Paleontology in China, 1979

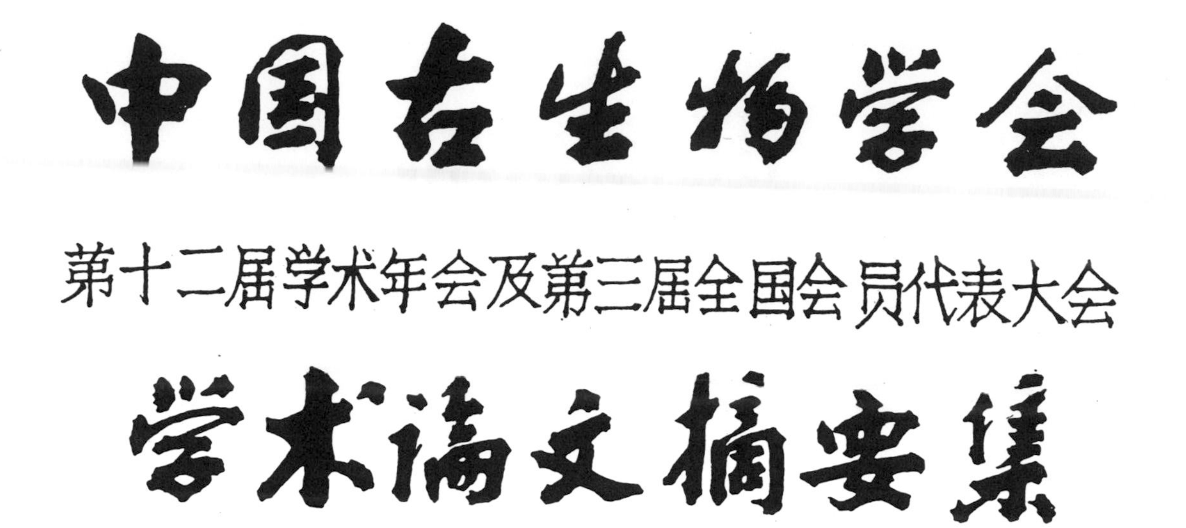

中国古生物学会编印

1979年4月 苏州

Cover of the program and abstracts of papers of the Third General Assembly and Twelfth National Meeting of the Palaeontological Society of China, April 1979. In the center is the seal of the Society printed in ancient Chinese characters.

Paleontology in China, 1979

Selected Papers Presented at the
Third General Assembly and Twelfth National Meeting
of the Palaeontological Society of China, April 1979

Edited by

CURT TEICHERT
Department of Geological Sciences
University of Rochester
Rochester, New York 14627
U.S.A.

LIU LU AND CHEN PEI-JI
Nanjing Institute of Geology and Palaeontology
Academia Sinica
Nanjing 210008
China

SPECIAL PAPER
187

THE GEOLOGICAL SOCIETY OF AMERICA
P.O. Box 9140 · 3300 Penrose Place · Boulder, Colorado 80301

Library of Congress Cataloging in Publication Data

Chung-kuo ku sheng wu hsüeh hui. Ch'üan kuo hui yüan
tai piao ta hui (3rd : 1979 : Su-chou shih, China)
Paleontology in China, 1979.

(Special paper / Geological Society of America ;
187)
Includes bibliographies.
1. Paleontology--China--Congresses. I. Teichert,
Curt, 1905- . II. Liu, Lu, 1937- . III. Chen,
Pei-Ji. IV. Chung-kuo ku sheng wu hsüeh hui. Hsüeh shu
nien hui (12th : 1979 : Su-chou shih, China) V. Title.
VI. Series: Special paper (Geological Society of
America) ; 187.
QE756.C6C553 1979 560.951 81-19997
ISBN 0-8137-2187-3 AACR2

Contents

Preface

At meetings held in Suzhou April 16 to 22, 1979, the Palaeontological Society of China celebrated the 50th anniversary of its foundation in Peking in 1929. In the 50 years of its existence the Society had met only eleven times, and the Suzhou convention was its Twelfth Annual Meeting and Third National Assembly. It was attended by 350 paleontologists from all parts of China who presented 264 papers covering most aspects of paleontology and related disciplines such as paleoecology, biostratigraphy, evolution, and paleobiogeography. Taken together, these papers gave a comprehensive picture of the state of the science of paleontology in the People's Republic of China (PRC) near the end of the eighth decade of the 20th century.

To mark the occasion, the Palaeontological Society of China had invited the following representatives of the International Palaeontological Association (IPA): Curt Teichert (U.S.A.), president; O. H. Walliser (German Federal Republic), secretary-general; and Fuyuji Takai (Japan), president of the Asian Branch of IPA. All were given the opportunity to visit some important centers of paleontologic research in the PRC and to meet with a large number of Chinese paleontologists.

Scientific work all over China had been severely disrupted from 1966 to 1976, in the period of the "Cultural Revolution," when most research work ceased, publication of scientific results was reduced to a trickle, and communications of Chinese scientists with their colleagues abroad were severely restricted. The removal of these repressive measures in 1976 triggered a tremendous surge of scientific activities, resulting in a veritable deluge of high quality publications in all branches of paleontology in an ever increasing number of publication outlets. On the systematic side, no field of study was neglected from micropaleontology and palynology to invertebrate and vertebrate paleontology and paleobotany. Great strides were made in biostratigraphy and stratigraphic correlation, including what may be the final step toward an understanding of the marine faunal succession at the Permian-Triassic transition. Large areas such as Xizang (Tibet) were opened up to scientific exploration on a large scale for the first time, and other, previously poorly known, parts of the country were extensively investigated.

The great bulk of this output, however, was printed in the Chinese language, occasionally with, though more commonly without, abstracts in English, and it became obvious to visitors that, mainly because of the language barrier, the renaissance of Chinese paleontology would be slow in being appreciated abroad. Before Teichert's departure from China, we, therefore, discussed plans for the organization of a volume of papers on Chinese paleontology in the English language, to be published in the United States. The suggestion to publish such a symposium was favorably received by the Geological Society of America, and the outcome is the present book which offers a choice from the 264 papers presented at the Suzhou meetings. Liu Lu and Chen Pei-ji were appointed coeditors who were responsible, in consultation with the President of PSC, for the selection of the papers, their translation into English, and transmission of the English manuscripts to Teichert. Teichert was responsible for preparation of the English texts from the standpoint of grammar and syntax, and he reviewed all reference lists and illustrations. Almost all manuscripts were in his hands by June 1980, and preliminary editing was completed that summer. Liu Lu was able to spend three months, from September to December 1980, at the University of Rochester working with Teichert on final editorial handling of the manuscripts. Liu also completed lists of all personal and geographic names used in the volume in Pinyin and Wade-Giles romanization, and these were edited and added to by Teichert after Liu's departure from Rochester.

In the fall of 1980 the Geological Society of America editors began editing the manuscripts; this was essentially completed by March 1981, by which time almost all remaining problems regarding text and illustrations had been cleared with the authors. In April 1981, Teichert visited the Geological Society headquarters in Boulder, Colorado, for a final review, meeting with Geological Society editors.

The present volume contains 24 articles authored by 48 paleontologists, 37 of whom are affiliated with the Nanjing Institute of Geology and Palaeontology. Obviously, this book could not have been compiled without the very active collaboration of this organization, for which thanks are due to its director, Professor Zhao Jinke [Chao King Koo], and its deputy director, Professor Lu Yanhao [Lu Yen-hao].

In the seventies, the government of the People's Republic of China officially enforced the change over from the old Wade-Giles system of romanization of Chinese names to the newer Pinyin system (see Teichert and Liu's introduction to the lists of names at the end of this volume). This move introduced many minor difficulties during the preparation of this book. Early on it was decided to cite all Chinese personal names in Chinese style, with the family name followed by the given name, both in Pinyin romanization, without intervening comma. First names that consist of two elements are hyphenated, although current tendency is to write such names as one word. (Example: An Taixiang, not An Tai-xiang.)

To assist the reader, throughout the text and reference lists the equivalents of many names in Pinyin are followed by Wade-Giles in square brackets and vice versa, although it has not been possible to be entirely consistent in this. If any problems arise from variant spellings, reference should be made to the lists at the end of the volume.

Generally, authors have been able to supply complete bibliographic documentation in their papers, and such reference lists are headed "References Cited." The heading "Selected References" indicates that not all references given in the text are included in the reference list, but the number of missing titles is usually few, except for the four introductory review papers which obviously could not list the complete literature of the past 30 years in their respective fields.

In all reference lists, square brackets around titles and publication sources indicate that the originals are in Chinese only and the texts between brackets are English translations supplied by the authors of the respective papers.

In order to avoid unacceptable delays, it was decided not to send proofs for correction to the authors in China. Instead proofs were read by Curt Teichert and Liu Lu who were assisted by Chen Jun-yuan and Margrit Gardner at the University of Rochester.

The editors regret their inability to provide subject and geographical indexes which would have been extremely valuable. Because of the wide separation of editors and authors, preparation of such indexes would have delayed publication of the book by at least another six months.

A grant from the National Science Foundation, EAR-8103339, contributed materially toward expenses incurred in retyping, photocopying, and mailing of manuscripts, and occasional travel.

CURT TEICHERT
Past President, International Palaeontological Association
YIN ZAN-XUN
President, Palaeontological Society of China

Geological Society of America
Special Paper 187
1981

Opening Speech of the Third Assembly and Twelfth National Meeting of the Palaeontological Society of China*

Yin Zan-xun
President of the Palaeontological Society of China, Institute of Geology, Academia Sinica, Beijing, People's Republic of China

Dear Delegates, Colleagues, and Honored Guests:

The Third Assembly and Twelfth National Meeting of the Palaeontological Society of China is now declared opened. In the name of the Council of the Society and Presidium of the Assembly, I warmly welcome the eminent guests and all our delegates and comrades who attend this meeting. The Second Assembly of our Society was held in Beijing in August 1962, more than 16 years ago. The Eleventh National Meeting took place in Nanjing in December 1964, more than 14 years ago. According to the decisions made at the Council meeting of the Society last October, the main tasks before the present Assembly are (1) to prepare an interim report of the Council activities since the second assembly, (2) to propose a revision of the statutes of the Society, (3) to elect new council members and to submit a new plan for future work of the Society, and (4) to organize multidiscipline symposia and to review the achievements in paleontology since 1949 and during the past 50 years since the founding of the Society.

Attending this meeting are distinguished guests from other countries. They are Professor Curt Teichert (U.S.A.), President of the International Palaeontological Association (IPA); Professor Otto H. Walliser (Federal Republic of Germany), Secretary-General of IPA; Professor Fuyuji Takai (Japan), President of the Asian Branch of IPA; Professor T. S. Westoll; Professor D. L. Dineley; Dr. A. J. Charig; and Dr. S. M. Andrews of the British delegation of vertebrate paleontologists. Their presence adds luster to our assembly. On behalf of the members of the Palaeontological Society of China, allow me to express heartfelt gratitude to our honored guests.

We also warmly welcome Professor Ge Ding-bang, a long-time member of our society who, residing abroad, returned to China to attend this meeting.

Just about a week ago, we had a discussion in Beijing with the authorities of the International Palaeontological Association concerning the application of the Palaeontological Society of China for affiliation with the international organization. We have arrived at a preliminary agreement on the matter and look forward to an early affiliation of our Society with IPA as one of its corporate members.

The National Assembly is the highest authority of our Society. Its duties are (1) to determine the policies and tasks of the Society, (2) to examine the working reports of the Council, (3) to elect the new Council, and (4) to regulate and revise the statutes of the Society. Therefore, we hope all our delegates and comrades will feel free to present their own views. Those who take the most active parts in scientific research and are able to unite the paleontological workers in good relations and enthusiasm for the work of the Society should be elected to the Council. The Council should also include members from different fields of industrial, educational, and scientific research organizations.

Two of the founders of our Society, Professor Sun Yun-zhu and Professor Yang Zhong-jian, passed away last winter. This was a great loss to the Society. We should continue to promote the cause they initiated and to carry on the paleontological work in China to ever higher levels.

During the last days before Professor Yang's death, while suffering from serious illness, he drew up an unfinished article to celebrate the 30th anniversary of the founding of the New China and the 50th anniversary of the Palaeontological Society. In the article, he recalled that in 1927 when he and Professor Sun were taking a geological field trip in the Harz Mountains, Germany, the plan to found a paleontological society in China came to their minds; they hoped that the establishment of the society would promote the development of paleontology in China.

Soon afterward, they both returned to their country and contacted a small number of colleagues, and on August 31, 1929, the founding ceremony of the Palaeontological Society of China took place in Peiping (now Beijing). At the meeting, the first statutes of the Society were adopted, Professor Sun Yun-zhu was elected chairman of the Society, and Professors Li Si-guang, Yang Zhong-jian, Zhao Ya-zeng, and Wang Gong-mu were elected councillors. Immediately after the founding ceremony, the first symposium was held on September 17, 1929, and three papers were delivered at the meeting. Among them was a paper on tetracorals by Professor Yue Sen-xun, who is present at this meeting.

*Abridged; delivered on April 16, 1979.

During the first decade of the Society's existence, works by members were published in the *Bulletin of Geological Society of China* and other journals. Since 1948, the Society issued two journals at irregular intervals: *Palaeontological Novitates,* which ceased after 8 issues, and *Proceedings of the Palaeontological Society of China,* which was interrupted after the 14th issue in 1964. Publication of that journal has been resumed, and the 15th issue is now in press. In addition, *Acta Palaeontologica Sinica* began publication in 1953, and up to 1979 18 volumes with 65 issues have been published. Eight issues of *Translated Papers on Palaeontology* have been published since 1955.

In the past half-century, members of our Society also made many other contributions to the scientific literature. Among these were (1) *Palaeontologia Sinica,* published since 1922, 155 numbers in 4 series, initially published by the National Geological Survey of China, now continued by the Academia Sinica; (2) *Vertebrata Palasiatica,* published since 1957 by the Institute for Vertebrate Palaeontology and Palaeoanthropology in Beijing, 17 volumes, 66 numbers; (3) *Acta Stratigraphica Sinica,* published since 1966, 2 volumes, 4 numbers (interrupted after 2 issues, resumed in 1978); and (4) *Main Fossil Groups of China* (in Chinese), published since 1962, 15 volumes; the last two series published by the Nanjing Institute of Geology and Palaeontology. Also, many papers and books for popular science readings were written by our members, in addition to the quarterly popular journal *Fossils* (in Chinese).

In the early days after the liberation of the country, we were working on ony 10 major fossil groups, whereas now 30 or more groups are under study; progress in research on microfossils is especially noteworthy. Personnel numbered only 20 to 30; now, the number of paleontologists in China is more than 1,000. For instance, there were only 2 or 3 persons engaged in research work on paleobotany and palynology, and now there are about 100 spread over the whole country. In addition to the basic descriptive work, investigations in systematics, evolution, ontogeny, and ecology have been gradually strengthened, and in recent years, progress has also been made in the study of paleobiogeography.

Members of the Society also have made many contributions in biostratigraphy. In addition to publication of a large number of papers, they have also been actively engaged in biostratigraphic work in connection with regional geological surveys, reconnaissance, and exploration work.

At this meeting, members of our Society are delivering a total of 352 papers on many fossil phyla with substantial results, which indicates that paleontological work in China has progressed considerably since the Eleventh National Meeting. New achievements in the study of such groups as fusulines, corals, brachiopods, gastropods, cephalopods, trilobites, and graptolites, and in macropaleobotany have increased our knowledge. Gaps and weak links in the study of spores and pollens, ostracodes, conodonts, tentaculites, estheriids, algae, stromatolites, charophytes, diatoms, and other microfossils have been filled in during past decades.

Despite 10 years of turmoil and destruction (mid-1966 to mid-1976), a great amount of paleontological work has been done in this country, a comparatively sound foundation has been laid, and significant contributions have been made to the national economic interest.

Now a period of modernization is ushered in, and the task of bringing Chinese paleontology to the highest level in a not too remote future is uppermost in the mind of every paleontologist. At the beginning of this "long march" we should be aware of our handicaps and exert maximum efforts to make up not only for our late start but also for a lost decade in recent years. It will be necessary to improve the teaching of paleontology at all levels. Research must be extended into theoretical and even speculative areas such as advanced bio-ecostratigraphy and the study of plate tectonics in paleobiogeographic reconstructions. Modern techniques should be introduced into our work by way of more frequent contacts with scholars from technically more developed countries. To facilitate the international exchange of ideas, so necessary for the modernization of China, language difficulties will have to be overcome. In our publications, longer summaries in one of the conventional scientific languages, preferably in English, will have to be given, and a better mastery of foreign languages, both in writing and the spoken word, has to be aimed at by our scientists, especially of the younger generation.

Manuscript Accepted by the Society April 22, 1981

Printed in U.S.A.

Geological Society of America
Special Paper 187
1981

Invertebrate paleontology in China (1949–1979)

Lu Yan-hao
Yu Chang-min
Chen Pei-ji
Nanjing Institute of Geology and Palaeontology, Academic Sinica, Nanjing 210008, People's Republic of China

INTRODUCTION

In China, the study of invertebrate paleontology has a long history. More than 50 years ago, the first paleontological work entitled *Contributions to the Cambrian faunas of North China* was published by Chinese paleontologists. But before this, foreign scholars had described and reported many invertebrate fossils. For example, Walcott's (1913) work was the first major publication on Cambrian fossils from China. A. W. Grabau, a famous American paleontologist, came to China in 1920 and by 1946 had completed a wealth of research work in the field of paleontology and stratigraphy. During the first half of the present century, the Chinese pioneer paleontologists had laid the foundations for the study of invertebrate paleontology. For example, Professors Li Si-quang [J. S. Lee] and Chen Xu [S. Chen] worked on fusulinids; Sun Yun-zhu [Y. C. Sun], Sheng Xin-fu [S. F. Sheng], Wang Yu [Y. Wang], and Zhang Wen-you [W. Y. Chang] on trilobites; Sun Yun-zhu, Yin Zan-xun [T. H. Yin], Xu Jie [S. C. Hsü], Zhang Xi-zhi [H. C. Chang], and Mu En-zhi [A. T. Mu] on graptolites; Yu Jian-zhang [C. C. Yü], Xue Sen-xun [S. S. Yoh], Huang Ji-qing [T. K. Huang], Ji Rong-sen [Y. S. Chi], Ma Ting-ying [T. Y. Ma], Zeng Ding-qian [T. C. Tseng], and Wang Hong-zhen [S. C. Wang] on corals; Zhao Ya-zeng [Y. T. Chao], Huang Ji-qing, Tian Qu-jun [C. C. Tien] and Wang Yu on brachiopods; Xu De-you [T. Y. Hsu] on bivalves; and Bing Zhi [C. Ping] on fossil insects. These paleontologists devoted themselves to the study of invertebrate fossils and their contributions drew the attention of paleontologists both at home and abroad. Since 1949, Chinese paleontologists have advanced with rapid strides in the study of invertebrate paleontology, as in other branches of paleontology. On the tenth anniversary of the founding of new China, progress in paleontological work was outlined in reports by Professors Yang Zhong-jian [C. C. Young] (1959) and Sze Hsin Chien (1960). At the present moment, it is therefore appropriate for us to sum up the achievements in the study of invertebrate paleontology during the past 30 years.

SCIENTIFIC PERSONNEL AND RESEARCH INSTITUTIONS

The past 30 years have witnessed the establishment of paleontological research institutions and the training of scientific personnel at a speed that was unimaginable in the old China. Before 1949, there were fewer than 30 paleontologists in China. Now their number has increased to more than 1,000, the majority of which are engaged in the study of invertebrate paleontology. Meanwhile, new research institutions and laboratories have been set up in rapid succession and are now found in almost every province, city, and autonomous region of this country. Among them are the Institute of Vertebrate Palaeontology and Palaeoanthropology, the Nanking Institute of Geology and Palaeontology, the Institute of Geology, Institute of Botany, Institute of Oceanology (in Qingdao), and the Institute of Oceanology of South China Sea (in Guangzhou), all belonging to the Chinese Academy of Sciences (Academia Sinica). Subordinate to the State Bureau of Geology are the Academy of Geological Sciences, geological institutes in large administrative regions and provinces, and laboratories with geological prospecting teams. In addition, similar laboratories have also been established in the production units of the petroleum and coal industry. What is more, paleontological teaching and research are being carried on in institutions of higher learning, such as Peking University, Nanking University, Chungshan University, Northwest-China University, Lanzhou University, Changchun Geological College, Wuhan Geological College, Chengtu Geological College, and others. Before 1949, only a few fossil faunas had been described, including fusulinids, corals, brachiopods, bivalves, cephalopods, trilobites, and graptolites. After 1949, we have filled up the research gaps in paleontology by starting to study such fossils as microfossils, stromatoporoids, Archaeocyatha, Scleractinia, bryozoans, Monoplacophora, gastropods, tentaculites, conulariids, Hyolitha, conodonts, Bradorida, Conchostraca, Malacostraca, Decapoda, Notostraca, echinoderms, insects, Chitinozoa, trace fossils, and other groups in order to complete the paleontological researches in our country. In addition, there exist fossils of unknown affinities that also have to be studied.

ACCUMULATION OF DATA

Beginning in 1962, Nanking Institute of Geology and Palaeontology issued a series of monographs entitled *All Groups of Fossils of China* that covers 15 major groups in 17 volumes. Among them, those dealing with invertebrate paleontology are *Graptolites of China* (Mu En-zhi and Chen Xu, 1962), *Bryozoa of China* (Yang Jing-zhi and Hu Zhao-xun, 1962), *Stromatoporoids of China* (Yang Jing-zhi and Dong De-yuan, 1962), *Fusulinids of China* (Sheng Jin-zhang, 1962), *Fossil ostracodes from China* (Hou You-tang and Chen De-qiong, 1962), *Fossil Corals from China* (Yu Chang-min, Wu Wang-shi, Zhao Jia-ming and Zhang Zhao-chang, 1963), *Fossil gastropods from China* (Yu Wen, Wang Hui-ji, and Li Zi-shun, 1963), *Fossil brachiopods from China* (Wang Yu, Jin Yu-gan and Fang Da-wei, 1964), *Fossil cephalopods from China* (Zhao Jin-ke, Liang Xi-luo, Zou Xi-ping, Lai Cai-gen, and Zhang Ri-dong, 1965), *Trilobites of China* (Lu Yen-hao, Zhang Wen-tang, Zhu Zhao-ling, Qian Yi-yuan, and Xiang Li-wen, 1965), *Fossil Lamellibranchia from China* (Gu Zhi-wei and others, 1976), *Fossil Conchostraca from China* (Zhang Wen-tang, Chen Pei-ji, and Shen Yan-bin, 1976). In addition, since 1959 this institute has published booklets on popular paleontology for beginners. They are *Fossils, Fusulinids, Graptolites, Ostracodes, Trilobites, Fossil Corals, Bryozoa,* and *Brachiopods.* In order to meet the needs of the geologists, especially those who wish to identify fossils in the field, we have published various handbooks concerning index fossils, for example, *Index Fossils of China* (five volumes) was published in 1953. In the late 1950s and early 1960s, regional handbooks on index fossils were published, for example, on the Central–South China region, the Yangtze region, and the Northwest China region. A significant publication in 1974 was the *Handbook of Stratigraphy and Paleontology in Southwest China,* in which are described about 1,000 species of different groups of fossils; this also includes more than 100 new genera and more than 700 new species. In recent years, the State Bureau of Geology has commenced compiling the multivolume *Fossil Atlas of China,* which already contains a profusion of materials obtained from various regions or provinces. It is expected that this atlas will soon be completed. In addition, we have published several volumes of *Palaeontologia Sinica* and other monographs. They are *New Materials of the Dendroid Graptolites of China* (Mu An-Tze, 1955), *Fusulinids from the Penchi Series of the Taitzeho Valley, Liaoning* (Sheng Jin-chang, 1958), *Ordovician Trilobite Faunas of Central and Southwestern China* (Lu Yan-hao, 1975), *Late Permian Cephalopods of South China* (Zhao Jin-ke, Liang Xi-luo, and Zheng Zhuo-guan, 1978), *Contribution to the Geology of the Mt. Quilianshan* (4 fascicles dealing with paleontology, 1960–1962), *Report of the Scientific Expedition to the Mt. Jolmo Lungma Region* (3 fascicules concerning paleontology, 1975–1976), *Palaeozoic Exposures in the Western Part of the Mt. Daboshan* (1975), *Contributions to the Paleontology of the Pohai Coastal Area* (1978), *Mesozoic and Cenozoic Red Beds of South China* (1979), *Devonian Symposia of South China* (1978), *Cretaceous-Tertiary Ostracodes from Liaohe-Songhuajiang Plains* (Hao Yi-chun and others, 1974), *Mesozoic and Cenozoic Fossils from Yunnan* (1976–1977), *Middle and Upper Triassic Brachiopods from Central Guizhou* (Yang Zun-yin and others, 1966), and others. Evidently, the data presented in the above mentioned volumes are indispensable for the preparation of paleontological researches at an advanced level, but much work remains to be done relative to the systematic description of fossils.

PROGRESS IN THE STUDY OF FOSSILS

After the birth of the new China, in the activities of geological prospecting and mineral exploration, large numbers of invertebrate fossils have been collected, which, in a broad way, facilitate the study of invertebrate palaeontology.

Archaeocyatha

Earlier, Ji Rong-sen described the Cambrian Archaeocyatha from the Yangtze Gorges district. More recently, reports have been published on the occurrence of the Cambrian Archaeocyatha in Xinjiang, North China, and Northeast China, especially in Central and Southwest China, which, biogeographically, are situated in the Yangtze region. According to the present knowledge, the early Cambrian Archaeocyathid faunas may be grouped into four assemblages with the lowest one found in the Chiungchussu Formation and probably belonging to the earliest Archaeocyatha assemblage of the world (Yuan Ke-xing and Zhang Sen-gui, 1979).

Coelenterata

The study of stromatoporoids has been carried out successfully mainly in the Devonian and Carboniferous strata of South China and partly in the strata the Chinghai-Tibet plateau (Yang Qing-zhi and others, 1975); this has resulted in the establishment of Devonian stromatoporoid sequences in South China. Moreoever, the Labechiida, being indigenous to China, have been found to occur in the basal part of the Lower Carboniferous.

Chinese paleontologists took early notice of Paleozoic corals. In 1960, it was estimated that about 681 species had been described in literature up to that time (Yu Chang-min and others, 1963). Since then, many additional reports have been published. It is on the basis of these reports that the establishment of coral sequences of Ordovician to Permian age has been made possible. Formerly, nothing was known of scleractinia. A scientific expedition to the Chinghai-Tibet plateau and field work in Yunnan has enabled us to study this group of fossils and we have already obtained some good results (Wu Wang-shi, 1975, 1977).

Brachiopoda

Brachiopoda, which have a long geological range and contain a considerable number of taxa, are of great importance. In the past 30 years, roughly 4,000 species in 1,000 genera have been described from China. In addition, about 100 brachiopod assemblages ranging from the Cambrian to the Quaternary have been established (Wang and others, 1979). Remarkably, the Early

Cambrian brachiopods are especially abundant and well preserved in this country. Recent studies of the Late Ordovician Hirnantian fauna has enabled us to solve, with much confidence, the problem of the boundary between the Ordovician and the Silurian (Rong Jia-yu, 1979). Lower Devonian brachiopods are extremely plentiful in South China.

Bryozoa

We began to study the bryozoa immediately after the liberation of this country. In 1960, it was calculated that about 122 species had been recorded (Yang Jing-zhi and Hu Zhao-xun, 1962). There have recently been new finds from the Ordovician to the Triassic. Detailed work has, by and large, been done on the late Paleozoic bryozoa with the result that the Devonian bryozoan sequences are preliminary set up in Guangxi and Hunan. (Yang Jing-zhi and Hu Zhao-xun, this volume). The Carboniferous bryozoans were found mainly in Weining of Guizhou (Xia Feng-shen, Lu Linhuang, and Li Wei-juan, 1978) and the northern slope of the Mount Qilianshan in Kansu, whereas the Permian bryozoans are especially numerous in South China.

Mollusca

The fossil Bivalvia—1,110 species (including subspecies) and 337 genera (including subgenera), ranging from Ordovician to Quaternary in age—have been described in *Fossil Lamellibranchia from China* (1976). Mesozoic nonmarine bivalves representing four faunal successions have been established in Jurassic and Cretaceous strata; this would seem to indicate that China is the place of origin of the earliest freshwater bivalves in Asia (Gu Zhi-wei, 1976, 1979). Marine bivalves of Triassic age are well known, and it has been possible to distinguish 12 assemblage zones in them. It should be pointed out that the study of marine bivalves from the Jurassic, Cretaceous, and Tertiary of Tibet is of special importance. In addition, attempts have been made to work on the Devonian bivalves with reference to their geographical distribution (Liu Lu, 1979).

More than 1,200 species of gastropods have been described so far. Recent studies of the Monoplacophora and gastropods from the lowest Cambrian in western Hubei provide substantial evidence bearing on the problem on the Precambrian-Cambrian boundary (Yu Wen, 1979).

In the early 1970s, we commenced to study Hyolitha and Tentaculites. Recent researches on the earliest Cambrian Hyolitha and other small shelly animals in South China demonstrated the possibility that the boundary between the Precambrian and the Cambrian may be drawn on a paleontological basis (Qian Yi, Chen Meng-e, and Chen Yi-yuan, 1977–1979). The tentaculites, a group of floating Mollusca, are so abundant in the Devonian deposits of South China that is has been possible to establish more than 10 tentaculite zones (Mu Dao-cheng, 1978), which are undoubtedly of great importance to the correlation and subdivision of the Devonian strata.

Among Cephalopoda, nautiloids, highly useful in the correlation and subdivision of early Paleozoic marine strata, occur abundantly in these deposits, particularly in the Cambrian. The study of the Late Cambrian nautiloids in North China suggests that at that time China may have been the place of origin of the Cephalopoda and the center of radiation of their early evolutionary stages (Chen Jun-yuan, Zou Xi-ping, Chen Ting-en, and Qi Dun-lun, 1979).

Ammonoids are generally believed to have great importance in the late Paleozoic and Mesozoic. Especially, they are fairly well developed in the Upper Permian of South China (1978), as described in the recent publication *Late Permian Cephalopods of South China* (1978) in which 155 species of ammonoids and 16 species of nautiloids are described. It seems that in the Late Permian a cephalopod fauna originated and developed in South China Sea; it migrated along the Tethys westward to Iran, Transcaucasia, and the southern part of Europe, for it is known to be distributed nearly in the same areas as that occupied by the Cathaysian flora and, therefore, it is called the Cathaysian ammonoid fauna. The ammonoids are associated with fusulinids and gastropods, and this Cathaysian fauna is generally considered to occur in the uppermost fossiliferous beds of the Upper Permian. This fauna is of considerable importance in the study of the defining of the boundary between the Triassic and the Permian, or in other words, the boundary between the Mesozoic and the Paleozoic.

Arthropoda

Chinese paleontologists took early notice of trilobites. As mentioned above, the late Sun Yun-chu, one of the founder of the Palaeontological Society of China, published in 1924 his monograph "Contribution to the Cambrian faunas of North China." Since then, many additional collections have been made. It was calculated in 1960 that about 1,233 species in 376 genera had been recorded. In the last 20 years, we have, in the main, finished the work on the systematic description of the Cambrian and Ordovician trilobites from Central and Southwest China and have summed up the work on the Cambrian and Ordovician biostratigraphy and zonation in North China and Northeast China. Recent studies on the trilobites from the upper part of the Fengshan formation in the Tangshan area indicate that the problem of the position of the Cambrian-Ordovician boundary in North China is expected to be solved soon (Zhou Zhi-yi and Zhang Jin-lin, 1978). Now we are intensifying our researches on Silurian and Devonian trilobites and are, to some degree, pursuing studies of trilobite ontogeny.

In earlier times, little was known of the fossil Conchostraca in China. More recently, large collections have been made from late Paleozoic, Mesozoic, and Cenozoic deposits, and 401 species and 63 genera have been described in *Fossil Conchostraca of China.* On the basis of materials available, a new proposal for the systematic classification of Conchostraca has been put forward, and, what is more, conchostracan assemblage sequences in Mesozoic and Cenozoic deposits have been established. Recently, the discovery of Tertiary Conchostraca in China has been reported (Wang Si-en, 1974).

Late in the 1950s, work was done on Bradorida collected mainly from Cambrian deposits in Yunnan, Guizhou, Szechuan, and Shaanxi; these fossils also occur in the Lower and Middle Cambrian deposits of North China. In general, the Bradorida are very rich both in number of individuals and of species, which may help to throw light on the origin, evolution, and systematic classification of this group once the materials have been fully described.

About 50 years ago Chinese paleontologists began to work on fossil insects (Bing Zhi, 1928), and, more recently, some papers dealing mainly with the Eocene insects from Fushun coal field (Hong You-chong, 1974, 1979), the Blattoidea of China, and fossil insects from Yunnan (Lin Qi-bin, 1978) have been published. In addition, research on the Miocene fossil insects from Shanwang and Shandong is now in progress.

In the past, few studies were made of such arthropods as crab, shrimp, and triopsids. Recently, however, it is reported that these arthropods have been found in some places. In Middle Asia, Kazacharthra were found in Lower Jurassic deposits, and they are known to be very richly represented in the Triassic of Xinjiang, even with their soft part well preserved, and are often found concurrently with the Yenchang flora.

Graptolites

In the past 30 years, much work has been done on graptolites and graptolite-bearing strata. Consequently, many graptolite zones have been established from the Cambrian to the Lower Devonian: 4 in the Upper Cambrian, 24 in the Ordovician, 19 in the Silurian, and 5 in the Lower Devonian. In Changning (Szechuan Province) the existence of two zones—the Arenigian and the Llanvirnian—containing pendent *Didymograptus*, is of special importance for the international correlation of Ordovician graptolite zones.

PALEOZOOGEOGRAPHY

Since the beginning of the 1970s, a new trend in the study of invertebrate paleontology has developed, and much attention is now being paid to the biogeographical division of different geological ages. One important result of this study is the application of the bio-environmental control hypothesis to Cambrian biostratigraphy (Lu Yen-hao and others, 1974) and to the distributional control of Cambrian mineral deposits. Important publications in this field are *Ordovician Graptolite Sequences and Biogeographical Regions in China* (Mu En-zhi, 1979) and *Preliminary Study on the Ordovician Corals with Their Stratigraphical Distribution and Biogeographical Provinces* (Yi Nong, 1974).

The biogeographical subdivisions in the late Paleozoic are based mainly on the fossil benthos. At the 12th annual meeting of the Palaeontological Society of China held in Suzhou in April 1979, several papers on paleobiogeography were presented, such as some on Silurian, Devonian, Carboniferous, and Permian corals (Wang Hong-zhen, He Xin-yi, 1979; Yu Chang-min, Kuang Guo-dun, 1979), on brachiopods (Wang Yu, Rong Jia-yu, 1979; Yang Shi-pu, 1979; Jin Yu-gan, 1979), and on Bryozoa (Yang Jing-zhi, Lu Lin-huang, 1979). Also, the Devonian nektonic tentaculites and nautiloids are of great help in the interpretation of Devonian biofacies and lithofacies.

In Mesozoic and Cenozoic marine deposits, biogeographic subdivisions are based mainly on bivalves and Foraminifera, as shown in the papers by (1) Lan Xiu (1979), (2) *A Preliminary Discussion on the Division of the Cenozoic Marine Lamellibranch Provinces of East China* by Lan Yi-chun and Zeng Xue-lu (1979), and (3) *Preliminary Analysis on the Tertiary Foraminifera and the Sedimentary Environment in Kashi of Tarim, Xinjiang* by Hao Yi-chun and Zeng Xue-lu (1979).

Here special attention is paid to the wide distribution of Mesozoic continental deposits and the geographical provinces of fresh-water animals of China, of which we knew nothing in the past. Recent papers on this subject include *Development of Mesozoic Nonmarine Lamellibranch Faunas in China and Its Significance* (Gu Zhi-wei, 1979), *Jurassic and Cretaceous Conchostracan Biogeographical Provinces in China* (Chen Pei-ji and Shen Yan-bin, 1979), *The Late Triassic and Early Jurassic Bivalve Assemblages and the Paleogeographic Characters of Hunnan, Jiangxi, and Guangdong Provinces* (Chen Jin-hua, 1979), and *Nonmarine Ostracode Biogeographical Provinces of the Early Cretaceous in China* (Ye Chun-hui, 1979). These papers make significant contributions to paleobiogeography and pave the way for future researches in biostratigraphy. Moreover, it is hoped that application of these biogeographical materials will contribute significantly to the study of important problems in earth sciences like plate tectonics and that it will provide a sound basis for the development of other branches of paleontology.

PALEONTOLOGICAL INVESTIGATION IN THE CHINGHAI-TIBET PLATEAU

The Chinghai-Tibet plateau is the youngest, the highest, and the largest plateau in the world. Mount Jolmo Lungma [Mount Everest] is a place famous as "the third pole of the Earth," along with the North and the South Poles, and has been visited by only a few people. However, this region has long attracted the attention of geologists because of its peculiar geologic development and its important geographic position. Beginning in the 1950s, under the auspices of the Chinese Academy of Sciences, five large-scale, multidiscipline expeditions were organized to investigate this region. Especially in 1966–1968 and 1975–1976, a considerable amount of data concerning lithology and fossils was obtained from there. In this area, deposits from the Cambrian-Ordovician to the Tertiary are exposed that have a total thickness of 11,207 m. Apart from the Cambrian, all the strata yielded invertebrate fossils, such as Foraminifera, Radiolaria, corals, Conularida, stromatoporoids, hydroids, bryozoans, brachiopods, bivalves, gastropods, nautiloids, ammonoids, belemnites, tentaculites, conodonts, trilobites, ostracodes, echinoids, crinoids, and graptolites. Among them are some fossil groups not previously reported from this region, for example: Ordovician nautiloids (Chen Jun-yuan, 1975) and trilobites (Qian Yi-yuan, 1976); Silurian and Devonian graptolites (Mu En-zhi and Ni Yu-nan, 1976); Devonian tentaculites (Mu Xi-nan, 1975); Early Carboniferous ammonoids (Liang Xi-luo, 1976); Permian corals (Wu Wang-shi, 1975) and brachiopods (Zhang Shou-xin and Jin Yu-gan, 1976); Late Triassic, Jurassic, Late Cretaceous, and Tertiary Foraminifera (He Yan, Zhang Bing-gao, Hu Lan-ying, and Sheng Jin-zhang, 1976); and Late Cretaceous and Early Tertiary ostracods (Huang Bao-ren, 1975). These finds are of great significance both stratigraphically and paleontologically.

MISCELLANEOUS

In China, application of statistical methods to paleontology, mainly to problems of ontogeny and bimorphism of animals, dates back to the 1960s. The mathematical methods often used are multivariate analysis, correlation analysis, discriminant analysis, cluster analysis, and factor analysis. During the 1970s these methods were being introduced into the study of paleogeography, paleoecology, paleoclimatology, paleogenetics, identification of taxa, division of faunal assemblages, and other problems.

Biomineralogy, as a new branch of paleontology, drew our attention in the early 1970s. Recent study of the calcareous structures of fossils leads to the recognition that environmental change has a great effect on the mineral components of the hard body parts of organisms.

Paleobiochemistry is a relatively new branch of paleontology. When carrying on geologic investigations in Precambrian and Cambrian strata, we attempted to study the geochemistry, particularly the amino acid and erythrophyll content of the rocks. These researches, however, need to be coordinated with traditional paleontological approaches. Only in this way can the paleobiochemical results find their full application.

Looking back to what has been done in the past 30 years, it would seem that considerable advances in invertebrate paleontology have been made. The number of research workers has increased considerably, new institutions have been founded and old ones enlarged, certain gaps in paleontological knowledege have been filled, and many new data have been accumulated. We have upgraded the level of research in theoretical fields and solved many problems of economic geology. Moreover, in certain taxonomic fields our progress has been comparable to that in the more advanced countries of the world.

SELECTED REFERENCES

Chang Wen You [Zhang Wen-you], 1937, Cambrian trilobite faunas of Anhui, Central China: Contributions from the National Research Institute of Geology, Academia Sinica, no. 6, p. 29–48, Pl. 1.

Chang Wen Tang [Zhang Wen-tang], 1949, Ordovician trilobites from Kaiping Basin, Hopei: Bulletin of the Geological Society of China, v. 29, p. 111–125, Pls. 1–2.

——1963, A classification of the Lower and Middle Cambrian trilobites from North and Northeastern China, with description of new families and new genera: Acta Palaeontologica Sinica, v. 11, no. 4, p. 447–488, Pls. 1–2 (in Chinese with English summary).

Chao Kingkoo [Zhao Jin-ke], 1959, Lower Triassic ammonoids from western Kwangsi: Palaeontologia Sinica, new ser. B, no. 9, p. 1–355, Pls. 1–45.

Chao Ya Tseng [Zhao Ya-zeng], 1927–1928, Productidae of China: Palaeontologia Sinica, ser. B, v. 5, no. 2, p. 1–244, Pls. 1–16, no. 3, p. 1–103, Pls. 1–6.

——1929, Carboniferous and Permian spiriferids of China: Palaeontologia Sinica, v. 11, no. 1, p. 1–133, Pls. 1–11.

Chen Jun-yuan, 1976, Advances in the Ordovician stratigraphy of North China with a brief description of nautiloid fossils: Acta Palaeontologica Sinica, v. 15, no. 1, p. 55–74, Pls. 1–3 (in Chinese with English abstract).

Chen Jun-yuan, Tsou Si-ping, Chen Ting-en, and Qi Dun-luan, 1979, Late Cambrian Cephalopoda of North China: Acta Palaeontologica Sinica, v. 18, nos. 1, 2, p. 1–23, 103–122, 8 pls. (in Chinese with English abstract).

Chen Pei-ji [Chen Pei Chi], 1975, [Tertiary fossil conchostracens from China]: Scientia Sinica, 1975, no. 6, p. 618–630, Pls. 1–2 (in Chinese).

——1979, An outline of palaeogeography during the Jurassic and Cretaceous Periods of China—with a discussion on the origin of Yangtze River: Acta Scientiarum Naturalium Universitais Pekinensis, no. 3, p. 90–109, Text-figs. 1–4 (in Chinese with English abstract).

Chen Su, 1934, Fusulinidae of South China: Palaeontologia Sinica, ser. B, v. 4, no. 2, p. 1–185, Pls. 1–16.

Chi Yung Shen, 1931, Weiningian (Mid. Carbon.) corals of China: Palaeontologia Sinica, ser. B, v. 12, no. 5, p. 1–70, Pls. 1–5.

Grabau, A. W., 1922–1928, Palaeozoic corals of China, Part 1: Palaeontologia Sinica, ser. B, v. 2, no. 1, p. 1–76, Pl. 1, 74 text-figs.

——1923–1928, Stratigraphy of China, Parts 1 and 2: Palaeontologia Sinica, 1302 p., 755 text-figs.

——1931–1933, Devonian Brachiopoda of China, Part 1: Palaeontologia Sinica, ser. B, v. 3, no. 3, 538 p., 54 pls.

——1934–1936, Early Permian fossils of China, Parts 1–2: Palaeontologia Sinica, ser. B, v. 8, nos. 3–4, 625 p., 42 pls.

Hao Yi-chun, Su De-ying, Li You-gui, Ruan Pei-hua, and Yuan Feng-dian, 1974, [Cretaceous-Tertiary ostracodes from Liaohe-Songhuajiang Plains]: Beijing, Geological Publishing House, 93 p., 30 pls. (in Chinese).

Hou You-tang and Chen De-qiong [Chen Te-chiung], 1962, [Fossil ostracodes from China]: Beijing, Science Press, 150 p., 29 pls. (in Chinese).

Hsü Singwu C. [Xu Jie], 1934, The graptolites of the Lower Yangtze Valley: Monographs of the National Research Institute of Geology, Academia Sinica, ser. A, v. 4, p. 1–106, Pls. 1–7.

Hsu Te You, 1937, Contribution to the Marine Lower Triassic fauna of southern China: Bulletin of the Geological Society of China, v. 16, p. 303–346, Pls. 1–4.

Ku Chih Wei [Gu Zhi-wei], 1948, Fauna of the Lower Triassic Tungkaitzu Formation of western Szechuan: Bulletin of the Geological Society of China, v. 28, nos. 3–4, p. 235–253, Pl. 1.

Lee Jonquei S., [Li Si-quang], 1927, Fusulinidae of North China: Palaeontologia Sinica, ser. B, v. 4, no. 1, p. 1–172, Pls. 1–24.

Li Yao-xi, Song Li-sheng, Zhou Zhi-qiang, and Yang Jing-yao, 1975, [The lower Paleozoic of western Mount Dabashan]: Beijing, Geological Publishing House, 232 p., 70 pls. (in Chinese).

Liu Hong-yun, 1959, [Paleogeographical maps of China]: Beijing, Science Press (in Chinese).

Liu Lu, 1979, [Distribution of Devonian bivalves of China]: Acta Stratigraphica Sinica, v. 3, no. 2, p. 127–130 (in Chinese).

Lu Yen-hao [Lu Yan-hao], 1941, Lower Cambrian stratigraphy and trilobite fauna of Kunming, Yunnan: Bulletin of the Geological Society of China, v. 21, no. 1, p. 71–90, pl. 1.

——1963, The ontogeny of *Hanchungolithus multiseriatus* (Endo) and *Ningkianolithus welleri* (Endo), with a brief note on the classification of the Trinucleidae: Acta Palaeontologica Sinica, v. 11, no. 3, p. 319–340, Pls. 1–3 (in Chinese with English summary).

——1975, Ordovician trilobite faunas of central and southwestern China: Palaeontologia Sinica, new ser. B, no. 11, 463 p., 50 pls. (in Chinese and English).

Lu Yen-hao, Zhang Wen-tang [Chang Wen Tang], Zhu Zhao-ling [Chu chao-ling], Qian Yi-yuan [Chien Yi-yuan], and Xiang Li-wen, 1965, [Trilobites of China]: Beijing, Science Press, 766 p., 135 pls. (in Chinese).

Lu Yen-hao, Zhu Zhao-ling, Qian Yi-yuan, Lin Huan-ling, Zhou Zhi-yi, and Yuan Ke-xing, 1974, [Bio-environmental control hypothesis and its application to the Cambrian biostratigraphy and paleozoogeography]: Memoirs of Nanking Institute of Geology and Palaeontology, Academia Sinica, no. 5, p. 27–116, 4 pls. (in Chinese).

Lu Yen-hao, Chu Chao-ling, Chien Yi-yuan, Zhou Zhi-yi, Chen Jun-yuan, Liu Geng-wu, Yü Wen, Chen Xu, and Xu Han-kui, 1976, [Ordovician biostratigraphy and paleozoogeography of China]: Academia Sinica, Memoirs of Nanking Institute of Geology and Palaeontology, no. 7 (in Chinese).

Mu An Tze [Mu En-zhi], 1955, The new materials of the dendroid graptolites of China: Palaeontologia Sinica, new ser. B, no. 5, 62 p., 10 pls. (in Chinese and English).

Mu En-zhi [Mu An Tze], 1974, Evolution, classification and distribution of Graptoloidea and graptodendroids: Scientia Sinica, v. 17, no. 2, p. 227–238.

Mu En-zhi [Mu An Tze] and Chen Xu, 1962, [Graptolites of China]: Beijing, Science Press, 170 p., 21 pls. (in Chinese).

Mu En-zhi [Mu An Tze], Ge Mei-yu [Geh Mei-yu], Chen Xu, Ni Yu-nan, and Lin Yao-kun, 1979, Lower Ordovician graptolites of Southwest China: Palaeontologia Sinica, new ser. B, no. 13, 192 p., 48 pls. (in Chinese with English abstract).

Ping Chih [Bing zhi], 1928, Cretaceous fossil insects of China: Palaeontologia Sinica, ser. B, v. 13, no. 1, p. 1–56, Pls. 1–3.

Qian Yi, 1978, The Early Cambrian hyolithids in Central and Southwest China and their stratigraphical significance: Memoirs of Nanjing Institute of Geology and Palaeontology, Academia Sinica, no. 11, p. 1–50, 7 pls. (in Chinese with English abstract).

Rong Jia-yu, 1979, [The Hirnantian fauna of China with comments on the Ordovician-Silurian boundary]: Acta Stratigraphica Sinica, v. 3, no. 1, p. 1–28, Pls. 1–2. (in Chinese).

Sheng Jing Chang [Sheng Jin-zhang], 1958, Fusulinids from the Penchi Series of the Tatzeho Valley, Liaoning: Palaeontologia Sinica, new ser. B, no. 7, 118 p., 16 pls. (in Chinese and English).

——1962, [Fusulinids of China]: Beijing, Science Press, 177 p., 27 pls. (in Chinese).

Sheng Si Fu [Sheng Xin-fu], 1934, Lower Ordovician trilobite fauna of Chekiang: Palaeontologia Sinica, ser. B, v. 3, no. 1, p. 1–28, Pls. 1–4.

Sun Yun Chu [Sun Yun-zhu], 1924, Contributions to the Cambrian faunas of North China: Palaeontologia Sinica, ser. B, v. 1, no. 4, p. 1–109, Pls. 1–5.

——1931, Ordovician trilobites of Central and Southern China: Palaeontologia Sinica, ser. B, v. 7, no. 1, p. 1–47, Pls. 1–3.

——1933, Ordovician and Silurian graptolites from China: Palaeontologia Sinica, ser. B, v. 14, no. 1, p. 1–52, Pls. 1–7.

——1935a, Lower Ordovician graptolite faunas of North China: Palaeontologia Sinica, ser. B, v. 14, no. 2, p. 1–14, Pls. 1–3.

——1935b, The Upper Cambrian trilobite faunas of North China: Palaeontologia Sinica, ser. B, v. 7, no. 2, p. 1–93, Pls. 1–6.

——1958, The Upper Devonian coral faunas of Hunan: Palaeontologia Sinica, new ser. B, no. 8, p. 1–28, Pls. 1–12.

Sze Hsin Chien [Si Xing-jian], 1960, Recent advances in Chinese palaeobotany and invertebrate palaeontology: Scientia Sinica, v. 9, no. 1, p. 1–13.

Tien Chi-chün, 1938, Devonian Brachiopoda of Hunan: Scientia Sinica, new ser. B, no. 4, p. 1–192, Pls. 1–22.

Tsao Rui-chi and Liang Yu-zhou, 1974, [On the classification and correlation of the Sinian System in China, based on a study of algae and stomatolites]: Memoirs of Nanking Institute of Geology and Palaeontology, Academia Sinica, no. 4, p. 1–26, Pls. 1–8 (in Chinese).

Walcott, C. D., 1913, The Cambrian faunas of China: Research in China, Volume 3: Carnegie Institution of Washington Publication 54, p. 1–230, 24 pls.

Wang Hong Chen [Wang Hong-zhen], 1948, The Silurian rugose corals of northern and eastern Yunnan: Bulletin of the Geological Society of China, v. 24, nos. 1–2, p. 21–32, Pl. 1.

Wang Yu, 1937, On a new Permian trilobite from Kansu: Bulletin of the Geological Society of China, vol. 16.

Wang Yu, Jin Yu-gan [Ching Yü-kan], and Fang Da-wei, 1964, [Fossil brachiopods from China]: Beijing, Science Press, 777 p., 136 pls. (in Chinese).

Xia Feng-sheng, Lu Lin-huang and Li Wei-juan, 1978, Carboniferous bryozoans from Weining, western Guizhou: Acta Palaeontologica Sinica, v. 17, no. 3, p. 319–342, 3 pls. (in Chinese with English Résumé).

Yang Jing-zhi [Yang King Chih], and Hu Zhao-xun, 1962, [Bryozoa of China]: Beijing, Science Press, 89 p., 28 pls. (in Chinese).

Yang Jing-zhi [Yang King Chih], and Dong De-yuan, 1962, [Stromatoporoids of China]: Beijing, Science Press, 40 p., 14 pls. (in Chinese).

Yang Tsun-yi [Yang Zun-yi], and Xu Geuiyong [Xu Gui-rong], 1966, Triassic brachiopods of Guizhou (Kueichow) Province, China: Beijing, 122 p., 151 pls. (in Chinese with English summary).

Yin Tsan Hsun [Yin Zan-xun], 1932, Gastropoda of the Penchi and Taiyuan Series of North China: Palaeontologia Sinica, ser. B, v. 11, no. 2, p. 1–53, Pls. 1–3.

——1933, Cephalopoda of the Penchi and Taiyuan Series of North China: Palaeontologia Sinice, ser. B, v. 11, no. 3, p. 1–46, Pls. 1–5.

——1935, Upper Palaeozoic ammonoids of China: Palaeontologic Sinica, ser. B, v. 11, no. 4, p. 1–44, Pls. 1–5.

Yin Tsan Hsun [Yin Zan-Xun] and Mu An Tze [Mu En-zhi], 1946, Lower Silurian graptolites from Tungtzu: Bulletin of the Geological Society of China, v. 25, p. 211–219, Pl. 1.

Yoh Sen Siung [Yue Sen-xin], 1961, On some new tetracorals from the Carboniferous of China: Acta Palaeontologica Sinica, v. 9, no. 1, p. 1–17, Pls. 1–3.

Yoh Sen Siung [Yue Sen-xin] and Huang, T. K. [Huang Ji-qing], 1932, The coral fauna of the Chihsia Limestone of the Lower Yangtze Valley: Palaeontologia Sinica, ser. B, v. 8, n. 1, p. 1–72, Pls. 1–10.

Yü Chien Chang, 1933. Lower Carboniferous corals of China: Beijing, Science Press, v. 12, no. 3, p. 1–133, Pls. 1–24.

Yu Chang-min [Yü Chang-ming], Wu Wang-shi [Wu Wang Shih], Zhao Jia-min, and Zhang Zhao-cheng, 1963, [Fossil corals from China]: Beijing, Science Press, 390 p., 98 pls. (in Chinese).

Yu Wen [Yü Wen], 1979, Earliest Cambrian monoplacophorans and Gastropoda from western Hubei with their biostratigraphical significance: Acta Palaeontologica Sinica, v. 18, no. 3, p. 233–266, 4 pls. (in Chinese with English abstract).

Yu Wen [Yü Wen], Wang Hui-ji, and Li Zi-shun, 1963, [Fossil gastropods from China]: Beijing, Science Press, 362 p., 66 pls. (in Chinese).

Zhang Wen-tang [Chang Wen Tang], Chen Pei-ji and Shen Yan-bin, 1976, [Fossil Conchostraca of China]: Beijing, Science Press, 325 p., 138 pls. (in Chinese).

Zhao Jin-ke [Chao Kingkoo], Liang Xi-luo [Ling Hsi-lo], Zou Si-ping, Lai Cai-gen, and Zhang Ri-dong, 1965, [Fossil cephalopods from China]: Beijing, Science Press, 389 p., 85 pls. (in Chinese).

Zhao Jin-ke [Chao Kingkoo], Liang Xi-luo, Zheng Zhuo-guan, 1978, Late Permian cephalopods of South China: Palaeontologia Sinica, new ser. B, no. 12, 194 p., 34 pls. (in Chinese with English summary).

Zhou Zhi-yi and Zhang Jin-lin, 1978, Cambrian-Ordovician boundary of the Tangshan area with descriptions of the related trilobite fauna: Acta Palaeontologica Sinica, v. 17, no. 1, p. 1–26, Pls. 1–4 (in Chinese with English abstract).

Miscellaneous References (Authors Unknown)

[Index Fossils of China] (Invertebrates, Parts 1–3): Beijing, Geological Publishing (in Chinese).

[Contribution to the geology of the Mt. Qilianshan: Beijing, Science Press, v. 4, nos. 1–5 (in Chinese).

[Handbook of stratigraphy and palaeontology in Southeast China]: Beijing, Science Press, 454 p., 202 pls. (in Chinese).

[Report of scientific expedition to the Mt. Jolmo Lungma region (Palaeontology, nos. 1–3)]: Beijing, Science Press (in Chinese).
[Devonian Symposia of South China]: Beijing, Geological Publishing House, 396 p., 53 pls. (in Chinese).
[Fossil lamellibrachia from China]: Beijing, Science Press, 521 p., 150 pls. (in Chinese).
[Mesozoic fossils from Yunnan]: Beijing, Science Press, nos. 1, 2 (in Chinese).
[Contribution to the palaeontology of the Bohai coastal area]: Beijing, Science Press, (in Chinese with English abstract).
1979, Abstracts of Papers, 12th Annual Converence and 3rd National Congress of the Palaeontological Society of China: Suzhou, 164 p. (Titles in English, texts in Chinese).
[Mesozoic and Cenozoic red beds of South China]: Beijing, Science 432 p., 62 pls. (in Chinese).

MANUSCRIPT ACCEPTED BY THE SOCIETY APRIL 22, 1981

Printed in U.S.A.

Geological Society of America
Special Paper 187
1981

Thirty years of micropaleontological research in China

Hao Yi-chun
Wuhan College of Geology, Xueyuan Road, Beijing, People's Republic of China

The study of micropalaeontology in our country has been developing rapidly since the birth of the People's Republic of China.

Before 1949, only a few groups of microfossils had been studied by a few palaeontologists, such as Li Si-guang (1924, 1927, 1934, 1936, 1942), Chen Xu (1934a, 1934b), and Lee and Chen (1930) on Carboniferous and Permian fusulinids and Middle Carboniferous small foraminifers; S. S. Joh (1932) on the discovery of a Bryozoa species, *Fistulipora sinensis,* in Lower Permian limestone; and Lu Yan-hao (1944, 1945) on Tertiary charophytes from Xinjiang and middle Devonian charophytes from Yunnan. Some of these works are well known and have been given high credit both at home and abroad.

After 1949, geological surveys were carried on all over the country to look for ores and minerals, and as one of the important approaches to stratigraphic correlation, the study of micropalaeontology received great attention and developed rapidly through the following stages:

1949 to 1957. Several groups of Paleozoic microfossils were studied more or less systematically on the basis of past biostratigraphical work. The following fossil groups were studied in some detail: (1) Carboniferous and Permian fusulinids from Yunnan, southeastern China, eastern part of Liaoning, Inner Mongolia, and Qinghia (Chen Xu, Sheng Jin-zhang, Zhang Lien-xin); (2) Ordovician, Devonian, Carboniferous, and Permian ostracodes from Guangxi, West Hubei, West Zhejiang, Jiangsu, East Liaoning, and Gansu (Hou You-tang, Chen De-quoing, Shi Cong-guang); (3) Ordovician, Silurian, Devonian, Carboniferous, and Permian bryozoans from Sichuan, Guangxi, Hubei, Hunan, Zhejiang, Shenxi and Jilin; and (4) Middle Devonian charophytes from Jiangyou and its vicinity in Sichuan province.

Students of micropalaeontology are still few in number; each group was studied by only two or three persons.

1958 to 1965. At the end of the 1950s a great many stratigraphic columns were described, and large numbers of microfossils were discovered by drilling in the prospecting for mineral resources. Rapid accumulation of abundant biostratigraphic materials accelerated the development of micropaleontologic study.

During the search for ore deposits in Mesozoic and Cenozoic strata, which are very well developed and widely distributed in China, overall attention was paid to the study of some important Mesozoic and Cenozoic groups of microfossils, such as ostracodes, charophytes, and small benthonic foraminifers. As a result, these became important in the subdivision and correlation of ore-bearing strata, and some important contributions were published during this period.

The Mesozoic and Cenozoic ostracodes and Charophyta collected mainly from nonmarine ore-bearing strata and red beds were studied by micropaleontologists at different localities in Jiangsu, Hubei, Hunnan, Guangdong, Xinjian, Qinghai, Kansu, and the Songliao Basin of Northeast China. Among the important publications may be mentioned *Jurassic and Cretaceous fresh-water ostracodes of Northwest and Northeast China* (Hou You-tang, 1958), *Cretaceous Ostracoda of the Songliao Basin* (Nechaeva and Liu Zong-yung, 1959), *Fossil Ostracoda from Kansen Region of Chaidamu Basin* (Huang Bao-ren, 1964), *Upper Triassic and Middle Jurassic Fossil Ostracoda of the Erdos Basin* (Zhong Xiao-chun, 1964); *Tertiary Fossil Charaphyta of the Chaidamu Basin, Qinghai* and *Mesozoic and Cenozoic Fossil Charophyta from the Jiuquan Basin of Kansu* (Wang Shui, 1961, 1965).

Two different faunas of small benthonic foraminifers were studied: a Triassic fauna from the Jialinjiang Limestone of southern Sichuan (He Yan, 1959) and a Quaternary fauna from the eastern part of Jiangsu (He Yan, Hu Lan-ying, and Wang Ke-liang, 1965).

Research on conodonts also began at the end of the 1950s. The first report on this group was on Permian conodonts from the Gufeng Formation in the vicinity of Nanjing (Jing Yu-gan, 1960).

On the basis of these researches, three generalized papers were published in 1962, namely *Fossil Bryozoa of China* (Yang Jing-zhi and Hu Zhao-xun), *Fusulinids of China* (Sheng Jin-zhang), and *Fossil Ostracodes of China* (Hou You-tang and Chen De-qiong), each of which included an introduction to morphology and structure, a taxonomic scheme (as well as systematic descriptions of all species known in China up to 1960), and lists of fundamental references.

1966 to 1975. During this period, the study of micropalaeontology, like scientific research in all other fields, had been seriously interrupted, and what little progress was made in this period may be summarized as follows:

First, serving as one of the important tools for stratigraphic correlation in the search for petroleum in the Paleozoic strata, the study of conodonts developed rapidly. Conodont faunas of Ordovician, Carboniferous, Permian, and Triassic ages were reported from different localities in Yunnan, Sichuan, Guangxi, Hubei, Shandong, Hobei, Xinjiang, and Xizang, while taxonomic and biostratigraphic studies were made on the Permian fauna of Sichuan and the Devonian fauna of Guangxi and Yunnan (Wang Cheng-yuan 1974, 1975).

Second, in reporting on brackish-water faunas of foraminifers from depositional basins in eastern and central China, some micropaleontologists (Wang Ping-zian and others, 1975) made the paleoecologic analyses of the faunas and interpreted the environments of sedimentation. This represented a new direction in foraminiferal research, proceeding from systematic description and biostratigraphic study to discussion of theoretical problems.

Third, five bryozoan species of *Paralioclema* were described by Yang Jing-zhi and Xia Feng-sheng (1975) in an article entitled "Fossil Bryozoa from the Chomolungma Region." These were the first Triassic bryozans reported from China.

1976 to 1979. In the course of a general drive for modernization, many new achievements were recorded in the study of micropaleontology.

1. Study of regional faunas or floras of different ages. For Ostracoda, the main contributions were concerned with the Mesozoic and Cenozoic faunas from nonmarine deposits, the important ones among which are the studies of Mesozoic and Cenozoic ostracodes from the Chaidamu basin (Yang Fan, Sun Zhi-chen, and Chen Tian-min, 1978) and the central to western part of Yunnan (Ye Chun-huei, Gou Yun-xian, and others, 1977), of Cretaceous ostracodes from the Songliao Basin of northeast China (Ye De-quan and Ding Lian-sheng, 1976), of Cretaceous and Tertiary ostracodes from the bordering regions of the Jianghan basin (Hou You-tang and He Jun-de, 1979), and of Early Tertiary ostracodes from the coastal region of the Bohai Sea.

In addition to these studies, taxonomic descriptions of Paleozoic, Mesozoic, and Cenozoic ostracodes from Hubei, Hunan, Guangdong, and Guangxi were published in 1976 (Guan Shaozeng, Sun Quan-ying, Jiang Yan-wen, and others).

Among foraminifers, abundant Mesozoic and Cenozoic faunas including planktonic and both large and small benthonic forms were reported from the Chomolungma region (Sheng Jin-zhang and others, 1976). Also of importance are taxonomic descriptions of Carboniferous, Permian, Triassic, and Tertiary foraminifers from Hubei, Hunan, Guangdong, and Guangxi (Lin Jia-xing, Zheng Yuan-tai, and Wang Nai-wen, 1976) and the study of Cenozoic foraminifers from the coastal region of the Bohai Sea.

For fusulinids, regional Carboniferous and Permian faunas of Guizhou, Guangxi, Hunan, Hubei, and Inner Mongolia were described taxonomically in 1976 and 1978 by Han Jian-xiu, Lin Kiaxing, Li Jia-xiang, Liu Chao-an, Xiao Xing-ming, and others.

For charophytes, Mesozoic and Cenozoic floras were reported from Yunnan, the Jianghan basin of Hubei, and the Sanshui basin of Guangdong, and systematic descriptions were published separately in 1976 and 1978 by Wang Zhen, Wuang Ren-jing, Wang Shui, Zhang Jie-fang and Lu Huei-ran. Later in 1978, new materials of Cretaceous and Early Tertiary charophytes were described, and subdivisions of floral assemblages as well as taxonomic problems were discussed (Wang Zhen, 1978a, 1978b).

The study of conodonts continues to speed up. On the basis of materials collected during the last period, two Triassic faunas were studied in 1976 and 1978, one from Yunnan, the other from western Hubei (Wang Cheng-yuan, Wang Zhi-hao, Cao Jin-yue). At the same time, more detailed studies of the subdivision of fossil zones have been made on Cambrian and Ordovician faunas from the coastal region of the Bohai Sea, on Ordovician faunas from Hubei and Jiangsu, on a Silurian fauna of Guizhou, on a Devonian fauna from Guangxi, on a Carboniferous fauna from Shanxi, and on Carboniferous and Permian faunas from Xinjiang as well as on Triassic faunas from Yunnan, Sichuan, Hubei, and Jiangsu.

2. Study of new subjects. In recent research on micropaleontology, more and more attention has been paid to new subjects, such as evolutionary trends, natural classification, paleoecological features, biogeographical distribution of different groups, and micropaleontological studies have advanced from systematic description and biostratigraphic analysis to discussion of more theoretical problems. A revision of the taxonomy of fusulinds was proposed by Zhang Lin-xin, the division of biogeographical provinces of Permian fusulinids in Xizang by Wang Yu-jing, and a revision of Lower Cretaceous nonmarine ostracodes by Ye Chunhui; the paleogeographical significance of Quaternary foraminiferal faunas from the coastal region of the Bohai Sea was discussed by Wang Nai-wen, the paleoclimatological significance of Mesozoic and Cenozoic freshwater ostracodes from Hainan Island was discussed by Geng Liang-yu (1979), and the origin and evolutionary trends of some Triassic charophyte genera were discussed by Wang Zhen and Ye Chun-hui (1979).

3. Use of new methods. Statistical methods (biometry) have been applied to identification and description of genera and species (Geng Liang-yu, 1979) as well as to diversity analyses (Wang Ping-xian and others, 1976, 1978); and the latter has been used in the study of the paleoecology of foraminifers and ostracodes (Wang Ping-xian and others, 1976, 1978).

4. Some groups previously unknown in China were discovered and studied. These include Triassic Radiolaria and Jurassic stromatoporoids from the Chomolungma region (Sheng Jin-zhang, 1977; Yang Jing-zhi and Wang Cheng-yuan, 1975); Late Cretaceous and Early Tertiary calcareous algae from the same region (Wang Yu-jing, 1976); Early Tertiary acritarchs and dinoflagellates from the Bose basin of Guangxi (He Cheng-quan and Qian Ze-shu, 1979) and from the coastal region of the Bohai Sea; and Upper Permian fossil Fungi from Guizhou (Mu Xi-nan, 1977).

Today, the study of micropaleontology in China covers about 15 groups of microfossils from almost all provinces and autonomous regions. About 500 students and technicians are now engaged in micropaleontologic research.

For the purpose of promoting scientific communication and raising the academic level, the Micropalaeontological Society of China, which is affiliated with the Palaeontological Society of China, was established and the first symposium was held in March 1979.

SELECTED REFERENCES

Chen, S. [Chen Xu], 1934a, Fusulinidae of South China: Palaeontologia Sinica, ser. B, v. 4, pt. 2, p. 1–185, Pls. 1–16.

——1934b, Fusulinids in the Huanglung and Maping Limestone of Kwangsi: National Research Institute of Geology, Memoirs, no. 14, p. 33–54, Pls. 6–8.

——1956, Fusulinidae of South China, Part II, Fusulinid fauna of the Permian Maokou Limestone in China: Palaeontologia Sinica, new ser. B, no. 6, p. 1–71, Pls. 1–14.

Chen, T. C. [Chen De-qiong], 1958, Permian ostracods from the Chihsia Limestone of Lungtan, Nanking: Acta Palaeontologica Sinica, v. 6, no. 2, p. 215–257, Pls. 1–8.

Ching Yü-kan [Jin Yu-gan], 1960, Conodonts from the Kufeng Suite (Formation) of Lungtan, Nanking: Acta Palaeontologica Sinica, v. 8, no. 3, p. 230–240, Pls. 1–2 (in Chinese with English abstract).

Geng Liang-yu, 1979, Some fresh-water ostracods from the Mesozoic and Cenozoic deposits in Hainan Island, Kwangtung: Acta Palaeontologia Sinica, v. 18, no. 1, p. 41–63, Pl. 1.

Hao Yi-chun, Su De-ying and others, 1974, [Cretaceous and Tertiary ostracods from Songliao Basin, Northeast China]: Peking, Geological Publishing House, p. 1–95, Pl. 1–30 (in Chinese).

He Cheng-quan and Qian Ze-shu, 1979, Early Tertiary dinoflagellates and acritarchs from the Bose Basin of Guangxi: Acta Palaeontologia Sinica, v. 18, no. 2, p. 171–188, Pls. 1–2.

He Yan, 1959, Triassic Foraminifera from the Chialingkiang Limestone of South Szechuan: Acta Palaeontologica Sinica, v. 7, no. 5, p. 393–418, Pls. 1–8.

He Yan, Hu Lan-ying, and Wang Ke-liang, 1965, [Quaternary Foraminifera from Eastern Jiangsu]: Memoirs of Nanjing Institute of Geology and Palaeontology, Academic Sinica, no. 4, p. 51–162, Pls. 1–16.

He Yan, Hu Lan-ying, 1978, [The Cenozoic Foraminifera from the coastal region of Bohai]: Science Press, p. 1–48, Pls. 1–10 (in Chinese).

He Yan, Zhang Bing-gao and others, 1976, [Mesozoic and Cenozoic Foraminifera from the Qomolungma region, Reports on Scientific Investigations in the Qomolungma region, 1966–1968]: Palaeontology, pt. II, p. 1–124, Pls. 1–36 (in Chinese).

Hou Y. T. [Hou You-tang], 1953, Some Tremadocian ostracods from Taitzeho Valley, Liaotung: Acta Palaeontologia Sinica, v. 1, no. 1, p. 40–49, Pl. 1.

——1955a, New Devonian ostracods from Hupeh: Acta Palaeontologica Sinica, v. 3, no. 3, p. 205–240, Pls. 1–9.

——1955b, On some new ostracods from Kwangsi: Acta Palaeontologica Sinica, v. 3, no. 4, p. 309–316, Pl. 1.

——1958, Jurassic and Cretaceous fresh-water ostracods from Northwest and Northeast China: Memoirs of the Institute of Geology and Palaeontology, Academia Sinica, v. 1, no. 1, p. 34–102, Pls. 1–13. (in Chinese and English).

Hou You-tang and Chen De-qiong, 1962, [Fossil Ostracoda of China]: Science Press, p. 1–150, Pls. 1–29 (in Chinese).

Hou You-tang, Ho Jun-de, and Ye Chun-hui, 1978, The Cretaceous-Tertiary ostracods from the marginal region of the Yangtze-Han River plain in Central Hubei: Memoirs of Nanjing Institute of Geology and Palaeontology, Academia Sinica, no. 9, p. 129–206, Pls. 1–20 (in Chinese with English abstract).

Hou You-tang, Li Ying-pei, Cai Zhi-guo, and others, 1978, [Early Tertiary ostracod fauna from the coastal region of Bohai]: Science Press, p. 1–205, Pls. 1–83 (in Chinese).

Huang Bao-ren, 1964, Ostracods from Gansen District, Chaidam Basin: Acta Palaeontologica Sinica, v. 12, no. 2, p. 241–260, Pls. 1–5 (in Chinese with English abstract).

Lee, J. S. [Li Si-quang], 1924, *Grabauina,* a transitional form between *Fusulinella* and *Fusulina:* Bulletin of the Geological Society of China, v. 3, p. 51–54, Figs. 1–2.

——1927, Fusulinidae of North China: Palaeontologica Sinica, ser. B, v. 4, pt. 1, p. 1–123, Pls. 1–24, Figs. 1–21.

——1931, Distribution of the dominant types of the fusulinoid Foraminifera in the Chinese seas: Bulletin of the Geological Society of China, v. 10, p. 273–290, Pl. 1.

——1934, Taxonomic criteria of Fusulinidae with notes on Permian genera: National Research Institute of Geology Memoirs, no. 14, p. 1–21, Pls. 1–5, Figs. 1–9.

——1942, Note on a new fusulinid genus *Chusenella:* Bulletin of the Geological Society of China, v. 22, p. 171–174, Figs. 1–3.

Lee J. S. and Chen S. [Chen Xu], 1930, The Haunglung Limestone and its fauna: National Research Institute of Geology Memoirs, no. 9, p. 85–144, Pls. 2–15.

Li You-qui, 1966, [Cenozoic ostracodes from Lantian of Shaanxi, south of the Weihe River: Symposium of on-the-spot meeting concerning the Cenozoic stratigraphy in Lantian, Shaanxi]: Science Press, p. 255–276, Pls. 1–2 (in Chinese).

Lu Yen hao [Lu Yan-hao], 1944, The Charophyta from the Kucha Formation near Kucha Sinkiang: Bulletin of the Geological Society of China, v. 24, no. 1–2, p. 33–36, Pl. 1.

——1945, Additional note on the Charophyta from the Kucha Formation of Sinkiang: Bulletin of the Geological Society of China, v. 25, p. 273–277, Figs. 1–2.

——1948, On the occurrence of *Sycidium,* a Palaeozoic Charophyta in the Lunghuashan Formation of Poshi, eastern Yunnan: Fiftieth Anniversary Papers of the National Peking University, p. 69–76, Pl. 1, Fig. 1.

Mu Xinan, 1977, Upper Permian fossil Fungi from Anshun of Guizhou: Acta Palaeontologica Sinica, v. 16, no. 2, p. 151–158, Pls. 1–2 (in Chinese with English abstract).

Sheng, J. C. [Sheng, Jin-zhang], 1949, On the occurrence of *Zellia* from the Maping Limestone of Chengkung, Central Yunnan: Bulletin of the Geological Society of China, v. 29, nos. 1–4, p. 105–109, Pl. 1.

——1951, *Taitzehoella,* a new genus of fusulinids: Bulletin of the Geological Society of China, v. 31, nos. 1–4, p. 79–85, Pl. 1.

——1958a, Some Upper Carboniferous fusulinids from vicinity of Beiyin obo, Inner Mongolia: Acta Palaeontologica Sinica, v. 6, no. 1, p. 35–50, Pls. 1–2.

——1958b, Fusulinids from the Maokou Limestone of Chinghai: Acta Palaeontologica Sinica, v. 6, no. 3, p. 268–291, Pls. 1–4.

——1958c, Fusulinids from the Penchi Series of the Taitzeho Valley, Liaoning: Palaeontologica Sinica, new ser. B, no. 7, p. 1–110, Pls. 1–16.

——1962, [Fusulinids of China]: Science Press, p. 1–177, Pl. 1–27 (in Chinese).

——1976, [Radiolarian fauna from the Jilong Group of Qomolungma region *in* Reports on scientific investigations in the Qomolungma region, 1966–1968, Palaeontology, Part 2]: Science Press, p. 125–136, Pls. 1–2 (in Chinese).

Sheng Jin-zhang and Chang, L. H. [Zhang Lin-xin], 1958, Fusulinids from the type-locality of the Changhsing Limestone: Acta Palaeontologica Sinica, v. 6, no. 2, p. 205–214, Pl. 1.

Shi Cong-guang, 1960, [Carboniferous ostracods from Oulungbuluk of Qinghai, geology of the Qilian Range, Volume 4, Part 1]: Science Press, p. 149–160, Pls. 1–11 (in Chinese).

Song Zhi-chen, He Cheng-quan, Zheng Guo-quang, and others, 1978, [On the Paleogene dinoflagellates and acritarchs from the coastal region of Bohai]: Science Press, p. 1–190, Pls. 1–49 (in Chinese).

Wang Chen-yuan and Wang Zhi-hao, 1976 [Triassic conodonts from the Qomolungma Region *in* Reports on scientific investigations in the Qomolungma region, 1966–1968, Palaeontology, Part 2]: Science Press (in Chinese).

Wang Chen-yuan and Wang Zhi-hao, 1978 [Early and Middle Devonian conodonts of Guangxi and Yunnan]: Beijing, Geological Press, Papers in Symposium on Devonian Stratigraphy of South China, p. 334–345, Pl. 39–41 (in Chinese).

Wang Ping-xian and Lin Jing-xing, 1974, Discovery of Paleogene brackish-water foraminifers in a certain basin, Central China, and its significance: Acta Geologica Sinica, no. 2, p. 175–181, Pl. 1.

Wang Ping-xian, Min Qiu-bao, Lin Jing-xing, and Cui Zhan-tang, 1975, [Discovery of brackish-water foraminiferal faunas in some Cenozoic basins of East China and its significance]: Beijing, Geological Press, Professional Papers of Stratigraphy and Palaeontology, No. 2, p. 1–36, Pl. 171–172 (in Chinese).

Wang Shui, 1961, Tertiary Charophyta from Chaidamu Basin, Qinghai: Acta Palaeontologica Sinica, v. 9, no. 3, p. 183–233, Pl. 1–7 (in Chinese with English abstract).

——1965, Mesozoic and Tertiary charophytes from Jiuquan Basin of Kansu Province: Acta Palaeontologica Sinica, v. 13, no. 3, p. 463–509, Pl. 1–5 (in Chinese with English abstract).

Wang Shui and Chang Shan-jean [Zhang Shan-zhen], 1956, On the occurrence of *Sycidium melo* var. *pskowensis* Karpinsky from the Devonian of Northern Szechuan: Acta PalaeontologicaSinica, v. 4, no. 3, p. 381–386, Pl. 1.

Wang Shui, Wang Zhen, and Huang Ren-jing, 1976, [Mesozoic and Cenozoic charophytes of Yunnan *in* Mesozoic fossils of Yunnan, Part 1]: Science Press, p. 65–86, Pl. 1–6 (in Chinese).

Wang Shui, Yang Chen-qiong, Li Hua-nan, and others, 1978, [Early Tertiary Charophytes from coastal region of Bohai]: Science Press, p. 1–49, Pl. 1–23 (in Chinese).

Wang Yu-jing, 1976, [Late Cretaceous and early Tertiary calcareous algae from the Qomolungma Region *in* Reports on scientific investigations in Qomolungma region, 1966–1968, Palaeontology, Part 2]: Science Press, p. 425–257, Pls. 1–12 (in Chinese).

Wang Zhen, 1978a, Cretaceous charophytes from the Yangtze-Han River Basin with a note on the classification of Porocharaceae and Characeae: Memoirs of Nanjing Institute of Geology and Palaeontology, Academia Sinica, no. 9, p. 61–100, Pls. 1–8 (in Chinese with English abstract).

——1978b, Paleogene charophytes from the Yangtze-Han River Basin: Memoirs of Nanjing Institute of Geology and Palaeontology, Academia Sinica, no. 9, p. 120–205, Pls. 1–5 (in Chinese with English abstract).

Wang Zhi-hao, 1978, Permian–Lower Triassic conodonts of the Liangshan area, southern Shaanxi: Acta Palaeontologica Sinica, v. 17, no. 2, p. 213–229, Pls. 1–2 (in Chinese with English abstract).

Yang Jing-zhi and Hu Zhao-xun, 1962, [Fossil Bryozoans of China]: Science Press, p. 1–82, Pls. 1–28 (in Chinese).

Yang Jing-zhi and Xia Feng-sheng, 1975, [Bryozoan fossils from the Qomolungma Region *in* Reports on scientific investigations of the Qomolungma region, 1966–1968, Palaeontology, Part 1]: Science Press, p. 39–70, Pl. 1–8 (in Chinese).

Yang Jing-zhi, and Wang Cheng-yuan, 1975, [Stromatoporoids and Hydrozoans from the Qomolungma Region *in* Reports on scientific investigations in the Qomolungma region, 1966–1968, Palaeontology, Part 1]: Science Press, p. 71–82, Pl. 1–14 (in Chinese).

Ye Chun-hui, Gou Yun-xian, and others, 1977, [Mesozoic and Cenozoic ostracod faunas from Yunnan *in* Mesozoic fossils of Yunnan, Part 2]: Science Press, p. 153–330, Pl. 1–24 (in Chinese).

Ye De-quan and Ding Lian-sheng, 1976, [Cretaceous ostracods from the Songliao Basin]: Science Press, p. 1–102, Pl. 1–37 (in Chinese).

Yoh, S. S. [Yue Sen-xun], 1932, Fossil corals in the Chihsia Limestone of the lower Yangtze region: Palaeontologica Sinca, ser. B, v. 8, pt. 1, p. 1–48, Pl. 1–5, Figs. 1–2.

Zhong Xiao-chun, 1964, Upper Triassic and Middle Jurassic ostracods from the Ordos Basin: Acta Palaeontologica Sinica, v. 12, no. 3, p. 426–465, Pl. 1–2 (in Chinese with English abstract).

MANUSCRIPT ACCEPTED BY THE SOCIETY APRIL 22, 1981

Printed in U.S.A.

Geological Society of America
Special Paper 187
1981

Vertebrate paleontology in China, 1949–1979

Zhou Ming-zhen
Institute of Vertebrate Palaeontology and Palaeoanthropology, Academia Sinica, Beijing, People's Republic of China

HISTORICAL REVIEW

In China the scientific study of fossil vertebrates probably dates back to the year 1846 when H. Falconer, an English scholar then working in India, reported some late Quaternary mammals that were collected by early British explorers from a site at Niti Pass on the northern slope of the Himalayas in southern Tibet. It is quite unexpected and strange that the starting point of our science was in this remote place, which was hardly accessible to scientists in general, instead of in a more developed coastal region as were other branches of natural science. Paleontologists generally consider the year 1870 as the beginning of the science of vertebrate paleontology in China. This year was marked by the publication of a paper entitled "On Fossil Remains of Mammals Found in China" by the great English anatomist Sir Richard Owen (1870). From then on up to 1927 occasional papers and a few monographs on Chinese fossil vertebrates appeared from time to time, but most of the materials studied were either purchased from Chinese drugstores or were merely surface material collected by amateur collectors, and the localities and horizons of many of these fossils were unknown. It was not until the early 1920s that fossil collections with authentic stratigraphical data were being assembled for the first time, notably through the explorations of the Central Asiatic Expeditions of the American Museum of Natural History in Inner Mongolia.

The publication in 1927 of the monograph "Fossile Nagetiere aus Nord-China" by C. C. Young [Young Chung-chien], one of the founders of the Palaeontological Society of China, is historically important in being the first paper published by a Chinese vertebrate paleontologist. Along with the work carried out simultaneously then at Chou-k'ou-t'ien, the subsequent founding of the Laboratory of Cenozoic Studies in 1929 may well be considered the birth of Chinese vertebrate paleontology. In the following two decades, work in vertebrate paleontology progressed steadily in this country, and scientifically, the formerly almost unexplored territories of China became the most active centers of this research, especially in the domain of fossil mammals and paleoanthropology and the relevant biostratigraphic problems of the Late Cenozoic. The studies of reptilian fossils played but a minor role then, and those of other vertebrate groups were hardly touched upon. Nevertheless, work done before 1949 laid the foundations for the study of fossil vertebrates in China.

The years 1949 and especially 1953 were turning points in the history of vertebrate paleontology in China. Soon after the founding of the People's Republic of China in the fall of 1949, excavation work at Chou-k'ou-t'ien was resumed after an interruption of some 10 years. In 1951, a team from the Laboratory for Vertebrate Palaeontology and Cenozoic Studies, then a subsidiary of the Palaeontological Institute of Nanking, Chinese Academy of Sciences, carried out extensive field excavations of dinosaur remains from the Cretaceous beds of Laiyang, Shantung. This was the first large-scale excavation of dinosaur fossils made in this country, resulting in the discovery of the skeletons of *Tsintaosaurus* and *Psitacosaurus,* now housed in the Museum of the Institute of Vertebrate Palaeontology and Palaeoanthropology.

Work actually began in full force in 1953 when the former laboratory became an independent institution, starting with a staff of about 20, including 5 paleontologists headed by Young and Pei. In the middle of the sixties, the research and technical staff in the new institute had increased to 100, out of a total of 150 persons. It had 4 research departments, a museum, and three field stations (Chou-k'ou-t'ien, Taiyuan, and Lantien) which were well equipped with laboratory and library facilities. The Institute published three serials including the *Vertebrata Palasiatica* (Quarterly) which was then the first and only periodical in the world devoted exclusively to the publication of papers on vertebrate paleontology. Despite its comparatively stagnant or "retarded" evolution during the period between 1966 and 1976, the Institute kept on growing, though slowly. It now has a staff of 260, including 150 scientific workers, and is expecting a further expansion in all aspects of its activities. Thirty years ago the Institute was essentially the only institution that "monopolized" research in vertebrate paleontology in China. Today the subject is being taught and studied by faculty members in five or six universities and technical colleges, and there are four museums of natural history in Peking, Shanghai, Tienkin, and Chungking, as well as a number of provincial museums with departments of paleontology that maintain exhibitions and take an active part in field exploration and research in the fields of vertebrate paleontology and paleoanthropology.

SOME MORE IMPORTANT ACHIEVEMENTS

Following is a brief survey of some of the more important aspects of progress of vertebrate paleontology in China, with emphasis on Devonian fish faunas, early Mesozoic reptiles, and early Tertiary mammals.

Paleoichthyology or Fossil Fishes

Devonian Fishes. The Devonian period has often been called "the Age of Fishes," mainly because early fishes and fishlike primitive vertebrates (agnathans) throw an important light on the origin, early radiation, and evolution of the vertebrates. Twenty-five years ago our knowledge of the Devonian fishes of China was almost a blank. Now there is plenty of concrete evidence showing that fossil remains of Devonian fishes are very rich and beautifully preserved and that the faunas contain many interesting endemic forms as well as other forms that are cosmopolitan in distribution.

Fossil Agnatha are known to occur extensively in the Lower Devonian of South China. Most of them have been allocated to the two subclasses Polybranchiaspida and Galeaspida. Both show high diversification at the generic level. The former group is represented by the genus *Polybranchiaspis,* and numerous other genera, which are the dominant fossil types in the nonmarine Lower Devonian of China.

The Arthrodires are the earliest of the jawed vertebrates, and their remains are also widespread and abundantly represented throughout the Devonian System. Increasing numbers of fossils of Dolichothoraci, such as *Arctolepis,* have been recovered from the southwestern provinces of Yunnan, Kweichow, and Kwangsi. The Brachythoraci first reported by Liu Hsien-ting from the Middle Devonian of northern Szechuan are predominantly marine and are now known to occur also in Yunnan and Kwangsi. Generally speaking, macropetalichthyiform arthrodires are rare as fossils and show little variety, but they appear to be rather common in China and also occur lower in the stratigraphic column. Their remains have recently been found in Silurian rocks, probably the earliest record so far known in the world.

The antiarchs are a large group of Devonian fishes, and considerable progress has been made in the last 30 years in their exploration and study. The question of the age of the *Bothriolepis*-bearing Tiaomachien Formation of Hunan has been debated since 1940. It is now known to be of Middle rather than Late Devonian age, through intensive field investigations and a series of discoveries of fossil antiarchs. In general, the stratigraphic horizons of these fishes, as in the case of the above-mentioned arthrodires, are apparently lower in the stratigraphic section than occurrences in other parts of the world. Although they occur also in the Upper Devonian beds, most of the Chinese antiarchs, as represented by the genus *Sinolepis* (Liu and Pan, 1958), are of Late Devonian age. Of greater interest is the recovery of rich faunas of Early Devonian antiarchs with the leading genus *Yunnanolepis.* These are the major components of the Early Devonian ichthyofauna of South China. *Yunnanolepis* and the other genera related to it show characters that are not only morphologically more primitive but also indicative of a closer relationship between antiarchans and arthrodires.

Interest in the study of the acanthodians has increased considerably in recent years, because they were thought to contain within them the stem group of all the osteichthyans. Their fossils are only meagerly represented by a few fragmentary specimens of problematic forms.

The Crossopterygians, especially the rhipidistians, are important in being the group containing forms of the ancestral group that gave rise to the earliest land-dwelling tetrapods. For a long time their fossil remains were practically unknown in China, and comparatively rare in the rest of the world as well. Great progress made in recent years is recovering a considerable number of good specimens from Yunnan and Kwangsi, some of them with well preserved endocrania.

Actinopterygians. The actinopterygians are mainly Late Paleozoic and Mesozoic fishes. During the field seasons of 1963 and 1974, members of the Institute of Vertebrate Palaeontology and Palaeoanthropology expeditions to Sinkiang collected a great number of well-preserved, Late Permian paleoniscoids and advanced genera of holosteans. The materials described by Liu, Su, and others indicated a strong element of endemism in many aspects of these faunas. The primitive acipenserid *Peipiaosteus* from the Jurassic of Peipiao, Liaoning, is among the earliest known and most primitive form of this group.

The early lycopteroid teleosts of North China are well known to vertebrate paleontologists, as well as to geologists, because of their wide geographical distribution and stratigraphical implications. The first important paper on this and related groups is that by Takai (1941), based on the materials collected from the northeastern provinces. Those of North China were studied by Liu and others (1963). Since then a great amount of data has been accumulated, but opinions still diverge as to their classification and phylogenetic relationships, as well as to the geological age of the faunas.

Since 1950 a large amount of teleost fossils has also been collected from the late Mesozoic beds in the southeastern provinces. The materials have been summarized in a recent paper by Chang and Chou (1977). Two distinct ichthyofaunas have been identified, one represented by the leading form *Mesoclupea showchangensis,* the other by *Paralycoptera wui, Pingolepis polyurocentralis,* and *Huashia gracilis.* The former fauna is considered to be Late Jurassic, the latter Early Cretaceous.

Other Important Finds. Other finds include those of *Petalodus, Sinohelicoprion, Cladodus,* and *Saurichtys,* and marine forms such as *Tibetodus, Sinorithoyes, Sinolepidotus,* and *Peltopleurus.* Cenozoic fishes ranging in age from Eocene upward have been found in abundance, but relatively few detailed descriptions have been published so far.

Lower Tetrapods

Investigations of the lower tetrapods (amphibians and reptiles), especially reptiles, have been particularly active in the past three decades. Thirty years ago our knowledge concerning the early reptiles of the Late Permian and earliest Triassic was mostly derived from the study of materials from South Africa and eastern Europe. Since the late 1960s, much new information has come to light from East Africa, India, and especially from South America. Progress made along this line in China is also quite impressive and significant. Many Gondwanan elements are known, indicating the presence, during these periods, of land connections between the northern and the southern continents, as well as the existence of a Permian-Triassic Pangaea. Land vertebrate faunas, in this regard, are important for their zoogeographical significance. Most of the discoveries in China were made after 1955, first in Shansi and later in Sinkiang and in other regions in the middle of the 1960s.

The *Sinokannemeyeria* faunal complex, known mainly from the Er-Ma-Ying Formation of Shansi and its neighboring districts, clearly shows a relationship with that of the South African Karroo Formation. The whole section is now divided in descending order into four horizons or faunal zones:

1. Upper Er-Ma-Ying, containing typical *Sinokannemeyeria* fauna, occurring in association with primitive thecodonts *(Shansisuchus), Neoprocolophon,* and *Sinognathus,* with labyrinthodont amphibians represented only by fragmentary material.
2. Lower Er-Ma-Ying, containing large dicynodonts, *Parakannemeyeria,* pseudosuchians, and procolophonids, differing from the above by the presence of different genera or species of the same group, by with *Scalophosaurus,* an important component in the Triassic of South Africa which is absent in the upper Er-Ma-Ying.
3. Hoshankou Formation, characterized by the absence of dicynodonts and increase in the proportionate number of procholophonids and probably *Protosaurus.*
4. Shichienfeng Formation and uppermost part of Shihotze Formation, containing *Pariesaurus* and deinocephalians in place of dicynodonts, procolophonids, and pseudosuchians.

To the west, in Sinkiang, where *Lystrosaurus* has long been known to occur, four faunal zones can be recognized. They are, in descending order:

1. Uppermost zone with labyrinthodonts and pseudosuchians.
2. *Kannemeyeria* zone (Karamayi Formation), a fauna with large dicynodonts as its dominating group and theocodonts.
3. *Lystrosaurus zone* (Chiutsaiyuan Formation) with *Chasmatosaurus, Shantaisaurus,* and possibly *Urumchia.*
4. *"Dicynodon"* zone.

Of those four faunal zones, the *Kannemeyeria* zone occurs in both regions and can be considered as being closely correlated. The Late Permian faunas of the two regions are distinctly different from each other, with a dominance of *Pariesaurus* in one and dicynodonts in the other. The material from Sinkiang has been described in a series of papers by Young and Sun, but the collections from North China, with the exception of those from the type locality in eastern Shansi, are still under investigation. It can be clearly seen, however, that the reptilian faunas of Sinkiang from the Upper Permian through the upper part of Middle Triassic are strikingly similar to those of South Africa in faunal sequence and composition. On the other hand, in North China the upper zones of the Triassic faunas are similar to those in Sinkiang, and also to those of South Africa, but the lower or Late Permian zones are quite different and more similar to the *Pariesaurus* fauna of northern European Russia.

A Middle Triassic reptilian fauna recently discovered in Shansi, Hunan, with the peculiar thecodont genus *Lotosaurus* (Chang, 1975) is of interest, but its real significance awaits further investigation. Formerly, fossils of marine and flying reptiles were essentially unknown in this country; now a fair amount of information has come to light with the discovery of quite a few beautiful specimens.

Another theromorphic reptile group to which important contributions have been made is the tritylodonts for which new data have been added from materials from Szechuan.

Dinosaurian Faunas

The study of dinosaurs has a comparatively longer history in China, but progress in the past 30 years is also impressive. Several important collections have been made, and many contain fairly complete skeletons, ranging in age from the latest Triassic through Late Cretaceous. Dinosaurian fossils are now known from many provinces, including the eastern part of Tibet (Changtu region). Twenty years ago dinosaurian remains were known only from the eastern part of northern China and a few isolated localities in the southwest. Now, all the major dinosaurian groups have been found, and it is impossible to give here even a brief account of them. Most of the materials have been described in the many monographs and papers by Young, Chao (H.C.), Tong, and others. In addition, the microhistological and biochemical studies made on dinosaurian eggshells and bones are worth mentioning. The taxonomic and phylogenetical classification proposed by Chao Tsu-kuei based on the microscopic structure of dinosaurian eggshells in coordination with stratigraphic data are not only of interest biologically but have proved to be valuable, at least for certain horizons and districts, in geological dating and paleontological studies. Other reptilian orders include the Testudinata (tortoises and turtles; studied mostly by Yeh), Squamata, including snakes and lizards (Sun, Hou, and others), and crocodiles (Young, Chow, Li, and others).

Finally, the recovery at Shantung of three well-preserved skeletons of Miocene birds from diatomaceous earth may be mentioned, one of which has been described by Yeh as a new phasianid.

PALEOMAMMALOGY

This branch of vertebrate paleontology has the longest history of study in China. Progress made in the past three decades has been uneven with regard to the different mammalian orders, as well as the special distribution of the fossils. Some important mammalian orders were unknown and even believed to be absent from the palearctic regions.

Paleocene

Little progress has been made with study of the earliest mammals, with the exception of the earliest Rhaetic, or Liassic, *Eozostrodon (= Morganucodon)* and its probable allies. However, highly significant is the discovery of an essentially complete stratigraphic sequence of Paleocene age, containing three or four mammalian assemblages, or faunal zones. Continental deposits of this epoch remained unknown for a long time and before 1959 were even assumed to be entirely absent in this country. In Asia outside of China only two nearby localities with mammal-bearing beds had been recorded in the southern Gobi of Mongolia. Essentially, they represent a single fauna (Gashato) which has been intensively investigated by American, Soviet, Polish, and Mongolian paleontologists since its first discovery in 1923. The fauna, which was formerly thought to be one of late or latest Paleocene, apparently shows characteristics of the Paleocene-Eocene transition.

In 1960, the presence of a primitive uintathere *(Prodinoceros)* from the Turfan basin of Sinkiang reported by Chow suggested the probable presence of continental sediments of the Gashatan (then considered to be of late Paleocene age) in Northwest China and Inner Mongolia. Several expeditions to Sinkiang in the middle 1960s and to Inner Mongolia (Erlien District) in the 1970s, organized by the Institute of Vertebrate Palaeontology with a view to verifying this, proved highly successful. It is of special interest that the age of the Gashato fauna was found to be probably earliest Eocene rather than late Paleocene.

Better documented and more important mammal-bearing formations are now known from South China in a number of localities in Kwantung, Kiangsi, Hunan, and Anhwei. Over 100 new species assigned to 20 families representing 17 mammalian orders have been described. The faunal zones or fossil assemblages from the Shanghu Formation of the Nanhsiung basin in Kwantung and the lower part of Wanghutun Formation of the Chienshan basin in Anhwei are of early and middle Paleocene age and represent the oldest known Cenozoic horizons anywhere in the world outside the Rocky Mountain regions of North America.

The faunas of the Nonshan Formation (Nanhsiung), Toumu Formation (Chienshan), and Shikiang Formation (Tayu, Kiangsi) include horizons roughly corresponding to the Gashato in the upper parts and, with one or two additional faunal zones in the underlying beds of late Paleocene age, are evidently older than the Gashatan. Study of the Paleocene mammals indicates that in China, as well as in other regions of northern Asia, the mammalian fauna of this period contains endemic groups, such as anagalids, Bemalamdidae, Archaeolambdidae, and Phenacolophidae. Members of these endemic groups in general consitute the overwhelming majority (about 80% or more) in each faunal complex at the generic level. This indicates that during most of Paleocene time, China and its adjacent regions were isolated from North America by seaways in the east and in the north and from Europe and western Asia by the "Trufai Strait" and Tethys seas. Besides, it is puzzling to find a primitive type of edentate, along with several forms of primitive notoungulates and other groups, generally thought to be confined to South America before Eocene time.

Eocene

Mammalian faunas of Eocene age had been documented earlier but were largely of later Eocene age. However, in recent years important progress has been made on the previously little known middle Eocene faunas and on early Eocene faunas, which were entirely unknown before.

The discovery of *Homogalax* (in Shantung), *Heptodon* (Shantung and Sinkiang), *Coryphodon* (widely distributed in the north, south, and west), and *Hyopsodus* (Sinkiang), all typical of the North American early Eocene (Wasatchian), are of considerable stratigraphic and zoogeographic significance.

Recently, the discovery of an early Eocene assemblage from beds containing fossils of the Nomogen and Naran Bulak faunas challenges the age assignment of this local fauna of the Gashatan horizon. From the same site and horizon where a fossil of "Propaleotherium" had been found in the late 1930s, some new forms, more or less typical of the early Eocene, were found. This clearly indicates that the barriers to free migration of mammals between eastern Asia and North America through Beringia were probably no longer in existence immediately after the very beginning of the Eocene.

Mammalian fossils of the middle Eocene were formerly poorly represented in China. Not until recent years were new finds made. The most important of these were by Qi Tao in 1977 who collected a large number of fossils from beds, formerly considered almost barren, of the Arshto Formation in Inner Mongolia. The fauna now seems to be slightly older than that of Kuan-chuang in Shantung.

Another new locality worth mentioning is that in Kweichow in Southwest China where primitive titanotheres and equids have recently been found, probably indicating the presence of middle Eocene.

Late Eocene faunas have had the longest history of study and remain the best known of the Chinese Paleogene. Formerly most of the work was done in Inner Mongolia in addition to a few sites in North China. In the past 25 years or so, considerable new material has been accumulated from various parts of this country. More important are those of Honan (Lushih), Lantian (Shensi), Base (Kwangsi), and Lunan and Likiang (Yunnan). In China and eastern Asia, the late Eocene assemblages were clearly differentiated into faunas of two substages, an upper one or Sharamuran, and a lower or Irdin Manhan. The local faunas of South China contain elements in common with those of Burmese Pongdong, and these extend farther south to Kalimantan, westward to Kazakhstan, and to the Maritime Province of the Soviet Far East, North Korea, and Japan to the east. The most characteristic features of the southern faunas is the presence of a large number of genera and species of anthracotheriid Artiodactyla, such as *Anthracokeryx, Anthracothema,* and others, and the helaletid Perissodactyla. The late Eocene mammals are urgently in need of systematic revision.

Oligocene

Recent work has been conducted mainly in the western parts of Inner Mongolia and in Kwangsi and Yunnan. Mammalian faunas of Oligocene age have been much less studied than those of the Eocene, though fossils are abundant, especially in the northern parts of the country.

Miocene

In spite of the fact that the Miocene is generally considered as the "golden age of mammals," not many localities and fossil forms are known in China. These include the fauna with *Anchitherium* in Nanking, the faunas of Hsiachaowan (Kiangsu), Zhihsien (Hopei), and Lantien (Shensi), and many isolated localities with a few but representative forms in North China in addition to that of Tunggur (Inner Mongolia). Most of these occurrences are of middle Miocene age.

Pliocene

Researches on mammals of this epoch were well advanced prior to the 1940s, but little further attention has been paid to them until quite recently. Recent discoveries made in Tibet (*Hipparion* fauna, at an altitude of 4560 m) and Yunnan (with *Ramapithecus*) and the Lower Yangtze Valley (*Hipparion* fauna) are among the most important ones and have attracted wide general interest.

Quaternary

Great progress has likewise been made on the study of Quaternary mammals, though most of the work in this field has been coordinated with paleoanthropological and paleolithic researches. The investigations carried out in Kwangsi on the *Gigantopithecus*-fauna (by Pei and others), at Yuanmo, Yunnan, at Lantien in Shensi, and at Tingtsen and Sihouto in Shansi, are most noteworthy.

CONCLUDING REMARKS

Despite the fact that a large amount of work has been done in the past, the study of the vertebrate paleontology in China may still be considered to be in the early state of its exploration and development. This is particularly true for early fishes, certain early tetrapods, and early mammals. So far, we have gained but a first dim glimpse of the true picture of vertebrate history of this part of the world.

So far as we can evaluate it, Devonian fishes, Triassic reptiles, especially theromorphs, and the earliest mammals, various early and late dinosaurian groups, and Paleogene mammals are particulary abundant and of great biological and geological significance. The study of these fossils deserves immediate close attention.

Past achievements, however great and successful they may appear to be, are only at the state of development reached in many developed countries before World War II.

Our work in the past five decades is for the most part in the nature of the accumulation of materials and basic data. In view of present worldwide progress of the science, more intensive research and systematic investigations should be emphasized, either in field exploration and stratigraphical and ecological studies or on the fossils and various vertebrate groups themselves. These should all be in accord with current concepts and take full advantage of modern equipment and techniques.

SELECTED REFERENCES

Black, D., Teilhard de Chardin, P., Young, C. C. [Young Chung-shien] Pei, W. C. [Pei Wen-chung], 1933, Fossil man in China: Memoirs of the Geological Survey of China, ser. A, No. 11, p. 1–166, 28 figs.

Chang Fa-kui [Zhang Fa-kui], 1975, A new Thecodont *Lotosaurus* from middle Triassic of Hunan: Vertebrate PalAsiatica, v. 13, no. 3, p. 144–147, 1 pl. (in Chinese with English summary).

Chang Mi-mann and Chou Chia-chien, 1977, On late Mesozoic fossil fishes from Zhejiang Province, China: Memoirs of the Institute of Vertebrate Palaeontology and Palaeoanthropology, no. 12, p. 1–59, 25 pls. (in Chinese with English summary).

Chi, Y. S. 1940, On the discovery of Bothriolepis in the Devonian of central Hunan: Bulletin of the Geological Society of China, v. 20, p. 57–72, 3 figs., 1 pl.

Chow, M. C. [Zhou Ming-zhen] and Qi Tao, 1978, Paleocene mammalian fossils from Nomogen Formation of Inner Mongolia: Vertebrata PalAsiatica, v. 16, no. 2, p. 77–85, 9 pls. (in Chinese with English summary).

Ding Su-yin, 1979, A new edentate from the Paleocene of Guangdong: Vertebrata PalAsiatica, v. 17, no. 1, p. 57–64 (in Chinese with English summary).

Hu Chang-kang and Qi Tao, 1978, Gongwangling Pleistocene mammalian fauna of Lantan, Shanxi: Paleontologia Sinica, new ser. C, no. 21, p. 1–65, 15 pls. (in Chinese with English summary).

Li Chuan-kuei, 1977, Paleocene eurymyloids (Anagalidlae, Mammalia) of Qianshan, Anhui: Vertebrata PalAsiatica, v. 15, no. 2, p. 103–118, 2 pls. (in Chinese with English summary).

Liu Hsien-t'ing, 1955, *Kiangyousteus*, a new arthrodiran fish from Szechuan, China: Acta Paleontologica Sinica, v. 3, no. 4, p. 261–274, 5 pls. (in Chinese with English summary).

Liu Hsien-t'ing, Su Te-tsao, Huang Wei-lung, and Chang Kuo-jui, 1963, Lycopterid fishes from North China: Memoirs of the Institute of Vertebrate Palaeontology and Palaeoanthropology, no. 6, p. 1–53, 19 pls. (in Chinese with English summary).

Liu Tung-sen and P'an Kiang, 1958, Devonian fishes from Wutung Series near Nanking, China: Paleontologia Sinica, new ser. C. no. 15, p. 1–41, 10 pls. (in Chinese with English version).

Liu Yu-hai, 1973, On the new forms of Polyranchiaspiforms and Petalichthyida from Devonian of southwest China: Vertebrata PalAsiatica, v. 11, no. 2, p. 132–143, 2 pls. (in Chinese).

Liu Yu-hai, 1975, Lower Devonian agnathans of Yunnan and Sichuan: Vertebrata PalAsiatica, v. 13, no. 4, p. 202–216, 4 pls. (in Chinese with English summary).

Owen, R. C., 1870, On fossil remains of mammals found in China: Quarterly Journal of the Geological Society of London, v. XXVI, p. 417.

Pei Wen-chung, 1965, Excavation of Liucheng *Gigantopithecus* Cave and exploration of other caves in Kwangsi: Memoirs of the Institute of Vertebrate Paleontology and Paleoanthropology, no. 7, p. 1–54, 11 figs. (in Chinese with English summary).

Qiu Zhan-xiang, Li Chuan-kuei, Huang Xue-shi, Tang Ying-jun, Xu Qin-qi, Yan De-fa, and Zhang Hong, 1977, Continental Paleocene stratigraphy of Qianshan and Xuancheng Basin, Anhui: Vertebrata PalAsiatica, v. 15, no. 2, p. 85–93 (in Chinese).

Sun Ai-ling, 1963, The Chinese kannemeyerids: Paleontologia Sinica, new ser. C, no. 17, p. 1–109, 59 figs., 9 pls. (in Chinese with English version).

Takai, F., 1943, A monograph on the lycopterid fishes from the Mesozoic of Eastern Asia: Journal of the Faculty of Science, University of Tokyo, sec. 2, v. 6, p. 207–270.

Teilhard de Chardin, P., and Leroy, P., 1942, Chinese fossil mammals (A complete bibliography analyzed, tabulated, annotated and indexed): Peking, Institut de Géo-Biologie, no. 8, p. 1–142.

Yeh Hsiang-k'uei, 1963, Fossil turtles of China: Paleontologia Sinica, new ser. C, no. 18, p. 1–112, 34 figs. 21 pls. (in Chinese with English version).

Young, C. C. [Young Chung-chien], 1927, Fossile Nagetiere aus Nord-China: Paleontologia Sinica, ser. C, v. 5, fasc. 3, p. 1–78, 3 pls.

Young, C. C., 1941, A complete osteology of *Lufengosaurus huenei* Young (gen. et sp. nov.) from Lufeng, Yunnan, China: Paleontologia Sinica, new ser. C, no. 7, p. 1–53, 25 figs., 6 pls.

Young, C. C., 1958, The dinosaurian remains of Laiyang, Shantung: Paleontologia Sinica, new ser. C, no. 16, p. 1–138, 61 fits. 6 pls. (in Chinese with English version).

Young, C. C., 1964, The Pseudosuchians in China: Paleontologia Sinica, new ser. C, no. 19, p. 1–205, 64 figs., 10 pls. (in Chinese with English version).

Young Chung-chien [Young, C. C.] and Dong Zhi-ming, 1972, Triassic marine reptiles in China: Memoirs of the Institute of Vertebrate Paleontology and Paleoanthropology, no. 9, p. 1–34, 13 pls. (in Chinese).

Young Chung-chien, Dong Zhi-ming, and Yeh Hsiang'k'uei, 1973, Pterosaurian fauna from Wuerho, Singkiang: Memoirs of the Institute of

Vertebrate Paleontology and Paleoanthropology, no. 11, p. 1–51, 7 pls. (in Chinese).

Young Chung-chien, Liu Hsian-t'ing, Sun Ai-ling, and Ma Feng-chen, 1973, [Permo-Triassic vertebrate fossil of Turfan Basin]: Memoirs of the Institute of Vertebrate Paleontology and Paleoanthropology, no. 10, p. 1–68, 9 pls. (in Chinese).

Zhai Ren-jie, Zheng Jia-jian, Tong Yong-sheng, Xu Yu-xuan, and Wang Jin-wen, 1978, [Tertiary stratigraphy and mammalian fossils of Turfan Basin]: Memoirs of the Institute of Vertebrate Paleontology and Paleoanthropology, no. 13, p. 1–140, 32 pls. (in Chinese).

Zhou Ming-zhen [Chow, M. C.] and Zhang Yu-ping, 1974, Fossil Proboscideans in China: Science Press, p. 1–74, 32 pls. (in Chinese).

Zhou Ming-zhen, Zhang Yu-ping, Wang Ban-yue, and Ding Su-yin, 1977, Mammalian fauna from the Paleocene of Nanxiong Basin, Guangdong: Paleontologia Sinica, new ser. C, no. 20, p. 1–100, 28 pls. (in Chinese with English summary).

MANUSCRIPT ACCEPTED BY THE SOCIETY APRIL 22, 1981

Printed in U.S.A.

Geological Society of America
Special Paper 187
1981

Thirty years of paleobotany in China, 1949–1979

Li Xing-xue
Zhou Zhi-yan
Song Zhi-chen
Ouyang Shu
Nanjing Institute of Geology and Palaeontology, Academia Sinica, Nanjing 21008, People's Republic of China

INTRODUCTION

The recognition of fossil plants in China may be traced back to the Sung Dynasty. One of the forerunners in this respect was a great Chinese scholar, Shen-gua [Sun-kou, 1029–1093]. He happened to witness a landslide in a region near Yenan, North Shaanxi. The landslide uncovered a fossil forest of what he interpreted as petrified bamboo. (The so-called fossil bamboos are probably *Neocalamites* of the Mesozoic of North Shaanxi). As bamboos did not at that time grow in North Shaanxi, he deduced that in the past the climate must have been different. This might be the earliest record of such an inference. This important discovery with interpretations was recorded in his famous book *Meng Chi Pi Tang,* Volume 21. In this connection, it should be pointed out that in Europe it was more than 400 years later that the well-known Italian philosopher Leonardo da Vinci (1452–1519) happened first to realize that fossils were actually remains of organisms and that they are not *lusus naturae.* In the past century and up to the 1920s, almost all paleobotanical research in China was done by foreign scholars; it was not until the 1930s that China's own paleobotanists, such as the late Professor Sze Hsin Chien and Professor Pan Zhong-xiang [C. H. Pan], began to work in this field. Before that, only a few paleobotanical studies were taken up by several Chinese geologists and botanists as a sideline.

Since 1949, the few paleobotanists who worked either in the former Institute of Geology, Academia Sinica, or the former Geological Survey of China, with very limited facilities, have been gathered together in the Nanjing Institute of Geology and Palaeontology (before 1958, under the name "The Institute of Palaeontology"), Academia Sinica, where a Palaeobotanical Section, the first paleobotanical organization in China, was established. In the same institute, the Laboratory of Palynology was founded in 1953. In 1961, the Laboratory of Palaeobotany (including Palynology) was set up in the Institute of Botany, Academia Sinica. Since then, much work has been done in the laboratories of Nanjing and Beijing. Nowadays the Laboratory of Palaeobotany and the Laboratory of Palynology in the Nanjing Institute have a staff of more than 50 research workers and are well equipped with sufficient facilities and some advanced instruments, such as a scanning electron microscope and a transmission electron microscope. Besides, much attention has also been paid to various research fields, such as Precambrian unicellular blue-green algae, charophytes, dinoflagellates, acritarchs, fungi, and fossil bacteria. In the Laboratory of Palaeobotany of the Institute of Botany, more than 15 research members are mainly engaged in studying plant remains from Devonian to Quaternary. They have also started to study Precambrian iron bacteria and Permian coal balls which are little known in China today. In addition, there are about 20 research workers in the Laboratory of Palaeobotany and Palynology of the Academy of Geological Sciences belonging to the Ministry of Geology of the People's Republic of China. The three organizations mentioned above are the main research centers of paleobotany and palynology in China. Since the 1950s, many paleobotanists and palynologists have been trained in these laboratories. Now they work either in universities, colleges, or in other geological organizations. It is estimated that there are about 400 workers in the field of palynology, nearly 100 in plant megafossils, and more than 100, including the Precambrian microbiologists, engaged in studying fossil algae. It is obvious that the research in paleobotany in China has made great headway during the past 30 years, and the main results are summarized below.

HIGHER PLANTS OR VASCULAR PLANTS

Vascular plant fossils ranging from Devonian to Quaternary were the chief subject of paleobotanical studies before 1950. Although up to that time more than 100 papers had been published, they were usually limited to a few fields of paleobotanical studies.

Since 1950, much progress has been made in the study of Devonian floras. As early as 1952, Sze published a monograph on the Late Devonian plants of China. Since then, Late Devonian plants have been found in various parts of China. During recent years, restudying the famous localities of Devonian plants in eastern Yunnan, Xu-Ren [formerly J. Hsü] discovered the genus *Zosterophyllum* in 1966. Subsequently, Li Xing-zue [H. H. Lee] and Cai Chong-yang made a large collection of Devonian plants in southern China between 1972 and 1975, from which they described more than 20 species of *Zosterophyllum*, a far greater number than known anywhere outside China. In 1978, these authors made a detailed study of a typical section of Lower Devonian deposits, the Tsuifengshan series in eastern Yunnan, and tentatively proposed a name for the *Zosterophyllum* flora with three plant assemblages as distinctive features in these deposits (Li and Cai, 1977). Last year, a general survey of the Devonian floras in China with a discussion of the correlations between the sequence of the Chinese megafloras and the sequence of those in other countries was also presented by the same authors (Li and Cai, 1978).

In the study of Carboniferous floras, a remarkable achievement has also been made. Although Early and Middle Carboniferous floras are widely distributed in China, only a few articles concerning plant remains of this age were published by Sze (1940, 1945). Later, Sze first made a restudy of some Culm plants from the Tseshui Formation in Hunan, and also described many plants ranging in age from Namurian to Stephanian from South Shaanxi, East Gansu, North Qinghai, North Jiangxi, and other localities. Among them, the Aolungpuluk Namurian flora from northern Qinghai is the most important, because this has proved to be the first documentation of a Namurian flora in eastern Asia. Another interesting discovery is a more characteristic Namurian flora found in the Tsingyuan Formation, eastern Gansu (Li and others, 1974). The flora has not been carefully studied yet. Despite all this, in addition to some forms seen in the Aolungpuluk flora, there is in the Tsingyuan Formation a very important plant, *Eleutherophyllum mirabile* (Nathorst), diagnostic of the early Namurian of Western Europe, which is especially important in association with *Eumorphoceras*, a leading goniatite genus of Namurian A. It is interesting to note also that several species of *Linopteris* were found nearly at the same horizon as the one with *Eumorphoceras*. Plants with network venation did not occur anywhere until the beginning of Westphalian, and the Tsingyuan Formation with its peculiar flora is thus referred to as early Middle Carboniferous or early Westphalian, despite the fact that the *Eumorphoceras*-bearing deposits are in general considered by many paleozoologists to be of late Early Carboniferous age.

Carboniferous Angara floras have recently been known to spread in the northern part of northeastern China and in the Junggar Pendi of Xinjiang. A few characteristic forms such as *Chacassopteris concinna* Radczenko were found in Early Carboniferous deposits, whereas species of *Angaropteridium* and *Angaridium* were of common occurrence in the Middle to Late Carboniferous, but none of them has been well studied.

In a monograph *Palaeozoic Plants from Central Shansi*, Halle (1927) added much to our knowledge of the general features and characteristics of the Permian-Carboniferous floras in eastern Asia. Much has been added to the information of these floras, owing to the restudy of some type sections of the Permian-Carboniferous and of important elements of the Cathaysia flora in North China by Sze, Li, and others in the past 30 years. The monograph *Fossil Plants of the Yuehmenkou Series, North China* (Lee, 1963) is one of the most important contributions not only from the paleobotanical but also from the biostratigraphical point of view. In it Lee first suggested that the late Paleozoic flora of northern China can be divided into six plant assemblages ranging from late Westphalian to the close of Permian in an unbroken succession which, though revised in later works, provides the best basis for subdivision and correlation of many similar plant-bearing strata in various places in eastern Asia. Another important contribution is the posthumous work of Sze, *Palaeozoic Plants from Tsingshuiho Region of Inner Mongolia and the Hokü District of Northwestern Shansi*, which will soon go to press. The material described in this monograph consists mainly of Permian plants and is of considerable importance for the study of the Cathaysian flora and its development under the influence of climate. In addition, in other papers and especially in the comprehensive book *Palaeozoic Plants from China* (Gu and Zhi, 1974), the emendation of the genus *Gigantopteris* Schenk, the discoveries of *Callipteris conferta* Brogniart, *C. changi* Sze, *Walchia bipannata* Gu and Zhi, *Discinites orientalis* Gu and Zhi, *Mariopteris* spp., *Fascipteris* spp., *Compsopteris (Proboblechnum)* spp., *Emplectopteridium alatum* Kawasaki, and others are of considerable significance both in plant taxonomy and in biostratigraphy. A research of many years on the Cathaysia flora from Yu Xian, Henan, by Zhang Shan-zhen and Mo Zhuang-guan will soon be accomplished.

Other paleobotanical researchers also described many Paleozoic vascular plants including some new forms from various parts of China. For example, the staff of the paleontological and stratigraphical lecture group of Wu Han College of Geology have paid much attention to the study of the Permian-Carboniferous flora and stratigraphy of Hulustai, Ningxia, Mentougou, Beijing, and also Yu Xian in Henan.

In the Permian period, the Cathaysian flora, characterized by many peculiar forms of gigantopterids, flourished and was widespread in China. Before 1960, research on the Permian Cathaysian flora in South China lagged somewhat behind that of North China. Since the beginning of the sixties, Zhou Zhi-yan (formerly T. Y. Chow) has tried to make a thorough study of the *Gigantopteris* flora of the Lungtan Formation in Jiangsu; it is regrettable that this important study was not formally published, but his main results were included in the book *Palaeozoic Plants from China* (Gu and Zhi, 1974). Following Zhou's work, others in recent years have made considerable progress in this field. The study of Zhao and others (1979) on *The Late Permian Flora Western Guizhou and Eastern Yunnan* is both of paleobotanical and biostratigraphical importance; Yao Zhao-qi (1978) became interested in the so-called *Gigantoperis*-bearing coal series in southern China, and found that this series, previously regarded as being of Early Permian or Late Permian age, has now proved to be ranging from late Early Permian to the end of Late Permian. Also of special paleophytogeographical importance are, among other new findings, a coal-bearing deposit with a typical late Cathaysian flora, both from northern and eastern Xizang (Tibet) (Li and Yao, 1979), as well as a *Glossopteris* flora in the Qubu Formation in southern Xizang (Hsü, 1973, 1976). The discovery of the *Glossopteris* flora in the northern slope of Qomolangma Feng indicates that the Himalayan Range was a part of the Gondwanaland, and that the conception of the so-called Himalayan Geosycline and the geosuture line between the Indian Plate and the Eurasia Plate occurring in the southern slope of Himalayas ought to be rejected.

In northeastern and northwestern China, the Permian, especially the Late Permian, Angara floras are found to be more dominant than floras of Carboniferous age. The one from the Lesser Khingan Range in northeastern China described by Huang Benhong (1978) is of interest, since this is the only carefully conducted research work on the Angara flora known in China.

Most recently, based on all available data and current studies, a paper "New Advances of Studies in the Cathaysia Flora" was written by Li (1979), and a sketch map showing "Carboniferous and Permian Floral Provinces of East Asia" was presented by him in collaboration with Yao (Li and Yao, 1979). Meanwhile, Zhao and Wu (1979) also dealt with the Carboniferous floral sequence and stratigraphy of South China.

The research on Mesozoic floras of China, as compared with that of Palaeozoic ones, lagged somewhat behind, but has advanced quite rapidly in recent years.

There was no authentic published record of Early Triassic plants in China before 1960. It is noteworthy that the most diagnostic Early Triassic species *Pleuromeia sternbergi* Corda and *P. rossica* Neuberg were reported for the first time by Wang Li-xin and others (1978) from the Qinshui basin of southeastern Shanxi, and later a similar florule was recorded in northeastern Hebei.

No less important than the above-mentioned results is a richer Early Triassic flora found from the Linwen series in east Hainan Island, Guandong (Zhou and Li, 1979). It consists of about 18 genera and more than 20 species, of which *Voltzia heterophylla* Brongiart, *Albertia elliptica* Schimper, *A. altifolia* Schimper and *Neuropteridium manrinatum* Zhou and Li, are almost all characteristic plants of the Buntsandstein of western Europe.

Middle Triassic plants were also little known in China before 1960. An important discovery of some plant remains from the Ermaying Formation of Nei Mongol seems to have filled this blank (Zhou Hui-qin and others, 1976). In general, the Ermaying flora resembles closely the Yenchang flora of North Shaanxi but may be distinguished by the appearance of several peculiar forms, such as *Pleuromeia wuziwanensis* Chow and Huang, *Nilssonia grandifolia* Chow and Huang, and *Leuthardia* sp.

In South China, although Middle Triassic sediments are mainly of marine origin, some vascular plant remains have been reported by Ye Mei-na (1979) and others from the Ladinian of W. Hupei, Anhui, and Jiangsu, of which the most important form is *Annalepis zeilleri* Flich, a lycopodiaceous plant originally described in the Upper Muschelkalk of France.

Since the publication of *Older Mesozoic Plants from North Shansi* by Pan (1936), the Keuper-Rhaetic flora in China, especially the Yenchang flora of North China, has received some of the attention which it deserves. A thorough investigation of the Yenchang flora of North Shaanxi by Sze in 1956 has revealed much information about the composition, botanical character, geological age, and geographical distribution of this flora, and a proposal for a major division of the Chinese Mesozoic strata on the basic four characteristic floral series was first presented in this important work. This division and/or the floral sequence might provide a fairly good basis for subdivision and correlation of other similar plant-bearing strata over vast areas. Since then, much research has been done on the Late Triassic floras both from North China (Sze, 1956; Chow and Chang, 1956; Wang Xi-fu, 1977; Sun Ge, 1979; and others) and from South China (Ao Zhen-kuan, 1956; Wu Shun-ching, 1966; Cao Zheng-yao, 1965; Hu Yu-fan and others, 1974; Hsü and others, 1974, 1975; Wang Xi-fu, 1977; Li Pei-chuan and others, 1976; Zhou Tong-shun, 1978; Chen Ye and others, 1979, and others). Besides, the study of the Keuper-Rahetic Hsuchiaho flora from Kwangyuan, North Sichuang (Le Pie-juan, 1964) and a recent work on the Late Triassic Baoding flora from Southwest Sichuan (Hsü and others, 1979), both dealing with the details of its floristic composition and stratigraphic relationships, are important contributions.

Other noteworthy research results (in prep. or in press) on Late Triassic plants may be briefly mentioned. (1) More than 10 localities with Late Triassic floras have been discovered in the eastern part of Xizang-Qinghai plateau. The floras from Anduo, Jiangda, and Gongjiao have been recently described by Wu Xiang-wu. These floras have a composition closely related to the Late Triassic floras of South China. (2) A flora characterized by some elements of the Yenchang flora of North Shaanxi, for example, *Neocalamites carconoides* Harris, *Danaeopsis* cf. *D. fecunda* Halle, *Cladophlebis grabauiana* Pan, *C. gracilis* Sze, *Glossophyllum shensiense* Sze, and others, were found from Hunjiang, South Jiling in northeastern China. (3) The finding of some dipterids, *Dictyophyllum nilssonii* (Brogniart) and *D. nathorsti* Zeiller from Kuche of South Xinjiang, is of special interest because they show a closer resemblance to the contemporaneous flora of South China, whereas those in Junggar Pendi of North Xinjiang have long been known to be nearly the same as the Yenchang flora of North Shaanxi.

As a great number of data of Late Triassic plants have been accumulated, two floral provinces seem to have been recognized in China: (1) the Northern province characterized by the *Danaeopsis-Bernoullia* flora which is similar to the Yenchang flora of North Shaanxi and is mainly distributed in North Xinjiang, North Qinghai, Gansu, Ningxia, Nei Mongol, and North China proper, and probably in the northern part of Northeast China; and (2) the Southern province characterized by the *Dictyophyllum-Clathropteris* flora which is represented by the Ipinglong flora in Yunnan, the Anyuan flora of Jiangxi, and other corresponding floras occurring in the vast areas of South China.

These two floras, though different in their first and last appearance, in geological distribution as well as to some extent in floral composition, seem contemporaneous mainly in being of Keuper-Rhaetic age, and may represent the provincial differences.

As compared with the increasing information on the Triassic floras, rather few contributions to the Jurassic flora of China have been made in the last two decades. A reinvestigation of the type section of the Hsiangchi series in West Hubei has, however, shown that (Wu Shun-ching and others, 1979) the plant-bearing horizons of this series, formerly all regarded as of Early Jurassic age, can be divided into two parts: (1) a lower part (Shacengxi Formation) characterized by some typical species such as *Pterophyllum sinense, Sinoctenis* sp., *Cvaedocarpidium erdmanni, Clathropteris mongugaica,* of Late Triassic age; and (2) an upper part (Hsingchi Formation, *s. s.*) yielding many plant fossils probably of an age ranging from Pliensbachian to Aalenian. Besides, two other coal-bearing formations, each with a characteristic plant assemblage probably corresponding to the *Lepidopteris* and *Thaumatopteris* zones of western Europe, may well be recognized in Hunan, Jiangxi, Guangxi, and Fujian.

Floras of the Jurassic coal series widely distributed in North and Northwest China were generally assigned tentatively to an Early to Middle Jurassic age (Sze, 1959, 1960; Shen Kuang-lung, 1961; Gu Da-yuan, 1978; Zheng Sao-ling and others, 1978). In recent years, however, it has been found that in some places, such as the area covered by the Tatung coal series of North Shanxi and the Metoukou series of the Western Hills in Beijian, a formation of moderate thickness occurs which contains plants of Early Jurassic aspect. This is overlain by a coal series containing a basal conglomerate which is tentatively considered to be of Middle Jurassic age.

Attempts have also been made to subdivide the Jurassic floras into stages, but their diagnostic characters have not yet been clarified.

The dating of the boundary of Late Jurassic and Early Cretaceous floras in China has presented difficulties because of a general lack of marine fossils in the plant-bearing sediments. However, some short papers on Early Cretaceous and/or Late Jurassic plants were published in recent years; for example, a preliminary study of Early Cretaceous plants from Qiaohe, Jiling (Lee and Yeh, 1964), Early Cretaceous plants from Xizang (Duan Shu-ying and others, 1977; Le Pie-juan, 1979), *Cephalotaxopsis* from Nei Mongol (Wang Zi-jian, 1976), and a new coniferous genus *Yanliaoia* from Liaoning (Pan Guang, 1976).

The Late Cretaceous plants are probably the least studied of the Chinese Mesozoic succession. No record was known of them before 1950. In 1959, Lee was the first to report an angiospermous plant *Trapa? microphylla* Lesqereux from the Fulungchuan Formation of Heilongjiang Province. Since then, additional angiospermous fossils have been found. Of greater importance are the Late Cretaceous plants from the Rikeze series of Xizang and the flora of the Bali Formation in Yongning, Guangxi, described by Guo Shuang-xing (1968). The finding of *Protophyllum* from the Hunchun series of Jilin, Northeast China, is of special interest, too, since it suggests a somewhat close floristic relationship between East Asia and North America.

In addition, in contrast with much progress in the study of grass morphology of fossil floras, there are few contributions to our knowledge concerning anatomical research on fossil plants from a botanical viewpoint. This might be due to the rare occurrence of well-preserved fossils in coal ball in the Chinese coal series. Nevertheless, Sze's paleobotanical research contains many important contributions not only to the knowledge of fossil foliages but also to that of fossil wood (Sze, 1951, 1952, 1954). Of special interest is *Phoroxylon scalariforme* Sze from the Cretaceous of Northeast China, which is a plant mixed with bennettitalean and araucarioid characters. The outstanding feature of this plant is that in the radial sections of its secondary wood, one may find that the tracheids are dominant with scalariform pittings and occasionally intermingle with the reticulated polygonal, more or less araucarioid pittings. Hsü also devoted himself to describing some petrified wood from Central Shanxi and East Shangdong (Hsü, 1952, 1953). A recent careful study on a rich collection of *Frenelopsis*-like plants from Early Cretaceous deposits of Eastern China (Zhou and Cao, 1977, 1979) established several new species and a new genus, *Suturovagina,* described with detailed cuticular structures. In these articles, they also discussed the taxonomical position and the evolutionary trends of *Frenelopsis*-like plants.

Furthermore, Sze and others (1963) compiled a comprehensive work *Mesozoic Plants from China* based mainly on materials described from China before 1960 in widely dispersed literature. They added descriptions of some new specimens and revised many previous identifications. The work is of paleobotanical and stratigraphical importance.

Summing up all the above-mentioned contributions, Zhou Zhiyan and Li Pei-juan (1979) have considered that there seem to have existed six floras in China during the Mesozoic Era: (1) the *Neuropteridium-Voltzia* flora (Early Triassic); (2) the *Danaeopsis-Bernoullia* flora (Middle to Late Triassic); (3) the *Dictyophyllum-Clathropteris* flora (Late Triassic-Aalenian); (4) the *Coniopteris-Phoenicopsis* flora (Early to Middle Jurassic); (5) the *Ruffordia-Onychiopsis* flora (Late Jurassic to Early Cretaceous); and (6) an angiospermous flora (late Early Cretaceous to Late Cretaceous).

The study of Cenozoic plants in China still lags considerably behind, because of the weak foundation for research on this field. In the earlier half of the 1950s, only few contributions to the study of Tertiary plants of China were made by Sze (1951a, 1951b, 1954), but several young researchers have been trained by Sze and Hsü since about 1956. They began to describe Cenozoic plants in 1965; for example, *Sabalites* from Guangdong and Guangxi by Guo Shang-xing, *Palibinia* from Hunan by Li Hao-min, and an Eocene florule from South Shaanxi by Tao Jun-rong (1965). Since then, studies of Cenozoic plants have developed more and more rapidly.

In 1978 an important book entitled *Cenozoic Plants from China* was compiled by paleobotanists of the Institute of Botany in Beijing and the Nanjing Institute of Geology and Palaeontology. This book seems to be very useful to our field geologists and paleobotanical lecturers and also seems to be a good guide to foreign colleagues for references on general aspects and chief characteristics of Cenozoic plants of China.

Recent studies in this field have also addressed themselves to the problems of vegetation and paleoclimatic changes. Among the more important articles are those on fossil *Quercus* from a locality near Xixiabanma Feng and on Quaternary angiosperms from the northern slope of Jolmolangma Feng (Hsu and others, 1973, 1976), a Late Tertiary flora from Eryuan, Yunnan (Tao and Kong, 1973), the Miocene Nanmuling flora from Xizang (Li and Guo, 1976), and Pliocene florulas from West Sichuan (Guo, 1979). Most of these researches provide important information about the problems of the history of the uplift of the Xizang-Qinghai plateau since the Late Tertiary.

LOWER PLANTS

Before 1950, the study of lower plant fossils seemed to be almost a blank. In the early years of the People's Republic of China, there was an urgent need for geological investigations. With the rapid development of socialist construction, some workers have come to be engaged in doing research work in such fields as Precambrian stromatolites, charophytes, calcareous algae, diatoms, dinoflagellates, acritarchs, and fungi remains. Following are some of the results gained in recent years.

Precambrian Stromatolites and Algae

Only a few stromatolites such as *Collenia* and *Gymnosolen* were described in the first half of this century in China (Grabau, 1922; Kao and others, 1934). Since 1960, Cao Rui-ji ([Tsao Rui-chi], 1964), Liang Yu-zhuo ([Lian Yu-tso], 1962), Cao and Liang (1974) and Zhu Shi-xing and others (1979), engaged in extensive geological field work in China, have described many stromatolites from the Sinian System or Suberathem. Through these studies, a tentative sequence of stromatolitic assemblages of North China can be recognized. It may be taken as a standard for a tentative subdivision and correlation of similar rock series covering extensive areas of China.

In the meantime, some of them have also begun to study other Precambrian microplants, such as bacteria, algae, and acritarchs, generally found in the stromatolites.

Recently, much attention has been paid to the search for very ancient organic remains essential for the understanding of the early evolutionary history of lower plants and the beginning of life. From the Anshan Group (2,300 to 2,500 m.y. old) in Liaoning, fossil bacteria and traces of other microorganisms were newly found by members of the Nanjing Institute of Geology and Palaeontology and the Institute of Botany in Beijing. Moreover, some possibly doubtful eucaryotic algae were described by Zhang-yun (1978) from Wumishan and the Hungshuichang Formation (about 1,300 to 1,400 m.y. old) of Hebei.

Charophytes

With the exception of several papers of fossil charophytes which were given by Y. H. Lu [Lu Yan-hao] (1944, 1945, 1948), all other contributions to this field were published after 1950.

The first is a study on *Sycidium* from the Middle Devonian, North Sichuan (Wang Shui and Chang Shan-zhen, 1956). The late Wang Shui made a great contribution to laying the foundations for study of fossil charophytes in China; most of the chief workers in this field today were trained by him. Of special interest among Wang-shui's other works are studies of Tertiary charophytes from the Tsaidam basin, Qinghai (1961), Mesozoic and Tertiary charophytes from Juiquan basin, West Gansu (1965).

Recently, several important works have been published by Wang Zhen (1976): (1) a taxanomical study of the Cretaceous and Early Tertiary charophytes from the Jiang-han basin of West Hubei; (2) a study of Middle Devonian charophytes from Southwest China conducted at first with a view to clarifying the growth process of enveloping cells of oogonia belonging to the genus *Sycidium;* (3) research on Early to Middle Triassic charophytes from North Shaanxi (in collaboration with Huang Ren-jin, 1978), which is noteworthy because of the first recognition of Triassic charophytes in east Asia and of providing evidence for the age of the Hoshankou and Ermaying Formations.

As charophytes have been found to be very useful for correlating strata devoid of megafossils, and especially for dating rocks from bore cores, a fossil charophytic training class of about 40 students organized by the Geological Department of Nanking University and the Nanjing Institute of Geology and Palaeontology, Academia Sinica, was held for two months in 1974. Later research in this field has developed at a rapid pace and has become more popular.

Calcareous and Noncalcareous Algae

Before 1950 no special attention was paid to calcareous fossil algae in China. Only a few paleontologists took an interest in them because of their abundance in nature.

From 1960 onward, in order to meet the needs mainly of petroleum-geological prospecting, research was first carried out on materials collected from the Permian of Sichuan (Zhang Lin-xin and others, 1974). More recently, some important papers concerning calcareous algae ranging in age from Ordovician to Eocene in various localities of Xizang were published (Wang Yu-jing, 1976; Mu Xinan, 1977). Mu has turned his interest to calcareous algae since 1975 and in the future will devote himself mainly to work in this field.

Apart from fossil calcareous algae, uncalcified algae remains have begun to be studied chiefly under the guidance of Chu Hao-jan. These studies seem to have advanced rapidly with the result that several contributions to Tertiary and Quaternary filamentous blue-green and green algae from North Jiangsu have been published in recent years (Zhang Chung-ying, 1977; Chu Hao-jan and others, 1978, 1979).

Diatoms

In contrast to other abundant plant-bearing beds, diatomaceous earth does not abound in China. However, in Shanwang, Linqu (Linchü) county of Shandong, well known for its Miocene megaflora (Hu and Chaney, 1940), the megafossil-bearing shales consist almost entirely of diatom remains, only a small number of species of which were previously described (Skvortzov, 1937). Since 1960, two Chinese workers have restudied the Shanwan diatom flora, adding some new materials and revising earlier information. An earlier paper dealing with Quaternary diatoms from Lantian, South Shaanxi, was also published by them (Li Qia-ying and Huang Cheng-nian, 1966).

Dinoflagellates and Acritarchs

The study of dinoflagellates, acritarchs, and other microplankton did not attract our special attention, and we did not begin to train specialists in this field until the 1970s. The first important contribution dealing with the description of 64 genera and 232 recognizable species (11 genera and 126 species described as new) of the Early Tertiary from the coast region of Bohai was made in 1968, by micropaleobotanists of the Nanjing Institute of Geology and Palaeontology, Academia Sinica, and the Academy of Petroleum Exploration of the Ministry of Petrochemical Industry. What is of special interest in this fossil microplankton flora is that the floral composition shows a pecular aspect quite different from those known in Europe and North America. Recently, some Early Tertiary dinoflagellates and acritarchs from the Bose basin, Guangxi, were described (He Chang-quan and Quian Ze-shu, 1979).

Fungi

Remains of parasitic fungi are commonly found in fossil wood (Hsü, J., 1953) and leaf impressions (Lee, H. H., 1963) of Early Devonian and later age.

It is significant that a few fossil fungi, well-preserved in cells of marine calcaerous algae or in shells of fusulinids, were found in Late Permian deposits of Guizhou. This appears to be the first record of fossil fungi occurring as parasites in marine biotas (Mu Xi-nan, 1977). It is of further interest to note that a few fungi were found in Eocene sediments of Xizang, as a by-product of macerating spore and pollen samples (Zhang Chungying, 1977). These fungi resemble closely the extant parasitic microthyriaceous fungi. Considering the present environment of the microthyriaceous fungus represented in the flora consisting mainly of angiospermous trees of subtropical forests, the conditions under which the fossil fungi grew might have been tropical to subtropical, with much moisture and stifling heat. This opinion is rather different from the conclusion reached by some authors (Wang Kai-fa and others, 1975) based on palynological analysis of the same material.

FOSSIL SPORES AND POLLEN

The study of fossil spores and pollen in China virtually began after the founding of the New China. Especially geological exploration provided a powerful impetus to the rapid development of palynology, and in this field Xu-ren [J. Hsü] has made a pioneering effort. In 1953, a palynological laboratory was first formed in the Nanjing Institute of Geology and Palaeontology, Academia Sinica. A little later, a palynology training class was organized under the direction of the Ministry of Geology of the People's Republic of China, its main aim being the stratigraphic application of palynology, especially in the correlation and subdivision of coal-bearing series based on spore and pollen analysis. About 30 students in the class received fundamental training in general palynological work. It resulted in setting up another palynological laboratory in the Institute of Geology and Mineral Resources of the Chinese Academy of Geological Sciences. The work carried out in the early stages of these laboratories was mainly under the guidance of Hsü.

In the second half of the 1950s, great progress was made in these institutions, and their membership was increased. Nowadays there are about 400 members in more than 100 laboratories or working groups engaged mainly in palynological research in different departments of national institutions all over China. The three research centers are (1) the Nanjing Institute of Geology and Palaeontology, Academia Sinica; (2) the Institute of Geology and Mineral Resources of the Chinese Academy of Geological Sciences; and (3) the Institute of Botany, Academia Sinica, in Beijing.

In mid-March, 1979, the First Palynological Symposium of China was held in Tianjin. There were nearly 200 attendants, including Dr. G. Norris, the Secretary-Treasurer of the International Commission for Palynology (I.C.P.), who was invited to lecture at this meeting. About 200 papers were presented dealing with palynological research and studies of various kinds of plant microfossils ranging from Precambrian to Quaternary in age. At the end of this symposium, the Palynological Association of China was established and Professor Hsü Jen was elected its first president.

Precambrian to Early Paleozoic

The earliest publications dealing mainly with the description of microorganisms from the Sinian, or late Proterozoic, rocks were presented by Yan Fu-hua (1965), Sin Yu-sheng and Liu Kui-zhi (1973). These papers provided evidence that plant microfossils seemed to be valuable and applicable to the study of Precambrian stratigraphy. Later on, other palynologists working on the phytoplankton remains from Southwest China (Ouyang Shu and others, 1974; Yin Lei-ming and others, 1978) expressed the opinion that the Toushantou Formation and the Tengyin Formation of the Sinian in Southwest China may roughly correspond to the Chingerhyu Formation of the Tsingpaikou Group of North China, the former being a little higher in stratigraphic position. Moreover, some rather well-preserved eucaryotic algae remains seen in slides of stromatolites from West Hubei are of considerable significance. Also, in regard to the plant microfossils and ultramicro-organisms discovered in Early Proterozoic and Archean rocks, for example, the Anshan Group and the Liaoho Group of east Liaoning (with a banded-iron formation, aged about 2,400 m.y.), some important papers (Ouyang Shu, 1979a, 1979b, Yin Lei-ming, 1979) have been presented. Generally, the Anshan Group has been found to be rich in various kinds of microfossils such as blue-green algae and iron-bacteria. If this discovery could be confirmed by further consistent evidence, its significance would be greatly enhanced.

Late Paleozoic

The study of Devonian spore assemblages has made progress in recent years. Kao Lian-da (1978) and Kao Lian-da and Hou Jin-peng (1978), and Lu Li-chang and Ouyang Shu (1976, 1978) described spore assemblages of Early and Middle Devonian age and discussed the stratigraphical significance of those from Southeast Guizhou, East Guangxi, and East Yunnan. With our present knowledge of the Devonian microflora in Southwest China, at least three spore assemblages can be recognized. On the whole, they resemble the contemporaneous spore assemblages of other countries in having a similar aspect, but some important elements seem to have made their first appearance earlier in China than anywhere else. For example *Archaeoperisaccus* (and *Nikitinsporites*), a type of Lycopsida, has been regarded to be diagnostic of the Frasnian in the United States, the Soviet Union, and Canada; however, it is found to occur in China in association with the *Protolepidodendron scharyanum* flora of Middle Devonian age. It may be mentioned further that Hou Jing-peng (1978) reported the first occurrence of fossil chitinozooans such as *Agnochitina* and *Ancyrochitina* from the Early Devonian of Guangxi.

The study of Permian and Carboniferous spore-pollen assemblages has increased since 1960. As early as the 1950s, Hsü, a teacher of the Palynology Training Class mentioned above, tried to describe some Permian and Carboniferous spores and pollen found in coal-bearing deposits from Pingdinshan of Heana, Qingshuiho of Nei Mongol, Leping of Jiangxi, and other localities, but these results have not been published. Thereafter, two Permian spore-pollen assemblages from Lungtan, Jiangsu, and Shanxi were described by Ouyang Shu (1962, 1964) in a summary report of the Carboniferous spore-pollen assemblages from Gansu and Shanxi by Kao (1962, 1976), and Permian spore-pollen assemblages of Hunan by Shen Jian-guo (1978). Recently, from the biostratigraphical viewpoint, Kao (1979) made attempt to sum up all the described spore-pollen assemblages of Late Paleozoic of China, while Ouyang Shu (1979a) presented an article on new miospore genera from the Permian and Carboniferous of China.

In North China an almost complete sequence of spore and pollen assemblages ranging from Visean to the latest Permian has now been established, while those in South China have only been recognized from the beginning of Permian to the latest Permian. This may be of great stratigrpahic significance, though some detailed work, especially the boundary between the Carboniferous and the Permian from a viewpoint of palynology, remains to be done.

Like those of megafloras, the spore-pollen assemblages, ranging from Late Carboniferous to the close of the Permian from the northern part of Northwest China as well as Northeast China, are characterized by the presence of many Angara floral elements and are essentially different from those of the Euramerican floral province. It should be pointed out further that some spore types such as *Crassispora, Knoxisporites,* and *Tripartites* generally considered to be restricted to the Permian and Carboniferous of the Euramerican floral province, might extend to the end of Late Permian, and even persist to the beginning of Early Triassic.

There are also a few contributions made by foreign authors, for example, the study of Permian and Carboniferous spores from the Kaiping basin, Hebei (Imgrund, 1969), and the study of the Permian microflora from Baode, Northwest Shanxi (Kaiser, 1976), based on materials obtained before 1950.

Mesozoic

Since the first paper on palynomorphs from the Lower Cretaceous of West Gansu by Hsü Jen and Chow Ho-i in 1956, more and more attention in China has been paid to the application of palynological remains to the study of Mesozoic terrestrial stratigraphy.

Two articles deserve to be mentioned. One is the study of the spore-pollen assemblage of the Kayitou Formation from east Yunnan; this assemblage is characterized by a peculiar palynological aspect which seems to present an intermediate link in the evolution of floral changes from late Paleozoic to early Mesozoic. The other is a preliminary study of the spore-pollen assemblage of the Liuchiakou Formation, that is, the middle part of the so-called Shihchienfeng series. Because of the abundant presence of *Lundbladispora* in this assemblage, an Early Triassic age ought to be assigned to the Liuchaikou Formation. This study, along with other investigations of fossil faunas and megaplants from the same formation, has made some important additions to our knowledge of the boundary between the Late Permian and the Early Triassic; in other words, the relation between the Liuchiakou Formation and the Shihchienfeng Formation (s.s.) has been clarified.

Rather little is known about the Middle Triassic spore-pollen assemblages in China, especially as compared with the increasing understanding of those of the Late Triassic in recent years. Many authors (Li Wen-ben, 1974, 1976; Zhang Zheng-lai, 1979; Lei Zuo-gi, 1978) have been devoted to describing the spore-pollen assemblages of the Late Triassic coal series such as the Hsuchiaho Formation of North Sichuan and the Anyuan Formation of East Jiangzi. Besides, studies of Late Triassic spore-pollen assemblages from East Xizang (Shang Yu-Ke, 1979) are also of interest. Through these efforts, it was found that the microfloras or palynological asemblages represented in the Hsuchiaho Formation are presumed to have been widespread in the same areas as those of the *Dictyophyllum-Clathropteris* megaflora, which grew near the littoral region in a tropical to subtropical environment. This is rather different from the more or less time-equivalent spore-pollen assemblages in North China, with a similar distribution to those of the *Danaeopsis-Bernoullia* flora of the Yenchange Formation. The latter flora is generally considered to have grown in a relatively arid to subarid zone.

In China, relatively few contributions have been made to the study of Jurassic spore-pollen assemblages. Zhang Lu-jin (1965, 1978) described some spore and pollen remains from the Yi-ma coal-bearing formation of Henan and the Shouchang and Guantou Formations of Zhejing, with special reference to the age of these plant-bearing formations. From a viewpoint of palynology, Zhang Dai-hua (1977) presented a paper on spore-pollen assemblages from the Wuchang basin of Nei Mongol, with a discussion about the boundary between the Late Jurassic and Early Cretaceous of North China. However, this problem is still defying solution, and it is the key to the solution of many other problems, such as the age of Jurassic-Cretaceous coal series in Northeast China, the problem of the various kinds of volcanic pyroclastic sediments spread widely in East China, and the problem of the Mesozoic "Red beds" covering a vast area of South China. Although many paleontologists (Yuan Yin-kuai, 1978; Li Weng-en, 1979; and others) have made some progress on this subject in the past 20 years, the problem probably cannot be solved until further efforts are made. One of the complex reasons involved in it is that there is considerable difference of opinion among authors in regard to the stratigraphic meaning of some important palynological genera such as *Cicatricosisporites* and *Classopollis.*

However, a number of authors (Hsü, J., 1958; Zhang Zheng-lai, 1978; Li Man-ying, 1978) have been making considerable progress along with other paleontological researchers in determining the age of the "Red beds" which are widely distributed in Hunan and Hubei. The age of the "Red beds" was formerly regarded to be entirely Early Tertiary, but now it has been proved that at least a part of them, for example, the Wulong Formation of West Hubei, ought to be assigned to the Early Cretaceous.

Recently, based on the studies mentioned above, together with many other available palynological data, Song Zhi-chen (1979) has briefly summarized the microfloral and paleoclimatic provinces extending from Late Cretaceous to Late Tertiary: (1) the northeastern province, represented mainly by the spore-pollen assemblages from the Fulungchuan Formation (Cenomanian) to the basal part of the Mingshui Formation (Senonian) of the Songliao basin, reveals a vegetation growing under subtropic climatic condition; (2) the Central China province, covering the main part of the Lower Huang-He (Yellow River) valley and a vast area to the south of North Jiangsu and the Jianghan plain of West Hubei, is composed of vegetation that might have existed in a tropical to a subtropical zone under arid or subarid climates.

Noteworthy also are two other papers, one on *Balmeisporites* from the Songliao basin (Zhao Chuna-ben, 1976), and the other dealing mainly with the evolutionary changes on the pollen morphology of Early Cretaceous angiosperms (Kao Rui-qi, 1979).

Cenozoic

Since the first study of Tertiary spore-pollen assemblages from the "Red beds" of West Gansu (Sung Tse-chen, 1958), a number of contributions have been accumulated. Among them, the most noteworthy, is a volume entitled *Early Tertiary Spore and Pollen Grains from the Coastal Region of Bohai*, which was published in 1978 by the Academy of Petroleum Exploration and Exploitation under the Ministry of Petrochemical Industry, P.R.C., and the Nanjing Institute of Geology and Palaeontology, Academia Sinica. The Bohai microflora comprises 152 genera, 470 recognizable species, and many forms of fungi remains, of which 7 genera and 165 species were treated as new. The stratigraphical and phytogeographical significance of the microflora was also discussed at some length in this article.

Other important papers worthy of mention here are studies of Cretaceous-Tertiary and pollen from Anhui (Wang Kai-fa and others, 1975), from Jianghan basin, West Hubei (Ma Jun-ying and others, 1976), from North Jiangsu (Song Zhi-chen and others, 1979), and the Tertiary palynological study from Beibu Gulf (Gulf of Tonkin) of South Guangdong (Zhang Yi-yong, 1977). Besides, similar work is being carried on by members of the Palaeobotany Laboratory of the Institute of Botany, Academia Sinica, from which notable contributions to Tertiary palynology may be expected.

Discovery of rocks of Paleocene age was first suggested by the study of vertebrate fossils in China, and these conclusions have been confirmed through palynological investigations such as the study of fossil spores and pollen from the Lingian Formation of the Qingjiang basin, Jiangxi (He Yue-ming and Sun Xing-jun, 1977), and from the Fushun coal field, Liaoning, (Sung Tse-chen and Tsao Liu, 1976).

Recently, the study of Tertiary palynomorphs seems to have turned from mainly descriptive work to the analysis of vegetation and paleoclimate. In the Tianjin Symposium held in mid-March, 1978, Song Zhi-chen gave a lecture on "The vegetation divisions of Late Cretaceous to Miocene Floras in China," and two papers with more or less the same contents, though with somewhat different conclusions, were presented by Sung Meng-rong and Sung Xiang-jun (1979), respectively.

The composition, characteristics, and stratigraphical relations of the Tertiary microfloras of China, especially in East China, have now been fairly well studied, and an almost complete standard sequence of Tertiary spore-pollen assemblages has been established. This seems to indicate that the study of Tertiary palynomorphs in this country has reached a new level.

The study of the Quaternary palynomorphs of China, though not as far advanced as that of the Tertiary forms, has made rapid progress in recent years, since it has a close bearing on many other problems such as the origin of loess, vegetation and climate changes, paleoanthropology, geomorphology, engineering geology, and in particular the Quaternary glaciations.

More than 30 articles on this subject have been published since 1965. Those dealing with the palynomorphs of North China are relatively systematic and detailed. It has been found that the spore-pollen assemblages of Early Pleistocene age are characterized on the one hand by the disappearance of some Pliocene plants generally growing in warmer climates and, on the other, by the abundance of some plants in colder climatic conditions. This is an obvious indication that the climate at this stage must have become colder than it was before. In the Middle Pleistocene, the spore-pollen assemblages indicated a succession of alternating colder and warmer stages; this knowledge would be useful for discussing the environment in which *Sinanthropus pekingensis* Black lived (Hsü Jen, 1965, 1966; Sung Meng-rong, 1965). According to evidence given by pollen of *Picea,* a cold temperate type is abundant in the contents of the Late Pleistocene spore-pollen assemblages. Some people (Zhou Kun-shu and others, 1978) considered this stage of colder climate to be roughly equivalent to the last glacial age in China. Others (Liu Jing-ling and Ye Pingyi, 1977), through studies of the Holocene spore-pollen assemblages of peat deposits from northern and northeastern China, found that there existed a regular fluctuation in temperate variations from the early through the middle to the late stages, and that the middle was the most favorable for plant growth.

Based on pollen spectrum analyses and study of paleoclimate changes, together with some isotopic age data, the search for the lower boundary of the Quaternary has made much progress.

Studies of the Quaternary spore-pollen assemblages from Shanghai and Zhejiang (Liu Jing-ling and Ye Ping-yi, 1977) showed that four stages of alternating warmer and colder temperatures existed in the valley of the Lower Yangzi Jiang (Yangtze River) and that these may roughly correspond to the four glacial and interglacial periods.

Of special interest are the studies of fossil spore-pollen assemblages of the Pliocene and Quaternary from Xizang (Hsü Jen and others, 1973, 1975; Zhou Kin-shu and others, 1976). On the basis of these works, it appears that the Himalayas and the southern part of the Xizang-Qinghai plateau have undergone an uplift of about 2,000 to 3,000 m since Pliocene time.

Although we have made great progress in the past 30 years, we should not overlook the shortcomings of our work as well as the gaps between the research of the advanced world and that of China in some fields. The tasks before us are arduous; there is a great deal of work to be done. We must remain modest, prudent, and strive to make greater contributions to the realization of the four modernizations of the People's Republic of China.

SELECTED REFERENCES

Chang Shan-chen [Zhang Shan-zhen], 1956, A Culm florule from eastern Kansu: Acta Palaeontologica Sinica, v. 4, no. 4, p. 641–646, 1 pl. (in Chinese and English).

Chen Ye, Duan Shu-ying, and Zhang Yu-cheng, 1978, New species of the Late Triassic plants from Yanbian, Sichuan: Acta Botanica Sinica, v. 21, no. 1–3, p. 57–63, p. 186–190, p. 269–273, Pls. 1–3 (in Chinese with English abstract).

Chu Hao-jan [Zhu Hao-ran], Tseng Chao-tsi, and Zhang Zhong-ying, 1978, Fossil *Pediastrum* algae from the Dainan Formation (Lower Tertiary) of Northern Jiangsu with note on their sedimentary conditions: Acta Palaeontologica Sinica, v. 17, no. 3, p. 233–243, 1 pl. (in Chinese with English abstract).

Grabau, A. W., 1922, The Sinian System: Bulletin of the Geological Society of China, v. 1, p. 44–88.

Guo Shuang-xing, 1965, On the discovery of fossil plants from the Tertiary formation of Kwangtung and Kwangsi: Acta Palaeontologica Sinica, v. 13, no. 4, p. 598–605, Pls. 1–2 (in Chinese and English).

—— 1975, [Plant fossils from the Rigeze Group of the Mt. Jolmo Lungma region, Xizang, in Report of scientific expedition to the Mt. Jolmo Lungma region (Palaeontology, Fasc. 1)]: Beijing, Science Press, p. 411–423, Pls. 1–3 (in Chinese).

He Cheng-quan and Qian Ze-shu, 1979, Early Tertiary dinoflagellates and acritarchs from the Bose Basin of Guangxi: Acta Palaeontologica Sinica, v. 18, no. 2, p. 171–187, Pls. 1–2 (in Chinese with English abstract).

Hou Jing-peng, 1978, [Devonian Chitinozoans from the Nakaoling Formation, Heng District, Guangxi: Devonian Symposium of South China, Nanning, 1974]: Beijing, Geological Publishing House, p. 359–373, Pls. 49–53 (in Chinese).

Hsü Jen [Xu Ren], 1953, On the occurrence of a fossil wood in association with fungous hyphae from Chimo of East Shantung: Acta Paleontologica Sinica, v. 1, no. 2, p. 80–83, 1 pl. (in Chinese and English).

—— 1966, The climatic condition in North China during the time of *Sinanthropus:* Scientia Sinica, v. 15, no. 3, p. 410–414.

—— 1976, On the discovery of a *Glossopteris* flora in southern Xizang and its significance in geology and palaeogeography: Scientia Geologia Sinica, 1976 (4), p. 323–331, Pls. 1–4 (in Chinese with English abstract).

—— 1978, On the palaeobotanical evidence for continental drift and Himalayan drift: Palaeobotanist, v. 25, p. 131–142, Pls. 1–3.

Hsü Jen [Xu Ren.] and Chow Ho-i [Zhou He-yi], 1956, Microflora and geological age of the Huihuipou Formation of the Chiuchuan Basin of western Kansu: Acta Palaeontologica Sinica, v. 4, no. 4, p. 491–524, pls. 1–4 (in Chinese with English summary).

Hsü Jen, Sung Tze-chen [Song Zhi-chen], and Chow Ho-i, 1958, Sporopollen assemblages from the Tertiary deposits of the Tsaidam Basin and their geological significance: Acta Palaeontologica Sinica, v. 6, no. 4, p. 429–440, Pls. 1–6 (in Chinese with English summary).

Hsü Jen, Tao Jun-rong, and Sung Xiang-jun, 1973, [On the discovery of a *Quercus semicarpifolia* bed in Mt. Shisha Panma and its significance in botany and geology]: Acta Botanica Sinica, v. 15, no. 1, p. 103–114, Pls. 1–4 (in Chinese).

Hu Hsen Hsu and Chaney, R. W., 1940, A Miocene flora from Shantung Province, China: Palaeontologia Sinica, new ser. A, no. 1 (whole series no. 112), p. 1–147, 57 pls., 15 figs., 11 tables.

Huang Ben-hong, 1977, [Permian flora from the southeastern part of the Xiao Hing'an Lin (Lesser Khingan Mt.), NE China]: Beijing, Geological Publishing House, 79 p., 43 pls. (in Chinese).

Imgrund, R., 1960, Sporae dispersae des Kaiping-Beckens, ihre paläontologische und stratigraphische Bearbeitung im Hinblick auf eine Parallelisierung mit dem Ruhrkarbon und dem Pennsylvanian von Illinois: Hannover, Geologisches Jarbuch, Bd. 77, p. 143–204.

Kaiser, H., 1976, Die Permische Mikroflora der Cathaysia-Schichten von Nordwest-Schansi, China: Palaeontographica, B., bd. 159, p. 83–157.

Kao Chen Si, Hsing Yong Hsien, and Kao Ping, 1934, Preliminary notes on Sinian stratigraphy of N. China: Bulletin of the Geological Society of China, v. 13, p. 243–276, Pls. 1–6.

Kao Lian-da, 1978, [Early Devonian spores and acritarchs from the Nakaoling Formation, Liujing, Guangxi: Devonian Symposium of South China, Nanning, 1974]: Beijing, Geological Publishing House, p. 346–358, Pls. 42–48 (in Chinese).

Kao Lian-da and Hou Jing-peng, 1975, [Early–Middle Devonian spore assemblages and their stratigraphical significance]: Professional Papers of Stratigraphy and Palaeontology, Beijing, Geological Publishing House, no. 1, p. 170–232, Pls. 1–13 (in Chinese).

Lee H[sing] H[süeh] [Li Xing-zue], 1959, *Trapa? microphylla* Lesq., the first occurrence from the Upper Cretaceous formation of China: Acta Palaeontologica Sinica, v. 7, no. 1, p. 33–40, 1 pl. (in Chinese and English).

—— 1963, Fossil plants from the Yuehmenkou Series, North China: Palaeontologia Sinica, new ser. A, no. 6, (whole ser. no. 148), 185 p., 45 pls. (in Chinese and English).

Lee H[sing] H[süeh] and Cai Chong-yang, 1978, Devonian floras of China: Papers read at the International Symposium on the Devonian System, Bristol, Nanjing Institute of Geology and Palaeontology, Academia Sinica, p. 1–9, Pls. 1–3.

Lee Pei-chuan [Li Pei-juan], 1964, Fossil plants from the Hsuchiaho Series of Kwangyuan, Northern Szechuan: Memoirs of the Institute of Geology and Palaeontology, Academia Sinica, no. 3, p. 101–178, Pls. 1–20. (in Chinese with English summary).

Lei Zuo-qi, 1978, The spore-pollen assemblage of Shezhe Formation of Yipinglang Coal Series in Luquan of Yunnan and its stratigraphical significance: Acta Botanica Sinca, v. 20, no. 3, p. 229–236, Pls. 1–2; no. 4, p. 361–372, Pls. 3–4 (in Chinese with English abstract).

Li Pei-juan [Lee Pei-chuan], Cao Zheng-yao [Tsao Cheng-yao], and Wu Shun-qing [Wu Shun-ching], 1976, [Mesozoic plants from Yunnan, *in* Mesozoic fossils from Yunnan, Fasc. 1]: Beijing, Science Press, p. 87–150, Pls. 1–47 (in Chinese).

Li Wen-ben, 1974, [Triassic and Early Jurassic spores and pollen, *in* Handbook of stratigraphy and palaeontology in Southwestern China]: Beijing, Science Press, p. 362–370, 378–379, Pls. 195–196, 202 (in Chinese).

—— 1976, [Mesozoic and Early Paleogene sporo-pollen assemblages from Yunnan, China. Part I: Late Triassic sporo-pollen assemblage from Nanping, Yunnan, *in* Mosozoic fossils of Yunnan, China, Book I]: Beijing, Science Press, p. 1–9, Pls. 1–2.

Li Xing-Xue and Cai Chong-yang, 1977, Early Devonian *Zosterophyllum*-remains from southwestern China: Acta Palaeontologica Sinica, v. 16, no. 1, p. 12–34, Pls. 1–5 (in Chinese with English abstract).

—— 1978, A type-section of Lower Devonian strata in SW China with brief notes of the succession and correlation of its plant assemblages: Acta Geologica Sinica, v. 52, no. 1, p. 1–12, Pls. 1–2 (in Chinese with English abstract).

Li Xing-xue [Lee Hsing Hsüeh], Yao Zhao-qi, Tsai Chung-yang [Cai Chong-yang], and Wu Siu-yuan, 1974, [Carboniferous biostratigraphy of Tsingyuan, eastern Kansu]: Memoirs of Nanking Institute of Geology and Palaeontology, Academia Sinica, no. 5, p. 99–117, Pls. 1–3 (in Chinese).

Li Xing-xue and Yao Zhao-qi, 1979, Carboniferous and Permian floral provinces in East Asia: Papers read at the 9th International Congress of Carboniferous Stratigraphy and Geology, Urbana, Illinois, 1979, p. 1–11.

Liu Jinling and Ye Pingyi, 1977, Studies on the Quaternary sporo-pollen assemblage from Shanghai and Zhejiang with reference to its stratigraphic and paleoclimatic significance: Acta Palaeontologica Sinica, v. 16, no. 1, p. 1–10, Pls. 1–2 (in Chinese with English summary).

Lu Lichang and Ouyang Shu, 1976, The Early Devonian spore assemblage from Xujiachong Formation at Cuifengshan, in Qujing of Yunnan: Acta Palaeontologica Sinica, v. 15, no. 1, p. 21–38, Pls. 1–3 (in Chinese with English abstract).

—— 1978, Devonian megaspores from the Zhanyi District, E. Yunnan:

Acta Palaeontologica Sinica, v. 17, no. 1, p. 69–79, Pls. 1–3 (in Chinese with English abstract).

Lu Yan-hao [Y.H. Lu], 1944, The Charophyta from the Kuche Formation near Kuche, Sinkiang: Bulletin of the Geological Society of China, v. 24, no. 1–2, p. 33–38, 1 pl.

Mu Xinan, 1977, Upper Permian fungi from Anshun of Guizhou: Acta Palaeontologica Sinica, v. 16, no. 2, p. 151–158, Pls. 1–2 (in Chinese with English abstract).

Ouyang Shu, 1964, A preliminary report on sporae dispersae from the Lower Shihhotse Series of Hokü District, NW Shansi: Acta Palaeontologica Sinica, v. 12, no. 3, p. 486–519, Pls. 1–8 (in Chinese with English summary).

——1979a, Notes on some new miospore genera from Permo-Carboniferous strata of China, *in* Papers for the 9th International Congress of Carboniferous Stratigraphy and Geology: Nanjing Institute of Geology and Palaeontology, Academia Sinica.

——1979b, Ultramicro- and micro-fossils from the Anshan Group and the Liache Group in E. Liaoning, NE China, *in* Selected works for a scientific symposium on iron-geology of China, sponsored by Academia Sinica 1977, Stratigraphy and Palaeontology: Beijing, Science Press, p. 1–32, 5 pls. (in Chinese with English abstract).

Ouyang Shu, Yin Lei-ming, and Li Zai-ping, 1974, [Sinian and Cambrian spores and acritarehs, *in* Handbook of Stratigraphy and Palaeontology in southwestern China]: Beijing, Science Press, p. 72–80, 114–123, Pls. 27–28, 45–48 (in Chinese).

Qu Li-fan, 1980, [Spores and pollen of the Triassic, *in* Mesozoic stratigraphy and palaeontology of the Shaan-Gan-Ning Basin]: Geological Institute, Academy of Geological Sciences, Beijing, Geological Publishing House, p. 105–204, Pls. 61–80 (in Chinese).

Shen Kuang-lung [Shen Guang-long], 1961, Jurassic plants from Meinhsien Series in the vicinity of Huicheng Hsien of S. Kansu: Acta Palaeontologica Sinica, v. 9, no. 2, p. 165–179, Pls. 1–2 (in Chinese with English abstract).

Sin Yu-sheng [Xing Yu-sheng] and Liu Kui-zhih [Liu Gui-zhi], 1973, On Sinian micro-flora in Yenliao region of China and its geological significance: Acta Geologica Sinica, no. 1, p. 1–64, Pls. 1–13 (in Chinese with English abstract).

Skvortzov, B. V., 1937, Neogene diatoms from eastern Shantung: Bulletin of the Geological Society of China, v. 17, no. 2, p. 193–204, Pls. 1–2.

Sung Meng-rong, 1965, [Sporo-pollen assemblages from the *Sinanthropus*-bearing horizons of Choukoutien, Peking]: Quaternaria Sinica, v. 4, no. 1, p. 84–96, Pls. 1–8 (in Chinese).

Sung Tze-chen [Song Zhi-chen], 1958, Tertiary spore and pollen complexes from the Red Beds of Chiuchuan, Kansu, and their geological and botanical significance: Acta Palaeontologica Sinica, v. 6, no. 2, p. 159–167, Pls. 1–7 (in Chinese with English summary).

Sung Tze-chen [Song Zhi-chen] and Tsao Liu [Cao Liu], 1976, The Paleocene spores and pollen grains from the Fushun Coalfield, Northeast China: Acta Palaeontologica Sinica, v. 15, no. 2, p. 147–162, Pls. 1–3 (in Chinese with English abstract).

Sung Tze-chen [Song Zhi-chen], Tsao Liu [Cao Liu], and Li Man-ying, 1964, Tertiary sporo-pollen complexes of Shantung: Memoirs of the Institute of Geology and Palaeontology, Academia Sinica, no. 3, p. 179–290, Pls. 1–28 (in Chinese with English abstract).

Sung Xiang-jun, 1979, Palynofloristical investigation on the Late Cretaceous and Paleocene of China: Acta Phytotaxonomica Sinica, v. 17, no. 3, p. 8–23 (in Chinese with English abstract).

Sze Hsin Chien [Si Xing-jian], 1949, Die mesozoische Flora aus der Hsiangchi Serie in Westhupeh: Palaeontologia Sinica, new ser. A, no. 2 (whole ser. no. 133), p. 1–68, Pls. 1–5 (in German with Chinese abstract).

——1952, Upper Devonian plants from China: Acta Scientia Sinica, v. 1, no. 2, p. 166–192, Pls. 1–6.

——1954, On the structure and relationship of *Phoroxylon scalariforme* Sze: Acta Palaeontologica Sinica, v. 2, no. 4, p. 347–354, Pls. 1–4 (in Chinese and English).

——1956, Older Mesozoic plants from the Yenchang Formation, Northern Shensi: Palaeontologia Sinica, new ser. A, no. 4 (whole ser. No. 139), 206 p., 56 pls. (in Chinese and English).

——1960, [The Namurian flora from Aolungpyluke region, Chinghai Province, *in* Contribution to the geology of the Mt. Qilianshan]: Beijing, Science Press, v. 4, no. 1, p. 1–11, Pls. 1–9 (in Chinese).

Sze Hsin Chien [Si Xing-jian] and Lee Hsing Hsüeh [Li Xing-xue], 1952, Jurassic plants from Szechuan: Palaeontologia Sinica, new ser. A, no. 3, (whole ser. no. 135), 38 p., 9 pls. (in Chinese and English).

Sze Hsin Chien [Si Xing-jian], Lee Hsing Hsüeh [Li Xing-xue], Li Pie-juan, Zhou Zhi-yan, Ye Mei-na, Wu Shun-qing, and Shen Guang-long, 1963, [Mesozoic plants from China]: Beijing, Science Press, 429 p., 118 pls. (in Chinese).

Tao Jun-rong, 1965, A late Eocene florule from the District Weinan of Central Shensi: Acta Botanica Sinica, v. 13, no. 3, p. 272–278, Pls. 1–3 (in Chinese with English abstract).

Tsao Cheng-yao [Cao Zheng-yao], 1965, Fossil plants from the Siaoping series in Kaoming, Kwangtung: Acta Palaeontologica Sinica, v. 13, no. 3, p. 510–528, Pls. 1–6 (in Chinese with English summary).

Tsao Rui-chi [Cao Rui-ji] and Liang Yu-zhou, 1974, [On the classification and correlation of the Sinian System in China, based on a study of Algae and Stromatolites]: Memoirs of Nanking Institute of Geology and Palaeontology, Academia Sinica, no. 4, p. 1–16, Pls. 1–8 (in Chinese).

Tuan Shu-yin, Chen Yeh: [Chen Ye], and Keng Kuo-chang [Geng Guo-chang], 1977, Some Early Cretaceous plants from Lhasa, Tibetan autonomous region, China: Acta Botanica Sinica, v. 19, no. 2, p. 114–119, Pls. 1–3 (in Chinese with English abstract).

Wang Kai-fa, Yang Jiao-wen, Li Zeng-rui, 1975, On the Tertiary sporopollen assemblages from Lungpola Basin of Xizang, China, and their palaeogeographic significance: Scientia Geologica Sinica, no. 4, p. 366–374, Pls. 1–4 (in Chinese with English abstract).

Wang Li-xin, Xie Zhi-min, and Wang Zi-qiang, 1978, On the occurrence of *Pleuromeia* from the Qinshui Basin in Shanxi Province: Acta Palaeontologica Sinica, v. 17, no. 2, p. 195–211, Pls. 1–4 (in Chinese with English abstract).

Wang Shui, 1961, Tertiary Charophyta from Chaidamu (Tsaidam) Basin, Qinghai (Chinghai) Province: Acta Palaeontologica Sinica, v. 9, no. 3, p. 183–219, Pls. 1–7 (in Chinese with English summary).

——1965, Mesozoic and Tertiary Charophyta from Jiuquan Basin of Kansu Province: Acta Palaeontologica Sinica, v. 13, no. 3, p. 463–499, Pls. 1–5 (in Chinese with English summary).

Wang Xi-fu, 1977, On the new genera of *Annularia*-like plants from the Upper Triassic in Sichuan-Shaanxi Area: Acta Palaeontologica Sinica, v. 16, no. 2, p. 185–190, Pls. 1–2 (in Chinese with English abstract).

Wang Yu-jing, 1976, [Cretaceous and Early Tertiary calcareous algae from the Mt. Jolmo Lungma Region, Xizang, *in* Report of scientific expedition to the Mt. Jolmo Lungma region (Palaeontology, Fasc. 2)], Beijing, Science Press, p. 425–457, Pls. 1–12 (in Chinese).

Wang Zhen, 1976, Middle Devonian *Sycidium* and *Chovanella* from Southwest China: Acta Palaeontologica Sinica, v. 15, no. 2, p. 175–185, Pls. 1–3 (in Chinese with English abstract).

Wang Zhen and Huang Ren-jin, 1978, Triassic charophytes of Shaanxi: Acta Palaeontologica Sinica, v. 17, no. 3, p. 267–274, Pls. 1–2 (in Chinese with English abstract).

Wu Shun-ching [Wu Shun-qing], 1966, Notes on some Upper Triassic plants from Anlung, Kweichow: Acta Palaeontologica Sinica, v. 14, no. 2, p. 233–241, Pls. 1–2 (in Chinese with English summary).

Wu Shun-ching [Wu Shun-qing], Ye Mei-na, and Li Bao-xian, 1980, Late Triassic and Lower and Middle Jurassic plants from western Hubei: Memoirs of Nanjing Institute of Geology and Palaeontology,

Academia Sinica, no. 14, p. 63–131, 39 pls. (in Chinese with English abstract).

Xu Ren [Hsü Jen], Zhu Jia-nan, Chen Ye, Duan Shu-yin, Hu Yu-fan, and Zhu Wei-qing, 1979, [Late Triassic Baoding flora, SW Sichuan, China]: Beijing, Science Press, 121 p., 75 pls. (in Chinese).

Yao Zhao-qi, 1978, On the age of "*Gigantopteris* Coal Series" and *Gigantopteris*-flora in South China: Acta Palaeontologica Sinica, v. 17, no. 1, p. 81–89 (in Chinese with English abstract).

Yan Fu-hua, 1965, Discovery of microfossils in Sinian rocks from eastern Yunnan and western Hupeh: Scientia Geologica Sinica, no. 4, p. 370–373, 1 pl. (in Chinese with English abstract).

Ye Mei-na [Yeh Mei-na], 1979, On some Middle Triassic plants from Hupeh and Szechuan: Acta Palaeontologica Sinica, v. 18, no. 1, p. 73–81, Pls. 1–2 (in Chinese with English abstract).

Yin Lei-ming, 1979, Microflora from the Anshan Group and the Liaohe Group in E. Liaoning with its stratigraphical significance, *in* Selected works for a scientific symposium on iron-geology of China, sponsored by Academia Sinica, 1977, Stratigraphy and Palaeontology: Beijing, Science Press, p. 39–58, Pls. 1–4 (in Chinese with English abstract).

Ying Lei-ming and Li Zai-ping, 1978, Precambrian microfloras from southwestern China: Memoirs of Nanjing Institute of Geology and Palaeontology, Academia Sinica, no. 10, p. 41–102, Pls. 1–9 (in Chinese with English abstract).

Zhang Chungying, 1977, On the discovery of the fossil blue-green algae from the Lower Tertiary of northern Kiangsu: Acta Palaeontologica Sinica, v. 16, no. 2, p. 159–164, Pls. 1–2 (in Chinese with English abstract).

Zhang Lujin [Chang Lu-chin], 1965, [Chzhan Lu-chin] K voprosu o znachenii sporo-pyultsevykh kompleksov Imaskoy uglenosnoy svity v zapadnoy chast provintsii Khenan. (Significance of sporo-pollen complexes from the Yima Coal-bearing formation of Western Honan Province): Acta Palaeontologica Sinica, v. 13, no. 1, p. 160–181, Pls. 1–6 (in Chinese with Russian abstract).

—— 1978, Mesozoic spores and pollen grains from the volcanic clastic sedimentary rocks in Zhejiang, with their stratigraphic significance: Acta Palaeontologica Sinica, v. 17, no. 2, p. 180–192, Pls. 1–4 (in Chinese with English abstract).

Zhang Yun, 1978, Eucaryotic unicellar microfossils in the mid-Proterozoic Wumishan Formation (Sinian System) from western Hopei, China: Acta Botanica Sinica, v. 20, no. 4, p. 293–304, Pls. 1–3 (in Chinese with English abstract).

Zhang Zhen-lai, 1979, [Cretaceous sporo-pollen assemblages of South-Central China, *in* Institute of Vertebrate Paleontology and Palaeoanthropology and Nanjing Institute of Geology and Palaeontology, Academia Sinica, eds., Mesozoic and Cenozoic red beds of South China], Beijing, Science Press, p. 132–140 (in Chinese).

Zhao Xiu-hu and Wu Xiu-yuan, 1979, Carboniferous macrofloras of South China: Papers read at the 9th International Congress of Carboniferous Stratigraphy and Geology, Urbana, Illinois, 1979, p. 1–8.

Zhao Xiu-hu, Zhang Shan-zhen [Chang Shan-chen], Yao Zhao-qi, and Mo Zhuang-guan, 1979, [Late Permian flora from West Guizhou and East Yunnan, *in* Late Permian stratigraphy and biota remains of West Guizhou and East Yunnan]: Beijing, Science Press, (in Chinese).

Zhou Hui-qin, Huang Zhi-gao, and Zhang Zhi-cheng, 1976, [Mesozoic plants, *in* Palaeontological atlas of North China, Nei Mongol no. 1]: Beijing, Geological Publishing House, p. 204–213, Pls. 105–120 (in Chinese).

Zhou Kun Shu, Chen Shuo-ming, Ye Yong-ying, and Liang Ziu-long, 1976, [On approaching some problems of the Quaternary paleogeography in the Mt. Jolmo Lungma region, Xizang, based on analyses of sporo-pollen materials, *in* Report of scientific expedition to the Mt. Jolmo Lungma region (Quaternary Geology)]: Beijing, Science Press, p. 79–92, Pls. 1–3 (in Chinese).

Zhou Tong-shun, 1978, [On the Mesozoic coal-bearing strata and fossil plants from Fujian Province]: Professional Paper of Stratigraphy and Palaeontology, no. 4, p. 88–128, Pls. 15–30 (in Chinese).

Zhou Zhi-yan [Chow Tse-yen] and Cao Zheng-yao, 1977, on eight new species of conifers from the Cretaceous of East China with reference to their taxonomic positions and phylogenetic relationship: Acta Palaeontologica Sinica, v. 16, no. 2, p. 165–181, Pls. 1–5 (in Chinese with English abstract).

—— 1979, [Some Cretaceous conifers from southern China and their geological significances, *in* Mesozoic and Cenozoic red beds of South China]: Beijing, Science Press, p. 218–222, Pls. 1–3 (in Chinese).

Zhou Zhi-yan and Li Bao-xian, 1979, A preliminary study of the Early Triassic plants from the Qianghai District, Hainan Island: Acta Palaeontologica Sinica, v. 18, no. 5, p. 444–462, Pls. 1–2 (in Chinese with English abstract).

Zhu Hao-ran [Chu Hao-jan], 1979, Microfossil algae from the Lower Tertiary Shanhejia Formation of Bin Xian, northern Shandong: Acta Palaeontologica Sinica, v. 18, no. 4, p. 327–344, Pls. 1–3 (in Chinese with English abstract).

Zhu Shi-xing, Cao Rui-ji [Tsao Rui Chi], Liang Yu-zuo and Zhao Wen-jie, 1979, [The studies on stromatolites from the stratotype of Sinian Suberathem in Chihsien County]: Biejing, Geological Publishing House, 94 p., 44 pls. (in Chinese).

Miscellaneous References (Many Authors, No Editors indicated)

1978, [Cenozoic plants of China]: Beijing, Science Press, 232 p., 149 pls. (in Chinese).

1978, [Early Tertiary spore and pollen grains from the Coastal Region of Bohai]: Contributions to the palaeontology of the Bohai Coastal Area: Beijing, Science Press, (in Chinese with English abstracts).

1974 [Paleozoic plants of China]: Beijing, Science Press, 226 p., 130 pls. (in Chinese).

MANUSCRIPT ACCEPTED BY THE SOCIETY APRIL 22, 1981

Printed in U.S.A.

Geological Society of America
Special Paper 187
1981

On the genus *Palaeofusulina*

Rui Lin
Sheng Jin-zhang
Nanjing Institute of Geology and Palaeontology, Academia Sinica, Nanjing 210008, People's Republic of China

ABSTRACT

This paper discusses the classification, characteristics, and geologic range of the fusulinid genus *Palaeofusulina.* Thirty-nine known species and subspecies referred to this genus are grouped into three types, each of which is treated as a genus. The first type, represented by *Palaeofusulina prisca* Deprat, the type species of *Palaeofusulina,* in *Palaeofusulina s. str.;* the other two are described as new genera, *Nanlingella* and *Parananlingella. Palaeofusulina* arose at the beginning of the Changhsingian of latest Permian age, flourished during the middle and late Changhsingian, and became extinct at the end of the Changhsingian. It is a short-ranging genus restricted to the Changhsingian which is referred to as the fusulinid faunal zone of *Palaeofusulina.* This zone is the highest fusulinid zone in the world.

INTRODUCTION

The genus *Palaeofusulina* is one of the most important fusulinaceans in the late Permian (Changhsingian). Since Deprat (1912, 1913) proposed *Palaeofusulina* based on P. prisca Deprat, about 39 species of this genus have been published; however, some of them are quite different from the type species. This procedure has not only brought about a confusion of the *Palaeofusulina s. str.* with related genera but also has obscured our understanding of its vertical distribution. It is necessary to revise the palaelofusulinids of the Changhsingian. The purpose of this paper is to discuss the classification, characters, and geological range of *Palaeofusulina* and its related genera.

DIAGNOSTIC CHARACTER OF *PALAEOFUSULINA* AND ITS RELATED GENERA

Deprat established the genus *Palaeofusulina* in 1912, and the type species was described in the next year. The diagnosis of the type species is as follows: shell large, inflated fusiform, having 4 volutions; normally coiled, expanding uniformly outward; spirotheca composed of a tectum and a diaphanothcca; septa highly and narrowly fluted throughout the length of the shell, with its nearly parallel-sided folds reaching almost to the ceilings of the chambers; chomata lacking; tunnel distinct; proloculus relatively large. To date, about 39 species have been assigned to this genus and these may be grouped into the following three types:

The first type is represented by the type species of *Palaeofusulina.* This is *Palaeofusulina s. str.,* including the following species: *Palaeofusulina prisca* Deprat, *P. sinensis* Sheng (=*P. convexa* Liem), *P. wangi* Sheng, P. *fusiformis* Sheng, *P. bella* Sheng, *P. pseudoprisca* (Colani), *P. mutabilis* Sheng, *P. pulla* Sheng, *P. ellipsoidalis* Sheng, *P. compacta* Sheng, *P. typica* Rui, *P. qinglongensis* Rui, *P. minima* Sheng and Chang, *P. pseudominima* Rui, *P. ampla* Rui, *P. nana* Likharev, *P. parafusiformis* Lin, *P. subcylindrica* Sheng, *P. kongdongshanica* Sheng, *P. simplicata* Sheng, *P. rhomboidea* Liem (=*P. kycungensis* Liem), *P.* sp. (Rui, 1979), *P. ovata* Sheng and Rui, and *P. chumipuensis* Sheng.

The second type is represented by a new genus *Nanlingella.* In the general profile, it is closely similar to the primitive species of *Palaeofusulina,* but its first two volutions are endothyroid with rudimentary chomata. The type species of this new genus is *Nanlingella meridionalis* Rui and Sheng, sp. nov. In addition to the type species, the following forms are now referred to this new genus: *Palaeofusulina? simplex* Sheng and Chang, *P. guizhouensis* Rui, *P. deltoidalis* Rui, *P. zhongyingica* Rui, *P. jiaozishanica* Rui, *P. abscondida* Lin, and *Dunbarula palaeofusulinaeformis* Sheng. This genus probably originated in the upper part of the *Codonofusiella* Zone and flourished in the lower and middle part of the *Palaeofusulina* Zone. It is similar in several features to *Codonofusiella;* however, the last uncoiled volution of *Codonofusiella* serves to distinguish the latter.

The third type is represented by *Palaeofusulina acervula* Sheng and Rui. This species is here designated as the type of species of new genus *Parananlingella.* Its first volution is endothyroid with a short axis coiled at a large angle to the outer volutions; its inner two volutions are tightly coiled, outer ones loosely coiled, and last ones strongly expanded but still coiled; its septa highly and intensely fluted throughout the length of the shell; in the inner volutions, parallel-sided septal loops reach to the ceilings of the chambers, are rather irregularly fluted, and look like spumes or cuniculi in the axial section of the last volution; chomata lacking. In addition to the type species, the following species are included in this genus: *Palaeofusulina laxa* Sheng, *P. oblata* Rui, *P. evoluta* (Chen), *P. xikouensis* Sun, and *P. shuanxiensis* Sun.

It is apparent that the genus *Palaeofusulina,* which arose at the beginning of the Changhsingian, flourished during the middle and late Changhsingian and became extinct at the end of that stage. It is a short-ranging genus whose entire history is encompassed in the Changhsingian Stage.

CHARACTERS AND STRATIGRAPHIC OCCURRENCES OF THE SPECIES GROUPS OF *PALAEOFUSULINA*

There are 25 species in the genus *Palaeofusulina* as defined by its type species. According to their shape, shell size, number of whorls, character of coiling, nature of septal fluting, and proloculus size, they may be subdivided into 6 subgroups.

Palaeofusulina minima group. Shell small, oval to inflated fusiform having 3 to 4 volutions with a form ratio about 1.5 to 1.7:1; septa loosely and broadly fluted throughout the length of the shell. This group contains *P. minima* Sheng and Chang, *P. ovata* Sheng and Rui, and *P. pseudominima* Rui and comprises the smallest and simplest palaeofusulinids. The typical form, *P. minima,* is found abundantly in the lower part of the Changhsingian.

Palaeofusulina simplicata group. Shell small to medium size, fusiform; adult shell possessing 6 volutions, with a form ratio about 1.8 to 2.2:1; volutions loosely coiled; septal folds loosely and irregulary arranged. This group which includes P. *simplica* Sheng and *P. kongdongshanica* is a more specialized branch and, except for *P. kongdongshanica,* may represent a primitive group of *Palaeofusulina,* based on the morphology and shell structures. It commonly occurs in the lower of middle part of the Changhsingian.

Palaeofusulina sinensis group. Shell medium size, fusiform, having 4 to 4½ volutions with a form ratio about 1.7:1; regularly coiled, expanding uniformly outward; septa strongly and narrowly fluted throughout the length of the shell. It comprises *P. sinensis* Sheng (=*P. covexa* Liem), *P. fusiformis* Sheng, *P. bella* Sheng, and *P. rhomboidea* Liem (=*P. kycungensis* Liem). These generally occur in the upper part of the Changhsingian.

Palaeofusulina ellipsoidalis group. Shell medium-size, ellipsoidal, possessing 5 to 6 volutions with a form ratio 1.3 to 1.6:1; septa strongly and narrowly fluted throughout the length of the shell. *Palaeofusulina ellipsoidalis* Sheng, *P. wangi* Sheng, *P. nana* Likharev, *P. compacta* Sheng, *P. pseudocompacta* Rui, and *P. chumipuensis* Sheng belong to this group and generally occur in the upper part of the Changhsingian.

Palaeofusulina typica group. Shell rather large, elongate fusiform to elongate ellipsoidal, having 5 to 6 volutions, with a form ratio over 2:1, all volutions tightly coiled; septa strongly and narrowly fluted throughout the length of the shell; proloculus minute. The following species are included in this group: *Palaeofusulina typica* Rui, *P. qinglongensis* Rui, *P. parafusiformis* Lin, *P. pulla* Sheng, *P. subcylindrica* Sheng, and *P. ampla* Rui, which occurs at a different level of the Changhsingian.

Palaeofusulina prisca group. Shell large, inflated fusiform, giving a form ratio 1.2 to 1.5:7, having 4 to 5 loosely coiled volutions; proloculus large. This group, the largest of *Palaeofusulina,* includes *P. prisca* Deprat, *P. pseudoprisca* (Colani), and *P. mutabilis* Sheng and commonly appears in the upper part of the Changhsingian.

VERTICAL DISTRIBUTION OF *PALAEOFUSULINA*

The strata bearing the *Palaeofusulina* fauna in South China have been called the *Palaeofusulina* Zone by Sheng (1955). This zone is now recognized as representative of the Changhsingian carbonate facies by most Chinese geologists and paleontologists. For the past decade, some authors (Wang and others, 1964; Han and others, 1977) found that few species of *Codonofusiella* are associated with *Palaeofusulina.* On the basis of this fact and the information from Yugoslavia and the Northern Caucasus, these authors suggested that the genus *Palaeofusulina* is not restricted to the Changhsingian and ranges from the Changhsingian downward to the Wuchiapingian and even into the Maokouan. According to the present study, during the Changhsingian, *Palaeofusulina* rapidly became dominant, flourishing both in number of individuals and in a variety of species, while the few species of *Codonofusiella* that are associated with *Palaeofusulina* were declining markedly in species and individuals at the end of its evolution. Near the middle or late Changhsingian, the genus *Codonofusiella* became entirely extinct. This fact suggests that the extinction of *Codonofusiella* did not occur near the end of Wuchiapingian Stage but possibly as late as the middle or even the late Changhsingian Stage. On the other hand, it also indicates that by the beginning of Changhsingian time, the rate of change of fusulinaceans was rather slow, though conspicuous.

Recent investigations show that the late Permian fusulinacean fauna that appeared in Lianxian, North Guangdong, consists mainly of *Gallowayinella, Nanlingella, Palaeofusulina,* and a few species of *Codonofusiella. Gallowayinella,* and *Nanlingella* are very dominant. *Gallowayinella* has a rather limited stratigraphic range, occurring only in the lower part of the Changhsingian. Together with *Palaeofusulina minima* and other species, these genera form a distinctive *Palaeofusulina minima-Gallowayinella* subzone, representing the lower Changhsingian of South China, while the new genus *Nanlingella* has been identified as *Codonofusiella* by some authors.

In Yugoslavia, the *Palaeofusulina* fauna has been found only in Bal, Montenegro, where it is found along with smaller foraminifers and as *Colaniella* and not with neoschwagerinids or verbeekinids. Kochancky-Devidě (1965) believed that this fauna and the *Neoschwagerina* and *Polydiexodina* fauna are contemporaneous in age, but are found in different limestone facies. However, this conclusion seems to be hardly acceptable for lack of enough substantial evidence.

In the northern Caucasus, Likharev (1926) and Miklukho-Maklay (1954) have reported the *Palaeofusulina* fauna from several localities. According to Miklukho-Maklay (1954), this fauna generally is found along with *Reichelina, Nankinella, Eoverbeekina, Codonofusiella, Colaniella,* and others, and not with neoschwagerinids or verbeekinids, except in Sahe where *Palaeofusulina* is associated with *Neoschwagerina, Parafusulina,* and

Pseudofusulina. Unfortunately, neither description nor illustrations accompanied Miklukho-Maklay's identification of *Palaeofusulina* in his monograph, and the reliability of this information remains doubtful. In the light of Likharev's report, the *Palaeofusulina* occurrence in the northern Caucasus is not in doubt, but no stratigraphic data are available.

In the Abadeh region of central Iran, *Palaeofusulina* was reported to be associated with *Chusenella* in the upper part of Unit 3 (Taraz, 1971). However, this identification was recently corrected by Ishii to be *Codonofusiella* (Kanmera and others, 1976).

Leven (1967) reported that *Palaeofusulina pamirica* Leven, which is not the only species of *Palaeofusulina* in the Pamir, is found in the middle part of the upper chert and limestone member of the Gansk Formation and in the Karaveles Formation. However, this species has been referred by Skinner (1969) to his new genus *Paradunbarula.*

To sum up, it is clearly known that the Changhsingian fusulinacean fauna is characterized by an abundance of the genus *Palaeofusulina.* Altough few species of *Codonofusiella* appear in this fauna, they have no actual stratigraphic significance, because the relics of *Codonofusiella* appearing in the *Palaeofusulina* Zone are of limited value for age determination. During the Changhsingian, *Palaeofusulina* diversified markedly and formed a distinctive fauna which characterized the *Palaeofusulina* Zone. This zone is the highest fusulinacean zone so far known in the world.

DESCRIPTION OF NEW GENERA AND NEW SPECIES

Subfamily BOULTONIINAE Skinner and Wilde, 1954
Genus *Nanlingella* Rui and Sheng (gen. nov.)
Type species: *Nanlingella meridionalis* Rui and Sheng (sp. nov.)

Shell small to medium-size, ellipsoidal, fusiform to elongate fusiform; mature shell consisting of 3 to 5 volutions; inner 1 or 2 volutions endothyroid, with the short axis coiled askew to the outer volutions, expanding uniformly outward; spirotheca thin, composed of a tectum and a diaphanotheca; septa strongly fluted, but weakly fluted in the median part; septal folds broad and low, having about one-half to two-thirds the height of the respective chambers; chomata weak, existing only in the first or second volution; tunnel distinct; proloculus small.

Discussion: This new genus closely resembles *Palaeofusulina* Deprat, but the juvenarium of the latter genus is not endothyroid, the septa are strongly and narrowly fluted, lateral sides of septal loops are nearly parallel, and chomata are absent. It is also similar to *Paradunbarula* Skinner, but the latter is larger, having a larger number of volutions, and the spirotheca is composed of tectum, diaphanotheca, and inner tectorium.

Age distribution: Late Permian of South China.

Nanlingella meridionalis Rui and Sheng nov.
Plate 1 (1–9)

Shell small, fusiform, one side of the median portion plane and the opposite side slightly vaulted; poles bluntly rounded; mature specimen possess 5 volutions; first volution endothyroid with the short axis coiled at large angles to the outer volutions; inner two volutions tightly coiled, expanding uniformly outward; spirotheca composed of a tectum and a diaphanotheca; septa strongly fluted, septal folds broad and loose, about one-half to two-thirds as high as the respective chambers; chomata weak, only seen in the first volution; tunnel distinct; proloculus minute. Measurements of representative specimens are given in Table 1.

Occurrence: This species occurs in the lower part of the Changhsingian at Liantang, Lianxian, Guangdong Province, and at Meitian, Yizhang, Hunan Province. It commonly is associated with *Gallowayinella meitienensis* Chen and *Palaeofusulina minima* Sheng and Chang.

Cat. Nos. 61465–61473.

Genus _Parananlingella_ Rui and Sheng (gen. nov.)
Plate 1 (10–17)
Type species: _Palaeofusulina acervula_ Sheng and Rui
(In Zhao Jin-ke and others, 1981, p. 49–50, 78–79; Pl. 4, figs. 32–38)

Shell of median size, quadrate, ellipsoidal to fusiform; adult shell possessing 4 to 5 volutions; first volution endothyroid, coiled at large angles to the coiling of the outer volutions; inner 2 or 3

TABLE 1. MEASUREMENTS (IN MM) FOR *NANLINGELLA MERIDIONALIS* RUI AND SHENG (SP. NOV.)

Specimen (Cat. Nos.)	L	W	F.R.	Diam. Prol.	Width of volutions					
					1	2	3	4	4½	5
61465	1.64	0.84	1.95	0.03	0.08	0.14	0.26	0.50	/	0.84
61467	1.67	0.80	2.09	0.04	0.08	0.14	0.25	0.47	/	0.80
61469	1.50	0.70	2.15	0.04	0.04	0.14	0.28	0.54	0.70	
61468	0.97	0.46	2.11	0.05	0.12	0.71	0.32	0.46		
61470	1.67	0.82	2.07	0.04	0.08	0.16	0.30	0.58	0.82	
61473	1.25	0.58	2.17	0.06	0.12	0.18	0.33	0.58		
61472	1.57	0.71	2.21	0.07	0.12	0.19	0.33	0.55	0.71	

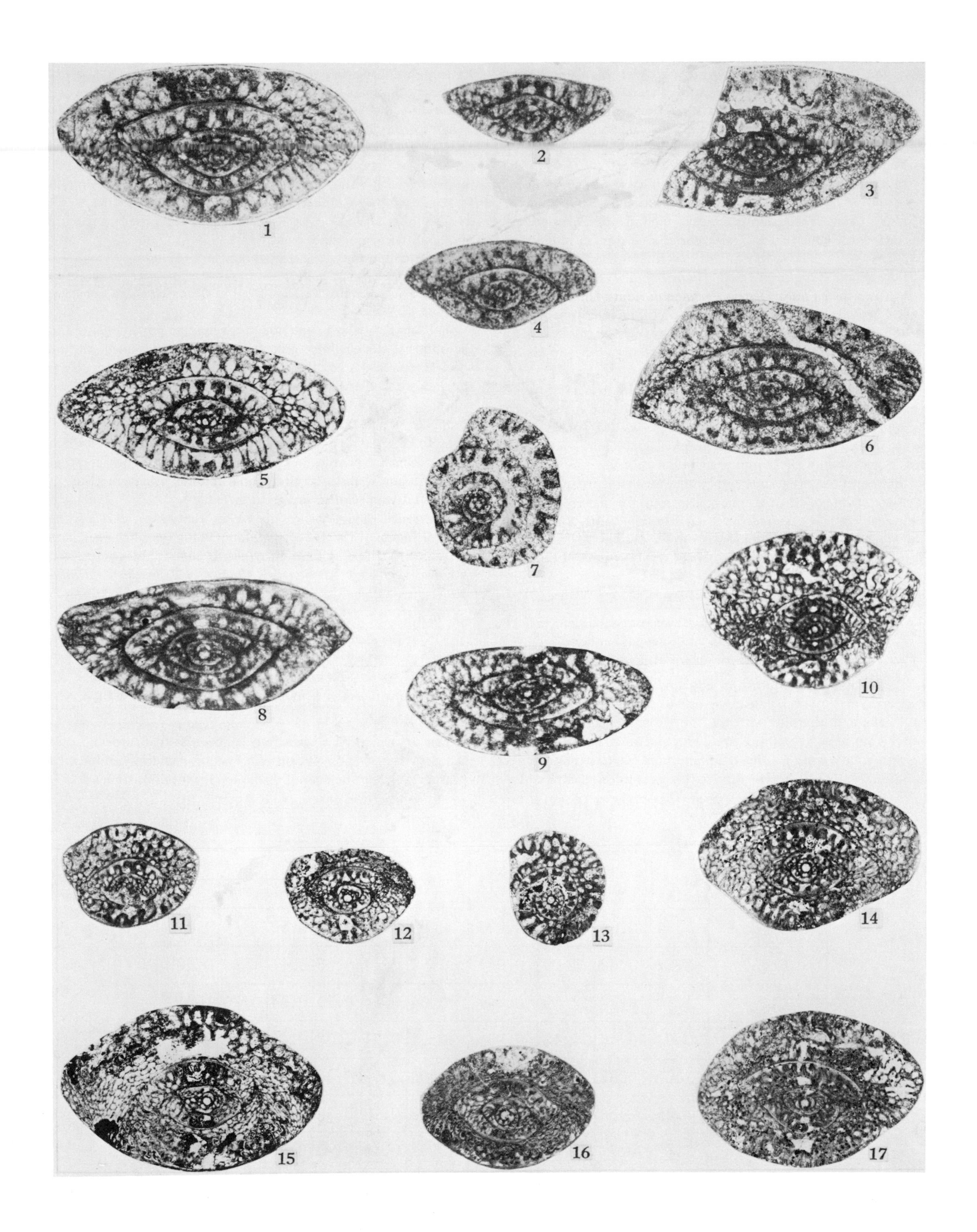

1
2
3
4
5
6
7
8
9
10
11
12
13
14
15
16
17

volutions tightly coiled, outer ones loosely coiled, last one highly expanding but not uncoiled; spirotheca composed of a tectum and a diaphanotheca; septa strongly fluted throughout the length of the shell, looking like cuniculi or spumes in axial section of the last volution; chomata lacking, tunnel indistinct in the outer variations; proloculus minute.

Discussion: In the shape of the endothyroid juvenarium, this new genus is closely related to *Nanlingella* Rui and Sheng (gen. nov.), but differs from that genus in its compactly coiled inner volutions and highly expanded last volution; its septa are more strongly fluted, looking like cuniculi or spumes in the axial section of the last volution.

Age and distribution: Latest Permian of South China.

ACKNOWLEDGEMENTS

We are much indebted to Dr. Charles A. Ross for his careful reading of the manuscript.

REFERENCES CITED

Colani, M., 1924, Nouvelle contribution a l'étude des Fusulinidés de l'extrême Orient: Mémoires du Service Géologique de L'Indochine, v. XI, fasc. 1, p. 1-191, Pls. 1-29.

Deprat, J., 1912, Etude des Fusulinidés de Chine et de l'Indochine et classification des Calcaires à Fusulines: Mémoires du Service Géologique de l'Indochine, v. 1, fasc. 3, p. 1-76, Pls. 6-9.

—— 1913, Les Fusulinidés des calcaires carbonifériens et permiens du Tonkin, du Laos et du Nord Annam: Mémoires du Service Géologique de l'Indochine, v. 2, fasc. 1, p. 1-74, Pls. 1-10.

Han Tong-xiang, Bai Qing-shao, and Xu Mao-yu, 1977, [Permian coal-bearing strata of Lianyan, Guangdong]: Beijing, Press of Coal Industry, p. 18-55 (in Chinese).

Kanmera, K., Ishii, K., and Toriyama, R., 1976, The evolution and extinction patterns of Permian fusulinaceans: University of Tokyo Press, Geology and Palaeontology of Southeast Asia, v. 17, p. 129-154.

Kochanskiy-Devidé, V., 1954, Permski foraminifera i vapnevacke Alge okolice Bara U Crnoj Gori: Geoloski Vjesnik Sv. V-VIII, God. 1951-1953, p. 295-298, Pls. 1-2.

Leven, E. Ya., 1967, Stratigrafiya i fuzulinidy permskikh otlozheniy Pamira: Trudy Geologicheskogo Institute Akademii Nauk SSSR, Vypusk 167, 223 p., 39 pls.

Likharev, D., 1926, *Palaeofusulina nana* sp. nov. iz antrakolitovykh otlozheniy severnogo Kavkaza: Izvestiya Geologicheskogo Komiteta, Tom 45, Vypusk 2, p. 59-66, Pl. 1.

Lin Jia-xing, Lee Jia-xiang, Chen Gong-xin, Zhou Zu-ren, and Zhang Bu-fei, 1977, [Palaeontological atlas of central-South China, Part 2]: Beijing, Geological Publishing House, p. 4-96, Pls. 1-30 (in Chinese).

Miklukho-Maklay, A. D., 1954, Foraminifery verkhnepermskikh otlozheniy severnogo Kavkaza: Trydy vsesoyuznogo Nauchno-issledovatel'skogo Geologicheskogo Instituta, 163 str., 19 tables.

Nguen Van Liem, 1974, Rod *Palaeofusulina* i ego novye vidy iz Vgetnama: Paleontologicheskiy Zhurnal, Vypusk 4, p. 11-17, Pl. 1-2.

Rui Lin, 1979, Upper Permian fusulinids from western Guizhou: Acta Palaeontologica Sinica, v. 18, no. 3, p. 271-300, Pls. 1-4 (in Chinese with English abstract).

Sheng Jing Chang [Sheng Jin-zhang], 1955, Some fusulinids from Changhsing Limestone: Acta Palaeontologica Sinica, v. 3, no. 4, p. 283-308, Pls. 1-4 (in Chinese with English summary).

—— 1963, Permian fusulinids of Kwangsi, Kueichow, and Szechuan: Palaeontologia Sinica, new ser. B, no. 10, 247 p., 36 pls. (in Chinese and English).

—— and Chang Lin Hsing [Zhang Lin-xin], 1958, Fusulinids from the type-locality of the Changhsing Limestone: Acta Palaeontologica Sinica, v. 6, no. 2, p. 205-214, 1 pl. (in Chinese with English abstract).

Skinner, J. W., 1969, Permian Foraminifera from Turkey: University of Kansas Paleontological Contributions, Paper 36, p. 1-14, Pls. 1-32.

Sun Xiu-fang, 1979, Upper Permian fusulinids from Zhenan of Shaanxi and Tewo of Gansu, NW China: Acta Palaeontologica Sinica, v. 18, no. 2, p. 163-170, Pls. 1-2 (in Chinese with English abstract).

Taraz, H., 1971, Uppermost Permian and Permo-Triassic transition beds in central Iran: Bulletin of the American Association of Petroleum Geologists, v. 55, no. 8, p. 1280-1294.

Wang Tzu-chuan, Wang Chang, and Wang Xi-zeng, 1964, New recognitions on Upper Permian subdivision of South China: Bulletin of the Coal Society of China, v. 1, no. 3, p. 94-106; no. 4, p. 1-32 (in Chinese with English abstract).

Zhao Jin-ke, Sheng Jin-zhang, Yao Zhao-qi, Liang Xi-luo, Chen Chu-zhen, Rui Lin, and Liao Zhou-ting, 1981, The Changhsingian and Permian-Triassic boundary of South China: Science and Technology Press of Jiangsu Province, Bulletin of Nanjing Institute of Geology and Palaeontology, Academia Sinica, no. 2, p. 1-95, Pls. 1-16.

MANUSCRIPT ACCEPTED BY THE SOCIETY APRIL 22, 1981

PLATE 1

All slides described in this paper are kept in Nanjing Institute of Geology and Palaeontology, Academia Sinica. All figures are unretouched photographs. Photo by Z. Y. Song.

Figure

1–9. ***Nanlingella maridionalis*** **Rui and Sheng (gen. sp. nov.)**
- **1.** **Axial section ('40) of holotype from the Changhsingian at Liantang, Lianxian of Guangdong Province. Cat. No. 61465.**
- **2–6.** **Five axial sections ('40) of paratypes from same locality. Cat. Nos. 61466–61470.**
- **7.** **Sagittal section ('40) of paratype from same locality. Cat. No. 61471.**
- **8, 9** **Two axial sections ('40) of paratypes from the Changhsingian at Meitian of Yizhang, Hunan Province. Cat. Nos. 61472, 61473.**

10–17. ***Parananlingella acervula*** **(Sheng and Rui)**
- **10.** **Axial section ('25) of holotype from the Changhsingian at Baoqing, Changxing of Zhejiang Province. Cat. No. 52868.**
- **11, 12 15, 16.** **Four axial sections ('25) of paratypes from same locality. Cat. Nos. 52866, 52867, 52972, 52869.**
- **13.** **Sagittal section ('25) of paratype from same locality. Cat. No. 52871.**
- **14, 17.** **Two axial sections ('25) of plesiotypes from Changhsingian at Gouwa, Guangde of Anhui Province. Cat. Nos. 61474, 61475.**

Printed in U.S.A.

Geological Society of America
Special Paper 187
1981

Lower Cambrian archaeocyathid assemblages of central and southwestern China

Yuan Ke-xing
Zhang Sen-gui
Nanjing Institute of Geology and Palaeontology, Academia Sinica, Nanjing 210008, People's Republic of China

ABSTRACT

The archaeocyathid faunas in Lower Cambrian of Central and Southwestern China may be divided into four assemblages. According to the archaeocyathids and their associated fossils, the assemblages of China may be correlated with the archaeocyathid faunas of Siberia and Altai-Sajan, Morocco, and South Australia, and the lower boundary of the Cambrian System in China may be drawn at the bottom of the Meishucun Stage.

BRIEF ACCOUNT OF THE LOWER CAMBRIAN BIOSTRATIGRAPHY IN CENTRAL AND SOUTHWEST CHINA

The Cambrian deposits in Central and Southwest China are well developed and have been under detailed investigation in recent years. There are many excellent and continuous Lower Cambrian sections, and the Precambrian-Cambrian is developed in uniform facies which may make this one of the best areas in the world for defining the lower boundary of the Cambrian System. Many groups of fossils were collected, among them archaeocyathids, poriferans, brachiopods, monoplacophorans, rostroconchs, gastropods, hyolithids, hyolithelminthids, trilobites, bradoriids, cambroscleritids, tommotiids, and microfloral fossils. These faunas include more than 500 species, in 150 genera and 40 families, of trilobites (Zhang and others, 1980) and about 100 species of earliest Cambrian shelly fossils (Qian, 1977, 1978; Qian and others, 1979; Yu, 1979; Liu, 1979). Recently, Lu and his colleagues divided the Lower Cambrian strata in Central and Southwestern China into stages and 13 fossil zones or assemblages in descending order as follows (Lu, in press):

Lower Cambrian
Lungwangmiao Stage

13. *Redlichia guizhouensis* Zone: *Redlichia (Redlichia) guizhouensis, R. nobilis, Antagmus, Eoptychoparia*
12. *Redlichia murakamii-Hoffetella* Zone: *Redlichia (Pteroredlichia) murakamii, R. (Redlichia), R. (Spinoredlichia), Hoffetella, Chuchiaspis, Panxinella, Eoptychoparia*

Tsanglangpu Stage

11. *Megapalaeolenus* Zone: *Megapalaeolenus, Redlichia (Redlichia), R. (Pteroredlichia), R. (Breviredlichia), Kootenia, Yuehsienszella, Xilingxia, Bonnia, Archaeocyathus, Retecyathus, Sanxiacyathus, Protopharetra*
10. *Palaeolenus* Zone: *Palaeolenus, Redlichia (Redlichia), R. (Conoredlichia), Kootenia, Shipaiella*
9. *Paokannia-Sichuanolenus* Zone: *Paokannia, Sichuanolenus, Pseudichangia, Ichangia, Redlichia, Shifangia, Shiqihepsis, Kootenia, Neocobboldia*
8. *Metaredlichioides-Changkouia* Zone: *Metaredlichioides, Chengkouia, Husaspis, Qingzhenaspis, Kootenia, Protolenella, Xiuqiella, Shifangia*
7. *Drepanuroides* Zone: *Drepanuroides, Drepanopyge, Redlichia, Mayiella, Paramalungia, Yiliangellina, Yinites, Yunnanaspidella, Coscinocyathus, Protopharetra, Archaeofungia, Rotundocyathus, Ajacicyathus, Dictyocyathus, Agastrocyathus*
6. *Yunnanaspis-Yiliangella* Zone: *Yunnanaspis, Yiliangella, Kueichowia, Shantania, Pseudoredlichia, Decerodiscus, Szechuananspis, S. (Zhenbadiscus), Guishoudiscus, Shizhudiscus, Hupeidiscus, Yinites*
5. *Malungia* Zone: *Malungia, Zhenbaspis, Chengyangia, Metaredlichia, Jingyangia, Hsuaspis, Szechuanaspis, Shizhudiscus, Guizhoudiscus, Hupeidiscus, Sinodiscus, Tsunyidiscus, Ajacicyathus, Rotundocyathus, Taylorcyathus, Coscinocyathus, Chengkoucyathus, Archaeofungia, Connanulofungia*

Chiungchussu Stage

4. *Eoredlichia-Wutingaspis* Zone: *Eoredlichia, Wutingaspis,*

Yunnanocephalus, Chaoaspis, Kuanyangia (Sapushania), Tsunyidiscus, Mianxiandiscus

3. *Parabadiella-Mianxiandiscus* Zone: *Parabadiella, Wutingaspis, Mianxiandiscus, M. (Liangshandiscus), Kunmingella, Liangshanella, Meishucunella*

Meishucun Stage

2. *Siphogonuchites-Zhjinites-Sachites* assemblage: *Siphogonuchites, Zhijinites, Sachites, Zeugites, Heraultipegma, Lapworthella, Turcutheca, Lenatheca, Allotheca, Trapezotheca, Quadratheca, Bemella, Tianzhushania*
1. *Anabarites-Circotheca-Protohertzina* assemblage: *Anabarites, Circotheca, Protohertzina, Shipaitubulus, Tiksitheca, Hyolithellus*

On the basis of gradual changes of the biofacies and lithofacies, from west to east, the Lower Cambrian may be subdivided into three regions, and the regional formational names corresponding to the above stages are given in Table 1.

ARCHAEOCYATHID ASSEMBLAGES IN CENTRAL AND SOUTHWEST CHINA

In the central and eastern regions of Central and Southwest China, archaeocyathids occur abundantly in Lower Cambrian calcareous deposits, such as limestone, muddy limestone, limy shale, and others, but, except in a few places, almost no archaeocyathids occur in the clastic sediments of the western region (Fig. 1), demonstrating the dependence of the distribution of the archaeocyathids on lithofacies; in other words, the distribution of archaeocyathids was strongly controlled by the nature of the marine environments.

Following are four significant districts (from east to west in Table 1) that have yielded many archaeocyathids.

Yichang District, Hubei Province

Archaeocyathids occur in the upper part of the Tianheban Formation. The rocks containing these fossils are gray limestones with argillaceous bands. Fossils are abundant and in most places occur as archaeocyathid reefs.

The most common species in these fossils are *Archaeocyathus hupehensis* Chi, *A. yichangensis* Yuan and Zhang [Pl. 4(5a,5b)], *A. tianhebanensis* Yuan and Zhang, *Retecyathus* cf. *comptophragma* Vologdin, *R. kuzmini* Vologdin, *R. Communis* Yuan and Zhang, *R. nitidus* Yuan and Zhang [Pl. 4(6a,6b)], *Retecyathus (Pararetecyathus) curvatus* Yuan and Zhang [Pl. 4(4a,4b)], *Protopharetra* sp. [Pl. 4(7)], and *Sanxiacyathus hubeiensis* Yuan and Zhang [Pl. 4(8a,8b)]. All of them are taenioids. This is consi-

TABLE 1. CORRELATION OF LOWER CAMBRIAN BIOSTRATIGRAPHY IN CENTRAL AND SOUTHWEST CHINA

Series	Stage	Substage	Zones or assemblages	Western Region: East Yunnan		Western Region: South Shaanxi (West)	Central Region: North Guizhou	Eastern Region: North Siehuan (Chengkou)	Eastern Region: West Hubei
Lower Cambrian	Lungwangmiao Stage		Redlichia guizhouensis Zone	Lungwangmiao Formation		Kongmingdong Formation	Qingxudong Formation	Shihlungtung Formation	
			Redlichia murakamii-Hoffetella Zone						
	Tsanglangpu Stage	Wulongjing Subst.	Megapalaeolenus Zone	Tsanglangpu Formation	Wulongjing Member	Yangwangbian Formation	Chingtingshan Formation	Tianheban Formation	
			Palaeolenus Zone					Yingzuiyan Formation	Shihpai Fm.
		Hongjingshao Substage	Paokannia-Sichuanolenus Zone		Hongjingshao Member				Shuijingtuo Formation
			Metaredlichioides-Chengkouia Zone						
			Drepanuroides Zone						
			Yunnanaspis-Yiliangella Zone				Minghsingssu Formation	Liangshuijing Formation	
			Malungia Zone			Xiannüdong Formation			
	Chiungchussu Stage		Eoredlichia Zone	Chiungchussu Formation		Kuojiaba Formation	Niutitang Formation		
			Parabadiella-Mianxiandiscus Zone						
	Meishucun Stage		Siphogonuchites-Zhijintes-Sachites assemblage	Meishucun Formation		Kuanchuanpu Formation		?	Huangshandong Member
			Anabarites-Circotheca-Protohertzina assemblage						
Late Precambrian				Töngying Formation					

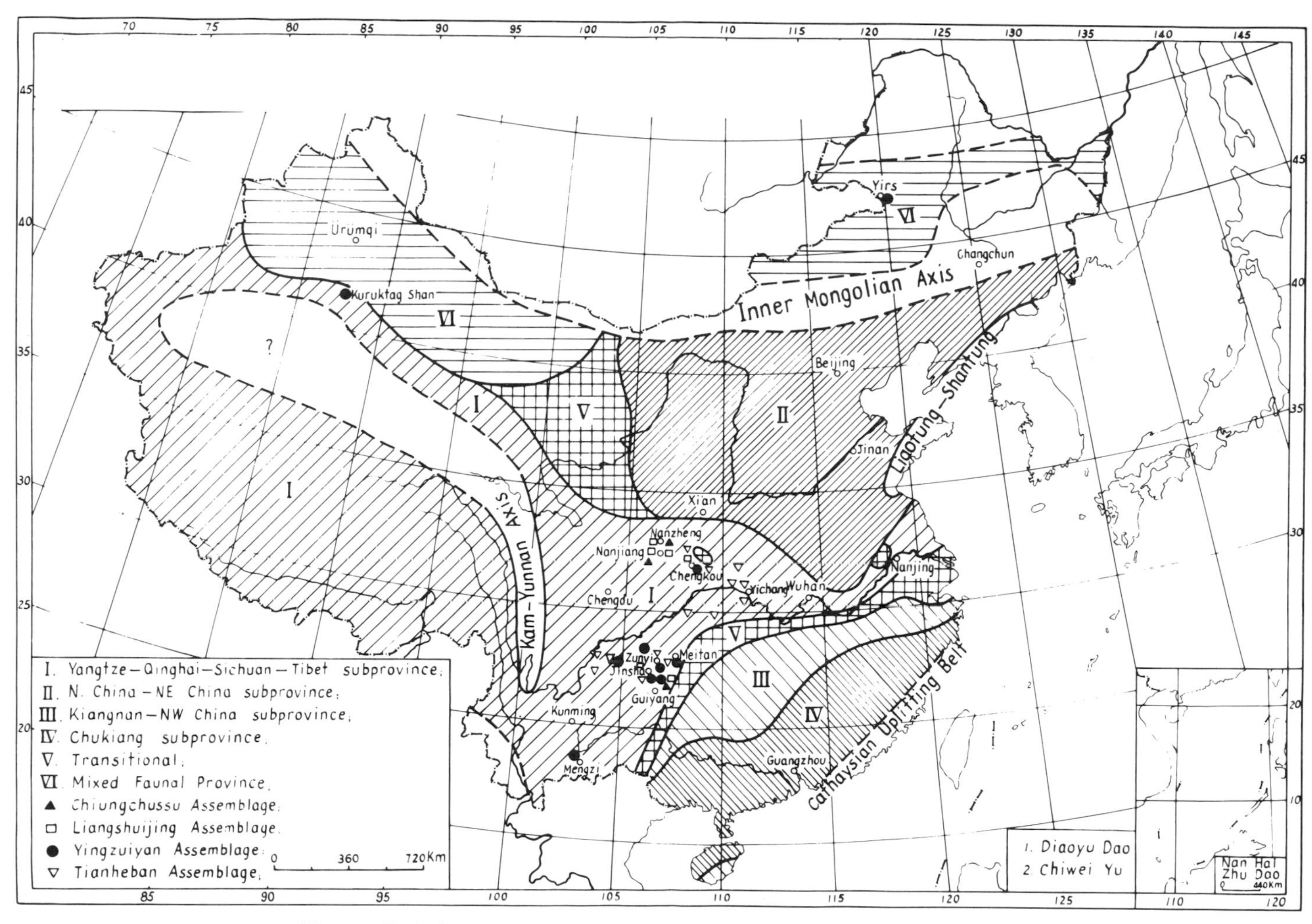

Figure 1. Early Cambrian paleobiogeography and distribution of Archaeocyatha in China.

dered to be a Tianheban assemblage. On top of the Tianheban Formation many trilobites are found, such as *Megapalaeolenus deprati* (Mansuy), *M.* sp., *Xilingxia ichangensis* (Chang), *Kootenia yui* Chang, *Redlichia kobayashii* Lu, and *R.* sp. These belong to the *Megapalaeolenus* Zone. At present, we are inclined to believe that both the trilobites and the Tianheban archaeocyathid assemblage are parts of the *Megapalaeolenus* Zone which is the youngest assemblage in Central and Southwest China.

Chengkou District, Sichuan Province

In this district three archaeocyathid-bearing horizons have been recognized. One of them is the middle of the Tianheban Formation. The archaeocyathids seen in dark gray, massive limestone with argillaceous bands are similar to those of the Tianheban assemblage in Yichang, Hubei. They are *Archaeocyathus validus* Yuan [Pl. 4(2a,2b)], *A. xuetangpingensis* Yuan, *A. cingyanzhaiensis* Yuan, *Retecyathus tubus* Yuan, *R. shixiqiaoensis* Yuan, *Protopharetra* sp. They formed the Tianheban assemblage which has the characteristics of a bioherm. This archaeocyathid bed is situated between two trilobite-bearing horizons belonging to the *Megapalaeolenus* Zone. One of them contains abundant *Xilingxia chengkouensis* Zhu, *Redlichia major* Lu, *R.* sp. B, *Megapalaeolenus* sp., *Dabashania minor* Qian and Yao, *Kootenia* spp., *Yuehsienszella* sp. B, and others. The other yields *Megapalaeolenus* cf. *M. Deprati* (Mansuy), *Redlichia hupenhensis* Hsü, *R.* sp., and *Kootenia* sp. Below the Tianheban assemblage of the Tianheban Formation lies the Yingzuiyan Formation which embraces three trilobite zones and one archaeocyathid assemblage—the Yingzuiyan assemblage—in descending order as follows:

4,3. The *Palaeolenus* Zone and the *Paokannia-Sichuanolenus* Zone containing *Palaeolenus lantenoisi* Mansuy, *Redlichia hupehensis* Hsü, and *Paokannia magna* Qian and Yao;

2. The *Metaredlichioides-Chengkouia* Zone with *Metaredlichioides constricts* Chien and Yao, *Chengkouia cylindrica* Chien and Yao, *C. pustulosa* Zhu, *Shifangia tumida* Chien and Yao, *S.* sp., *Protolenella angustilimbata* Qian and Yao, *Xiuqiella rectangularis* Chien and Yao, and *X.* sp.; and

1. The Yingzuiyan archaeocyathid assemblage, a gray limestone facies bearing *Protopharetra chengkouensis* [Pl. 3(6a,6b)] Yuan, *Coscinocyathus honghuaensis* Yuan, *C. erzishanensis* Yuan, *Erugatocyathus yingzuiyanensis* Yuan and Zhang [Pl. 3(4a,4b)], *Agastrocyathus grandis* Yuan and Zhang [Pl. 4(1a,1b)], *Archaeofungia* cf. *A. dissepimentalis* Taylor, and *Taylorcyathus* sp., belonging partly to the Irregulares and partly to the Regu-

lares. The archaeocyathid remains are so densely crowded that in places they form typical archaeocyathid reefs.

Below the Yingzuiyan assemblage in the Chengkou District is the Liangshuijing Archaeocyathid assemblage of the Liangshuijing Formation. Most of the archaeocyathid remains are fragmental and distributed sporadically in oolitic limestone. They are mainly regular archaeocyathids, including *Ajacicyathus sichuanensis* Yuan [Pl. 1(7a,7b)], *Rotundocyathus xiuqiensis* Yuan [Pl. 3 (2a,2b)], *Archaeofungia shixiensis* Yuan [Pl. 2(1a,1b)], *Taylorcyathus shifangensis* Yuan [Pl. 2(6a,6b,6c)], *Chengkoucyathis shabaensis* Yuan [Pl. 1(8a,8b)], *Coscinocyathus liangshuijingensis* Yuan [Pl. 2(3a,3b)], *C. zhuyuanensis* Yuan, and *Clathricoscinus dabashanensis* Yuan [Pl. 3(1a,1b)]. The part of the Liangshuijing Formation above the archaeocyathid beds contains a few trilobites, including *Yunnanaspis-Yiliangella* Zone. Therefore, the Liangshuijing Archaeocyathid assemblage may be correlated with the *Malungia* Zone of the lower part of the Tsanglangpu Formation.

Zunyi District and Jinsha District
Guizhou Province

Three archaeocyathid assemblages occur in these districts. One of them is the Tianheban assemblage which corresponds to the upper part of the Chintingshan Formation. The Tianheban assemblage contains *Protopharetra yanjiaoensis* Yuan, *Retecyathus wanfusiensis* Yuan [Pl.4(3a,3b)], *R.* sp., *Archaeocyathus* sp., *Sanxiacyathus* sp., and other taenioid genera in gray, massive or thick-bedded limestone. The upper part of the Chintingshan Formation also yields *Megapalaeolenus deprati* (Mansuy), *M. fengyangensis* (Chu), *Redlichia* spp., *Kuehsienszella* sp., and *Kootenia* sp. of the *Megapalaeolenus* Zone. Hence, this assemblage is assignable to the *Megapalaeolenus* Zone of the Tsanglangpu Stage. The second is named as the Chintingshan assemblage located in the lower part of the Chintingshan Formation, which consists of gray, massive or thick-bedded limestone and green-yellowish shale. There may be as many as eight limestone beds with rich archaeocyathid faunas which locally may develop into archaeocyathid reefs. *Conannulofungia jinshaensis* Yuan [Pl. 2(2a,2b,2c)], *Rotundocyathus shilixiensis* Yuan [Pl. 3(3a,3b,3c)], *Archaeofungia meitanensis* Yuan, *Clathricoscinus zunyiensis* Yuan [Pl. 3(5a,5b,5c)], *Dictyocyathus jindingshanensis* Yuan [Pl. 3(7a,7b,7c)] are the most important members of the assemblage, most of which are regular archaeocyathids. At some distance above the archaeocyathid-bearing bed, abundant trilobites such as *Pseudichangia damiaoensis* (Chang), *Redlichia* sp., and *Paokania* sp. are recorded. These trilobites are the members of the *Paokannia-Sichuanolenus* Zone. *Drepanuroides* and *Yinites* occur together with the archaeocyathids, and we consider that this assemblage is coeval with the *Drepanuroides* Zone. In the Jinsha District, we found a third archaeocyathid assemblage which includes *Connanulofungia jinshaensis* Yuan, some species of *Ajacicyathus* and *Rotundocyathus,* and coscinocyathids. This is the Jinsha assemblage. Most of them are regular archaeocyathids. These fossils occur in a dark gray reef limestone bed more than 1 m above the base of the Minghsingssu Formation. Above this bed many significant trilobites such as *Kueichowia liui* Lu and *Pseudoredlichia* sp. were collected which are included in the *Yunnanaspis-Yiliangella* Zone. It appears that stratigraphically the Jinsha assemblage belongs to the *Malungia* Zone.

South Shaanxi–Northwest Sichuan District

Here is found a well-developed archaeocyathis limestone called "Xiannüdong Limestone." The Xiannüdong sequence consists of carbonate rocks of shallow marine facies, including oolitic and argillaceous limestone and dolomite. The oolitic limestone beds contain many well-preserved archaeocyathids, which are temporarily named "Xiannüdong assemblage." Important fossils are *Conannulofungia nanzhengensis* Yuan and Zhang, *Taylorcyathus clarus* Yuan and Zhang [Pl. 2(4a,4b,4c)], *Dictyocyathus intextus* Yuan and Zhang, *D. Lepidus* Yuan and Zhang [Pl. 1(6a,6b)], *Rotundocyathus shaanxiensis* Yuan and Zhang [Pl. 1(5a,5b)], *Ajacicyathus* cf. *A. sichuanensis* Yuan, *Archaeofungia biseriatus* (Chi), *A. abnormis* Yuan and Zhang, *Coscinocyathus angustus* Yuan and Zhang [Pl. 2(5a,5b,5c)], *C. nanzhengensis* Yuan and Zhang, *C.* cf. *C. spatiosus* Vologdin, and *Protopharetra polymorpha* Bornemann. The associated trilobites include *Malungia granulosa* Zhou and *Micangshania gracilis* Zhou. The Xiannüdong assemblage certainly belongs to the *Malungia* Zone and should be subdivided into several subzones after a more detailed investigation has been made. It is worth mentioning that in Shatan of Nanjiang District, Sichuan Province, and in Mount Yuanshan and Mount Huashan of Nanzheng District, Shaanxi Province, an archaeocyathid fauna is found which is considered to be of Chiungchussu Stage. This fauna is characterized by the association of *Yunnanocephalus yunnanensis* Mansuy of the *Eoredlichia* Zone. The fauna occurs in the marls below the oolitic limestones. *Ajacicyathus spinosus* Yuan and Zhang [Pl. 1(1a,1b)], *Rotundocyathus distinctus* Yuan and Zhang [Pl. 1(3a,3b)], *Taylorcyathus annuliformis* Yuan and Zhang [Pl. 1(2a,2b)], and *Dictyocyathus(?)* sp. [Pl. 1(4)] have been identified. It represents the oldest archaeocyathid assemblage in China, a fauna consisting of species characterized by small size, simple structure, and by varying conditions of preservation. We call it provisionally "Chiungchussu assemblage."

CORRELATION OF THE ARCHAEOCYATHID HORIZONS AND THE PROBLEM OF THE LOWER BOUNDARY OF THE CAMBRIAN FROM THE ARCHAEOCYATHID POINT OF VIEW

The worldwide correlation of the Lower Cambrian strata and the problem of the lower boundary of the Cambrian System have not yet been satisfactorily settled. As suggested by Lu and others (1974), the Cambrian of the globe may be divided into the Oriental, the Occidental, and the Transitional Realms, each Realm having its distinct faunas. Some of them were not widely distributed and never played an important role in stratigraphic correlation. The Lower Cambrian of Central and Southwest China belongs to the Oriental (or Redlichiid) Realm.

Our tentative correlation is given in Table 2 and the distribution of archaeocyathid genera is shown in Figure 2. In the Siberia platform and the complementary Altai-Sajan region, four important periods of archaeocyathid evolution have been recognized and consequently used to divide the Lower Cambrian into four

TABLE 2. TENTATIVE CORRELATION OF LOWER CAMBRIAN OF SOME AREAS OF THE WORLD

	China		Siberia	South Australia	Morocco	North America
Lower Cambrian	Lungwangmiao Stage		Lenian	No. 7	Niveau d'Ouriken d'Ourmast	Bonnia-Olenellus Zone
	Tsanglangpu Stage	Wulongjing Substage	Botomian	Clastics	Asrir	
					Issafence	
		Hongjingshao Substage	Atdabanian	No. 4 Ajax Limestone	Amouslekien	Nevadella Zone
					Ouneinien	Fallotaspis Zone
	Chiungchaussu Stage		Tommotian	No. 2 No. 1 Parachilna Fm.	Adoudounien Supérieur	Pre-trilobite Zone
	Meishucun Stage			Uratanna Fm.	Adoudounien	
Late Pre-cambrian	Töngying Formation		Judoma Suite	Pound Quartzite	Inférieur	

Note: Units not drawn to scale.

stages in the Soviet Union. According to similar and related forms of the archaeocyathids, associated trilobites and small shelly fossils, the Tianheban assemblage corresponds to the early Lenian, because both contain most Irregulares, whereas the Yingzuiyan and the Liangshuijing may be correlated with the Atdabanian, which represents the second period in development of the archaeocyathids in Siberia. The Chiungchussu assemblage is considered to be Tommotian. In Siberia the protolenid trilobites flourished in the Botomian, but in Central and Southwest China the protolenids were well developed following the Yingzuiyan assemblage (Fig. 2). The archaeocyathids of the Yingzuiyan and the Liangshuijing assemblages are mainly typical Atdabanian. Moreover, in the associated fauna with the Xiannüdong assemblage (same as the Liangshuijing in age), many tommotiids have been found. Tommotiids are the representatives of the Tommotian with the exception of *Tommotia kozlowskii* (Missarzhevsky) which may range into early Atdabanian. In the Chiungchussu assemblage no coscinocyathids and nochoroicyathids are unknown. It is an important and significant feature of the Chiungchuss assemblage that the archaeocyathids are small in size and simple in structure. In Siberia and the Altai-Sajan region, coscinocyathids and nochoroicyathids appeared, respectively, in the middle and early Tommotian. This broad correlation as determined at higher taxonomic levels suggests that parts of the Chiungchussu Stage of China may be approximately correlatives of the Tommotian of the Soviet Union.

Recently, a new archaeocyathid fauna has been found at Tiout (Anti-Anlas, Morocco). Debrenne and Debrenne (1978) stated that the oldest archaeocyathid fauna in Tiout contained 15 species belonging to 11 genera, and was equivalent to the Atdabanian Stage. Comparing the Moroccan archaeocyathid fauna with that of the Yingzuiyan and Liangshuijing assemblages, we find that two genera (*Coscinocyathus* and *Agastrocyathus*) are common to the Moroccan and the Yingzuiyan assemblages, and two genera (*Coscinocyathus* and *Ajacicyathus*) are common to the Moroccan and the Liangshuijing assemblages. Thus, the age of the Moroccan fauna may perhaps be somewhere between that of the Liangshuijing and the Yingzuiyan faunas. In South Australia, archaeocyathid faunas are also well known. The oldest archaeocyathid bed bearing coscinocyathids in South Australia might be younger than the Chiungchussu assemblage of China, and this bed may correspond to the Tommotian of Siberia.

From what has been stated above, it may be concluded that the Chiungchussu Stage with trilobites quite probably coresponds to part of the Tommotian, and that the underlying Meishucun Stage, or at least part of it, which yields many shelly fossils, is older than the Tommotian. This opinion has been considered by some workers on shelly fossils. In fact, Daily (1972, p. 22) agreed with this point of view by stating that "the overlying *Eoredlichia* Zone may span the Tommotian-Atdabanian boundary, in which case the appearance of trilobites within parts of the Realm would have been slightly in advance of that for olenellid Realm." In other words, the Chiungchussu, bearing *Eoredlichia,* may be older than the base of Atdabanian with *Protallotaspis* and *Fallotaspis.* Evidence from absolute age dating also supports this view. The age of the base of the Tommotian is 578 m.y., the age of the base of the Shuijingtao Formation (at Shuijingtao of Yichang, Hubei Province) is 613 m.y., and the age of the rocks from the lower part of the Meishucun Formation without trilobites (at Meishucun of Jinning, Yunnan Province) is 612 or 603 m.y. It is worthy of notice that Savitsky (1978) has pointed out that a lithologic and stratigraphic unconformity is present between the Judoma and the

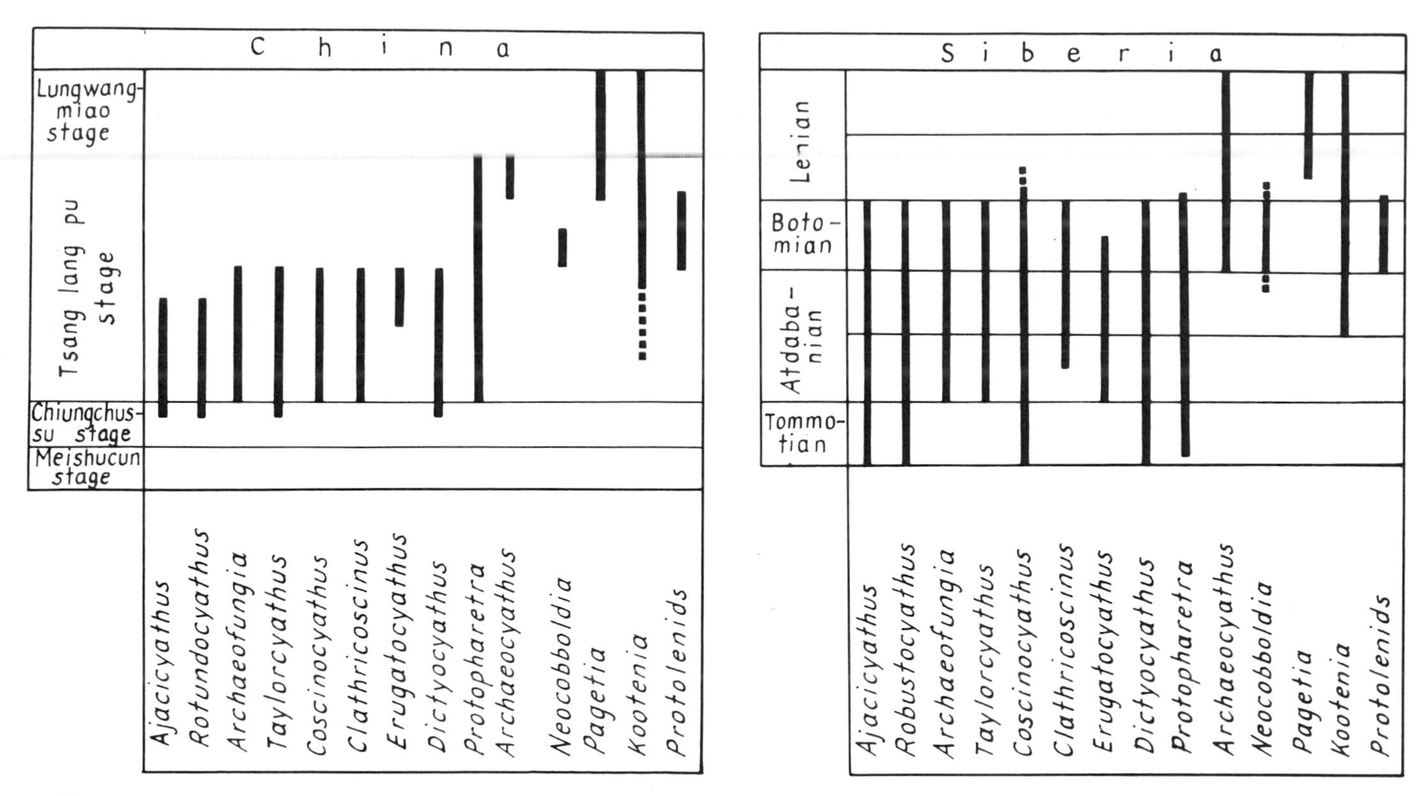

Figure 2. Comparison and range of important Lower Cambrian archaeocyathids and trilobites of Central and Southwest China (left) and of Siberia (right).

Pestrotsvet Formations (the Tommotian); that is, that a hiatus existed below the Tommotian.

CONCLUSIONS

1. The Lower Cambrian archaeocyathid faunas in Central and Southwest China may be divided into four assemblages: the Tianheban, the Yingzuiyan (or the Chintingshan), the Liangshuijing (or Xiannüdong or Jinsha), and the Chiungchussu.

2. A tentative correlation of the Lower Cambrian of Central and Southwest China and of Sibera suggests that the Tianheban assemblage corresponds to the early Lenian, the Yingzuiyan and the Liangshuijing assemblages coincide with the Atdabanian, and the Chiungchussu assemblage is to be considered as Tommotian. Moreoever, the oldest archaeocyathid fauna of Morocco may perhaps be situated between the Yingzuiyan and the Liangshuijing assemblages.

3. It is possible that the Chiungchussu Stage with trilobites corresponds to a part of the Tommotian, and that part of the Meishucun Stage is older than the Tommotian.

ACKNOWLEDGMENTS

Professor Lu Yan-hao critically read and modified in detail the manuscript, for which we thank him very much. The illustrated specimens were photographed by Zhou Si-san, Song Zhi-yao, Huang Zhao-qi, Zhang Fu-tian, and Liang Xiae-yun. All are unretouched photographs.

REFERENCES CITED

Chi Yung Shen, 1940, Cambrian Archaeocyathina from the Gorge District of the Yangtze: Bulletin of the Geological Society of China, v. 20, no. 2, p. 121–146, 3 pls.

Cowie, J. W., and Glaessner, M. F., 1975 (With contribution by A. Boudda, G. Choubert, A. Faure-Muret, W. B. Harland, A. Yu. Rozanov and V. E. Savitsky) The Precambrian-Cambrian Boundary: A Symposium: Earth-Science Reviews, v. 11, p. 209–251.

Daily, B., 1956, The Cambrian in South Australia: *in* Rodgers, John, ed., El Sistema Cambrico, su Paleogeografis y el Problema de su Base: International Geological Congress, 20th, Mexico, Tomo 2, p. 91–148.

—— 1972, The base of the Cambrian and the first Cambrian Faunas, *in* Jones, J. B., and McGowran, B. (eds.), Stratigraphic problems of the later Precambrian and Early Cambrian: Centre for Precambrian Research, University of Adelaide, Special Paper, no. 1, p. 13–37, 2 pls.

Debrenne, F., 1964, Archaeocyatha. Contribution à l'étude des faunes Cambriennes du Maroc, de Sardaigne et de France: Notes et Mémoires du Service Géologique du Maroc, no. 179, v. 1 Texte, 265 p., 69 text-figs., 29 tables; 2 Planches, 52 pls.

Debrenne, F., and Debrenne, M., 1978, Archaeocyathid fauna of the lowest fossiliferous levels of Tiout (Lower Cambrian, Southern Morocco): Geological Magazine, v. 115, no. 2, 101–120, 4 pls.

Handfield, R. C., 1971, Archaeocyatha of the Mackenzie and Cassiar Mountains, Northwestern Canada: Geological Survey of Canada Bulletin 201, 119 p., 16 pls.

Hill, D., 1965, Archaeocyatha from Antarctica and a review of the Phylum: Trans-Antarctic Expedition 1955–1958, Scientific Reports no. 10, Geology 3, 151 p., 12 pls.

—— 1972, Archaeocyatha (second edition), *in* Teichert, C., ed., Treatise on invertebrate paleontology, Part E, Volume 1: Boulder, Colorado, Geological Society of America (and University of Kansas Press), 158 p.

Khomentovski, V. V., and Repina, L. N., 1965, Nizhniy kembriy stratotipicheskogo razreza Sibiri: Moskva, Izdatel'stvo "Nauka," str. 3–196, 14 tables.

Kruse, P. D., 1978, New Archaeocyatha from the Early Cambrian of the Mt. Wright area, New South Wales: Alcheringa, v. 2, no. 1, p. 27–48, 12 figs.

Li Yao-xi, Song Li-sheng, Zhou Zhi-qiang, and Yang Jing-yao, 1975, [The Lower Palaeozoic of Western Mt. Dabashan]: Beijing, Geological Publishing House, 232 p., 70 pls. (in Chinese).

Liu Di-yong, 1979, Earliest Cambrian brachiopods from Southwest China: Acta Palaeontologica Sinica, v. 18, no. 4, p. 503–511, 2 pls. (in Chinese with English abstract).

Lu Yan-hao [Lu Yen-hao], 1962, [The Cambrian System of China]: Beijing, Science Press, 117 p. (in Chinese).

Lu Yan-hao [Lu Yen-hao], Zhu Zhao-ling [Chu Chao-ling], Qian Yi-yuan [Chien Yi-yuan], Lin Huan-ling, Zhou Zhi-yi, Yuan Ke-xing, 1974, [Bio-environmental control hypothesis and its application to the Cambrian biostratigraphy and palaeozoogeography]: Memoirs of Nanking Institute of Geology and Palaeontology, Academia Sinica, no. 5, p. 27–116, 4 pls. (in Chinese).

Lu Yan-hao, Zhu Zhao-ling, Qian Yi-yuan, Lin Huan-ling, and Yuan Jin-liang, (1981) [Correlation of Cambrian System in China], *in* Nanjing Institute of Geology and Palaeontology, ed., Stratigraphical correlations in China (I): Beijing, Science Press (in Chinese) (in press).

Nelson, C. A., 1978, Late Precambrian–Early Cambrian stratigraphic and faunal succession of eastern California and the Precambrian–Cambrian boundary: Geological Magazine, v. 115, no. 2, p. 121–126.

Qian Yi, 1977, Hyolithia and some problematica from the Lower Cambrian Meishucun Stage in Central and SW China: Acta Palaeontologica Sinica, v. 16, no. 2, p. 255–275, 3 pls. (in Chinese with English abstract).

——1978, The Early Cambrian Hyolithids in Central and Southwest China and their stratigraphical significance: Memoirs of Nanjing Institute of Geology and Palaeontology, Academia Sinica, no. 11, p. 1–50, 7 pls. (in Chinese with English abstract).

Qian Yi, Chen Meng-e, and Chen Yi-yuan, 1979, Hyolithids and other small shelly fossils from the Lower Cambrian Huangshandong Formation in the Eastern Part of the Yangtze Gorge: Acta Palaeontologica Sinica, v. 18, no. 3, p. 207–230, 4 pls. (in Chinese with English abstract).

Repina, L. N., Khomentovsky, V. V., Zhuravleva, I. T., and Rozanov, A. Yu, 1964, Biostratigrafiya nizhnego kembriya Sayano-Altayskoy skladchatoy oblasti: Moskva, Izdatel'stvo "Nauka," 365 str., 48 tables.

Rozanov, A. Yu, 1967, The Cambrian lower boundary problem: Geological Magazine, v. 104, no. 5, p. 415–434.

——1969, Nekotorye voprosy sistematiki arkheotsiat, *in* Zhuravleva, I.T., ed., Biostratigrafiya i paleontologiya nizhnego kembriya Sibiri i Dalnego Vostoka: Moskva, Izdatel'stvo "Nauka," str. 106–113, tables 40–42.

——1976, Granitsa kembriya i dokembriya, *in* Granitsy Geologicheskikh sistem: Moskva, Izdatel'stvo "Nauka," str. 31–53.

Rozanov, A. Yu, and Debrenne, F., 1974, Age of Archaeocyathid assemblages: American Journal of Science, v. 274, p. 833–848.

Rozanov, A. Yu, and Missarzhevsky, V. V., 1966, Biostratigrafiya i fauna nizhnikh gorizontov kembriya: Trudy Geologicheskogo Instituta Akademii Nauk SSSR, vpy. 148, str. 1–120, 13 tables.

Rozanov, A. Yu, Missarzhevsky, V. V., Volkova, H. A., Voronova, L. G., Krylov, I. N., Keller, B. M., Korolyuk, I. D., Lendzion, K., Mikhnyak, R., Pykhova, N. G., and Sidorova, A. D., 1969, Tommotksiy yarus i problema nizhney granitsy kembriya: Trudy Geologicheskogo Instituta Adademii Nauk SSSR, vyp. 206, str. 1–405, 55 tables.

Savitsky, V. E., 1978, The Precambrian-Cambrian boundary problem in Siberia and some general problems of stratigraphy (preliminary report): Geological Magazine, v. 115, no. 2, p. 127–130.

Yu Wen, 1979, Earliest Cambrian monoplacophorans and gastropods from Western Hubei with their biostratigraphical significance: Acta Palaeontologica Sinica, v. 18, no. 3, p. 233–266, 4 pls. (in Chinese with English abstract).

Yuan Ke-xing, 1974, Archaeocyatha, *in* [A handbook of the stratigraphy and palaeontology of Southwest China]: Beijing, Science Press, p. 80–82, Pls. 29–30 (in Chinese).

Yuan Ke-xing and Zhang Sen-gui, 1977, Archaeocyatha, *in* [Atlas of palaeontology in Central and Southern China]: Beijing, Geological Publishing House, pt. 1, p. 4–8, Pls. 1–2 (in Chinese).

——1978, Archaeocyatha, *in* [The stratigraphy and palaeontology of Sinian to Permian in the eastern part of the Yangtze Gorge]: Beijing, Geological Publishing House, p. 138–140, Pls. 16–17 (in Chinese).

——1980, Lower Cambrian Archaeocyatha of Central and Southwestern China: Acta Paleontologica Sinica, v. 19, no. 5, p. 380–392, 4 pls. (in Chinese with English abstract).

——1981, [Archaeocyatha], *in* Xi'an Institute of Geology and Mineral Resources, ed., Atlas of palaeontology in Shaanxi, Gansu and Ningxia. Part 1: Beijing, Geological Publishing House (in Chinese) (in press).

Zhang Wen-tang [Chang Wen Tang], Lu Yan-hao, Zhu Zhao-ling, Qian Yi-yuan, Lin Huan-ling, Zhou Zhi-yi, Zhang Sen-gui, and Yuan Jin-liang, 1980, Cambrian trilobite faunas of Southwestern China: Palaeontologia Sinica, new ser. B, no. 16, whole series no. 159, Beijing, Science Press (in Chinese with English summary).

Zhuravleva, I. T., 1960, Arkheotsiaty Sibirskoy platformy: Moskva, Izdatel'stvo Akademii nauk SSSR, 344 str., 33 tables.

MANUSCRIPT ACCEPTED BY THE SOCIETY APRIL 22, 1981

PLATE 1

Figure

Chiungchussu Archaeocyathid assemblage

1a,1b. ***Ajacicyathus spinosus*** **Yuan and Zhang. 1a. Transverse section, ×6. 1b. Longitudinal section, ×6. Top of Kuojiaba Formation, Lower Cambrian, Mount Huashan of Nanzheng District, Southern Shaanxi.**

2a,2b. ***Taylorcyathus annuliformis*** **Yuan and Zhang, 2a. Transverse section, ×6. 2b. Longitudinal section, ×6. Same as 1a,1b.**

3a,3b. ***Rotundocyathus distinctus*** **Yuan and Zhang. 3a. Transverse section, ×8. 3b. Longitudinal section, ×8. Horizon and locality: Same as 1a,1b.**

4. ***Dictyocyathus(?)*** **sp. Transverse and oblique sections, ×6. Horizon and locality: The top of Kuojiaba Formation, Lower Cambrian, Shatan of Nanjiang District, Northern Sichuan.**

Liangshuijing or Xiannüdong Archaeocyathid assemblage

5a,5b. ***Rotundocyathus shaanxiensis*** **Yuan and Zhang. 5a. Transverse section, ×10. 5b. Longitudinal section, ×10. Horizon and locality: The Xiannüdong Formation, Lower Cambrian, Moujiaba of Nanzheng District, Southern Shaanxi.**

6a,6b. ***Dictyocyathus lepidus*** **Yuan and Zhang. 6a. Transverse section, ×6. 6b. Longitudinal section, ×6. Horizon and locality: Same as 5a,5b.**

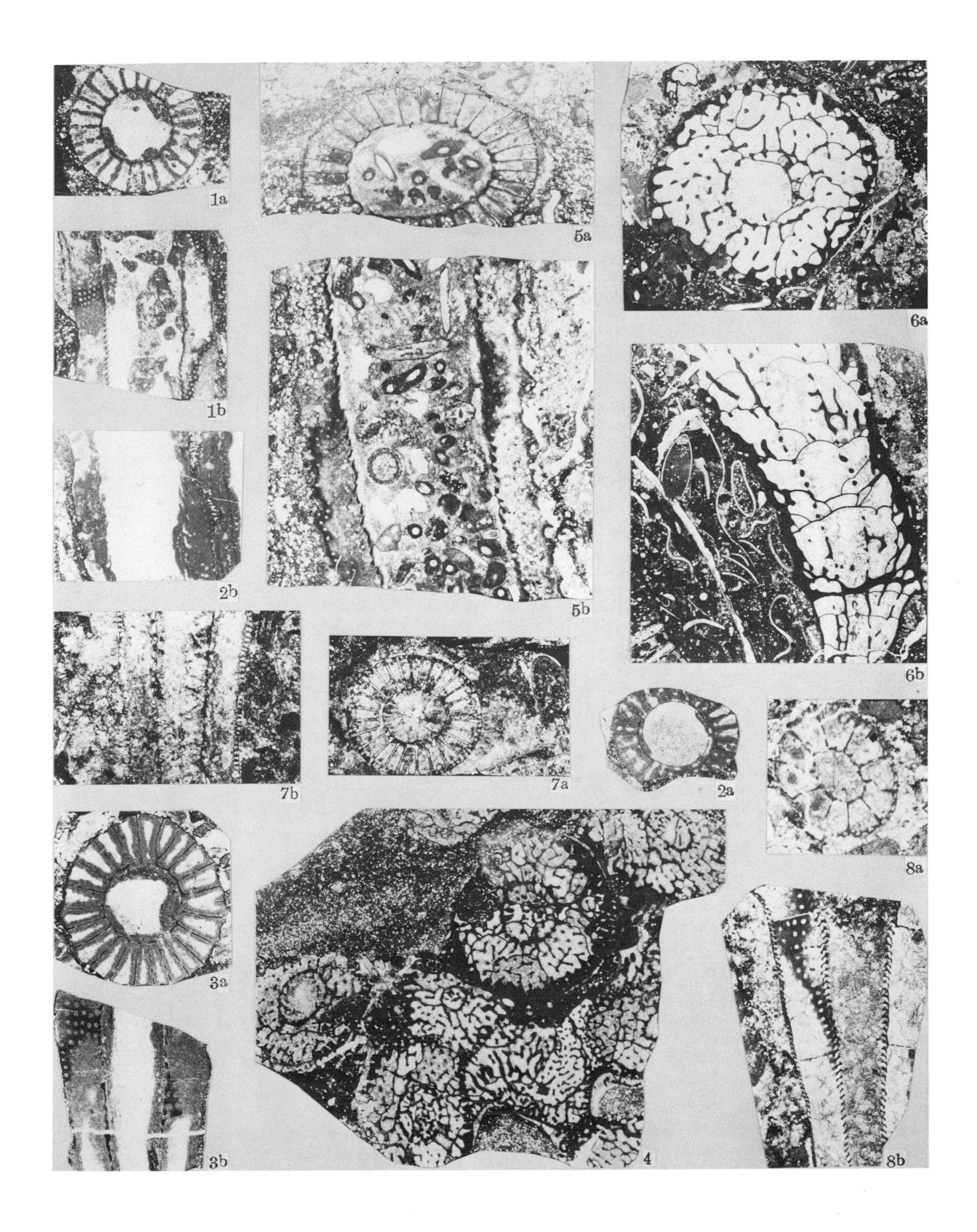
1a
5a
6a
1b
2b
5b
6b
7b
7a
2a
8a
3a
4
3b
8b

PLATE 2

Figure

Liangshuijing or Xiannüdong Archaeocyathid assemblage

1a,1b ***Archaeofungia shixiensis*** **Yuan. 1a. Transverse section, ×4. 1b. Longitudinal section, ×4. Horizon and locality: The upper part of Liangshuijing Formation, Lower Cambrian, Shixihe of Chengkou District, Northern Sichuan.**

2a,2b,2c. ***Conannulofungia jinshaensis*** **Yuan. 2a. Transverse section, ×4. 2b. Longitudinal section, ×4. 2c. Tangential section, ×4. Horizon and locality: The Minghsingssu Formation, Lower Cambrian, Yankong of Jinsha District, Guizhou.**

3a,3b. ***Coscinocyathus liangshuijingensis*** **Yuan, 3a. Transverse section, ×5. 3b. Longitudinal section, ×5. Horizon and locality: Same as 1a, 1b.**

4a,4b,4c. ***Taylorcyathus clarus*** **Yuan and Zhang. 4a. Transverse section, ×10. 4b. Longitudinal section, ×10. 4c. Tangential section, ×10. Horizon and locality: The Xiannüdong Formation, Lower Cambrian, Moujiaba of Nanzheng District, Southern Shaanxi.**

5a,5b,5c. ***Coscinocyathus angustus*** **Yuan and Zhang. 5a. Transverse section, ×5. 5b. Longitudinal section, ×5. 5c. Tangential section, ×5. Horizon and locality: Same as 4a–4c.**

6a,6b,6c. ***Taylorcyathus shifangensis*** **Yuan. 6a. Transverse section, ×5. 6b. Longitudinal section, central cavity on right, ×5. 6c. Tangential section, ×5. Horizon and locality: Same as 1a, 1b.**

7a,7b. Ajacicyathus sichuanensis Yuan. 7a. Transverse section, ×10. 7b. Longitudinal section, ×10. Horizon and locality: The upper part of Liangshuijing Formation, Lower Cambrian, Shixihe of Chengkou District, Northern Sichuan.

8a,8b. ***Chengkoucyathus shabaensis*** **Yuan. 8a. Transverse section, × 16. 8b. Longitudinal section, ×10. Horizon and locality: Same as 7a, 7b.**

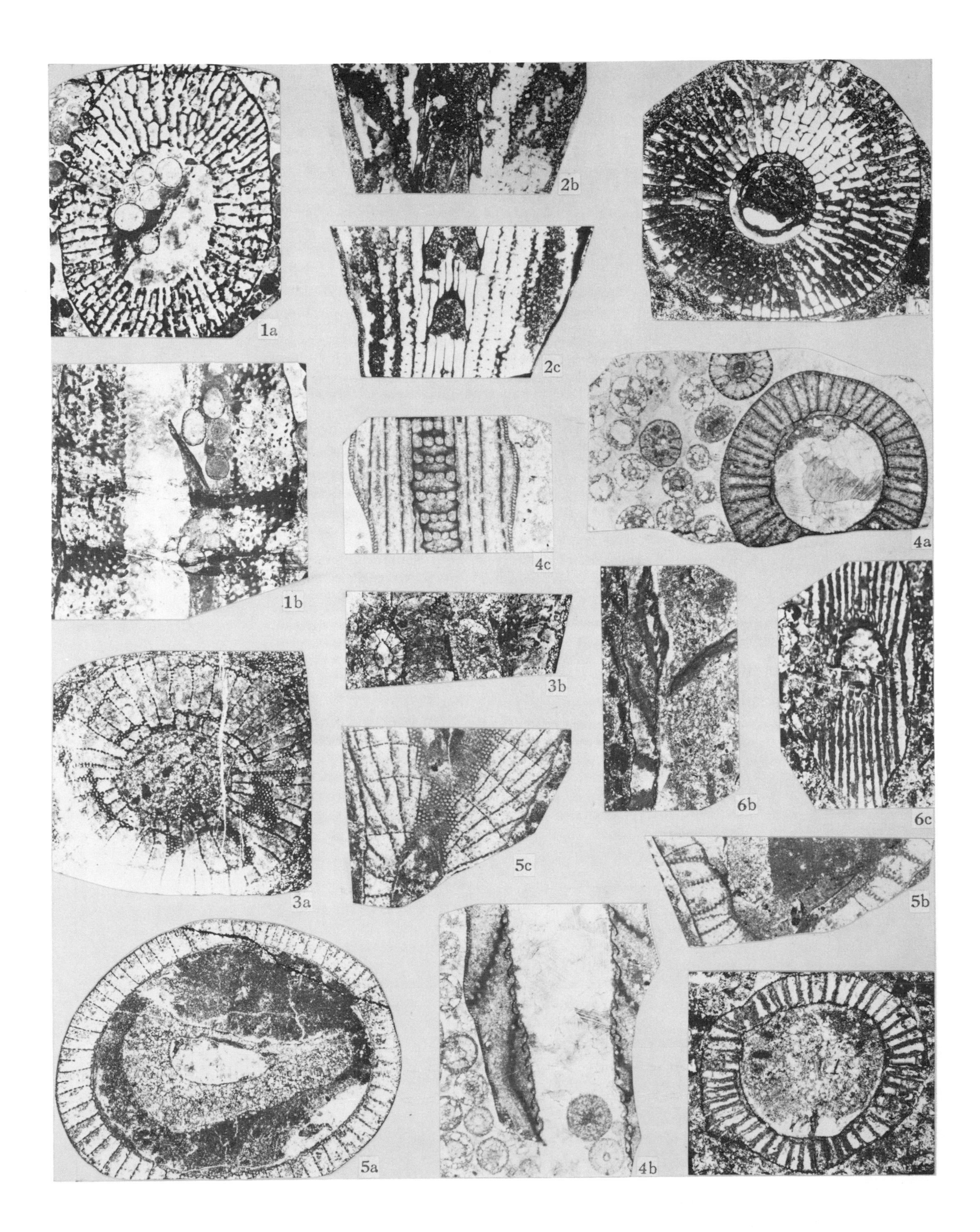
1a
2b
2c
1b
4a
4c
3b
6b
6c
3a
5c
5b
5a
4b

PLATE 3

Figure

Liangshuijing or Xiannüdong Archaeocyathid assemblage

1a,1b. *Clathricoscinus dabashanensis* Yuan. 1a. Transverse section, ×8. 1b. Longitudinal section, ×8. Horizon and locality: The Liangshuijing Formation, Lower Cambrian, Shixihe of Changkou District, Northern Sichuan.

2a,2b. *Rotundocyathus xiuqiensis* Yuan. 2a. Transverse section, ×10. 2b. Longitudinal section, ×10. Horizon and locality: Same as 1a, 1b.

Yingzuiyan or Chintingshan Archaeocyathid assemblage

3a,3b,3c. *Rotundocyathus shilixiensis* Yuan. 3a. Transverse section, ×10. 3b. Longitudinal section, ×10. 3c. Tangential section, ×10. Horizon and locality: The lower part of Chintingshan Formation, Lower Cambrian, Niuchang of Meitan District, Guizhou.

4a,4b. *Erugatocyathus yingzuiyanensis* Yuan and Zhang. 4a. Transverse section, ×6. 4b. Longitudinal section, ×6. Horizon and locality: The lower part of Yingzuiyan Formation, Lower Cambrlan, Shixihe of Chengkou District, Northern Sichuan.

5a,5b,5c. *Clathricoscinus zunyiensis* Yuan. 5a. Transverse section, ×6. 5b. Longitudinal section, ×6. 5c. Tangential section, ×10. Horizon and locality: The lower part of Chintingshan Formation, Lower Cambrian, Jinpenshui of Meitan District, Guizhou.

6a,6b. *Protopharetra chengkouensis* Yuan. 6a. Transverse section, ×6. 6b. Longitudinal section, ×6. Horizon and locality: Same as 4a, 4b.

7a,7b,7c. *Dictyocyathus jindingshanensis* (Yuan). 7a. Transverse section, ×4. 7b. Longitudinal section, ×4. 7c. Tangential section, ×4. Horizon and locality: The base of Chintingshan Formation, Lower Cambrian, Jindingshan of Zunyi District, Guizhou.

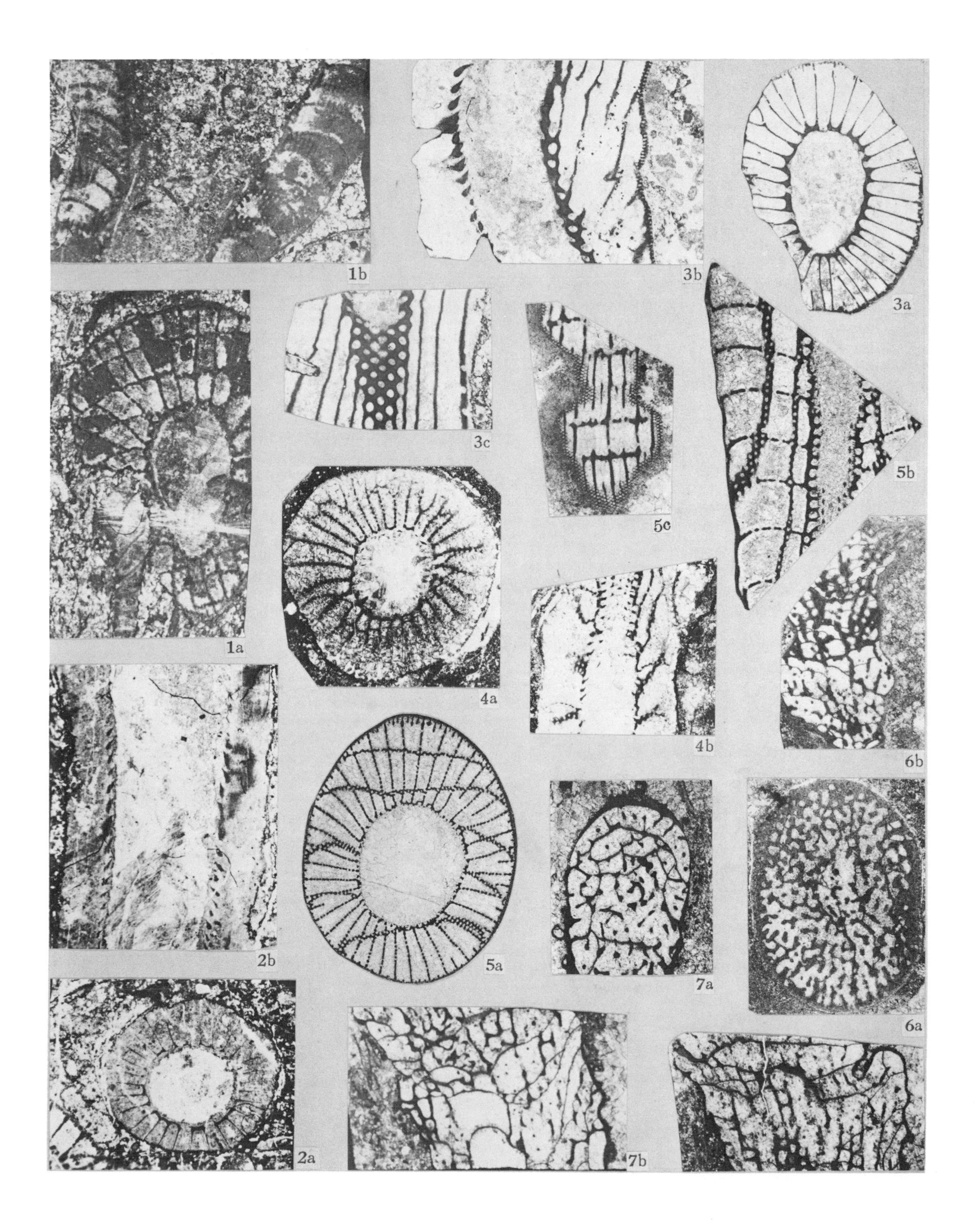
1b
3b
3a
3c
5b
5c
1a
4a
4b
6b
2b
5a
7a
6a
2a
7b

PLATE 4

Figure

Yingzuiyan Archaeocyathid assemblage

1a, 1b. *Agastrocyathus grandus* Yuan and Zhang. 1a. Transverse section, ×5. 1b. Longitudinal section, ×5. Horizon and locality: The lower part of Yingzuiyan Formation, Lower Cambrian, Shixihe of Chengkou District, Northern Sichuan.

Tianheban Archaeocyathid assemblage

2a,2b. *Archaeocyathus validus* Yuan. 2a. Transverse section, ×6. 2b. Longitudinal section, ×6. Horizon and locality: The Tianheban Formation, Lower Cambrian, Shixihe of Chengkou District, Northern Sichuan.

3a,3b. *Retecyathus wanfusiensis* Yuan. 3a. Transverse section, ×5. 3b. Longitudinal section, ×5. Horizon and locality: The upper part of Chintingshan Formation, Lower Cambrian, Jindingshan of Zunyi District, Guizhou.

4a,4b. *Retecyathus (Pararetecyathus) curvatus* Yuan and Zhang. 4a. Transverse section, ×2. 4b. Longitudinal section, ×2. Horizon and locality: The Tianheban Formation, Lower Cambrian, Shilongdong of Yichang District, Hubei.

5a,5b. *Archaeocyathus yichangensis* Yuan and Zhang. 5a. Transverse section, ×4. 5b. Longitudinal section, ×4. Horizon and locality: Same as 4a, 4b.

6a,6b. *Retecyathus nitidus* Yuan and Zhang. 6a. Transverse section, ×4. 6b. Longitudinal section, ×4. Horizon and locality: Same as 4a, 4b.

7. *Protopharetra* sp. Oblique transverse section, ×4. Horizon and locality: Same as 4a, 4b.

8a,8b. *Sanxiacyathus hubeiensis* Yuan and Zhang. 8a. Transverse section, ×4. 8b. Longitudinal section, ×4. Horizon and locality: Same as 4a, 4b.

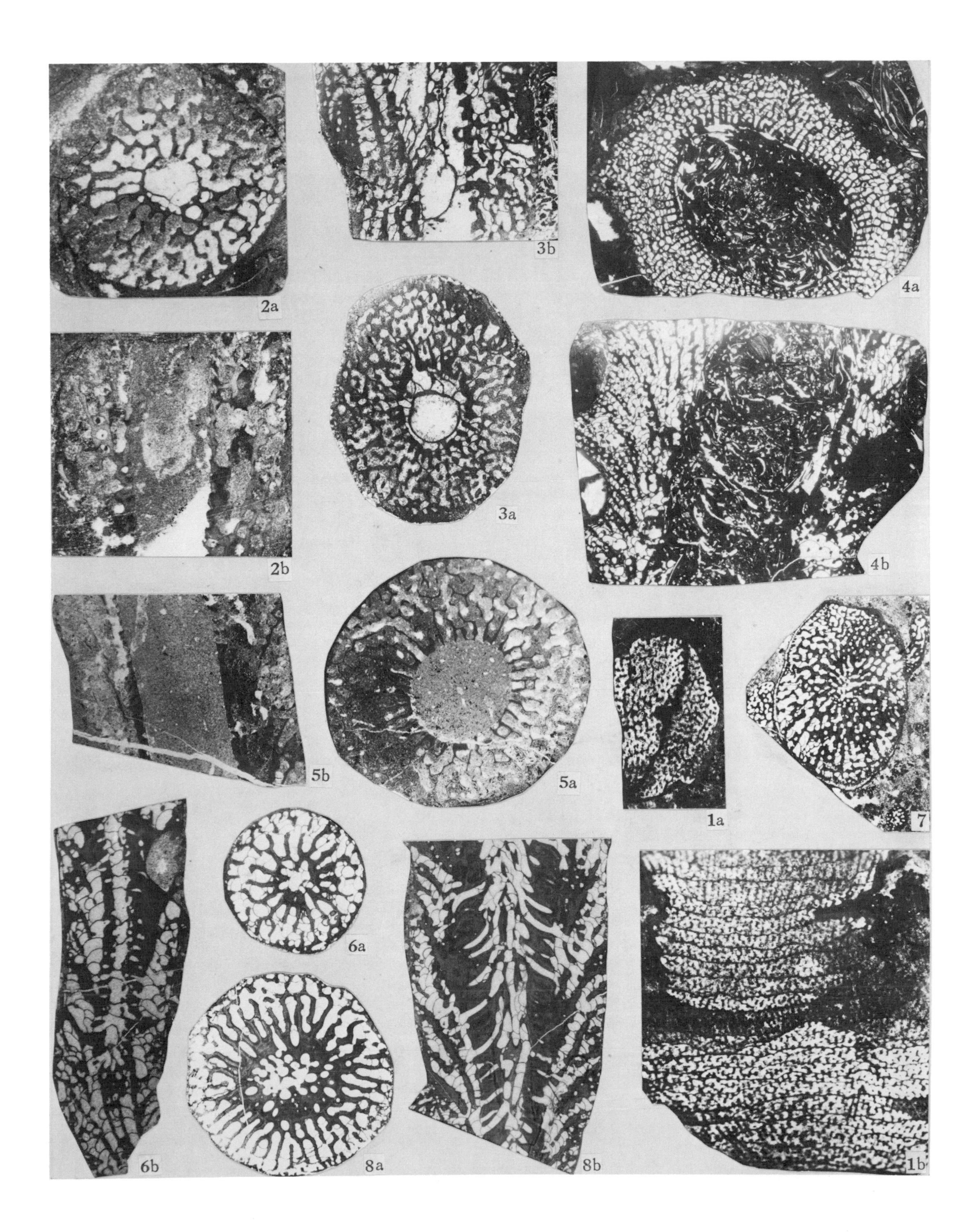
3b
2a
4a
3a
2b
4b
5b
5a
1a
7
6a
6b
8a
8b
1b

Printed in U.S.A.

Geological Society of America
Special Paper 187
1981

Silurian rugose coral assemblages and paleobiogeography of China

Wang Hong-zhen
Ho Xin-yi
Beijing Graduate School, Wuhan College of Geology, Xueyuan Road, Beijing, People's Republic of China

ABSTRACT

The first part of this paper describes seven Late Ordovician to Silurian rugose coral assemblages and their stratigraphic position. They are named in ascending order: (1) *Borelasma-Sinkiangolasma,* (2) *Dinophyllum-Rhabdocyclus,* (3) *Kodonophyllum-Maikottia,* (4) *Kypohophyllum-Idiophyllum,* (5) *Micula-Ketophyllum,* (6) *Weissermelia-Ataja,* and (7) *Mucophyllum-Pseudomicroplasma* assemblages.

The second part includes a discussion of the Silurian biogeography and types of sedimentation in China. Six sedimentation domains consisting of 15 sedimentation regions with various associations are shown in accompanying sketch maps. Four biogeographical realms, the Boreal, the Tethyan, the Australo-Pacific, and the Gondwanan(?), are recognized worldwide. On the basis of rugose coral faunas, four provinces, the Tianshan-Beishan and the Inner Mongolia-Jiliao (belonging to the Boreal) and the Yangtze and the Qinling-Sanjiang (belonging to the Paleotethyan but also related to the Australo-Pacific realms), are distinguished in China. A table is compiled to show the distribution of coral assemblages in some of the representative sedimentation regions.

INTRODUCTION

After Lindström's description of some Chinese corals in 1883, recent research on Silurian rugose corals of China began in the early 1940s. The recent publication of regional stratigraphic tables and index fossils organized by the Chinese Academy of Geological Sciences and the Nanjing Institute of Geology and Palaeontology, Academia Sinica, have greatly increased our knowledge of Silurian stratigraphy and palaeontology (see recent summaries by Yin, 1966, and by Lin, 1979).

This paper is an attempt to correlate the rugose coral assemblages and to estimate their paleobiogeographic significance.

SILURIAN RUGOSE CORAL ASSEMBLAGES OF CHINA AND THEIR STRATIGRAPHIC DISTRIBUTION

Silurian, including latest Ordovician, rugose corals may be grouped in seven assemblages which are distributed in different parts of China (Fig. 1; Table 1). They are, in descending order:

Upper Silurian
7. *Mucophyllum-Pseudomicroplasma* Assemblage, Erdaogou Formation
6. *Weissermelia-Altaja* Assemblage, Xibiehe Formation
Middle Silurian
5. *Micula-Ketophyllum* Assemblage, Gaundi Formation
4. *Kyphophyllum-Idophyllum* Assemblage, Ningqiang Formation
Lower Silurian
3. *Kodonophyllum-Maikottia* Assemblage, Shiniulan Formation, upper part
2. *Dinophyllum-Rhabdocyclus* Assemblage, Shiniulan Formation, lower part
Uppermost Ordovician
1. *Borelasma-Sinkiangolasma* Assemblage, Guanyinqiao beds

Borelasma-Sinkiangolasma Assemblage. The stratigraphic horizon and type locality of this assemblage is the Guanyinqiao beds of northern Guizhou, on the Yangtze platform. Rugosa are abundant, among which *Borelasma* is the most characteristic and representative. *Borelasma* is widely distributed in the uppermost Ordovician *Dalmanitina* bed of Sweden, in bed 5b of Norway, and in bed F2 of Esthonia, which can be adequately correlated with each other. *Sinkiangolasma* is comparatively rare in the Guanyinqiao beds, but is known extensively in the Upper Orodovician of China, including northwestern and northeastern China. These horizons may, however, be a little lower than the Guanyinqiao beds.

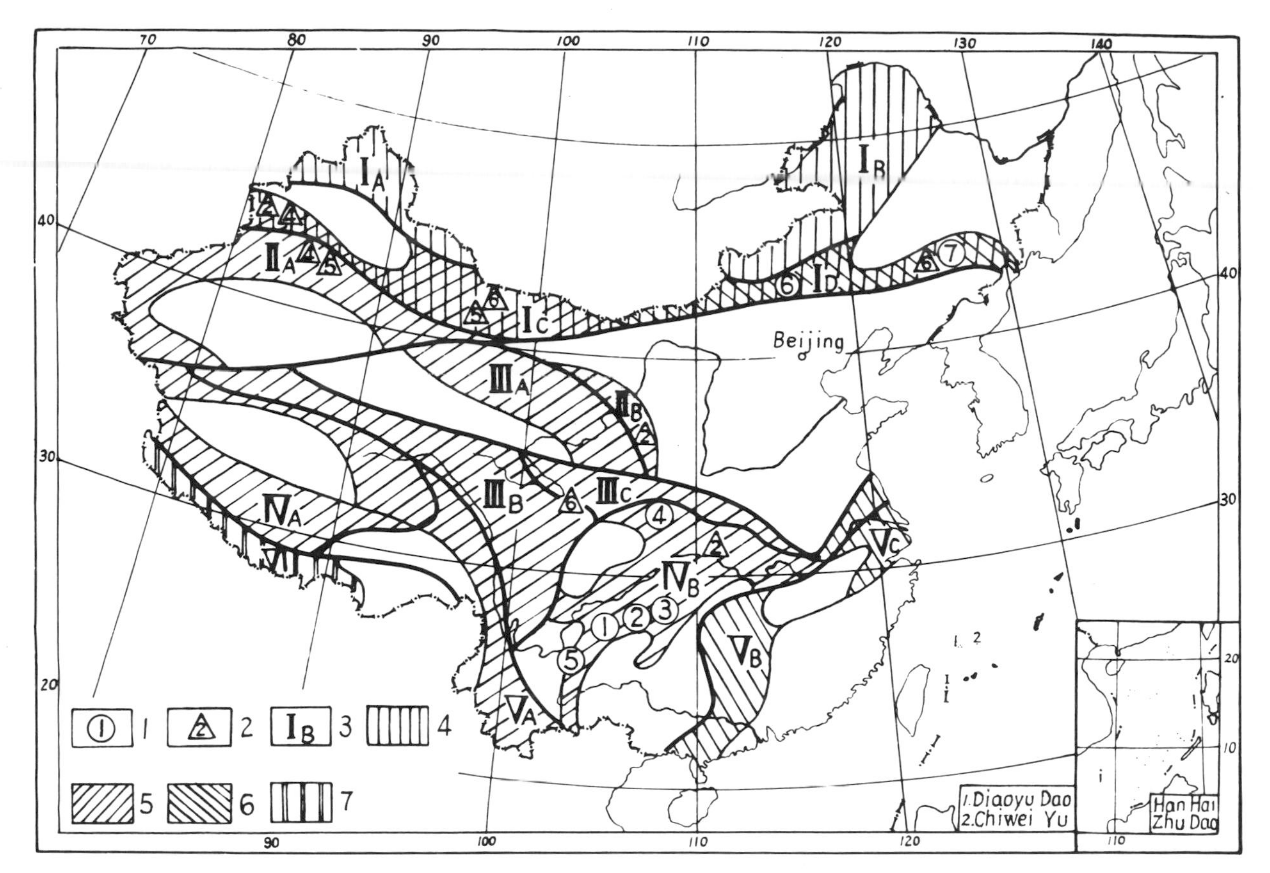

Figure 1. Silurian sedimentation regions, rugose coral assemblages, and biogeography of China. 1. Type localities of coral assemblages. 2. Other localities of coral assemblages. 3. Sedimentation regions: I. Northern mobile domain: IA, Altai region; IB, Hingan region; IC, Tianshan-Beishan region; ID, Inner Mongolia–Jiliao region. I, North stable domain; IIA, Tarim (South Tianshan) region; IIB, Shaanganning region. III. Median mobile domain; IIIA, Qilian region; IIIB, Kunlun–West Sichuan region; IIIC, Qinling region; IV. Southern stable domain: IVA, South Tibet region; IVB, Yangtze region. V. Southern mobile domain: VA, West Yunnan region; VB, Central Hunan-Qinfang region; VC, Anhui-Zhejiang region. VI. Gondwana (?) domain, Himalaya Region. 4–7. Biogeographical provinces: 4, Boreal realm; 5, Tethyan realm; 6, Australia-Pacific realm; 7, Gondwana (?) realm.

Among the constituents of this assemblage are *Brachylasma primum, Streptelasma, Crassilasma, Kenophyllum, Pycnactis, Siphonolasma* Ho, and *Paramplexoides* Ho (1978). All are simple corals, with Streptelasmatidae the most abundant; Pycnactidae and Cystiphyllidae also are well represented, and *Proheliolites* was also found in the Guanyinqiao beds. The presence of *Lambeophyllum* is worth noting, as it is generally known from the Middle Ordovician of North America and Europe. The recent discovery of this genus in the early Late Ordovician Sanjushan Limestone in western Zhejiang and in the Upper Ordovician of northeastern Kazackstan indicates that its time range may extend to Late Ordovician. The age of the Guanyinqiao beds has been much discussed lately. As many of the Rugosa are typical Late Ordovician forms, we are included to refer the Guanyinqiao beds to the latest Ordovician.

Although strata bearing the *Hirnantia–Dalmanitina mucronata* fauna are widely known in China, the *Borelasma–Sinkangolasma* fauna seems to be comparatively more restricted, as it has been found only in Yanzikou and Bijie (northwestern Guizhou) and in Renhuai, Sinan, and Yinjiang (northern Guizhou), and was probably controlled by paleogeographic and paleoecologic conditions. The Wufengian in the Yangtze region is mostly graptolite facies deposited under restricted and reducing conditions and thus unsuitable for coral growth. Normal epeiric seas with mixed facies or shell facies deposits appeared in Guanyinqiao time, and consequently corals abounded under locally stable conditions of calcareous deposition. The *Borelasma-Sinkiangolasma* fauna is closely related to the European Baltic forms, and sea ways connecting Europe and South China via Central Asia might have been present.

Early Silurian Ruogse Coral Assemblages

In the Upper Yangtze region, early and middle Llandoverian are represented by the Lungmachi Formation of graptolite facies: which is followed by a shelly facies mainly composed of limestone and marl; this is known as the Lojeping Formation in northwestern Hubei and the Shiniulan Formation in northern Guizhou. The Shiniulan Formation as used here comprises the original Shiniulan Formation and strata below the lower red beds of the original Hanjiadian Formation, thus conforming to the Xiangshuyuan Formation and the Leijiatun Formation (*senso lato*) of the Nanjing Institute of Geology and Palaeontology. As the lateral facies change is very pronounced, we regard it adequate to retain the original name Shiniulan Formation, which contains two members characterized by two rugose coral assemblages.

Dinophyllum-Rhabdocyclus Assemblage. This assemblage is typically developed in northern Guizhou and occurs in the lower member of the Shiniulan Formation of late Llandoverian age. Small simple corals are prevalent, including the following: *Brachylasma, Sibiricum, Crassilasma, Cymatelasma, Amplexoides, Tryplasma, Cystiphyllum, Cysticonophyllum, Palaeophyllum, Ceriaster,* and *Fletcheria.* Among tabulates and heliolitids, *Troedssonites, Syringoporinus,* and *Helioplasmolites* are the most common. *Troedssonites,* constant in stratigraphical position and widespread in Guizhou, is a good index fossil of the lower member of the Shiniulan Formation.

Some of the constituent members of the assemblage in question are typical Late Ordovician forms, such as *Leolasma. Dinophyllum yunnanense* Wang and *Holmophyllum conicum* Wang, originally found in the Daguan Formation of northeastern Yunnan and referred to the Middle Silurian, are now believed to be Lower Silurian, as both species occur in the Lower Silurian of northern Guizhou. The coral beds containing the *Dinophyllum-Rhabdocyclus* assemblage are probably equivalent to the lower member of the Shiniulan Formation. *Dinophyllum yunnanense,* though well known in northern Guizhou and northeastern Yunnan, is not found in the Lojeping Formation of western Hubei. It is probable that *Densiphylloides* Ge and Yu (1974) represents a dinophyllid with moderately dilated and partly contiguous inner septal ends.

The general aspect of the rugose faunas found in the Lojeping Formation bears a basic resemblance to that of the lower part of the Shiniulan Formation. Forms common to both formations comprise *Brachylasma sibiricum, Palaeophyllum hubeiense, Ceriaster minor, Rhizophyllum minor, Amplexoides, Rhabdocyclus, Cantrillia,* and *Tryplasma.* In addition, the Lojeping fauna contains special forms such as *Onychophyllum pringlei* and *Pseudophaulactis?,* known respectively from the Lower Silurian of Europe and Siberia. Mention should also be made here that the coral referred to *Pycnactis hubeiensis* from the Lojeping Formation has attenuate septa in the counter parts and well-developed tabulae in the adult stage. It may, therefore, be referred to *Pseudophaulactis* rather to *Pycnactis.*

Kodonophyllum-Maikottia Assemblage. This assemblage is typically represented in the upper member of the Shiniulan Formation. The generic name *Maikottia* is used here, for it seems to us that *Qianbeilites* Ge and Yu (1974) is a junior synonym of *Maikottia* Lavrusevich (1967).

The associated genera are *Gyalophyllum, Hedstroemophyllum, Zelophyllum, Pilophyllia, Tabularia,* and *Stauria,* mostly continuing upward from the lower assemblage. *Ketophyllum* and *Pycnostylus* occur mainly in the Middle and Upper Silurian of Europe and Australia, but are also found in the lower Silurian of Guizhou. Stauriids are also frequent in the upper member of the Shiniulan Formation, including advanced forms with stable dissepiments and axial structures, such as *Ceriaster columellatus* and *Stauria normola. Grewingkia* was also discovered. As *Grewingkia* was found in recent years in the lower and middle Llandoverian of Iran and Australia, it would seem that the Early and Middle Silurian Rugosa of the Yangtze region bear a rather close resemblance to contemporaneous forms found in Europe and Australia.

Tabulates are also abundant. *Meitanopora, Baikitolites, Eoroemerites, Thecostegites,* and favositids were reported, among which *Meitanopora* and *Eoroemerites* are the most diagnostic, widespread, and characteristic of the upper member of the Shiniulan Formation. In addition, dizonal Rugosa seem to have been abundant in this assemblage, especially in southern Sichuan and northeastern Yunnan, where *Entelophyllum* (Ho Yuan-xiang, unpub. dat), *Stombodes,* and *Yassia* are frequently met with at the horizon corresponding to the upper part of the Shiniulan Formation.

The Early Silurian Rugosa in the Wangjiawan Formation of southern Shaanxi contain two assemblages: the lower one, which includes *Amplexoides, Tryplasma, Cysticonophyllum, Onychophyllum,* and *Ceriaster,* may be correlated with the lower member, whereas the upper, which contains *Kodonophyllum, Pilophyllum,* and *Yassia,* may be correlated with the upper member of the Shiniulan Formation.

Middle Silurian Rugose Coral Assemblages

Kyphophyllum-Idiophyllum Assemblage. This assemblage is referred to the Wenlockian and is represented in the Ningqiang Formation of southern Shaanxi. In addition to *Idiophyllum* and *Kyphophyllum,* the genera frequently met with are *Gyalophylloides, Pilophyllia, Shensiphyllum, Miculiella,* and *Ceriaster,* followed by *Tryplasma, Amplexoides, Chonophyllum,* and *Calostylis* in abundance. The features distinguishing *Idiophyllum* Cao (in Li Yao-xi and others, 1975) from *Nanshanophyllum* are the usual occurrence of the third order septa, the split inner septal ends, and the bilateral septal symmetry in the cardinal parts. *Idiophyllum* is mainly found in the Ningqiang Formation of Dabashan and also in the Xiushan Formation in Shimen, northern Hunan.

Shensiphyllum is widely distributed in the Ningqiang Formation and its equivalents. Although Ge and Yu did not mention the presence of horseshoe dissepiments in the diagnosis of their new genus in 1974, we deem that *Shensiphyllum* does include such forms, and we consider *Stereoxylodes phacelloides* Cao from the Ningqiang Formation and *Neopetrozium hubeiense* Wu (in Jia and Wu, 1977) from the Shamao Formation in Changyang, western Hubei, to be members of *Shensiphyllum.*

i In the Ningqiang Formation, apart from the typical Wenlockian Rugosa *Rhyzophyllum gotlandicum, Micula, Miculiella,* and *Chonophyllum,* there occur abundant tabulates and heliolitids among which *Antherolites, Subalveolites,* and *Thaumatolites* are restricted to the Middle Silurian.

In northeastern Yunnan, Middle Silurian corals comparable to tose found in the Ningqiang and Xiushan Formations remain so far unstudied. Formerly reported were *Pseudocystiphyllum, Zelophyllum, Pilophyllum,* and *Gyalophyllum* among the Rugosa, and *Alveolites, Subalveolites,* and *Pilophyllum* were also reported from the Middle Silurian of Ailaoshan, western Yunnan, but their correlation with adjacent regions is not clear.

In the northern mobile domain, Middle Silurian Rugosa are found at many localities. In the Qilianshan region, *Nanshanophyllum, Kyphophyllum, Tryplasma,* and *Cystiphyllum* occur in the Quannaogoushan Formation, in association with *Sichuanoceras* and *Coronocephalus,* and are thus probably correlatable with the *Kyphophyllum-Idiophyllum* assemblage of the Ningqiang and

TABLE 1. SILURIAN RUGOSE CORAL ASSEMBLAGES IN SOME REPRESENTATIVE

Stages of western usage	North Mobile Domain			
	IB Hingan region	IC Northern Tianshan and Beishan region		ID Inner Mongolia-Jiliao region
D_1 Gedinnian	D_1	D_1		D_1 Erdaodou Fm.
S_3^2 Pridolian	Gulanhe Fm.	Boluohuoluo-shan Fm.	Upper Gongpoquan Fm. *Schlotheimophyllum* *Kodonophyllum* *Holmophyllum* *Zelophyllum* *Pilophyllum*	Zhaganhabu Fm. *Entelophyllum* *Songophyllum* *Tryplasma* *Mucophyllum* *Pseudomicroplasma* *Tryplasma* *Microplasma* *Spongophyllum* -------------------- *Warburgella* *Encrinurus*
S_3^1 Ludlovian	Woduhe Fm. *Enterolasma* *Tryplasma* -------------------- *Tuvaella* *Tannuspirifer*			Sibiehe Fm. *Weissermelia* *Altaja* *Entelophyllum* *Spongophyllum* *Ptychophyllum* *Strombodes* *Conchidium*
S_2^2 Upper Wenlockian	Bashilixiaohe Fm. *Tuvaella*	Jifuke Fm. *Kyphophyllum* *Strombodes* *Yassia* *Holmophyllum*	Lower Gongpoquan Fm. *Tabularia* *Ketophyllum* *Ceriaster*	Halayan Fm.
S_2^1 Lower Wenlockian				
S_1^3 Upper Llandoverian	Huanghuogou Fm. *Hindella*	*Gyalophyllum* *Dentilasma* *Rhabdocyclus* *Microplasma* *Zelophyllum*	Heijianshan Fm.	Bulogshan Fm.
S_1^{1-2} Lower & Middle Llandoverian O_3–S_1^1 Hirnantian		Nileikehe Fm.		
O_3	O_2	O_3	O_3	O_3

SEDIMENTATION REGIONS OF CHINA

Median Mobile Domain		South Stable Domain	South Mobile Domain	Gondwana (?) Domain
IIIA Qilian region	IIIC Qinling region	IVB Upper Yangtze region	VB Central Hunan-Qinfang region	VI Himalayan region
D_1	D_1	D_1	D_1	D
Hanxia Fm.	Beilongjiang Group *Entelophyllum* *Spongophyllum* *Holmophyllum* *Strombodes*	Yulongsi Fm. *Warburgella* *Athyrisina* *Protathyris*	Fangcheng Group	
		Miaogao Fm. *Howellella tingi* *Protathyris*		
Quannaogoushan Fm. *Kyphophyllum* *Tryplasma* *Dinophyllum* *Nanshanophyllum*	Zhouqu Fm. *Stereoxylodes* *Holmophyllum* *Gyalophyllum*	Guandi Fm. *Micula-Ketophyllum* *Holmophyllum* *Ketophyllum* *Gyalophyllum* *Mucophyllum* *Squameofavosites*	Hepu Fm.	Shiqipo Group: Upper Fm.
		Ningqiang Fm. *Kyphophyllum-Idiophyllum* *Shensiphyllum* *Miculiella* *Nikiforovaena* *Pilophyllia* *Sichunoceras* *Gyalophylloides* *Coronocephalus* *Chonophyllum* *Salopina*		
Angzanggou Fm.	Diebu Group: Jiannigou Fm.	Upper Shiniulan Fm. *Kodomophyllum-Maikottia* *Gyalophyllum* *Hedstroemophyllum* *Dentilasma* *Pilophyllia* *Paraconchidium* *Stauria* *Subalveolites*	Wentoushan Fm.	Shiqipo Group: Lower Fm.
		Lower Shiniulan Fm. *Dinophyllum-Rhabdocyclus* *Brachyelasma* *Cymatelasma* *Pycnactis* *Cantrillia* *Crassilasma* *Cystiphyllum* *Stricklandinia*		
Xiaoshihugou Fm.	Diebu Group: Anzigou Fm.	Lunmachi (Lungmachi) Fm.	Liantan Fm.	
O_3	6-0	O_3 Guanyinqiao beds *Borelasma-Sinkiangolasma* *Streptelasma* *Grewingkia* *Brachyelasma* *Pycnactis* *Crassilasma* *Lambeophyllum*	O_3	

Xushan Formations. Again, the Middle Silurian Jifuke Formation of the central Tianshan contains *Kyphophyllum, Holmophyllum, Zelophyllum, Kodonophyllum, Amplexoides, Strombodes,* and *Yassia.* The Middle Silurian Shaiwusu Formation of Liaoning Province contains *Streptelasma, Dinophyllum, Palaeophyllum,* and *Mucophyllum.* All are comparable with the assemblage under discussion, but precise correlation with the Rugosa fauna found in the Yangtze region is difficult.

Micula-Ketophyllum Assemblage. This assemblage is typically represented by the Rugosa found in the Guandi Formation, with *Holmophyllum, Gyalophyllum,* and *Mucophyllum* as the main associated forms. Tabulate corals include *Squameofavosites* and *Syringopoda. Mucophyllum* is known in the lower Middle and Upper Silurian of China. *Squameofavosites* was formerly believed to occur only in the Upper Silurian. But, according to Kaljo and Klaaman (1973), it was also found in the Middle Silurian of Kazakhstan and the Sayan region. As the fauna of the Guandi Formation is closely related to those of the upper Xiushan Formation bearing *Sichuanoceras* and *Coronocephalus,* this assemblage may be referred to the late Middle Silurian.

Late Silurian Rugose Coral Assemblages

Weissermelia-Altaja Assemblage. The representative locality and horizon is the Ludlovian Xibiehe Formation of Inner Mongolia. Associated forms consist of *Spongophyllum, Strombodes, Kyphophyllum, Mucophyllum,* and some cystimorphs. *Weissermelia* is an Upper Silurian genus in the Urals and Europe. *Altaja* was originally reported from the Middle Silurian of Kazakhstan and Siberia. It seems to occur in a slightly higher horizon in China.

The Xingshuwa Formation of western Liaoning, an equivalent of the Xibiehe Formation, contains as main forms *Weissermelia* and *Tryplasma hedstoremi.* The coral bed in the upper part of the Bailongjiang Group of western Qinling may also represent the same horizon, in which were found *Entelophyllum, Strombodes, Spongophyllum, Pycnostylus, Tabularia, Tryplasma, Cystiphyllum, Ketophyllum, Holmophyllum, Gyalophyllum, Gukoviphyllum, Nipponophyllum, Kyphophyllum, Pseudamplexus,* and *Schlotheimophyllum.* It may be mentioned in this connection that the Rugosa found in the Koktiekdaban Group of the southern Tianshan bear a resemblance to those of the Bailongjiang Group and thus may be of the same age. Common genera in the two groups are *Tabularia, Ketophyllum, Holmophyllum, Kyphophyllum,* and *Pycnostylus.*

Mucophyllum-Pseudomicroplasma Assemblage. The type locality and horizon is the Erdaogou Formation of central Jilin. Associated genera are *Tryplasma, Cystiphyllum, Microplasma, Rhizophyllum, Diplochone, Entelophyllum, Spongophyllum,* and *Lyrielasma,* with the tabulates *Squameofavosites* and *Mesofavosites. Lyrielasma* and *Pseudomicroplasma* have long been known as exclusively Devonian. But the latter genus was recently reported from the uppermost Silurian in the Urals. Though the Erdaogou fauna contains some Early Devonian forms, the prevailing constituents are Late Silurian. We refer here the Erdaogou Formation tentatively to the Pridolian to Gedinnian. The Xiashipai Formation of western Liaoning is equivalent to the Erdaogou Formation.

PALAEOBIOGEOGRAPHY OF THE SILURIAN RUGOSE CORALS OF CHINA

Silurian Paleogeography and Sedimentation

Before entering into discussions of the Silurian paleobiogeography, it is advisable to give a brief review of the sedimentation types and sedimentation regions of China in Silurian times, which should be of help in the understanding of the distribution and significance of the coral faunas.

According to sedimentation types and geotectonic settings, the Silurian of China includes 6 sedimentation domains. They are, from north to south: I, northern mobile domain; II, northern stable domain; III, median mobile domain; IV, southern stable domain; V, southern mobile domain; and VI, Gondwanan (?) stable domain (Figs. 1, 2).

Figure 2 is a sketch map of China showing the Early to Middle Silurian sedimentation types and paleogeography that covers the time span from Shiniulan to Xiushan age. During the Silurian, North China was an old land of denudation, and the main parts of Tarim were also above sea level. To the north of this vast latitudinal stable domain is the complicated "northern geosynclinal domain." The boundary between these two major units is the northern boundary fault zone of the Inner Mongolian Axis in the east and the southern boundary fault zone of the central Tianshan in the west. The Hingan region is characterized by graywacke associations rich in volcanic material. Andesite prophyrites and tuffs are also abundant in central and northern Tianshan and in western Junggar. Sediments representative of the north stable domain are the Kopingtagh Formation of northern Tarim and the Zhaohuajing Group fo Ningxia, the latter containing a rich coral fauna.

The median mobile domain that essentially divided China into a northern and a southern part in Silurian time includes the Qilian, Kunlun, and Qinling geosynclines. The Middle Silurian Angzanggou and Quannaogoushan Formations form the latest sedimentary record of strong volcanic activities and subsidence in the northern Qilian trough, which came to an end in the late Silurian. The Kunlun-Qinling troughs were the main latitudinal mobile belt of the time. The Silurian in the southern slope of Burhanbuda of eastern Kunlun is rich in volcanic units, though variable both in facies and in thickness.

The southern stable units are represented by the Yangtze massif, which is also the best studied. As compared with the Ordovician, Silurian rocks are more restricted in distribution, more complicated in facies differentiation, and generally thicker, thus reflecting a much stronger topographic relief in the border regions of the massif. Early Middle Silurian rugose coral assemblages are best developed here. It may be pointed out that the north Tibet region might have been another large median massif in Silurian time (Wang, 1978), as a stable type of Silurian sediments was recently reported from Bange, Tibet.

Because of the general uplift of the southeastern maritime provinces, the south mobile domain of marine sedimentation in the Silurian was confined to the central Hunan and the Anhui-Zhejiang basins and the Qinfang trough. All these regions were converted to areas of denudation or sites of terrestrial or paralic deposition after the Middle Silurian, except the Qinfang trough, which continued to exist until the end of the late Paleozoic.

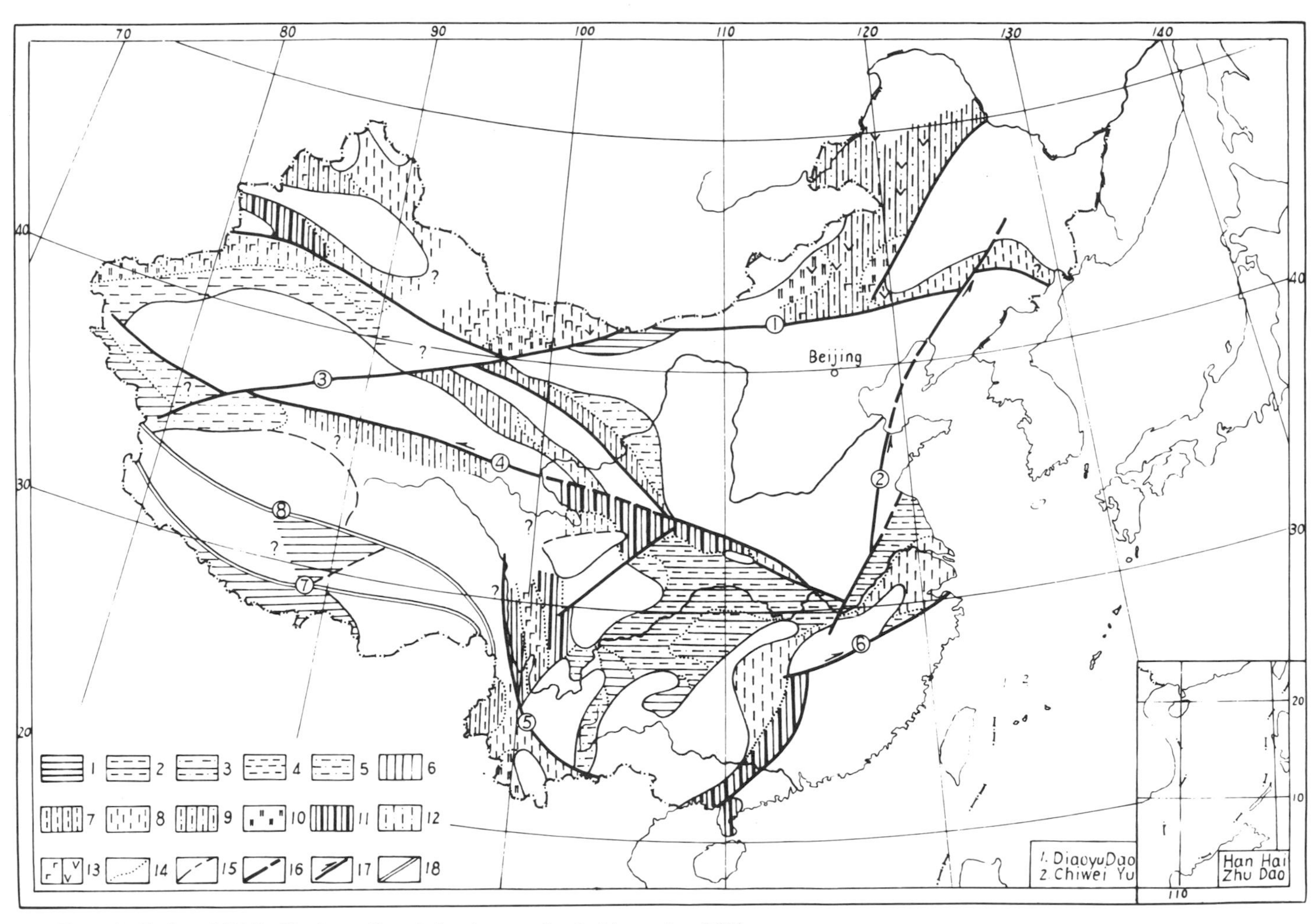

Figure 2. Early to Middle Silurian sedimentation types and paleobiography of China.

1–5. Marine stable type: 1, mainly carbonates; 2, carbonates and argillaceous rocks; 3, clastic and argillaceous rocks; 4, mainly argillaceous rocks, mostly Lower Silurian only; 5, mainly clastic and argillaceous rocks.

6–12. Marine mobile type: 6, mainly carbonates; 7, mainly calcareo-argillaceous rocks; 8, argillaceous flysch and paraflysch; 9, clastic and argillaceous flysch and paraflysch; 10, argillaceous flysch with silicolites; 11, carbonaceous rocks with silicolites; 12, mainly clastic and argillaceous rocks, overlapping Middle Silurian.

13. Median to basic (left) and median to acid (right) volcanic rocks.

14. Sedimentation type and overlap boundaries.

15. Boundaries of sedimentation region.

16. Syndepositional faults and boundary faults.

17. Later transcurrent faults.

18. Crustal consumption zones.

1–8. (in circles). Principal fault zones: 1, Inner Mongolia northern border; 2, Tanlu-Yitong; 3, Altyn; 4, Kolamilan-Xiugou; 5, Jinsha-Honghe; 6, Yichuan-Shaoxing; 7, Yarlungzangbo; 8, Bangonghu-Nujiang.

Finally, the Silurian Shiqipo Formation in the Himalayan region is a stable type deposit, but its relation to the Gondwana supercontinent remains unclear. In all probability it represents the northern margin of the southern continent, and forms the southern shallow-water tract of the Paleotethys, which was then of a much vaster extent.

Palaeobiographical Provinces of Chinese Silurian Rugose Corals

Kaljo and Klaamann (1973), among others, have regarded the Late Ordovician and Silurian coral faunas as mostly cosmopolitan. In fact, Middle and Late Silurian corals display a considerably marked provinciality. Oliver (1977) pointed out the evident endemism of the North American Late Silurian Rugosa. In China, and in the Eastern Hemisphere in general, three realms— the Boreal, Tethyan, and Australo-Pacific—may be recognized in the Silurian faunas as a whole. All three realms may be recognized among the Chinese late Ordovician and Silurian coral faunas, the late Middle and Upper Silurian being more prominent. They are briefly discussed below.

1. Latest Ordovician. According to Ho (1978), the *Borelasma-Sinkiangolasma* Assemblage of the Yangtze region includes 13 genera and 34 species, most genera being known in western Europe and the Baltic area, thus belonging to the northern tract of the Tethys realm *sensu lato*. Although most genera are cosmopolitan, there are 25 new species, about 74% of the total, that show an evidently endemic character. The Late Ordovician *Favistella* and

Amsassia fauna found in the northern regions may be attributed to the Siberia-Mongolian province of the Boreal realm.

2. Early Silurian. In the late Early Silurian assemblages, the lower one is mainly composed of simple monozonal streptalasmatids and some cystimorphs. Dizonal corals appear mainly in the upper assemblage. The monozonal *Cymatelasma, Pycnactis mitratus, Streptelasma whittardi,* the cystimorph *Cysticonophyllum cylindricum, Zelophyllum, Rhabdocyclus,* and the dizonal *Strombodes* are all European forms. But at the same time there are the endemic stauriids and *Pilopyllia* that are peculiar to the Yangtze region. The synchronous Zhaohuajing coral fauna on the northern side of the Qilian trough is also mainly European but includes the Boreal genus *Tungussophyllum.* The more or less contemporaneous coral faunas found in Tianshan and Beishan seem to represent the epeiric marine population along the northern margin of the North China and the Tarim massif. They bear a close resemblance to the Uralian and north European forms and may be regarded as transitional between the Tethyan *(senso lato)* and the Boreal realms (see Table 1).

3. Middle Silurian. The differentiation of the Middle Silurian Rugosa is remarkable, as shown by the presence of such endemic genera as *Idiophyllum* and *Shensiphyllum* in the Yangtze region. The same assemblage is known from the Zhouqu Formation of western Qinling. Similar forms also occur in the Quannaogoushan Formation of Qilianshan, where the endemic *Nanshanophyllum* is found. It is worth noting that the diagnostic *Tuvaella* and *Tqunuspirifer* brachiopod fauna of the Boreal realm is widespread from Hingan and Mongolia to Altai, the Sayan Mountains, and Siberia. The Middle Silurian *Altaja* found from Siberia and the Sayan region has not been discovered in contemporaneous beds of China.

The *Micula-Ketophyllum* assemblage, so far only known in eastern Yunnan, is peculiar for its minor simple cystimorphs *Ketophyllum* and *Gyalophyllum.* They represent probable local faunas of inland seas (Fig. 2).

4. Late Silurian. Late Silurian rugose coral assemblages are typically developed in the Inner Mongolia-Jiliao sedimentation region. Corals of the same age were also found in western Qinling and Beishan. According to the material of Guo (1976), Rugosa in the Xibiehe Formation total 20 genera and 28 species. Apart from the 10 new species, 5 are Boreal, 5 are akin to Australian forms, and the remaining 10 are cosmopolitan columnariids, mostly *Entelophyllum* and *Spongophyllum.* From the uppermost Silurian Erdaogou coral fauna, 10 genera and 12 species have been described. Among them, two are typically Asian, three are probably Australian. All cosmopolitan and some new species are related to northwestern European forms. It would seem that the Inner Mongolia-Jiliao region was the juncture of the north Asian and the Australian coral faunas, with a notable imprint of the Australo-Pacific realm.

Coral beds correlatable with the *Weissermelia-Altaja* assemblage occur in the upper part of the Gongpoquan Formation of Beishan, which contains *Schlotheimophyllum, Holmophyllum, Pilophyllum,* and *Kodonophyllum* (Cao Xuan-duo and Cai Tu-si, unpub. data) evidently related to contemporary northern European forms.

The Rugosa found in the upper part of the Bailongjiang Group of the Qinling region are also characterized by European and cosmopolitan genera such as *Entelophyllum, Spongophyllum, and Holmophyllum.* The difference is the absence of the north Asian *Altaja* and the presence of the European species *Pilophyllum keyserlingi* along with some Australian forms (Cao Xuan-duo, unpub. data). In Silurian time, western Qinling was situated at the northern end of the Sanjiang-Burmese-Malayan geosyncline and was connected with the European seas via Kunlun, central Asia, and the middle east. The Late Silurian Rugosa of Qinling and western Sichuan belong to the Tethys realm which was, to a certain extent, also connected with Australia. Marine transgressions were then much restricted in the Yangtze region and no coral faunas were reported.

To summarize, the Silurian Rugosa in Tianshan, Qilian, and Beishan are related to Uralian and western European forms and occupy a mixed belt of the Tethyan and the Boreal realms, here called the Tianshan-Beishan Province. Corals in eastern Inner Mongolia and the Jiliao region are related to both the north Asian and the Australian faunas and form the Inner Mongolia-Jiliao Province. The Qining and Sanjiang geosynclinal region is typical of the Tethyan realm, and forms the Qinling-Sanjiang Province. The Yangtze region had long been occupied by epicontinental seas connected with the Sanjiang and Qinling troughs, thus forming a subprovince with pronounced endemism within the Tethyan realm.

Judging from the relative abundance and distribution of the corals in various stages, the Early and Middle Silurian belt of profuse coral growth from the Yangtze to southern Tianshan via Qinling has a northwest-southeast direction and may represent regions of the northern low latitudes of the time. In the Late Silurian, the rich coral belt seemed to have shifted northward to the Inner Mongolia-Jiliao and Beishan regions, which might have extended to the Urals and northern Europe. The remarkable Qilian-Tianshan belt of arid deposition might represent the arid climatic zone north of the Equator. A problem needing further research is whether the coral faunas in the Sanjiang-Burmese-Malaya region and in Australia represent the southern low-latitude belt of coral growth. As the Silurian faunas in the Himalayan region and in Nepal are comparatively meager, they may possibly represent the southern high-latitude region on the northern border of the Gondwana supercontinent. The present position is the result of the tremendous northward shift of the Indian massif since the Mesozoic. The apparent similarity between some of the faunas from the Himalaya and from Hingan-Mongolia may be only a reflection of similar environmental conditions of the two opposite high-latitude regions.

ACKNOWLEDGMENTS

We are especially thankful to Mrs. Ho Yuan-xiang, Cao Xuan-duo, Guo Sheng-zhe, and Cai Tu-si for reference to some of their results of research in advance of their publication.

REFERENCES CITED

Ge Zhi-zhou and Yu Chang-min, 1974. [Silurian corals, *in* Handbook of stratigraphy and palaeontology of Southwest China"], Beijing, Science Press, p. 165–173, Pls. 72–79 (in Chinese).

Guo Sheng-je, 1976, [Rugosa, *in* Palaeontological atlas of North China, division for Inner Mongolia]: Beijing, Geological Publishing House, p. 63–100, Pls. 24–29 (in Chinese).

Ho Xin-yi, 1978, [Tetracoral fauna of the Late Ordovician Guanyinqiao Formation, Bijie, Guizhou Province]: Beijing, Geological Publishing House, Professional Papers of Stratigraphy and Palaeontology, no. 6, p. 1–45, Pls. 1–13 (in Chinese).

Jia Hui-zhen and Wu Jin-zhu, 1977, [Anthozoa, *in* Palaeontological atlas of Central and South China]: Beijing, Geological Publishing House, p. 9–24, Pls. 4–9 (in Chinese).

Kaljo, D., and Klaamann, E., 1973, Ordovician and Silurian corals, *in* Hallam, A., ed., Atlas of palaeobiogeography: Elsevier, Amsterdam, p. 37–46.

Lavrusevich, A. I., 1967, Late Silurian rugose corals from Central Tadzhikistan: Paleontological Journal, no. 3, p. 18–24.

Li Yao-xi, Cao Xuan-duo, Song Li-sheng, Zhou Zhi-qiang, and Yang Jing-yao, 1975, [Silurian corals, *in* The lower Palaeozoic of western Dabashan]: Beijing, Geological Publishing Hoiuse, p. 179–221, Pls. 34–70 (in Chinese).

Lin Bao-yu, 1979, The Silurian of China: Acta Geologica Sinica, vol. 53, no. 3, p. 173–189 (in Chinese with English abstract).

Oliver, W. A., Jr., 1977, Biogeography of late Late Silurian and Devonian rugose corals: Palaeogeography, Palaeoclimatology, Palaeoecology, v. 22, p. 85–135.

Wang Hong-chen [Wang Hong-zhen], 1978, [On the subdivision of the stratigraphical provinces of China]: Acta Stratigraphica Sinica, v. 2, no. 2, p. 81–104 (in Chinese).

Yi Nung, 1974, A preliminary study of the stratigraphical distribution and zoogeographical provinces of the Ordovician corals of China: Acta Geologica Sinica, 1974, no. 1, p. 1–22 (in Chinese with English abstract).

Yin Tsan-hsun, 1966, China in the Silurian Period: Journal of the Geological Society of Australia, v. 13, pt. 1, p. 277–297.

Manuscript Accepted by the Society April 22, 1981

Printed in U.S.A.

Geological Society of America
Special Paper 187
1981

A preliminary study of the rugose corals from the Wase Group of Dali, western Yunnan

Liao Wei-hua
Nanjing Institute of Geology and Palaeontology, Academia Sinica, Nanjing 210008, People's Republic of China

ABSTRACT

The subdivision and correlation of the Wase Group of Dali, western Yunnan, are reviewed. The following rugose corals are described and figured from the lower and middle parts of the Wase Group: *Cystiphyllum* sp., *Hedstroemophyllum* sp., *Holmophyllum* sp., *Entelophyllum daliense* sp. nov., *Tabularia erhaiensis* sp. nov., *Stereophyllum floriforme* Soshkina and *Chalcidophyllum zashipengense* sp. nov. Coral faunas indicate a Late Silurian age for the lower part and an Early Devonian age for the middle part of the Wase Group.

INTRODUCTION

The Wase (Waseh) Group of Dali (Tali) was first proposed by Sun Yun-zhu [Y. C. Sun] in 1945. Later, Yang Qi [Yang Chi'i], in the winter of 1945–1946, visited the type locality in the area east of the Erhai (Tali lake), western Yunnan, and brought back a large collection of fossils. In 1948, Yang published his well-known paper "The Silurian Waseh Formation of Western Yunnan and Its *Favosites* Faunas." The succession given by him is as follows (in descending order):

- S_6 Chert and limestone, unfossiliferous
- S_5 Chert with subordinate shale; tentaculites in the intercaleted greenish shales
- S_4 Gray, limy shale with subordinate argillaceous limestone, containing brachiopods, bivalvia, tentaculites, gastropods, trilobites
- S_3 Black limestone with subordinate shale, yielding brachiopods, tentaculites, gastropods, and cephalopods
- S_2 Dark gray, crystalline limestone, highly fossiliferous, yielding Tabulata, Rugosa, stromatoporoids, brachiopods, and gastropods
- S_1 Gray, massive dolomitic limestone, unfossiliferous

Yang claimed that the *Favosites*-bearing crystalline limestone (S_2) is of Middle Silurian age. The faunas in the upper units of the Wase Group (S_3 to S_5) indicate Late Silurian age and can be correlated with the Zebingyi beds in the Northern Shan States, Burma, while unit S_6 (chert and limestone) of the Wase Group might belong to a younger age.

Concerning the age of the Zebingyi Formation in the Northern Shan States, Burma, formerly assigned to the Late Silurian, more recently, K. J. Müller (1967) considered it to be Silurian, but ranging upward into the Lower Devonian. Most probably, the formation is a transitional sequence straddling the Silurian-Devonian boundary.

Since 1958, many important biostratigraphical data have been accumulated by several geological surveys, and knowledge regarding the Wase Group has been considerably increased.

In 1973, Xiao Ying-wen, a member of the first regional geological surveying team of Yunnan, studied the Wase Group and divided it into five formations, the succession of which is as follows:

Upper Devonian
- (5) Shaziqing Formation, 120 to 130 m
- (4) Changyucun Formation, 610 m

Middle Devonian
- (3) Lianhuaqu Formation, 680 m
- (2) Qingshan Formation, 970 m
- (1) Kanglang Formation, 1,100 m

Of these formations, the Kanglang, Qingshan, Changyucun, and Shaziqing correspond, respectively, to Yang's units S_1, S_2, S_5, and S_6, while the Lianhuaqu Formation may be essentially an equivalent to Yang's S_3 and S_4.

In the summer of 1972, I had an opportunity to make a field trip to the classical area of the Wase Group and brought back a rich collection of fossil corals which contained seven species assigned to as many genera. Among them, three species are considered as new and three as indeterminable. The specimens were obtained from the following horizons: (1) top of the Kanglang Formation,

(2) base of the Qingshan Formation, (3) lower portion of the Qingshan Formation, and (4) lower part of Lianhuaqu Formation. In addition to the four horizons mentioned above, in Zashipeng, a village about 10 km east of Wase, a fossiliferous stratum which includes *Chalcidophyllum zashipengense* Liao, sp. nov., has been found.

The following is a general account of the essential result of my preliminary study which, it is hoped will be followed by more detailed work.

TENTATIVE CLASSIFICATION AND CORRELATION OF WASE GROUP

As mentioned above, Xiao divided the Wase Group into five formations. There exists between the Wase Group and the Yulongfeng granite a quartzitic sandstone bed about 13 m thick intercalated with conglomerate beds. This bed contains some brachiopods: for example, *Metorthis, Orthis, ?Aporthophyla,* and *?Toquimia.* Judging from the fossil aspect, I think this quartzitic sandstone bed may belong to the Lower Ordovician.

The Kanglang Formation is composed mainly of dolomitic limestone, attaining 1,100 m in thickness, containing only a few "cystimorph" corals at the top, such as *Cystiphyllum* sp. and *Hedstroemophyllum* sp. These two genera have been found in the Upper Silurian Guandi [Kuanti] Formation of Qujing [Chutsing], East Yunnan. The first genus is also known from the Silurian of New South Wales, Estonia, and the Urals; the second genus is recorded from the Upper Silurian of the Donets River and the Island of Gotland. Thus, at least the upper horizon of this formation may be referable to the Upper Silurian.

The Qingshan Formation consists chiefly of massive, crystalline limestone, measuring 970 m in thickness. Its basal part contains *Entelophyllum daliense* Liao (sp. nov.), together with *Holmophyllum* sp. The lower part is characterized by *Tabularia erhaiensis* Liao (sp. nov.).

Entelophyllum daliense Liao (sp. nov.) shows an affinity with *E. articulatum* (Washlenberg) and *E. medium* Ivanovsky from the Silurian of the Siberian platform; *Holmophyllum* sp. is known from the Silurian of Quebec and Maine; *Tabularia erhaiensis* Liao (sp. nov.) bears similarity to *Tabularia turiensis* Soshkina and *Zelophyllum ludlovense* Zheltonogova from the Upper Silurian of the Urals and the Altai Mountains. Therefore, the lower part of Qingshan Formation may be referrable to the Upper Silurian.

The Lianguaqu Formation, 680 m thick, is composed of shale, argillaceous limestone, and limestone. Its lower part has yielded *Stereophyllum floriforme* Soshkina which is known to occur in the Lower Devonian of the Urals. For this reason, this present formation may represent the Lower Devonian.

Directly resting on the Lianhuaqu Formation is the Changyucun Formation with a thickness of 610 m. It consists chiefly of siliceous shale, which contains only tentaculites (*Styliolina* sp.) which are restricted to the lower part. It lies comfortably on the lower Devonian Lianhuaqu Formation and is overlain by the Shaziqing Formation, which is characterized by early Carboniferous conodonts, *Polygnathus bischoffi.* Taking into consideration the lithological character and the stratigraphical position, I regard the Lianhuaqu Formation as Middle and Upper Devonian.

BIOGEOGRAPHY AND FAUNAL PROVINCES OF LATE SILURIAN AND EARLY DEVONIAN RUGOSE CORALS IN SOUTH CHINA

As far as we know, the Early Devonian and Late Silurian rocks of South China may be divided into two different provinces; namely, the South China province and the Qinghai-Tibet province which are divided by the Longmenshan and Ailaoshan Mountains.

In the former province, the Devonian deposits are of platform type, yielding rich faunas belonging to diverse phyla, whereas Late Silurian rocks are distributed only in some local areas. Generally in South China, the Upper Silurian is overlain by Lower Devonian strata with remarkable disconformity, although conformable relationships between Silurian and Devonian deposits are known in two places, the Qinzhou District of southern Guangxi [Kwangsi] and the Qujing District of eastern Yunnan. In the rest of the South China province, the Devonian unconformably overlies the pre-Devonian rocks, with a prominent clastic series at its base.

The Qinghai-Tibet province covers a vast extent of area lying to the south of the Kunlun Mountains and to the west of Longmenshan Mountain, Ailaoshan Mountains. Most of the area belongs to the Qinghai-Tibet Plateau with an average height of over 3,000 m, some peaks even exceeding 6,000 to 8,000 m above sea level. The marine Lower Devonian strata of the Qinghai-Tibet province which immediately overlie the Upper Silurian are essentially of miogeosynclinal type characterized by the occurrence of planktonic and nectonic faunas, principally graptolites, tentaculites, ammonoids, and others, which are associated with the benthic faunas, mostly corals and brachiopods.

In the Qinghai-Tibet province, the best biostratigraphical section of Lower Devonian rocks was seen at Alengchu, northeast of Lijiang. Recently, Professor Yu Chang-min and I studied the rugose corals, recognizing five assemblages in ascending order as follows: (1) *Tryplasma* cf. *T. subcruciatum* (Zheltonogova) Assemblage; (2) *Spongophyllum pseudofritchi* (Soshkina) Assemblage; (3) *Pseuodochonophyllum lijiangense* Yu and Liao Assemblage; (4) *Enterolasma strictum* Simpson Assemblage; and (5) *Siphonophrentis alengchuensis* Yu and Liao Assemblage.

These coral assemblages belong to the Lochkovian, Pragian, and Zlichovian Stages.

It should be pointed out that during the period from Late Silurian to Early Devonian, most coral genera in the Qinghai-Tibet province are noticeably cosmospolitan. They obviously bear a very close relationship to those of New South Wales, the Altai Mountains, Salair, Siberia, the Urals, Estonia, the Donets River, Bohemia, the Island of Gotland, Alaska, and the Yukon, as well as New York State.

DESCRIPTION OF SPECIES

Family COLUMNARIIDAE Nicholson, 1879
Genus *Tabularia* Soshkina, 1937
***Tabularia erhaiensis* Liao, sp. nov.**
Plate 1(6–8); Plate 2(6)

Corralum simple, cylindrical, 4 to 7 mm in diameter; epitheca about 0.4 mm thick; septa acanthine, with blunt, thickened spines;

tabulae complete, flat over a broad axial area, 1.5 to 2 mm apart.

Remarks: This new form differs from *Tabularia turiensis* Soshkina, 1937, in its smaller diameter. It also differs from *Tryplasma* sp. (Oliver and others, 1976, Pl. 9, figs. 5–6) in its less numerous tabulae.

Occurrence: Lower part of the Qingshan Formation.

Cat. Nos. 62102, 62103 (holotype); 62104–62106, 62112 (paratype).

Stereophyllum floriforme Soshkina, 1937
Plate 1(9–11); Plate 2(1–3)

1937 *Stereophyllum floriformis* Soshkina, p. 22–23, Pl. 1, figs. 3–4
1937 *Stereophyllum massivum* Soshkina, p. 20–21
1949 *Columnaria floriformis* Soshkina, p. 105–107, Pl. 3

Corallum massive, composed of cerioid or angulo-circular corallites, ranging in diameter from 4 to 5 mm; septa of two orders, numbering 30 to 50 in each, major septa acanthine, appearing as peripheral ridges, extending only about one-fourth to two-fifths the distance to axis; minor septa not extending beyond stereozone. All septa greatly thickened and in lateral contact peripherally.

Tabulae typically horizontal, mostly complete, a few incomplete, spaced an average distance of 1 mm apart; no dissepiments.

Remarks: Our specimens are similar to the holotype of this species but have more septa and shorter major septa.

Occurrence: Lower part of Lianhuaqu Formation.

Cat. Nos. 62107–62117.

Family _ENTELOPHYLLIDAE_ Hill, 1940
Genus _Entelophyllum_ Wedekind, 1927
Entelophyllum daliense Liao, sp. nov.
Plate 2(4–5); Plate 3(1–4); Plate 4(1–4)

Corallum phaceloid, composed of cylindrical corallites averaging 4 to 7 mm in diameter, corallites separated by distances ranging from 1 to 5 mm, but may connected laterally; epitheca thin; transverse section shows 40 thin septa, radially arranged and relatively straight. Major septa extend to or nearly to the axis, slightly twisted in their axial parts; minor septa are only one-half to one-third as long as the major septa.

Dissepimentarium composed of 3 or 4 rows of small globose dissepiments; tabulae incomplete, vesicular, boundary between tabularium and dissepimentarium indistinct.

Remarks: This new species differs from both *E. articulatum* (Wahlenberg) and *E. medium* Ivanovskiy by its cystose tabulae.

It differs also from *Disphyllum siluriense* Wang 1947, from the Upper Silurian of Qujing (Kutsing), eastern Yunnan, in its smaller corallites and thinner septa.

Occurrence: Base of the Qingshan Formation.

Cat. Nos. 62123, 62124 (holotype); 62118–62121, 62125–62135 (paratype).

Family CYTHOPHYLLIDAE Dana, 1846
Genus _Chalcidophyllum_ Pedder, 1965
Chalcidophyllum zashipengense Liao, sp. nov.
Plate 1(3–4)

Simple or weak compound coralla, maximum diameter of cross section 22 mm; major septa rather long, but not extending to axis; minor septa short and variable, generally less than one-half as long as major ones, occasionally appearing as peripheral ridges. In transverse section, dissepiments display a herringbone pattern.

In longitudinal section, the dissepimentarium composed of 5 to 7 rows of small globose dissepiments, inclined inward; tabulae incomplete, strongly depressed axially.

Remarks: This new form is similar to the Australian species *C. discorde giandarrense* Pedder (1971), but has a smaller diameter and more septa. *C. campanense* Pedder (1965) differs from the present species in its weakly compound corallum and larger diameter.

Occurrence: Lower Devonian, Zashipeng.

Cat. Nos. 62094–62096 (holotype); 62097–62099 (paratype).

Family CYSTIPHYLLIDAE Edwards and Haime, 1850
Genus _Cystiphyllum_ Lonsdale, 1839
Cystiphyllum sp.
Plate 1(1a–1b)

Corallum simple, cylindrical, about 15 to 20 mm in diameter; calyx filled with irregular vesicles; septal crests present.

Remarks: This species seems close to *Cystiphyllum* cf. *C. bohemicum* described by Hill (1940, p. 397, Pl. 2, fig. 13) from New South Wales, Australia, but it has a smaller diameter.

The specimen figured as *Cystiphyllum* ex gr. *siluriense* by Ivanovskiy (1965, p. 91, Pl. 34, fig. 1) has more septal crests.

Occurrence: Top of Kanglang Formation.

Cat. Nos. 62090, 62091.

Genus _Hedstroemophyllum_ Wedekind, 1927
Hedstroemophyllum sp.
Plate 1(2a–2b)

Corallum simple, about 9 mm in diameter, having numerous discontinuous longer septal spines, extending almost to the axis, numbering about 26 in each order. In longitudinal section, tabulae incomplete and irregularly distributed; disspeimentarium composed of 3 to 4 rows of small globose dissepiments steeply inclined and pierced by acanthine septa directed upward and toward the axis.

Remarks: The present specimens differ from *H. weissermeli* (Wedekind, 1927, p. 66, Pl. 2, figs. 3–4) in its much smaller diameter.

H. stollevi sinense Wang can be distinguished from our specimens by its larger diameter and greater number of septal spines.

Occurrence: Top of the Kanglang Formation.

Cat. Nos. 62902, 62903.

Genus _Holmophyllum_ Wedekind, 1927
Holmophyllum sp.
Plate 1(5a–5b)

1962 *Holmophyllum* sp. 1, Oliver, p. 15, Pl. 7, figs. 8–9

Corallum simple, cylindrical, about 10 mm in diameter; major septa number about 35, extending one-half of distance to axis; minor septa variable in length.

In longitudinal section, the tabularium occupies about one-half the diameter; tabulae widely spaced and complete.

Remarks: The present specimen is close to, and possibly conspecific with, *Holmophyllum* sp. 1 (Oliver, 1962, p. 15, Pl. 7, figs. 8, 9), but the Quebec specimens are smaller than the Chinese ones, and a larger collection will be needed to determine the exact relationships.

Occurrence: Base of the Qingshan Formation.
Cat. Nos. 62100, 62101.

ACKNOWLEDGMENTS

All specimens are from the Wase Group of Dali, western Yunnan, and are deposited in the Nanjing Institute of Geology and Palaeontology, Academia Sinica.

REFERENCES CITED

Hill, D., 1940, The Silurian Rugosa of the Yass-Bowning District, New South Wales: Proceedings of the Linnean Society of New South Wales, v. 65, pts. 3 and 4, p. 388–420, Pls. XI–XIII.

Ivanovskiy, A. B., 1963, Rugozy i Silura Sibirskoy Platformy: Akademiya Nauk SSSR, Sibirskoe otdel, Institut Geologii i Geofiziki, Nauka, Moscow, p. 1–152, Pls. 1–33.

—— 1965, Drevneyshie Rugozy. Akademiya Nauk, SSSR, Sibirskoe otdel, Institut Geologii i Geofiziki, Nauka, Moscow, p. 1–144, Pls. 1–39.

Merriam, C. W., 1973, Silurian rugose corals of the central and southwest Great Basin: U.S. Geological Survey Professional Paper 777, p. 1–66, Pls. 1–16.

Mu An Tze [Mu En-zhi], 1962, [Silurian system of China]: Beijing, Science Press (in Chinese).

Müller, K. J., 1967, Devonian of Malaya and Burma, *in* Oswald, D. H., ed. International Symposium on the Devonian System, Calgary, 1967: Calgary, Alberta, Canada, Alberta Society of Petroleum Geologists, p. 565–568.

Oliver, W. A., Jr., 1962, Silurian rugose corals from the Lake Têmiscouata area, Quebec: U.S. Geological Survey Professional Paper 430, p. 11–19.

Oliver, W. A., Jr., Merriam, C. W., and Churkin, M., Jr., 1976, Ordovician, Silurian and Devonian corals of Alaska: U.S. Geological Survey Professional Paper 823–B, p. 1–44, Pls. 1–25.

Pedder, A.E.H., 1965, A revision of the Australian Devonian corals previously referred to *Mictophyllum*: Proceedings of the Royal Society of Victoria, v. 78, pt. 2, p. 201–220, Pls. XXX–XXXIV.

—— 1971, Lower Devonian corals and bryozoa from the Lick Hole Formation of New South Wales: Palaeontology, v. 14, pt. 3, p. 371–386, Pls. 67–68.

Soshkina, E. D., 1937, Korally verkhnego Silura i nizhnego Devona vostochnogo i zapadnogo sklonov Urala: Trudy Paleozoologicheskogo Instituta, v. 6, pt. 4, p. 1–112, Pl. 1–21.Moskva, Leningrad. Izdatel'stvo Akademii Nauk SSSR.

—— 1949, Devonskiye korally Rugosa Urala: Trudy Paleontologicheskogo Instituta Tom XV, vypusk 4, p. 1–160, Pls. 1–58. Moskva, Leningrad, Izdatel'stvo Akademii Nauk SSSR.

Stumm, E. C., and Oliver, W. A., Jr., 1962, Silurian corals from Maine and Quebec: U.S. Geological Survey Professional Paper 430, p. 1–9, Pls. 1–14.

Sun Yun Chu [Sun Yun-zhu], 1945, The Sino-Burmese geosyncline of early Palaeozoic time with special reference to its extent and character: Bulletin of the Geological Society of China, v. 25, p. 1–7.

—— 1948, Problems of the Palaeozoic stratigraphy of Yunnan: 50th Anniversary Papers of the National Peking University, Geological Series, p. 1–28.

Wang Hong Chen [Wang Hong-zhen], 1944, The Silurian rugose corals of northern and eastern Yunnan: Bulletin of the Geological Society of China, v. 24, nos. 1–2, p. 21–32, Pl. 1.

—— 1947, New material of Silurian rugose corals from Yunnan: Bulletin of the Geological Society of China, v. 27, p. 171–192, Pls. 1–2.

Wedekind, R., 1927, Die *Zoantharia Rugosa* von Gotland (bes. Nordgotland): Sveriges Geologie Undersökning, ser. Ca, v. 19, p. 1–95, Pls. 1–30.

Wu Wang-shih [Wu Wang-shi], 1958, Some Silurian corals from the vicinity of Beiyin Obo, Inner Mongolia: Acta Palaeontologica Sinica, v. 6, no. 1, p. 66–82, Pl. 1 (in Chinese and English).

Yang Ch'i [Yang Qi], 1948, The Silurian Waseh Formation of western Yunnan and its *Favosites*-faunas: 50th Anniversary Papers of the National Peking University, Geological Series, p. 129–140, Pl. 1.

Yin Tsan Hsun [Yin Zan-xun], 1949, Tentative classification and correlation of Silurian rocks of south China: Bulletin of the Geological Society of China, v. 29, p. 1–62.

Yu Chang-min [Yü Chang-ming], 1956, Some Silurian corals from the Chiuchüan basin, western Kansu: Acta Palaeontologica Sinica, v. 4, no. 4, p. 610–636, Pls. 1–2 (in Chinese and English).

Zheltonogova, W. A., and Zinchenko, W. G., 1967, Corals and brachiopods of the Siluro-Devonian boundary beds of western Siberia, *in* Oswald, D. H., ed., International Symposium on the Devonian System, Calgary, 1967: Calgary, Alberta, Canada, Alberta Society of Petroleum Geologists, v. 2, p. 885–892.

MANUSCRIPT ACCEPTED BY THE SOCIETY APRIL 22, 1981

PLATE 1

All figures are unretouched photographs.

Figure

1a–1b. ***Cystiphyllum* sp.**
Occurrence: Top of Kanglang Formation. Coll. No. K-1-1-6.
1a. **Transverse section ×2. Cat. No. 62090.**
1b. **Longitudinal section ×2. Cat. No. 62091.**

2a–2b. ***Hedstroemophyllum* sp.**
Occurrence: Same as preceding species. Coll. No. DL-7030-2.
2a. **Transverse section ×3. Cat. No. 62092.**
2b. **Longitudinal section ×3. Cat. No. 62093.**

3–4. ***Chalcidophyllum zashipengense* Liao, sp. nov.**
Occurrence: Lower Devonian.
3a.–3c. **Coll. No. Z-2-5-1 (holotype).**
3a. **Transverse section ×2. Cat. No. 62094.**
3b. **Transverse section ×. Cat. No. 62095.**
3c. **Longitudinal section ×2. Cat. No. 62096.**
4a.–4c. **Coll. No. Z-2-5-2 (paratype).**
4a. **Transverse section ×2. Cat. No. 62097.**
4b. **Transverse section ×2. Cat. No. 62098.**
4c. **Longitudinal section ×2. Cat. No. 62099.**

5a.–b. ***Holmophyllum* sp.**
Occurrence: Base of Qingshan Formation. Coll. No. Q-1-1-3.
5a. **Transverse section ×2. Cat. No. 62100.**
5b. **Longitudinal section ×2. Cat. No. 62101.**

6–8. ***Tabularia erhaiensis* Liao, sp. nov.**
Occurrence: Lower part of Qingshan Formation.
6a.–b. **Coll. No. Q-3-7 (holotype).**
6a. **Transverse section ×3. Cat. No. 62102.**
6b. **Longitudinal section ×3. Cat. No. 62103.**
7a.–b. **Coll. No. Q-3-10.**
7a. **Transverse section ×3, Cat. No. 62104.**
7b. **Longitudinal section ×3. Cat. No. 62105.**
8. **Coll. No. Q-3-6. Longitudinal section ×3. Cat. No. 62106.**

9–11. ***Stereophyllum floriforme* Soshkina 1937**
Occurrence: Lower part of Lianhuaqu Formation.
9a.–9b. **Coll. No. DL-7016-4.**
9a. **Transverse section ×3. Cat. No. 62107.**
9b. **Longitudinal section ×3. Cat. No. 62108.**
10a.–10b. **Coll. No. DL-7016-5.**
10a. **Transverse section ×3. Cat. No. 62109.**
10b. **Longitudinal section ×3. Cat. No. 62110.**
11. **Coll. No. DL-7016-6.**
Longitudinal section ×3. Cat. No. 62111.

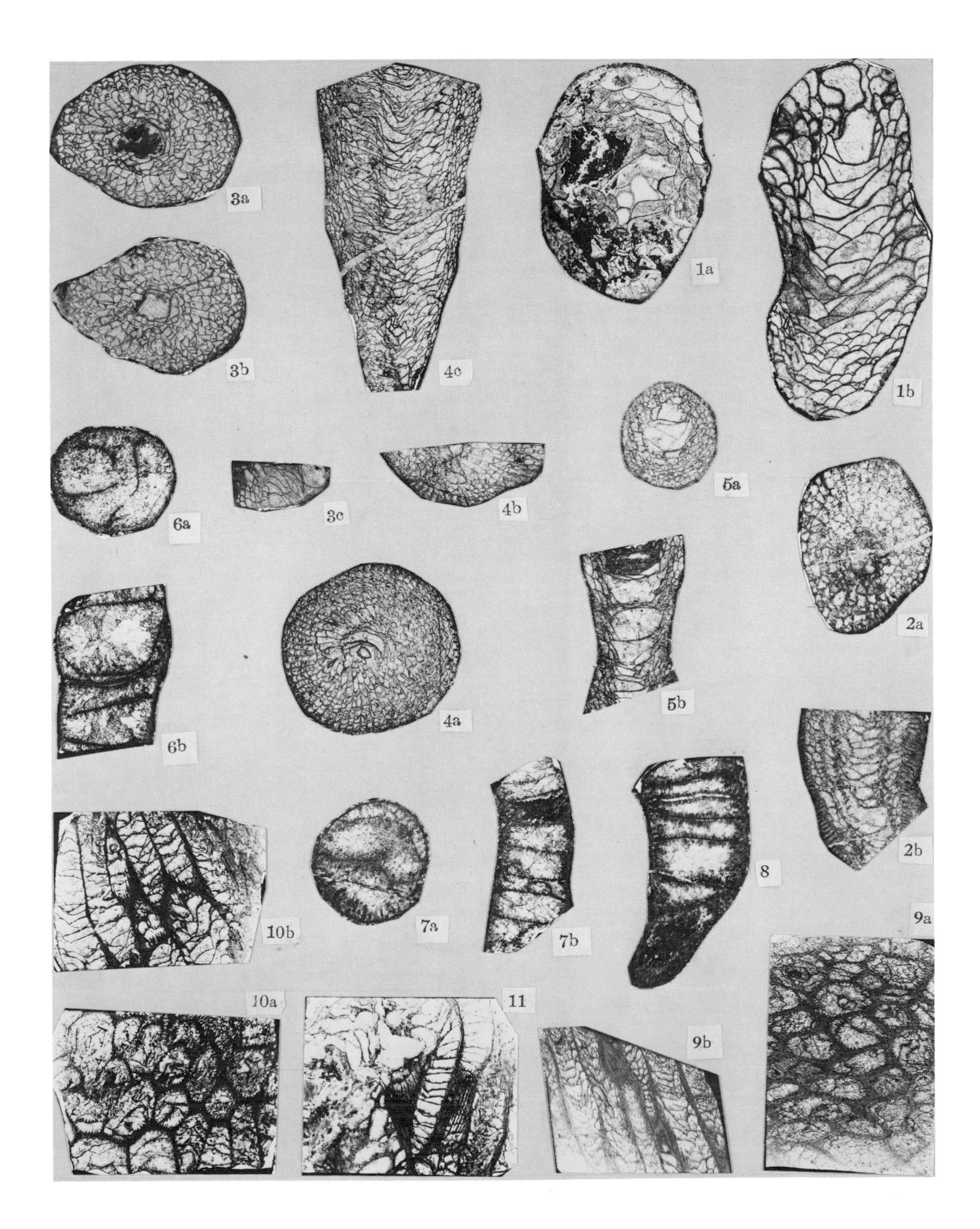
3a
3b
4c
1a
1b
6a
3c
4b
5a
2a
6b
4a
5b
2b
8
10b
7a
7b
9a
10a
11
9b

PLATE 2

All figures are unretouched photographs.

Figure		
1–3.	***Stereophyllum floriforme*** **Soshkina 1937**	
	Occurrence: Lower part of Lianhuaqu Formation.	
	1a.–1b.	**Coll. No. DL-7016-1.**
	1a.	**Transverse section ×3. Cat. No. 62112.**
	1b.	**Longitudinal section ×3. Cat. No. 62113.**
	2a.–2b.	**Coll. No. DL-7016-3.**
	2a.	**Transverse section ×3. Cat. No. 62114.**
	2b.	**Longitudinal section ×3. Cat. No. 62115.**
	3a.-3b.	**Coll. No. DL-7016-3.**
	3a.	**Transverse section ×3. Cat. No. 62116.**
	3b.	**Longitudinal section ×3. Cat. No. 62117.**
4–5.	***Entelophyllum daliense*** **Liao, sp. nov.**	
	Occurrence: Base of Qingshan Formation.	
	4.	**Coll. No. Q-1-1-9 (paratype).**
		Longitudinal section ×3. Cat. No. 62119.
	5.	**Coll. No. Q-1-1-10 (paratype).**
		Longitudinal section ×3. Cat. No. 62118.
6.	***Tabularia erhaiensis*** **Liao, sp. nov.**	
	Occurrence: Lower part of Qingshan Formation.	
	Coll. No. Q-3-13 (paratype).	

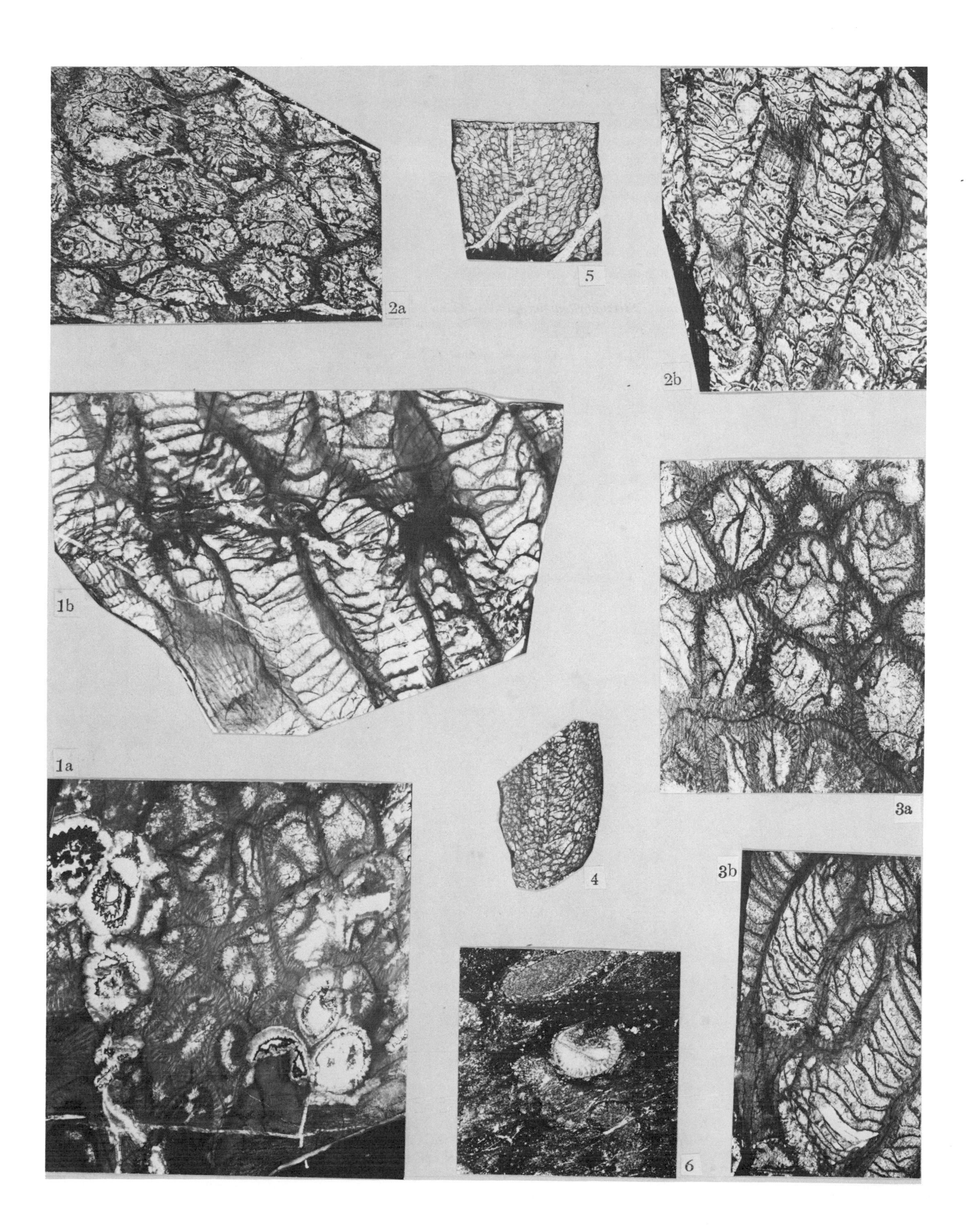
2a
5
2b
1b
1a
3a
4
3b
6

PLATE 3

All figures are unretouched photographs.

Figure

1–4. ***Entelophyllum daliense*** **Liao, sp. nov.**

Occurrence: Base of Qingshan Formation.

1a.-2b.	**Coll. No. Q-1-1-11 (holotype).**
1a.	**Transverse section ×3. Cat. No. 62124.**
1b.	**Longitudinal section ×3. Cat. No. 62124.**
2a.–2b.	**Coll. No. Q-1-1-1 (paratype).**
2a.	**Transverse section ×3. Cat. No. 62125.**
2b.	**Longitudinal section ×3. Cat. No. 62126.**
3.	**Coll. No. Q-1-1-9 (paratype).**
	Transverse and longitudinal sections ×3.
	Cat. No. 62120, 62121.
4a.–4b.	**Coll. No. Q-1-1-4 (paratype)**
4a.	**Longitudinal section ×3. Cat. No. 62127.**
4b.	**Longitudinal section ×2. Cat. No. 62128.**

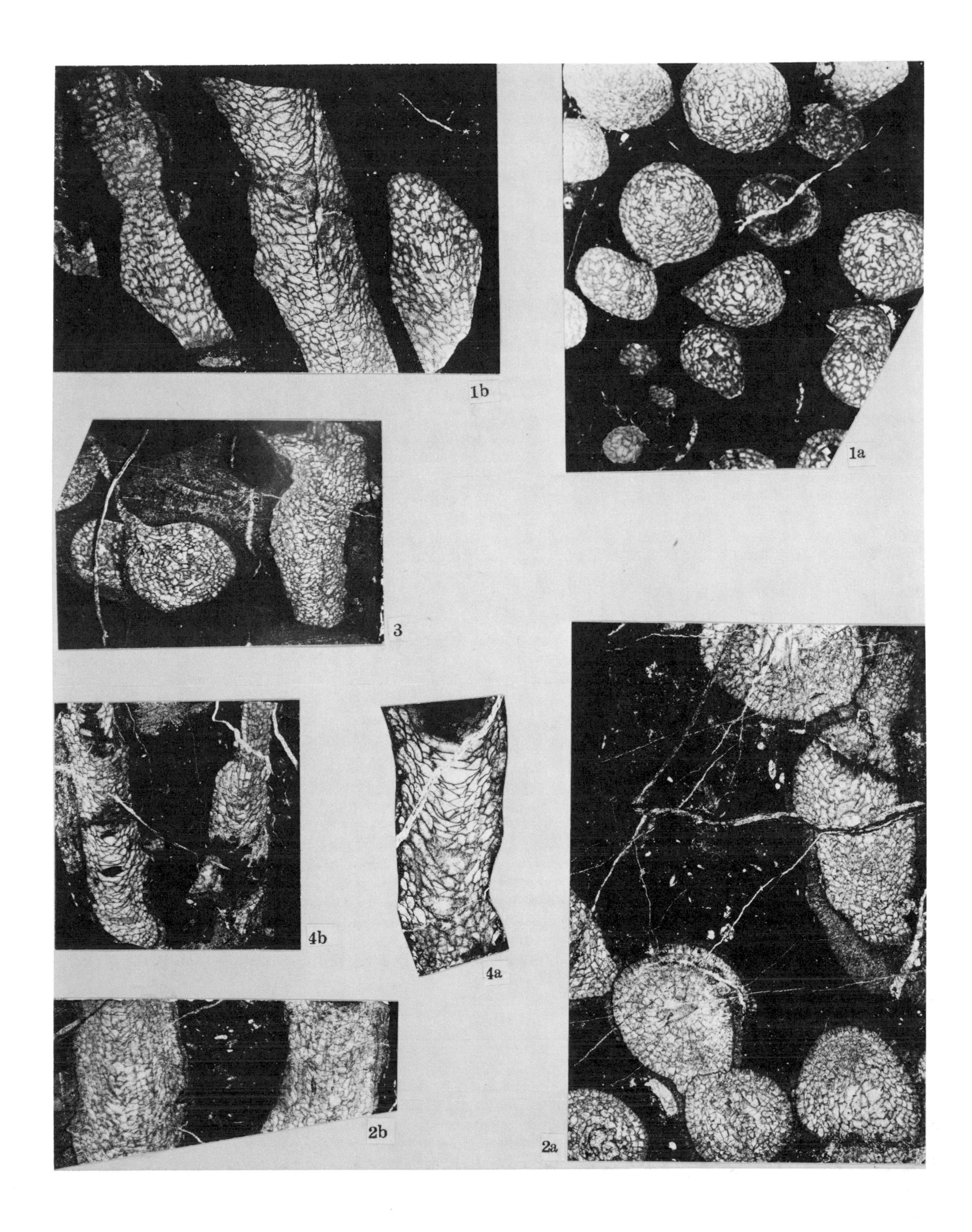
1b
1a
3
4b
4a
2b
2a

PLATE 4

All figures are unretouched photographs.

Figure		
1–4.	***Entelophyllum daliense*** **Liao, sp. nov.**	
	Occurrence: Base of Qingshan Formation.	
	1a.–1b.	**Coll. No. Q-1-1-2 (paratype).**
	1a.	**Transverse section ×3. Cat. No. 62129.**
	1b.	**Longitudinal section ×3. Cat. No. 62130.**
	2a.–2b.	**Coll. No. Q-1-1-4 (paratype).**
	2a.	**Transverse section ×3. Cat. No. 62131.**
	2b.	**Longitudinal section ×3. Cat. No. 62132.**
	3.	**Coll. No. Q-1-1-10 (paratype).**
		Transverse section ×3. Cat. No. 62133.
	4a.–4b.	**Coll. No. Q-1-1-7 (paratype).**
	4a.	**Transverse section ×3. Cat. No. 62134.**
	4b.	**Longitudinal section ×3. Cat. No. 61235.**

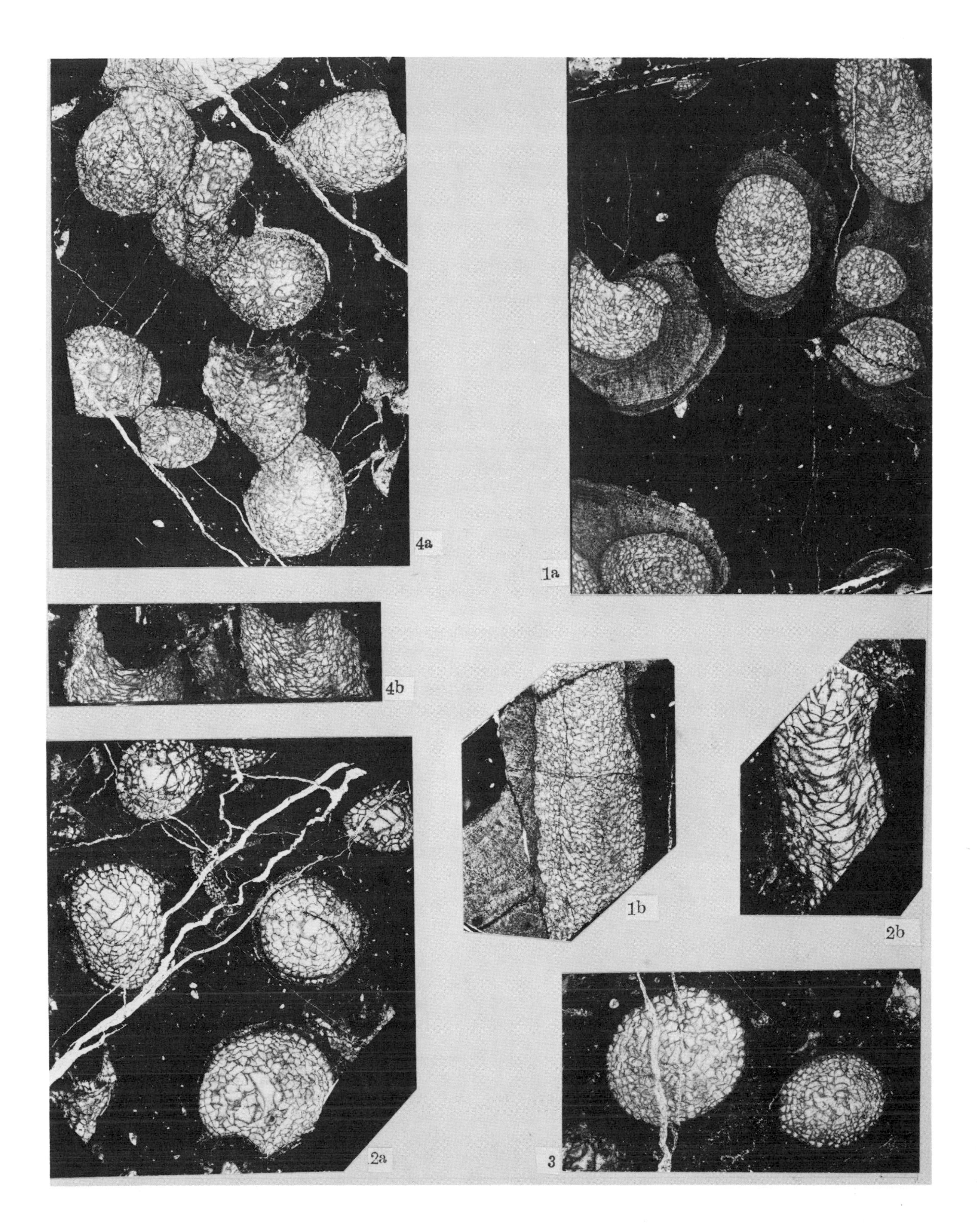

4a
1a
4b
1b
2b
2a
3

Printed in U.S.A.

Geological Society of America
Special Paper 187
1981

Some new Trepostomatous Bryozoa from the Sikuangshan Formation (Famennian) of central Hunan, China

Yang Jing-zhi
Hu Zhao-xun
Nanjing Institute of Geology and Palaeontology, Academic Sinica, Nanjing 210008, People's Republic of China

ABSTRACT

The Sikuangshan Formation of central Hunan in China approximately corresponds to the Famennian of Europe. The fauna described in this account consists of 4 new genera (*Multiphragma, Sinoatactotoechus, Polyspinopora,* and *Armillopora*) and 10 new species that were collected from the lower middle part of the Sikuangshan Formation.

The new bryozoan genera bear their own peculiar characteristics. We suggest that the *Multiphragma multiseptatum–Sinoatactotoechus hunanensis* assemblage is an assemblage representing the lower middle Sikuangshan Formation of Famennian age.

INTRODUCTION

The Upper Devonian Bryozoa of China studied and reported by Yang King Chih during the past three decades include *Atactotoechus hunanensis, A. lui,* and *Monotrypa hsui* from Lingling, Qiyang, and Xikuangshan (Sikuangshan) in Hunan (Yang, 1950); *Anomalotoechus carinatus, Leptotrypa mui,* and *Rhombopora maanshanensis* from Changyang in Hubei (Yang, 1964); and *Shulgina sinensis* in Sinjiang (Yang, 1963).

The late Upper Devonian rocks named the Sikuangshan Formation by previous authors correspond approximately to the Famennian of Europe and are well developed and widely distributed in the central part of Hunan Province which comprises one of the important regions for Upper Devonian biostratigraphy. The fauna consists largely of bryozoans, brachiopods, ostracodes, conodonts, and a few small corals.

This paper deals with only some of the Bryozoa obtained by us during stratigraphic investigation in 1962 and 1964. The whole work concerning the detailed complete study of the bryozoan fauna collected in this formation in this region will be published in a separate paper. This account deals only with 4 new genera and 10 new species, all of which come from the Tutzetang Limestone of the lower middle part of the Sikuangshan Formation.

The new bryozoan genera bear their own peculiar characteristics. In *Multiphragma,* the diaphragms are suddenly increased in the mature region of the zooecial tubes, with as many as 25 or more occupying the distance of a tube diameter. The cystoidal diaphragms which commonly occur in Ordovician and Silurian species of the family Monticuliporidae, are rarely detected in the Devonian forms. But in the Devonian genus *Sinoatactotoechus,* the cystodial diaphragms are also very well developed and crowded on one side of the zooecial tubes. These peculiar structures were used to lever up the zooid so as to keep up with the distal growth of the zooecial walls and to strengthen the support of the zoarium. In other words, it is assumed that they were adapted to turbulent conditions.

Another new genus, *Armillopora,* which we place in the family Trematoporidae, bears a close relationship to the genus *Abokana,* and *Abokana* is closely related to *Neotrematopora.* The phylogenetic relationships of these genera could be represented thus: *Trematopora*→*Neotrematopora*→*Abokana*→*Armillopora.* The evolutionary trend is clearly shown by the mesopores and diaphragms changing from more to less and acanthopores varying from more and large to less and small.

Some of the Famennian seas, especially near the coast, were enclosed or semi-enclosed by barriers and separated into lagoons or basins. The waters between different basins or lagoons could not freely interchange, and the same is true for their biotas. Bryozoa are benthic animals. Their distribution is controlled by depositional environments and the nature of the sea bottoms; therefore, the endemic characters of the Famennian bryozoan fauna were very obvious. Different regions bear their own local faunal elements. Thus, in the Fitzroy basin of Western Australia, the Famennian bryozoan fauna is characterized by *Fitzroyopora, Percyopora,* and *Granivallum* (Ross, 1961); in the central part of Khazakhstan S.S.R. the Famennian bryozoan fauna is represented by *Triplopora* and *Eodyscritella* (Trotskaya, 1970); whereas in central Hunan the Sikuangshan Formation corresponds to the Famennian of Europe as characterized by the new genera *Multiphragma, Sinoatactotoechus, Polyspinopora,* and *Armillopora.* We suggest that the *Multiphragma multiseptatus–Sinoatactotoechus hunanensis* assemblage is an assemblage representing the lower middle Sikuangshan Formation of Famennian age.

SYSTEMATIC DESCRIPTIONS

Order TREPOSTOMATA Ulrich, 1882
Family ATACTOTOECHIDAE Duncan, 1939
Genus *Multriphragma* Yang and Hu (gen. nov.)
Types species: *Multiphragma multiseptatum* Yang and Hu (gen. et sp. nov.)

Definition: Zoarium ramose; in tangential section, zooecia oval, subcircular, or subpolygonal in shape; zooecial walls usually integrated with light, transparent zooecial boundaries; inner parts of zooecial walls wide and armillary. Acanthopores conspicuous, large, granular, usually with fine central tubes, located at zooecial corners. Mesopores often absent. In longitudinal sections, mature region wide; walls irregularly thickened; zooecial boundaries lightly shaded, the laminae appearing continuous from adjacent zooecia. Diaphragms very numerous, spacing close and regular, slightly inclined and parallel to each other. Nonparallel diaphragms relatively small, triangles commonly formed by the intersection of two adjacent diaphragms with the zooecial wall.

Remarks: *Multiphragma* is closely related to the genus *Atactotoechus* Duncan and is placed in the same family with that genus. Both are characterized by the presence of atactotoechid wall structures and spacing of diaphragms. In *Multiphragma* the diaphragms are spaced, more numerous, and arranged parallel. Acanthopores are moderately numerous. *Sinoatactotoechus* Yang and Hu (gen. nov.) differs from *Multiphragma* in having a series of strong, curved cystoidal diaphragms in the mature region.

Occurrence: Middle part of Sikuangshan Formation, Hunan.

***Multiphragma multiseptatum* Yang and Hu (sp. nov.)**
Plate 1(1–8)

Description: Zoaria ordinarily ramose, 4.5 to 5.1 mm in diameter, surface smooth, without monticules. In tangential section, zooecial apertures oval, subcircular or subpolygonal, regularly arranged, long diameter 0.15 to 0.17 mm, short diameter 0.11 to 0.15 mm; with the longer diameters of the large ones ranging from 0.21 to 0.25 mm and the shorter diameters from 0.16 to 0.17 mm, usually 7 to 8, rarely 9 to 9.5, in a space of 2 mm. Zooecial walls integrated, moderately thickened, 0.07 to 0.12 mm in thickness; zooecial boundaries transparent. Inner parts of zooecial walls wide, armillarily laminated, and slightly raised above the apertures. Acanthopores conspicuous, granular, generally hollow in center, moderate in size, 0.03 to 0.07 in diameter, 2 to 4 surrounding each aperture, located at the zooecial corners.

In longitudinal sections, zooecial tubes bend gradually from immature region toward mature region and obliquely to surface of zoarium. Zooecial walls thin, granular, slightly crenulated in immature region and irregularly thickened in mature region; walls or adjacent zooecia sharply divided by zooecial boundaries. Laminae form a V-shaped pattern using the boundary line as the apex and trend nearly parallel to the boundary. In immature region, 4 to 7 diaphragms occur in each tube, about 1.5 to 2.5 tube diameters apart. Tyically, in mature region, diaphragms arranged closely and regularly, horizontal, slightly inclined or curved, and parallel to each other, ordinarily 25 to 36 in a tube, with 4 to 5 in a distance of one tube diameter. Small, triangular spaces commonly formed by the intersection of two adjacent nonparallel diaphragms. Mature region wider than the immature region.

Remarks: This new species is distinguished from all species of the genus *Atactotoechus* by the presence of closely, regularly arranged and more numerous diaphragms in the mature region, and well-defined, numerous, and larger acanthopores.

Occurrence: Middle part of Sikuangshan Formation of Shaodong, Hunan.

Cat. No. 62190 (holotype).

***Sinoatactotoechus* Yang and Hu (gen. nov.)**
Type species: *Sinoatactotoechus hunanensis* Yang and Hu (sp. nov.)

Definition: Zoarium ramose; immature region and nondiagnostic, displaying thin granular walls that range from straight to slightly crenulated. Zooecial walls regularly thickened, displaying an atactotoechid wall structure in the mature region. As seen in longitudinal section, the traces of many of the wall laminae approaching the zooecial boundaries generally remain straight and intersect the boundaries at angles of less than 90°. Diaphragms few, thin, and straight in immature region. Markedly early and middle mature region characterized by closely and regularly spaced cystoidal diaphragms that are thin, few flat, and slightly curved. In tangential section, zooecia polygonal or subpolygonal in outline. Zooecial walls generally integrated with light, transparent zooecial boundaries. Narrowly amalgamate structure rare. Acanthopores granular, large, and their numbers and distributions vary within some species. Mesopores usually absent.

Remarks: *Sinoatactotoechus* is most closely related to the genus *Homotrypa.* Both genera are characterized by closely spaced and strongly curved cystoidal diaphragms arranged in parallel series within a zooecium. *Homotrypa* possesses numerous and conspicuous mesopores and larger acanthopores. The new genus differs from *Atactotoechus* in having closely and regularly spaced cystoidal diaphragms.

Occurrence: Middle part of Sikuangshan Formation, Hunan.

***Sinoatactotoechus hunanensis* Yang and Hu (sp. nov.)**
Plate 1(9); Plate 2(1–7)

Description: Zoaria ramose and overgrown, or a combination of both forms, 3.8 to 5.3 mm in diameter. In tangential sections, zooecia subpolygonal, a few subcircular in outline, long diameter 0.21 to 0.25 mm and short one 0.15 to 0.17 mm; some large zooecia have a longer diameter, 0.37 mm, and a shorter one that is 0.20 mm; they average 8 to 8.5 in a space of 2 mm. Zooecial walls about 0.05 to 0.06 mm in thickness, integrated, and with light transparent zooecial boundaries. Acanthopores having diameters ranging from 0.03 to 0.06 mm, conspicuous, large, granular, average 4 to 5 to a zooecium. Mesopores absent.

In longitudinal section, zooecial tubes bend gradually outward from immature region, oblique, sometimes perpendicular to surface in outer mature region. In immature region, zooecial walls thin, slightly crenulated, and composed of granular tissue that gradually thickens in mature region. Zooecial boundaries narrow and defined by the intersections of laminae, are continuous across the boundary as seen in mature region. Diaphragms few, flat in immature and in early mature region, with only 2 or 3 in each zooecium. Cystoidal diaphragms well developed, strongly curved, more numerous and closely arranged in parallel series in the zooecium. Both diaphragms and cystoidal diaphragms ordinarily 6 to 9 in each zooecium, about 3 in a distance of one tube diameter. The mature region is 0.75 to 1.20 mm wide, and the immature region is 2.30 to 3.20 mm wide.

Remarks: *Sinoatactotoechus hunanensis* may be readily distinguished from all species of *Atactotoechus* by the presence of closely arranged and strongly curved cystoidal diaphragms. This species differs from all species of *Homotrypa* in having much thicker walls and more numerous larger acanthopores.

Occurrence: Middle part of Sikuangshan Formation, Shaodong, Hunan.

Cat. No. 62191 (holotype).

Sinoatactotoechus obliquus Yang and Hu (sp. nov.)
Plate 3(2–11)

Description: Zoaria slender ramose, with smooth surface, without monticules. In tangential section, zooecia polygonal, apertures subcircular or oval in shape, arranged longitudinally in alternating rows; long diameters 0.21 to 0.24 mm and short ones 0.10 to 0.13 mm; diameter of smaller zooecia from 0.16 to 0.19 mm, and 0.07 to 0.10 mm, about 6 to 6.5 in a space of 2 mm, regularly distributed. Zooecial walls 0.07 to 0.12 mm thick, integrate with light, transparent zooecial boundaries. Acanthopores large, moderate in number, located at the angles of junctions, and sometimes on the walls of zooecia; 0.05 to 0.07 in diameter, generally 3 to 4, rarely 5, to a zooecium. Mesopores usually absent.

In longitudinal sections, zooecial tubes in the immature region strongly curved toward the mature region, where they obliquely meet the zoarial surface. In immature region, zooecial walls thin, granular, straight, or slightly crenulated, regularly thickened, and composed of laminated tissue in mature region. Flat diaphragms about 2 to 2.5 tube diameters apart in immature region; in mature region they become closely spaced, 4 to 5 in each zooecium, with 2 in a distance of one tube diameter. Typically, cystoidal diaphragms also observed 1 to 3 in each zooecium in late mature region. Immature region wider than mature region.

Remarks: In comparison with *Minussira grandis* Morozova (1961) from the upper Frasnian of the Soviet Union (Kuznetsk), the new species lacks mesopores and has conspicuously larger acanthopores and closely arranged cystoidal diaphragms. The present species differs from *Sinoatactotoechus hunanensis* in having larger zooecia, thicker walls, and fewer cystoidal diaphragms.

Occurrence: Same as the preceding species.

Cat. No. 62192 (holotype); 62193 (paratype).

Sinoatactotoechus rarispinus Yang and Hu (sp. nov.)
Plate 2(7–10); Plate 3(1)

Description: Zoaria slender ramose, overgrowth common, 4.5 mm in diameter. Surface of the zoarium smooth without monticules. In tangential section, zooecia oval or subpolygonal, long diameter 0.11 to 0.15 mm and short diameter 0.06 to 0.08 mm; in some larger specimens the long diameter is 0.19 to 0.21 mm and the short diameter is 0.08 to 0.12 mm, about 7 to 7.5 in a space of 2 mm, regularly distributed. Walls narrow, integrated, with light zooecial boundaries. Acanthopores few, irregularly distributed, absent in some parts, diameter ranging from 0.05 to 0.06 mm, average 2 to 3 in a zooecium. Mesopores absent, occasionally smaller zooecia observed.

In longitudinal sections, zooecial walls thin, granular in immature region; gradually thickened in mature region and composed of laminated tissue. Diaphragms thin, planar and widely spaced, composed of growth stages of zooecial tubes in immature region. Diaphragms in inner layer of mature region planar to slightly curved, irregularly spaced, 1 to 3 in each tube; in mature region near the zoarial surface, both diaphragms and cystoidal diaphragms more numerous, closely arranged 2 to 3 tube diameters apart. Outside layer (0.58 to 0.64 mm) wider than inner (0.25 mm) of the mature region.

Remarks: *Sinoatactotoechus rarispinus* differs from *S. hunanensis* in having fewer cystoidal diaphragms, smaller acanthopores and zooecia, and from *S. obliquus* in having zoaria with conspicuous overgrowth.

Occurrence: Same as the preceding species.

Cat. No. 62194 (holotype).

Sinoatactotoechus densiseptatus Yang and Hu (sp. nov.)
Plate 4(1–7)

Description: Zoaria ramose, average diameter 4.8 mm, surface smooth. In tangential section, zooecia subpolygonal in outline, long diameter 0.17 to 0.19 mm, short diameter 0.11 to 0.12 mm; in some parts, long diameter is 0.24 to 0.28 mm, and short diameter is 0.16 to 0.19 mm, 6.5 to 7.5 in a distance of 2 mm. Walls integrate, occasionally amalgamate, about 0.06 to 0.09 mm thick. Typically, acanthopores larger, conspicuously more numerous, tissue granular, diameter ranges from 0.06 to 0.07 mm, usually at junction of zooecial walls; slightly inflected zooecia also observed, 3 to 4 around each zooecium. Mesopores or smaller zooecia absent.

In longitudinal section, zooecial tubes bend gradually from axial region to periphery, oblique to surface of zoarium. Zooecial walls thin, slightly curved in axial region; in mature region, walls thickened and fused together in some parts. In axial region, diaphragms widely spaced, 1 to 2 in each tube. Usually, cystoidal diaphragms present in middle or late parts of zooecial tube, imbricate, 4 to 6 occurring in each tube and 3 occupying a distance of a tube diameter. Immature region (2.80 to 3.20 mm) wider than mature region (0.57 to 0.89 mm).

Remarks: Characteristic features of *Sinoatactotoechus densiseptatus* are the usually larger and numerous acanthopores, the more numerous diaphragms in the immature region, and the imbricate arrangement of the cystoidal diaphragms in the middle or late parts of the mature region. This new species differs from *S. obliquus* in having numerous diaphragms in the axial region and a longer mature zone; it differs from *S. rarispinus* in having larger and more numerous acanthopores and different zoarial growth.

Occurrence: Same as the preceding species.

Cat. No. 62159 (holotype).

Polyspinopora Yang and Hu (gen. nov.)
Type species: _Polyspinopora shaodongensis_ Yang and Hu (sp. nov.)

Definition: Zoaria ramose, overgrowth rare, or a combination of the two growth habits. Monticules or clusters formed by groups of zooecia larger than average. Apertures oval or subpolygonal in shape. Zooecial walls generally integrate, but a few of them appear narrowly amalgamate. Acanthopores usually well defined, large, numerous, with granular tissue. Many small granules arranged in the walls of large zooecia, sometimes absent in some species. Mesopores absent. Walls thin, crenulated in immature region, gradually thickening from beginning of mature region; wall structure resembles atactotoechid aspect. Zooecia oblique, slightly bent toward surface in mature region. Diaphragms in immature region rare or absent. In mature region diaphragms numerous, thick, rarely curved; conspicuous, thicker compound diaphragms commonly occur on the distal end of zooecial tubes. Mature region very short.

Remarks: *Polyspinopora* differs from *Sinoatactotoechus* mainly in developing more numerous, large acanthopores, in having narrow mature region, and in having thick compound diaphragms, that developed into cystoidal diaphragms, closely aranged; the new genus differs from *Atactotoechus* in having conspicuous wide mature region, closely spaced diaphragms, and rare or absent acanthopores.

Occurrence: Middle part of Sikuangshan Formation, Hunan.

Polyspinopora shaodongensis Yang and Hu (sp. nov.)
Plate 5(5–10); Plate 6(1–12)

Description: Zoaria slender, ramose, overgrowth rare, or a combination of these two kinds of growth; diameter 4.7 to 5.0 mm. In tangential sections, zooecia commonly oval, rarely subpolygonal, average long diameter 0.13 to 0.16 mm, short diameter 0.08 to 0.10 mm, in some parts the long diameter is 0.21 to 0.25 mm, and short diameter is 0.12 to 0.15 mm, 7.5 to 8 zooecia in 2 mm of longitudinal rows. Zooecial walls generally integrate, less commonly narrowly amalgamate, 0.07 to 0.11 mm wide. Acanthopores conspicuous, large, numerous, with granular tissue, 0.03 to 0.05 mm in diameter, distributed at zooecial corners, 5 to 6 surrounding a zooecium. Small granules commonly occur at junction of the zooecial walls, about 0.01 to 0.02 mm in diameter. Mesopores absent.

In longitudinal section, zooecial walls thin in immature region and thick in mature region, fused together in some parts, with zooecial tubes bending gradually from immature region toward mature region and oblique to surface of zoarium. Zooecial walls thin, slightly crenulated in immature region, gradually thickened from beginning of mature region. Wall structure atactotoechid. Zooecial boundary normally marked by dark line made by the overlapping laminae. In immature region, diaphragms thin, planar, generally 3 to 4 in zooecial tube, approximately 1.5 tube diameters apart. Usually only 1 compound diaphragm near early parts of mature region; inner layer of mature region with 2 to 3 diaphragms, commonly 4 to 5 in each zooecium in outer layer, average 3 in a distance of one tube diameter. Immature egion wider than mature region.

Remarks: The larger and more numerous acanthopores, the small number of compound diaphragms, and the absence of cystoidal diaphragms are characteristics that distinguish *P. shaodongensis* from *Sinoatactotoechus*.

Occurrence: Middle part of Sikuangshan Formation, Shaodong, Hunan.

Cat. No. 62196 (holotype).

Polyspinopora ampliata Yang and Hu (sp. nov.)
Plate 6(3–11)

Description: Zoaria ramose, 4.9 to 5.2 mm in diameter with slightly elevated maculae 2.9 mm apart, measured from center to center. In tangential section, zooecia oval to subpolygonal in outline, regularly arranged 7 to 7.5, rarely 6 in a distance of 2 mm measured longitudinally; average long diameter ranges from 0.15 to 0.17 mm, short diameter from 0.11 to 0.12 mm; in the maculae region, larger zooecia with long diameter ranges from 0.29 to 0.37 mm, and short diameter ranges from 0.25 to 0.30 mm. Zooecial walls generally amalgamate, sometimes integrate, about 0.07 to 0.11 mm wide. Acanthopores present but not prominent, tissue granular, usually 0.04 mm in diameter, absent in some parts. Many small conspicuous and well-developed granules arranged usually near junctions of zooecial walls, with about 20 or more surrounding each zooecial aperture. Mesopores generally absent; a few small zooecial occasionally present.

In longitudinal section, interzooecial space of immature region filled with calcareous material. Zooecial tubes nearly straight, very few oblique to surface in mature region. In mature region, zooecial walls thickened with laminated tissue and atactotoechid pattern. Only 1 or 2 diaphragms in early parts of mature region; in mature region, they are more numerous, planar, thin, average 6 to 8 each zooecial tube, and one tube diameter apart. Mature region 1.25 to 1.75 mm wide; immature region 2.0 to 2.5 mm wide.

Remarks: *Polyspinopora ampliata* may be compared with *P. shaodongensis.* It differs in having a smaller number of acanthopores, a greater number of diaphragms, and conspicuous, closely arranged smaller granular tissue.

Occurrence: Same as the preceding species.

Cat. No. 62197 (holotype); 62198 (paratype)

Polyspinopora regularis Yang and Hu (sp. nov.)
Plate 4(8–11); Plate 5(1–4); Plate 7(8–10)

Description: Zoaria ramose and overgrowth, or a combination of both habits, 5.3 to 5.5 mm in diameter. Slightly elevated maculae 2.7 mm apart, measured from center to center. In tangential section, zooecia oval to subcircular, long diameter 0.12 to 0.16 mm, short diameter 0.08 to 0.10 mm; in the maculae, long dimater ranges from 0.21 to 0.28 mm, short diameter from 0.19 to 0.21 mm, regularly arranged 7.5 to 9, or in some parts 6.5 to 7 in a distance of 2 mm measured longitudinally. Zooecial walls integrate, 0.08 to 0.11 m wide. Inner part of zooecial walls composed of concentric laminae, slightly elevated above apertural surface. Acanthopores well developed, regularly spaced, generally 5 to 6 surrounding each zooecium with dark dense tissue; diameter ranges from 0.03 to 0.05 mm. Small granular tissue and mesopores absent.

In longitudinal sections, zooecial tubes parallel to longitudinal direction of zoarium in immature region, curved gradually outward and forming a large angle with surface. Zooecial walls thickened gradually from immature to mature regions, and boundary between these two regions indistinct. In mature region, zooecial walls thick, with some parts fused. Zooecial walls typically atactotoechid, and zooecial boundaries marked by black overlapping laminae. Usually, 1 to 3 diaphragms disposed in early part of mature region, thin, complete and planar. In both middle and peripheral parts of mature region are thicker and curved compound diaphragms with laminate tissue, average 5 to 8 inserted in a zooecial tube and usually one tube diameter apart. Mature region 1.25 to 1.50 mm wide; immature region 2.5 to 2.7 mm wide.

Remarks: The present species can be easily distinguished from *Leptotryella (Leptotrypella) mesostana* Boardman from the Hamilton Group of the United States (Boardman, 1960, p. 55) by the well-developed and more numerous acanthopores and by the thicker, oblique, compound diaphragm.

Occurrence: Same as *Polyspinopora shaodongensis* Yang and Hu (sp. nov.)

Cat. No. 62199 (holotype); 62200 (paratype).

Family TREMATOPORIDAE Miller, 1889
Genus ARMILLOPORA Yang and Hu (gen. nov.)
Types species: *Armillopora sinensis* Yang and Hu (sp. nov.)

Definition: Zoaria slender ramose; walls thin, granular tissue, straight to slightly crenulated in immature region, gradually thickening from beginning of mature region to surface, sometimes fused. Zooecial walls integrate with light, transparent zooecial boundaries in tangential section. Diaphragms very few, thin and straight, Hemisepta prominent, usually short, strong, in early mature region. Inner part of zooecial walls (or peristomes) very conspicuous, wide, armillary-shaped laminate tissue. Mesopores very rare or absent. Acanthopores well developed, with concentric laminae or smaller tube.

Remarks: The distinctive characteristics of *Armillopora* are the well-developed peristomes, conspicuous hemisepta, and more numerous but smaller acanthopores. This new genus differs from *Abakana* Morozova (1961, p. 114) in having wide peristomes and conspicuous hemisepta and from *Pseudobatostomella* Morozova (1961, p. 102) in lacking mesopores and in having more numerous acanthopores and conspicuous hemisepta. *Armillopora* is most closely related to the genus *Trematopora* and is therefore placed in the same family with that genus.

Occurrence: Middle part of Sikuangshan Formation, Hunan.

***Armillopora sinensis* Yang and Hu (sp. nov.)**
Plate 7(1–3)

Description: Zoarium slender ramose, 6 to 8 mm long, 2.7 to 3.3 in diameter. Maculae not observed. In tangential sections, zooecia oval or subcircular, longer diameter of oval zooecial 0.12 to 0.15 mm and shorter diameter 0.07 to 0.10 mm, of subcircular zooecia 0.10 mm, 8 to 9 in a distance of 2 mm, regularly arranged. Conspicuous, peristomes well developed and surrounded by dark concentric laminae, 0.05 to 0.10 mm wide. Acanthopores more numerous, small, black, composed of granular tissue, usually 2 to 4 surrounding each zooecium, average diameter 0.02 to 0.04 mm. Mesopores very small and few in number, resembling smaller tubes, diameter 0.07 mm. Zooecial walls integrate.

In longitudinal section, zooecial tubes bend gradually outward from immature region, perpendicular to surface in central mature region. Zooecial walls thin, slightly curved, composed of granular tissue, gradually thickening in immature region, fused in mature region. Diaphragms very few, can be observed at early and middle parts of immature region, 2 in a distance of one tube diameter. Usually only one hemiseptum disposed near the mature region, comparatively short, about one-fourth to one-third tube diameter long. Mature region wider than immature region.

Remarks: In comparison with *Abakana macrospina* (Schoenmann) from the Givetan of the Soviet Union (Minusinsk) (Morozova, 1961, p. 114), the mesopores of the new species are less developed or even absent, yet hemisepta are developed.

Occurrence: Middle part of Sikuangshan Formation, Lianyuan, Hunana.

Cat. No. 62201 (holotype).

***Armillopora spinosa* Yang and Hu (sp. nov.)**
Plate 7(4–7)

Description: Zoaria slender ramose, 5 to 6 mm in length, 1.75 to 2.10 mm in diameter. Maculae absent. In tangential section, zooecial oval or polygonal, small, irregularly arranged; longer diameter averages 0.07 to 0.10 mm and shorter diameter averages 0.05 to 0.07 mm; usually 6 to 7, sometimes 9, zooecia in a length of 2 mm. Peristomes conspicuous, composed of concentrically laminated tissue, armillary in shape, generally 0.07 to 0.12 mm wide. Conspicuous acanthopores well defined and more numerous, located near junction of zooecia, sometimes inflect the zooecia, diameter 0.02 to 0.05 mm, 6 to 6 surrounding each zooecium. Zooecial walls of medium thickness, usually amalgamate in pattern.

In longitudinal section, zooecial tubes bend gradually outward from immature region, perpendicular to surface in mature region. Zooecial tubes irregularly arranged. Walls thin, slightly crenulated in immature region, regularly thickened and fused in some parts in mature region. Diaphragms and hemisepta very few, generally 1 to 2 in early parts of mature region. Width of mature region 0.85 mm and of immature region 0.50 mm.

Remarks: This new species differs from *Armillopora sinensis* in having more numerous acanthopores and very few diaphragms and hemisepta.

Occurrence: Same as the preceding species.

Cat. No. 62202 (holotype).

ACKNOWLEDGMENT

All specimens described in this paper are deposited in the Nanjing Institute of Geology and Palaeontology, Academic Sinica, People's Republic of China.

REFERENCES

Boardman, R. S., 1960, Trepostomatous Bryozoa of the Hamilton Group of New York State: U.S. Geological Survey Professional Paper 340, p. 1–83, Pls. 1–22.

Duncan, H., 1939, Trepostomatous Bryozoa from the Traverse Group of Michigan: Contributions from the Museum of Paleontology, University of Michigan, v. 5, no. 10 p. 171–270, Pls. 1–16.

Fritz, M. A., 1944, Upper Devonian Bryozoa from New Mexico: Journal of Paleontology, v. 18, no. 1, p. 35–38, Pls. 1–2.

Morozova, I. P., 1961, Devonskie mshanki Minusinskikh i Juznetskoi Kotlovin: Trudy Paleontologicheskogo Instituta, Akademiya Nauk SSSR, Tome 86, p. 1–207, Pls. 1–34.

Ross, J. P., 1961, Ordovician, Silurian and Devonian Bryozoa of Australia: Bureau of Mineral Resources, Geology and Geophysics, Bulletin 50, p. 1–111, Pls. 1–28.

Troitskaya, T. D., 1968, Devonskie mshanki Kazakhstana: Geologicheskiy Fakultet Gosudarstvennogo, Moskovskogo Universiteta, p. 1–239, Pls. 1–35.

——1970, Nekotorye Famenskie mshanki tsentralnogo Kazakhstana: Novye vidy Paleozoyskih Mschanok i Korallov, Akademiya Nauk SSSR, p. 40–47, Pls. 16–18.

Yang King Chih (Yang Jing-zhi), 1950, Some Bryozoa from Upper Devonian and Lower Carboniferous of Hunan: Palaeontological Society of China, Palaeontological Novitiates, no. 6, p. 1–14, Pls. 1–3 (in Chinese and English).

——1963, Bryozoa, *in* [Index fossils of Northwest China]: Beijing, Science Press, p. 73, Pl. 35 (in Chinese).

——1964, Some bryozoans from the Upper Devonian of Chang-yang, Western Hupeh: Acta Palaeontologica Sinica, v. 12, no. 1, p. 26–31, Pl. 1 (in Chinese with English abstract).

Yang King Chih and Hu Zhao-xun, 1965, Bryozoa of the Tungkangling Formation of Xiangzhou, Kwangsi: Memoirs of Institute of Geology and Palaeontology, Academia Sinica, no. 4, p. 1–50, Pls. 1–12 (in Chinese with English abstract).

MANUSCRIPT ACCEPTED BY THE SOCIETY APRIL 22, 1981

PLATE 1

Figure

1–8. ***Multiphragma mutiseptatum*** **Yang and Hu (gen. et sp. nov.) Cat. No. 62190 (holotype).**

1. **Tangential section showing the shape, size, and arrangement of the zooecia, and the characteristics of acanthopores, ×50.**

2. **Tangential section, ×50.**

3. **Tangential section of another part showing the characteristics of zooecial walls, ×100.**

4. **Longitudinal section showing the shape of the zoarium, ×10.**

5. **A part of figure 4 showing the characteristics of the mature region and distribution of the diaphragms, ×20.**

6. **A part of figure 4, ×20.**

7. **A part of the same longitudinal section showing atactotoechid structure of thick zooecial walls and the arrangement of the diaphragms, ×100.**

8. **Transverse section, ×10.**

9. ***Sinoatactotoechus hunanensis*** **Yang and Hu (gen. et sp. nov.) Cat. No. 62191 (holotype).**
Transverse section, ×10.

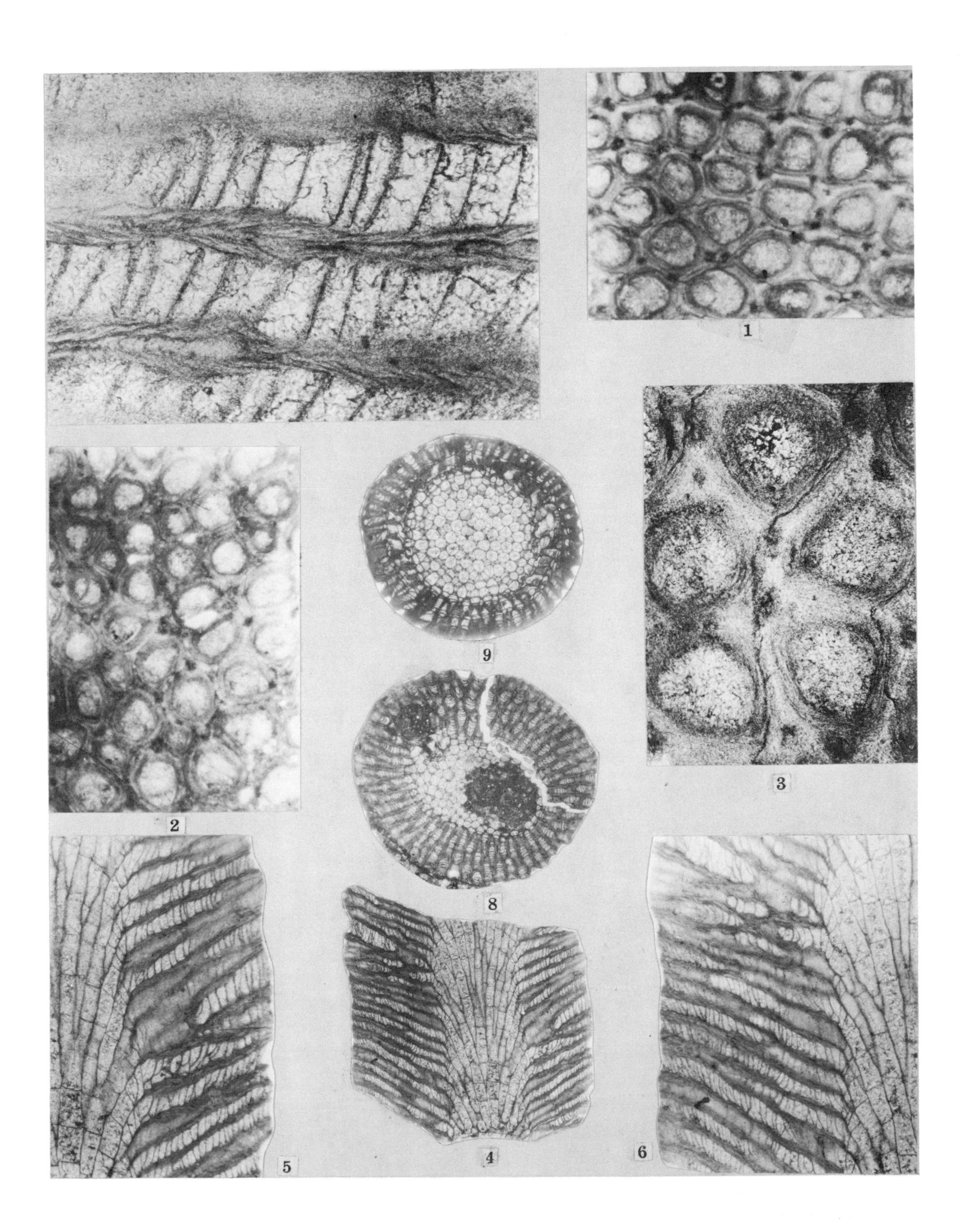
1
9
2
3
8
5
4
6

PLATE 2

Figure	
1–7.	*Sinoatactotoechus hunanensis* Yang and Hu (gen. et sp. nov.) Cat. No. 62191 (holotype).
1.	Tangential section showing the characteristics of the zooecial walls and distribution of the acanthopores, ×40.
2.	A part of tangential section showing light zooecial boundaires of integrate walls, ×100.
3.	Longitudinal section showing branching part of zoarium, ×10.
4.	Longitudinal section showing cystoidal diaphragms in mature region.
5.	A part of longitudinal section, ×20.
6.	A part of longitudinal section displaying many cystoidal diaphragms, ×100.
7–10.	*Sinoatactotoechus rarispinus* Yang and Hu (gen. et sp. nov.) Cat. No. 62194 (holotype).
7.	Transverse section, ×10.
8.	Tangential section showing the shape, size, and arrangement of the zooecia, ×20.
9.	Longitudinal section displaying well-preserved cystoidal diaphragms, ×20.
10.	Longitudinal section of the same specimen, ×20.

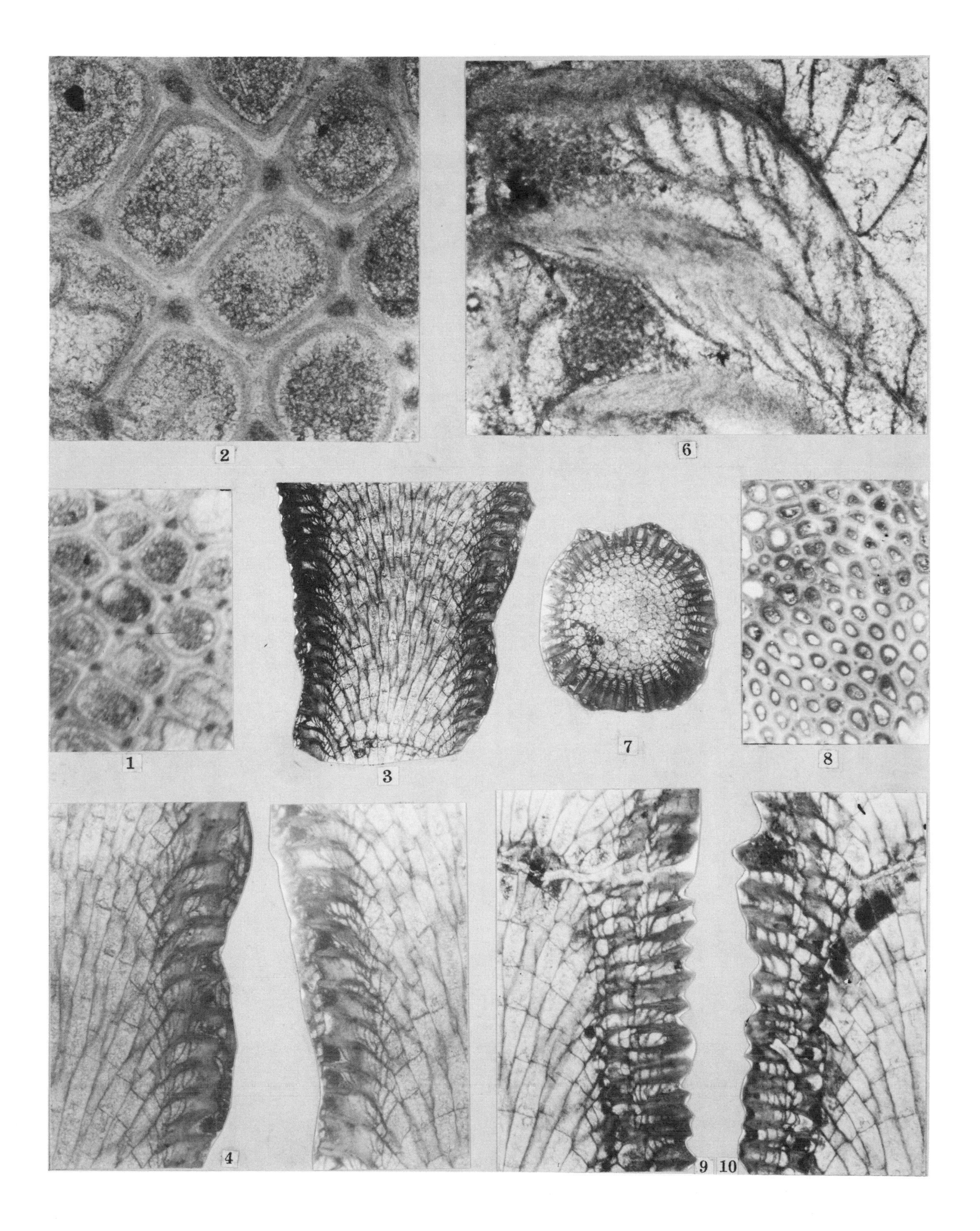

2
6
1
3
7
8
4
9
10

PLATE 3

Figure	
1.	***Sinoatactotoechus rarispinus*** **Yang and Hu (gen. et sp. nov.).** **Cat. No. 62194 (holotype).** **Longitudinal section, ×20.**
2–11	***Sinoatactotoechus obliquus*** **Yang and Hu (gen. et sp. nov.).** **2–7, Cat. No. 62192 (holotype).**
2.	**Tangential section showing the shape, size, and arrangement of the zooecia, ×20.**
3.	**Tangential section, a part of figure 2, displaying thick zooecial walls, ×50.**
4.	**Tangential section, ×50.**
5.	**Longitudinal section showing branching of zoarium, ×10.**
6.	**Longitudinal section displaying cystoidal diaphragms, ×50.**
7.	**Longitudinal section of another part, ×50.** **8–11, Cat. No. 62193 (paratype).**
8.	**Tangential section, ×20.**
9.	**Tangential section of the same specimen, ×50.**
10.	**Longitudinal section, ×20.**
11.	**Longitudinal section, ×50.**

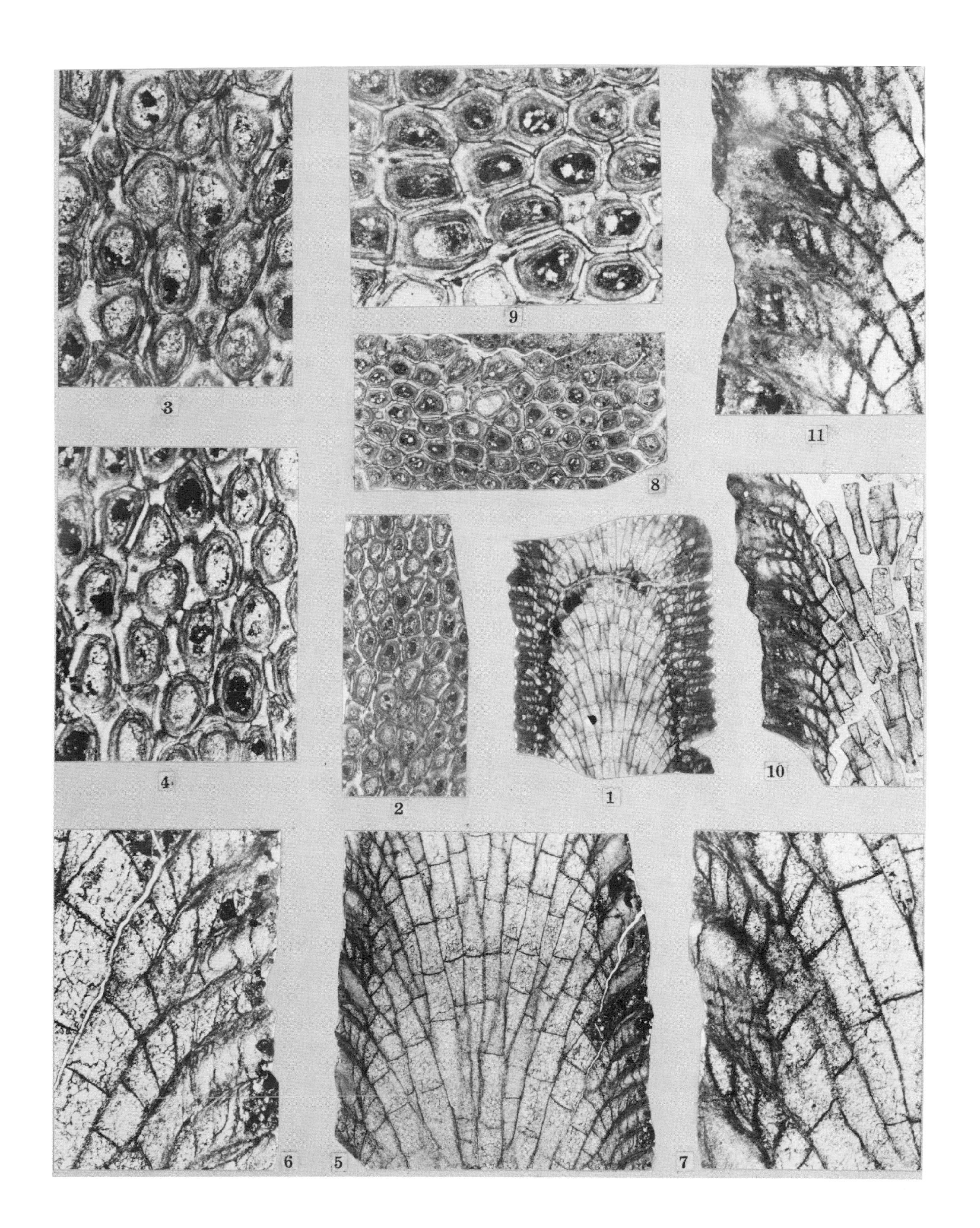
9
3
11
8
10
1
2
4
6
5
7

PLATE 4

Figure

1–7. *Sinoatactotoechus densiseptatus* Yang and Hu (gen. et sp nov.). Cat. No. 62195 (holotype).

1. Tangential section, showing numerous acanthopores, ×20.
2. Tangential section displaying well-preserved and numerous acanthopores, ×50.
3. Tangential section, ×50.
4. Longitudinal section showing cystoidal diaphragms in mature region, ×20.
5. Longitudinal section, ×20.
6. Longitudinal section showing the arrangement of cystoidal diaphragms, ×50.
7. Longitudinal section of the same specimen, ×50.

8–11. *Polyspinopora regularis* Yang and Hu (gen. et sp. nov.). 8–11, Cat. No. 62199 (holotype).

8. Tangential section showing the shape, size, and arrangement of the zooecia; small and numerous acanthopores, ×20.
10. Longitudinal section displaying well-preserved compound diaphragms in mature region, ×20.
11. Longitudinal section, ×20.

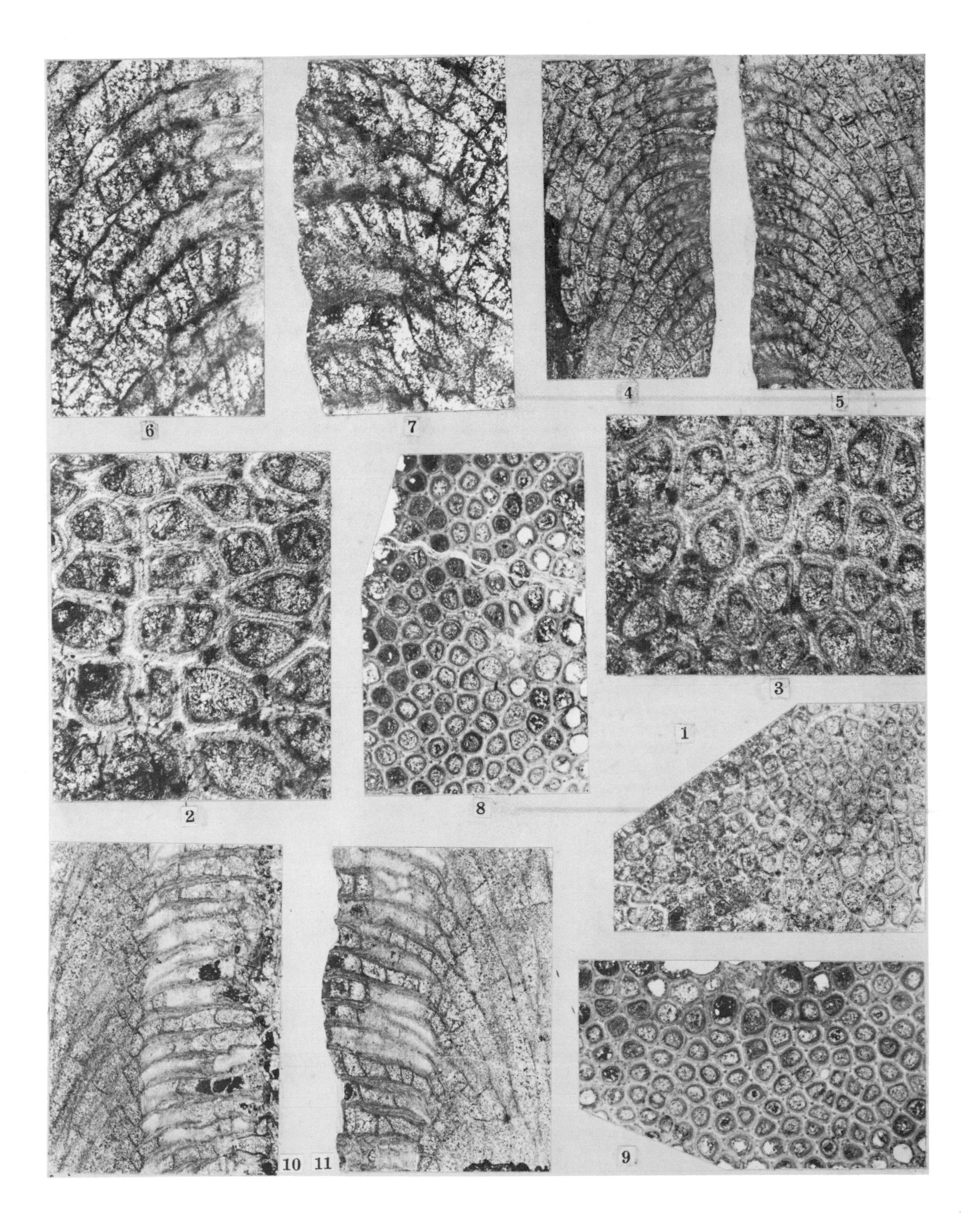

6
7
4
5
3
2
8
1
10
11
9

PLATE 5

Figure	
1–4	*Polyspinopora regularis* Yang and Hu (gen. et sp. nov.). Cat. No. 62199 (holotype).
1.	Tangential section showing well-developed acanthopores, ×50.
2.	Tangential section, ×50.
3.	Longitudinal section showing well-preserved compound diaphragms in tubes of mature region, and detailed well structures, ×50.
4.	Longitudinal section showing the detailed wall structures, ×50.
5–10.	*Polyspinopora shaodongensis* Yang and Hu (gen. et sp. nov.). Cat. No. 62196 (holotype).
5.	Longitudinal section showing characteristics of the tubes of mature region, ×20.
6.	Longitudinal section, ×20.
7.	Tangential section showing well-developed and numerous acanthopores, ×20.
8.	Tangential section, ×20.
9.	Tangential section showing the shape, size, and arrangement of the zooecia, and well-developed acanthopores, ×50.
10.	Longitudinal section showing detailed wall structures, ×50.

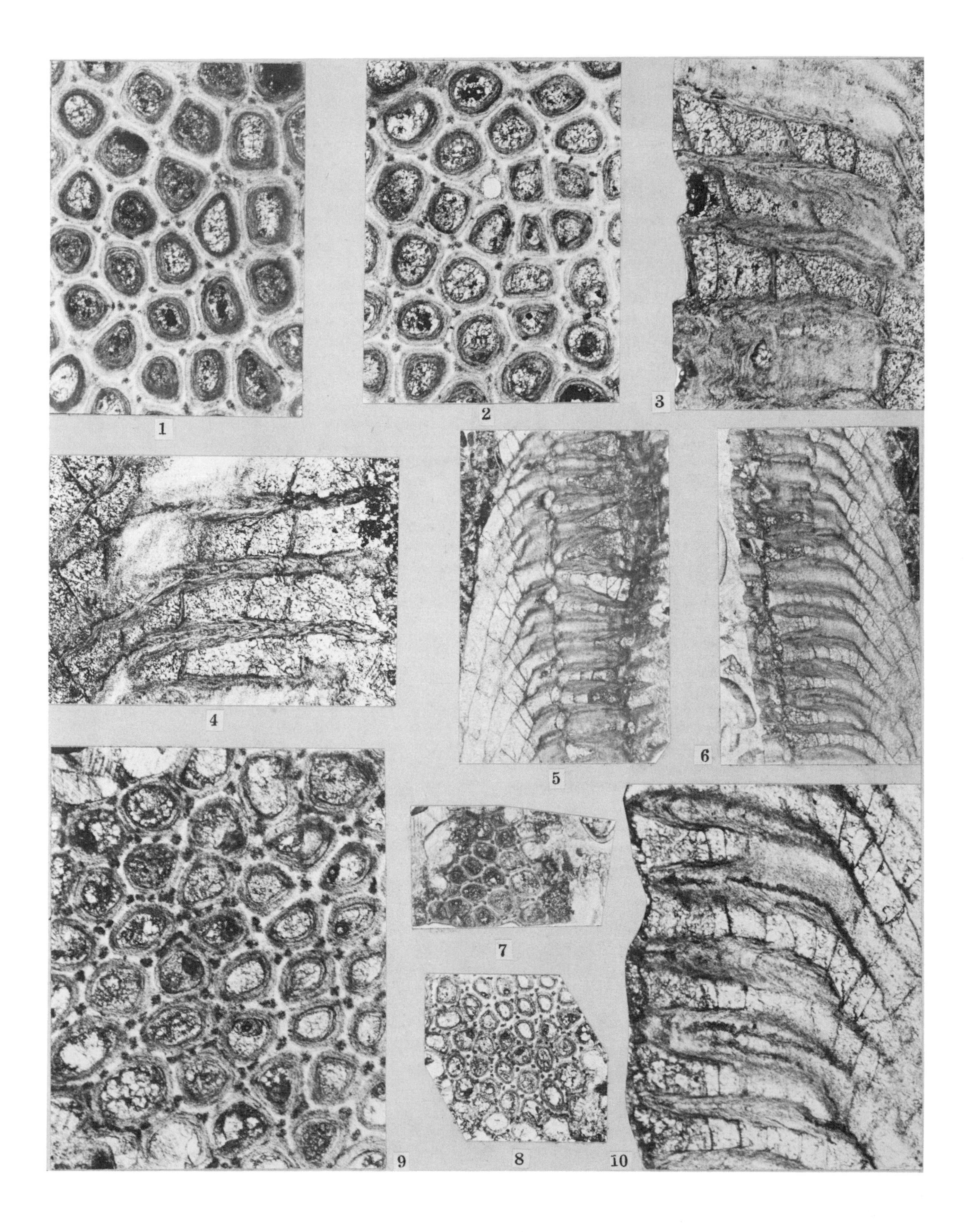
1
2
3
4
5
6
7
8
9
10

PLATE 6

Figure

1–2. *Polyspinopora shaodongensis* Yang and Hu (gen. et sp. nov.). Cat. No. 62196 (holotype).

1. Tangential section showing distribution of the acanthopores, ×50.
2. Longitudinal section showing Characteristics of the zooecial walls, ×50.

3–11. *Polyspinopora ampliata* Yang and Hu (gen. et sp. nov.). 3–9, Cat. No. 62197 (holotype). 10–11, Cat. No. 62198 (paratype).

3. Tangential section of another part of same specimen showing the shape and size of zooecia, ×20.
4. Tangential section of another part of same specimen, showing distribution of the acanthopores, ×20.
5. Longitudinal section showing distribution of diaphragms, ×20.
6. Longitudinal section, ×20.
7. Tangential section, displaying well-preserved acanthopores, ×50.
8. Tangential section showing a few small acanthopores, ×50.
9. Part of longitudinal section, ×50.

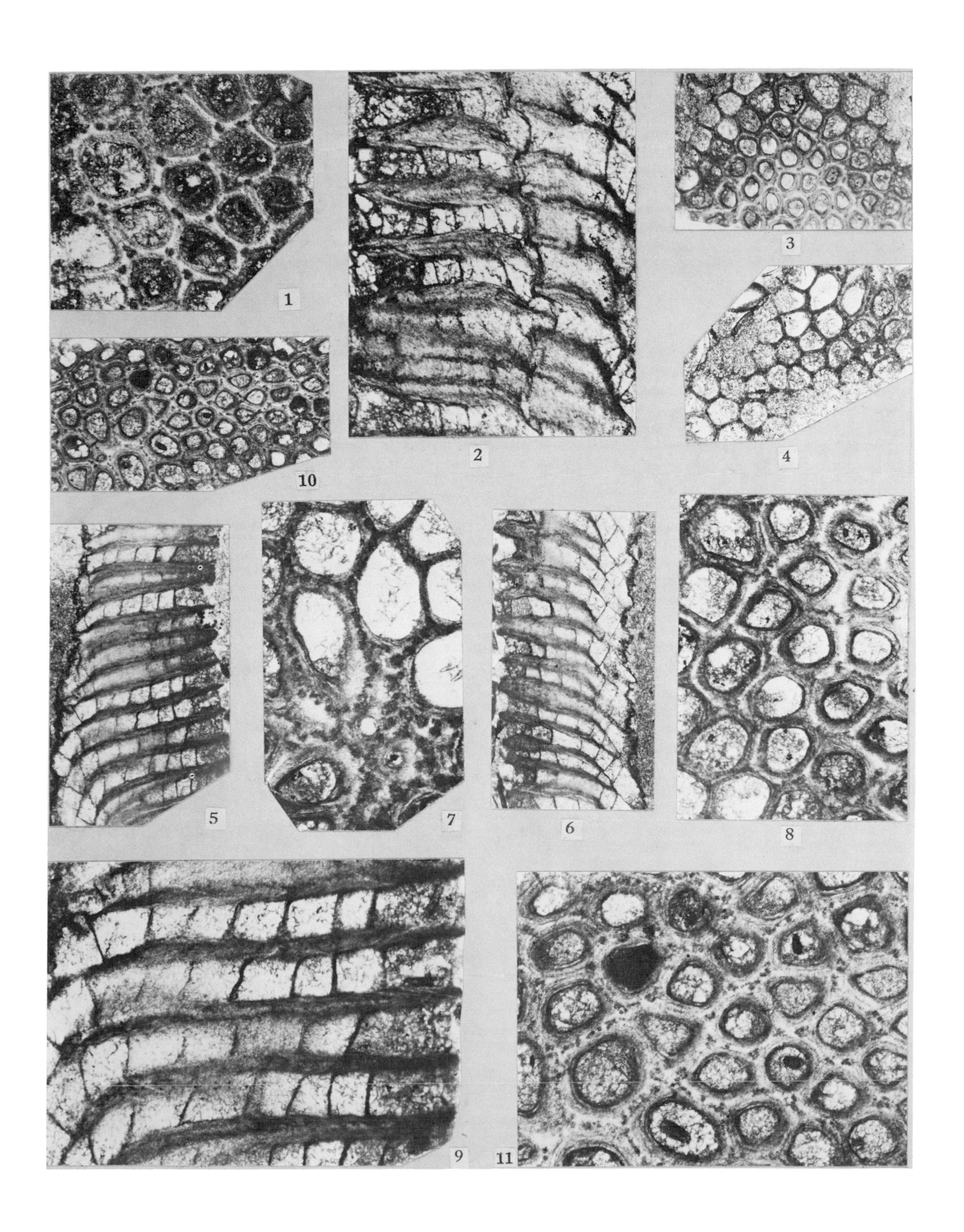

1
2
3
4
10
5
7
6
8
9
11

PLATE 7

Figure

1–3. *Armillopora sinensis* Yang and Hu (gen. et sp. nov.).
Cat. No. 62201 (holotype).

1. Tangential section showing the shape and arrangement of the zooecia and acanthopores, ×20.
2. Tangential section displaying well-preserved armillary perisetoma, ×50.
3. Longitudinal section showin hemisepta in early mature region, ×20.

4–7. *Armillopora spinosa* Yang and Hu (gen. et sp. nov.).
Cat. No. 62202 (holotype)

4. Tangential section showing the shape and size of the zooecia, ×20.
5. Tangential section displaying well-developed small acanthopores, ×50.
6. Longitudinal section showing a few hemisepta in early mature region, ×20.
7. Transverse section, ×20.

8–10. *Polyspinopora regularis* Yang and Hu (gen. et sp. nov.).
Cat. No. 62200 (paratype).

8. Longitudinal section showing the wall structure, ×50.
9. Tangential section showing numerous acanthopores, ×50.
10. Part of longitudinal section showing arrangement of the diaphragms, ×50.

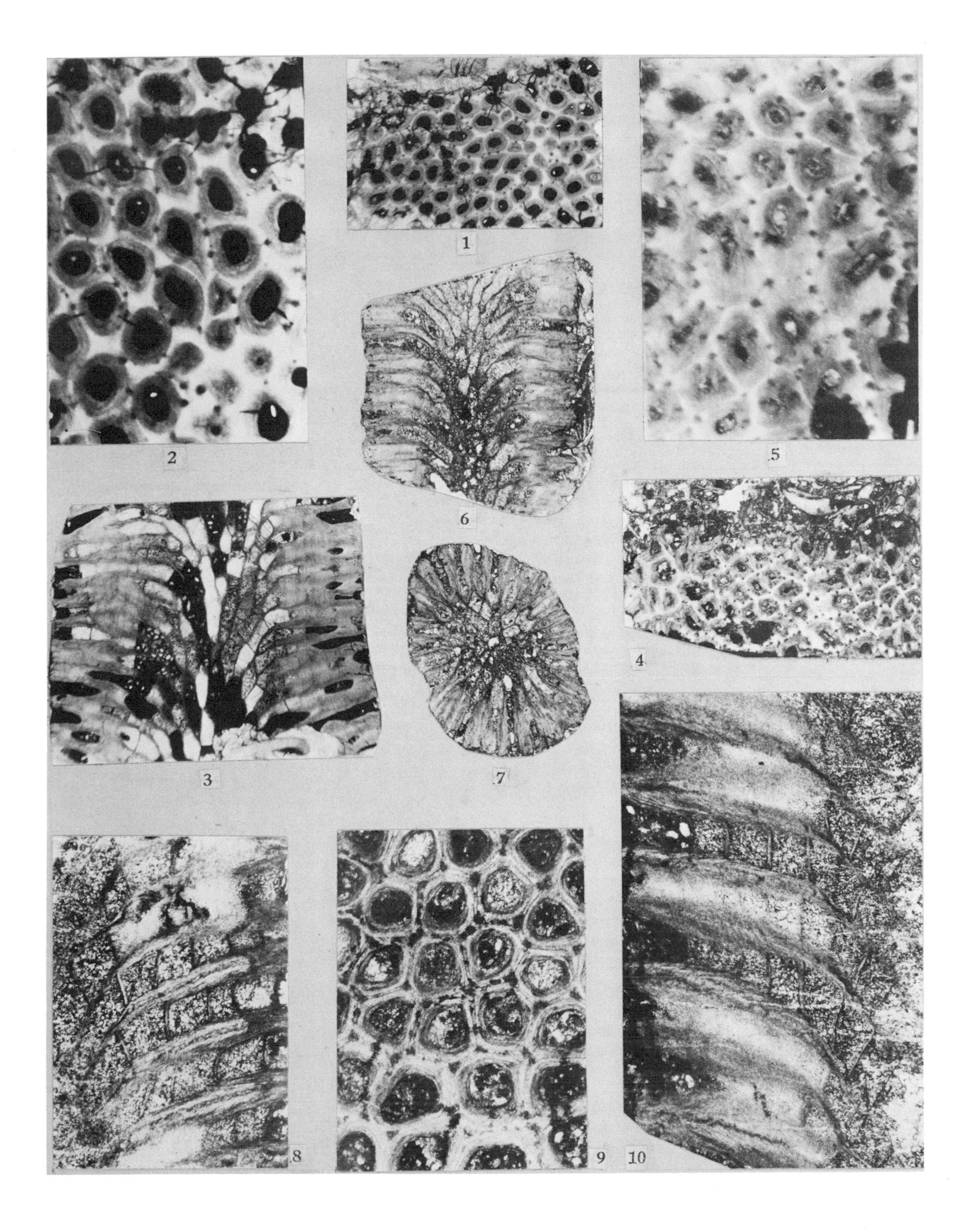
1
2
3
4
5
6
7
8
9
10

Printed in U.S.A.

Geological Society of America
Special Paper 187
1981

Stratigraphic distribution of Brachiopoda in China

Wang Yu
Jin Yu-gan
Liu Di-yong
Xu Han-kui
Rong Jia-yu
Liao Zhuo-ting
Sun Dong-li
Yang Xue-chang
Nanjing Institute of Geology and Palaeontology, Academic Sinica, Nanjing 210008, People's Republic of China

ABSTRACT

A brief description of current developments of the study of Brachiopoda in China is presented. The succession of brachiopod assemblages and the general geographic distribution of brachiopods from Cambrian to Tertiary are described. The main factors that contribute to difficulty in correlating successions of brachiopod faunas are tentatively discussed.

INTRODUCTION

The wide attention given in China to the study of the stratigraphic and geographic distribution and chronologic usefulness of Brachiopoda is demonstrated by the fact that during the past three decades, nearly 120 Chinese workers have devoted themselves to the study of Brachiopoda, with a corresponding increase in the number of publications. Progress in this field is attributable mainly to the discovery of new taxa (Fig. 1). According to Wang and others (1964), about 253 genera and 1,400 species of brachiopods were recorded in China prior to 1959. Since then, 240 new genera and about 1,200 new species have been described. Such progress is also manifested by the extension of knowledge of stratigraphic and geographic occurrences. Apart from South and North China, where the great majority of brachiopod faunas was collected, huge collections have been made in some virgin areas. For example, the number of localities of brachiopods in Tibet has increased from a few to more than 85. So far, only a few papers have been published dealing with brachiopods of Silurian, Triassic, and Jurassic age, a subject almost unknown to scientists some years ago.

Because of the fact that large numbers of brachiopods recorded in fossil handbooks have not been studied stratigraphically and that some of the important studies on brachiopods carried out in the past few years have not yet been published, the stratigraphic distribution of brachiopod faunas in China is still quite imperfectly known. Consequently, it is time both to summarize what is known about the stratigraphic distribution of brachiopods in China and to discuss several factors that make it difficult to correlate the succession of faunas.

THE SUCCESSION OF BRACHIOPOD FAUNAS

The distribution of brachiopods in China is marked by provincialism in many stages. The faunal sequences established in South China are more complete than those in other regions. Synthesis of regional sequences of each province permits recognition of about 90 assemblages (Fig. 2), including 8 Cambrian, 15 Ordovician, 9

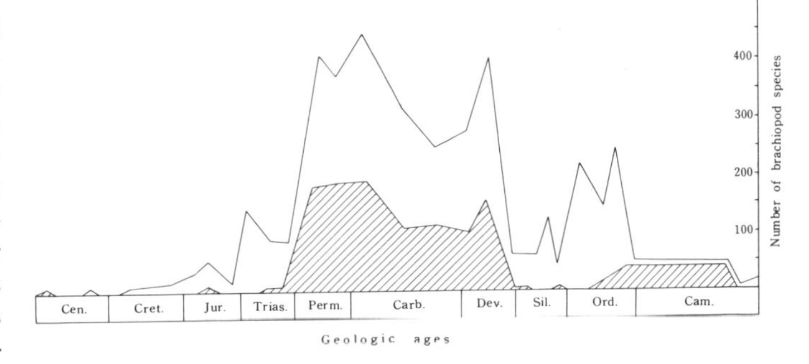

Figure 1. Diagram showing number of species in various geological periods in China. Shaded areas show number of species described prior to 1959.

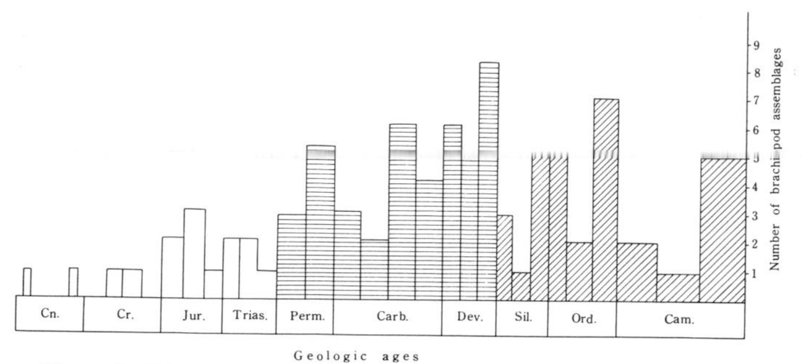

Figure 2. Diagram showing number of brachiopod assemblages in various geological periods in China.

Silurian, 19 Devonian, 15 Carboniferous, 8 Permian, 5 Triassic, 6 Jurassic, 2 Cretaceous, and 2 Tertiary ones. However, most of the assemblages discussed here should not be interpreted as assemblage zones until more details of the stratigraphic and geographic distribution of brachiopods are available.

Cambrian

The earliest brachiopod fauna was found in the Meishucun Member in Yunnan and its equivalents in Sichuan and Hubei. It contains varied forms among which 11 genera such as *Ocruranus, Scambocris, Protobolus,* and *Psamathopalas* were described by Zhong (1977) and Liu Di-Yong (1979). Most of them possess a pseudointerarea and a homoedeltidium with no pedicular groove, a unique morphological character of the earliest brachiopods. This assemblage is approximately coeval with that from Tommotian rocks of Siberia and the basal bed of the Comley Sandstone in England.

From the lower Lower Cambrian Chiungchussu Formation in Yunnan, Wang and others (1974) described a new genus *Diandongia.* The Tsanglangpu Formation in Xinjiang and Anhui yielded abundant *Kutorgina* and *Israelaria.*

The other five Cambrian brachiopod assemblages are named on the basis of materials from North China, all being dominated by inarticulate brachiopods as well as some primitive articulate brachiopods of potential stratigraphic value included (Table 1).

Ordovician

Based largely on unpublished data, the genera of Ordovician articulate brachiopods from China number more than 240. A succession consisting of 14 brachiopod assemblages has been recognized and is listed in Table 2.

In the Lower and Middle Ordovician of China a complete and easily recoverable sequence of brachiopod assemblages occurs only in southwestern China, where the sedimentary facies is mainly calcareous. The Lower and Middle Ordovician brachiopod faunas found in China outside this area, although stratigraphically very important, are few and sparse and are dealt with below.

A brachiopod fauna of Middle Ordivician age from the Mount Jolmolungma region (Liu, 1976) is dominated by *Aporthophyla, Aporthophylina,* and *Xizangostrophia* and is closely related to that found in central Jilin and Ejne Qi of northern Gansu within the Paleozoic geosynclinal region of northern China.

TABLE 1. CAMBRIAN BRACHIOPOD ASSEMBLAGES

Stage	Assemblage
Upper Cambrian	
Fengshan Stage	Mesonomia-Huenella Assemblage
Changshan Stage	Eoorthis-Palaeostrophia Assemblage
Middle Cambrian	
Hsüchuang Stage	Wimanella-Wynnia Assemblage
Lower Cambrian	
Maochuang Stage	Paterina-Cambrotrophia Assemblage
Manto Stage	Nisusia-Eoconcha Assemblage
Tsanglangpu Stage	Kutorgina-Israelaria Assemblage
Chiongchussu Stages	Diandongia Assemblage
Meischucun Stage	Ocruranus-Scambocris Assemblage

TABLE 2. ORDOVICIAN BRACHIOPOD ASSEMBLAGES

Upper Ordovician	Ashgillian	Changwu Stage
		1. Hirnantia-Kinnella Assemblage
		2. Eoconchidium-Rhynchotrema Assemblage
	Caradocian	Beiguoshan Stage
		3. Trimerellina-Ovalospira Assemblage
		Taoqupo Stage
		4. Ovalospira-Hallina Assemblage
		Pagoda Stage
		5. Ptychopleurella-Anisopleurella Assemblage
Middle Ordovician	Llanvirnian	Dashaba Stage
		6. Porambonites-Leptestia Assemblage
		7. Horderleyella-Hesperina Assemblage
Lower Ordovician	Arenigian	Dawan Stage
		8. Diorthelasma-Anechophragma Assemblage
		9. Lepidorthis-Martellia Assemblage
		Hungshiang Stage
		10. Orthis-Schedophyla Assemblage
		11. Diparelasma-Tritoechia Assemblage
	Tremadocian	Fenhsiang Stage
		12. Oligorthis-Syntrophira Assemblage
		Nantsinkuan Stage
		13. Archaeorthis-Imbricatia Assemblage
		14. Finkelnburgia-Apheorthis Assemblage

The brachiopod faunas of the Dashaba Stage collected from some localities in North China, such as Zhouzishan in western Neimongol and Mianyang County of Shaanxi, are closely comparable to those from Southern Xizang. The common elements of those faunas consist of *Glyptorthis, Mimella, Hesperorthis, Lep-*

telloidea, and others. As a rule, their diversity is evidently lower than that of the contemporaneous faunas from southwestern China.

Upper Ordovician brachiopod-bearing deposits are developed in a relatively restricted region. The 11th and 12th assemblages were recognized in Mianxian County of Shaanxi and the 13th and 14th assemblages in Jiangshan County of Zhejian.

The unusually distinctive and widespread *Hirnantia* fauna has been recorded (Rong, 1979) from the Kuanyinchiao Beds and its equivalents in China. This fauna contains more than 20 genera, of which the most common taxa are *Kinnella, Paromalomena, Aphanomena, Plectothyrella,* and *Hindella,* in addition to *Hirnantia,* usually found in association with *Dalmanitina*. It is proposed to divide this fauna into three assemblages: (1) *Kinnella-Plectothyrella,* (2) *Draborthis-Toxorthis,* and (3) *Paromalomena-Aegiromena*. Their correlation with graptolite zones suggests that the *Hirnantia* faunas of China and other countries is markedly diachronous, although it spread fairly rapidly, and that the acme of its development was in the latest Ashgillian.

Silurian

The Silurian brachiopod faunas of South China contain various elements widespread in the Northern Silurian Realm, such as *Stricklandia, Pentamerus, Zygospiraella, Eospirifer. Striispirifer, Atrypoidea,* and *Protathyris*. These faunas also include some genera restricted to Asia or South China, such as *Xinanospirifer, Nalivkinia, Protathyrisina, Pleurodium,* and *Paraconchidium* (Table 3).

Assemblages collected from such geosynclinal areas as northern Xinjiang, Neimongol, and Heilongjiang are part of the famous *Tuvaella* faunas. Though the correlation of this fauna with its contemporaneous faunas remains a topic of debate, its age is believed to be from late Early to early Late Silurian. Apart from abundant *Tuvaella* and *Tannuspirifer,* all the other genera of this fauna are cosmopolitan elements of the northern hemisphere. It is, therefore, reasonable to believe that the *Tuvaella* fauna represents one of the communities in the Silurian Cosmopolitan Realm.

TABLE 3. SUCCESSION OF SILURIAN BRACHIOPOD ASSEMBLAGES IN SOUTH CHINA

Series	Formation	Assemblage		
Pridolian	Yulongsi Formation	Unnamed assemblage		
	Miagao Formation	*Protathyrisina plicata-Schizophoria hesta* Assemblage		
Ludlovian	Kuandi Formation	*Protathyrisina uniplicata-Atrypoidea qujingensis* Assemblage		
Wenlockian				
	Xiushan Formation	*Salopina-Xinanospirifer* Assemblage		
	Hanchiatian Formation	*Nalivkinia-Nucleospira* Assemblage		
Llandoverian	Shiniulan Formation	*Paraconchidium-Virgianella* Assemblage	*Pentamerus-Eospirifer* Assemblage	*Stricklandia-Merciella* Assemblage
		Borealis-Kritorhynchia Assemblage		
	Lungmachi Formation	Graptolitic facies		
	Wulipo Beds	"*Protatrypa*"-*Hindella* Assemblage		

The results of the studies of brachiopods show that these have played an important role in the progress of Silurian stratigraphy of South China during recent years. For example, Silurian strata of South China were traditionally divided into three ascending units: the Lungmachi, Lojoping, and Shamao Groups, correlated respectively with the Llandoverian, Wenlockian, and Ludlovian series in Great Britain. Based on the data of brachiopod fauna, particularly the discovery of a number of Llandoverian taxa, such as *Stricklandia, Kulumbella, Merciella* and *Zygospiraella* from the Lojoping formation, the age of the formation and its equivalents has been assigned to the late Early Silurian.

Devonian

Two distinctive Devonian biogeographical provinces, the South China Province and the Junggar-Hingan Province, are separated by the northern China uplift. The brachiopod faunas of the South China Province are part of the Old World Realm and may be subdivided into two major facies, corresponding to the Rhenish community and the Bohemian community proposed by Boucot and others (1969). One of them, the Xiangzhou facies, is characterized by the dominance of benthic fauna, including highly diversified large and thick-shelled brachiopods. Most Devonian brachiopods described previously were obtained from rocks of this facies, with a faunal succession better defined than that of any other facies. During recent years, however, it has been a matter of controversy whether the *Euryspirifer* and the *Zdimer* assemblages represent two distinctive contemporaneous communities or whether they occur at different levels. Because the earliest Devonian brachiopod faunas are found only in western Qinlin, it is evident that marine deposits of this period are entirely lacking in South China. *Lanceomyonia-Linguopugnoides* and *Spirigerina-Gypidula* are two assemblages that have been recognized.

The Nandan facies also is dominated by sparse, pelagic forms of small and thin-shelled brachiopods. Only one assemblage is given here, though it is reportedly found to encompass several assemblages.

The brachiopod faunas of the Junggar-Hingan Province have close affinities to those of Siberia and the Appalachians. The succession is shown in Table 4, but it should be remembered that until now it has been difficult to correlate with that of the South China Province.

Carboniferous

Three Carboniferous brachiopod provinces are recognized: the southern, the central, and the northern (Table 5). The southern province consists of southern Xizang (Tibet) and western Yunnan. The Early Carboniferous brachiopod faunas are characterized by the common occurrence of syringothyrids, while the latest Carboniferous *Stepanoviella-Trigonotreta* assemblage shows a strong affinity with that of the Umaria marine bed of India. The northern province covers a vast area mainly north of lat 35° N. Syringothyrids, the characteristic group of Tournaisian and early Visean brachiopod faunas, disappear generally in the late Visean when the faunas are dominated by gigantoproductids and yield fewer megachonetids and straitiferids than the corresponding faunas from South China. The Late Carboniferous brachiopod faunas are essentially the same in generic composition as those of the central province except the one from Northern Great Hingan, where a brachiopod fauna with boreal affinities has been partly described. The third province that includes the remaining areas of China is characterized by the rich presence of gigantoproductids, megachonetids, striatiferids, and some endemic genera, such as *Weiningia* and *Lochengia,* while syringothyrids are found to be quite sporadic.

TABLE 4. SUCCESSION OF DEVONIAN BRACHIOPOD ASSEMBLAGES IN CHINA

Stage	South China Province		Junggar-Hingan Province
Famennian	Nayunnella supersynplicata-Tenticospirifer triplisinosus Assemblage; Nayunnella synplicata - T. gortani Assemblage; Yunanellina hanburyi-Cyrtiopsis davidsoni Assemblage		
Frasnian	Cyrtospirifer hunanensis (sp. nov.)-Ptychomaletoechia Assemblage; Hypothyridina hunanensis Assemblage; Cyrtospirifer sinensis-Leiorhynchus Assemblage		Cyrtospirifer; Spinatrypa
Givetian	Emanuella takwanensis-Stringocephalus burtini Assemblage; Stringocephalus-Bornhardtina spp. Assemblage; Rensselandia circularis-Bornhardtina uncitoides Assemblage		Mucrospirifer; Mediospirifer; Proleptostrophia
Eifelian	Acrospirifer fongi-Eospiriferina lachrymosa Assemblage	Zdimir-Megastrophia Assemblage	Alatiformia; Brevispirifer
Delejan	Euryspirifer-Indospirifer Assemblage	(Zdimir-Megastrophia Assemblage)	Acrospirifer; Leptaenopyxis
Zlichovian	Howellella-Reticulariopsis Assemblage; Vagrania-Leptathyris Assemblage		Fimbrispirifer; Megastrophia
Zlichovian	Dicoelostrophia-Rostrospirifer Assemblage	Eosophragmophora-Parathyrisina Assemblage; Sinoatrypa (gen. nov.) - Parachonetes Assemblage	Leptostrophia (Rhytistrophia); Gladiostrophia
Pragian	Orientospirifer Assemblage		Leptocoelia; Coelospira
Lochkovian			Strophochonetes; Protathyris; Rugoleptaena

The Late Carboniferous faunas are chiefly composed of dictyoclostids, choristitids, and some terbratulids. In this province, the lowest *Sphenospira-Mesoplica* assemblage is dominated by cyrtospiriferids and *Mesoplica* sp., similar on the whole to the Famennian fauna of the Hsikuanshan Formation. The remarkable difference between them lies in the disappearance of *Nayunnella* and *Yunnanellina* in this assemblage, which are common in the Hsikuanshan Formation. The recent discovery of *Polygnathus normalis, P. obliquicostatus,* and other conodonts in association with this assemblage suggests that it may be correlated with the K zone in England. It is also worth noticing that a direct correlation of the Visean/Tournaisian boundary between those two regions cannot be made on the basis of brachiopods because *Levitusia hurmosa,* a key element of the C_2 zone in England, is missing from the Lower Carboniferous in China. The age of the *"Fusella"*

TABLE 5. SUCCESSION OF CARBONIFEROUS AND LOWER PERMIAN BRACHIOPOD ASSEMBLAGES IN CHINA

Stage	Southern Province	Central Province	Northern Province
Sakmarian	Stepanoviella-Trigonotreta Assemblage	Tyloplecta richthofeni-Choristites pavlovi Assemblage; Dictyoclostus uralicus Assemblage	Dictyoclostus taiyuanfuensis Chonetes pygmae
Asselian		Protanidanthus Choristites jigulensis Assemblage	
Moscovian		Buxtonia grandis Assemb. Choristites mansyi-Semicostata panxianensis Assemblage	Brachythyris abnormalis Choristites mansuyi borohoensis
Bashkirian	Gigantoproductus-Balakhonia Assemblage	Gigantoproductus edelburgensis-Gondolina Assemblage	Productus concinnus Striatifera striata
Visean	Marginirugus-Syringothyris Assemblage	Gigantoproductus gigantoides Assemblage; Gigantoproductus moderatus Assemblage; Vitiliproductus groeberi-Pubilis hunanensis Assemblage; Delepinea subcarinata-Megachonetes zimmermani Assemblage	Gigantoproductus sarsimbai; Antiquatonia antiquata; Dictyoclostus crafordsvillensis; Grandispirifer mylkensis
Tournaisian	Fusella-Ovatia Assemblage	Fusella ohaoyangensis Assemblage; Eochoristites-Martiniella Assemblage; Schuchertella-Yanguania Assemblage	Dictyoclostus robustus; Syringothyris cf. S. texta
Etroeungt		Sphenospira-Mesoplica Assemblage	

shaoyangensis assemblage and its corresponding brachiopod faunas in the other two provinces is still a matter of dispute. These faunas are characterized by the presence of *Schuchertella magna, Syringothyris* cf. *S. taxa,* and *Dictyoclostus crafordovillensis* in western Yunnan.

Permian

During the Permian, faunal provinciality was very clearly defined, and three provinces have been distinguished (Jin in Zhang and Jin, 1976). The northern province with its southern boundary at about lat 42°N evidently communicated with the Boreal realm. The southern province is restricted within the limits of southern Xizang, at approximately lat 32°N around the Nu River in the east, with a fauna related to those in the northern marginal sea of the Gondwana continent. The remaining areas in China are included in the central province and are characterized by typical faunas of the Tethyan realm (Table 6).

In addition to the assemblages listed below, there are a few contemporaneous assemblages from unusual facies. For example, the *Peltichia zigzag–Spinomarginifera chengyaoyanensis* assemblages is fairly common in carbonate facies of South China and is always replaced by *Paryphella sulcatifera–Paracrurithyris pigmae* in siliceous facies. In the southern province, the facies changes of the Permian deposits are marked by diminishing terrestrial materials along with increasing amounts of carbonate rocks northward and upward, reflecting an environmental change from onshore to offshore in a marginal sea. Corresponding to the lithological transition, there appear in succession the *Taeniothaerus* assemblage with abundant *Spiriferella* and *Costiferina,* the *Calliomarginatia* assemblage dominated by some productids, and the *Chonetella* assemblage mainly consisting of cemented and pedicular brachiopods.

Triassic

Triassic brachiopods of China were mainly collected from the Tibetan plateau and the Yunnan-Guizhou plateau; yielding a total number of 177 described genera and 256 species, of which 44 genera and 153 species are introduced as new. Liao (1980) tentatively divided the faunas into nine asemblages belonging to five stages which may be summarized as follows:

Griesbachian Stage

Fusichonetes pigmae Paryphella triquetra Neowellerella pseudoutah Assemblage

Anisian Stage

Nurdirostralina griesbachi–Tulungospirifer stracheyi Assemblage (Himalya region)

Nudirostralina subtrinodosa–Diholkorhynchia sinensis Assemblage (Yunnan-Guizhou region)

Qilianoconcha opima–Aequispiriferina multiplicata Assemblage (Qilianshan region)

Ladinian Stage

Volirhynchia multicostata–V. himaica Assemblage (Himalaya region)

Neoretzia tibetensis–Oxycolpella oxycolpos–Sangiaothyris elliptica Assemblage (Yannan-Guizhou region)

Norian Stage

Himalayirhynchia media–Eoseptaliphoria tulungensis Assemblage (Himalaya region)

Halorella donqoensis–Septamphiclina qinghaiensis–Sacothyris sinosa Assemblage (Tanggula region)

Orientospira cf. *O. gregaria* Assemblage (Boreal region)

The Griesbachian brachiopods were collected from the basal bed in 15 localities of South China (Liao, 1979). This fauna contains 17 species in 9 genera, but most of them are restricted to a bed no more than 1 m in thickness; only two genera, *Crurithyris* and *Neowellerella,* extend upward into the higher horizons.

The area of distribution of Anisian brachiopod faunas covers three main regions, each containing many endemic forms, such as *Tulunspirifer* of the Himalayas region, *Diholkorhynchia* and *Septaliphoroidea* of the Guizhou-Yunnan region, and *Qilianoconcha* of the Qilianshan region.

The Ladinian brachiopod faunas are transitional with those of the Anisian and Carnian. Relatively characteristic forms are *Volirhynchia, Nudispiriferina,* and *Sangiaothyris.*

The Carnian brachiopod faunas of China contain a large number of endemic genera, for example, *Dierisma, Yidunella, Saccorhynchia, Septamphiclina,* and *Zidothyris.* Most of these are found in reef facies in southern Qinghai, eastern Xizang, and western Sichuan. The faunas dominated by *Halorella* and *Halorelloidea* were collected only from northern Xizang. Moreover the discovery of the genus *Orientospira* in northeastern Heilongjiang testifies to the fact that a Norian fauna of the Boreal Province is also present in China.

Jurassic

A sequence including three Jurassic brachiopod assemblages has been described from the Mount Jolmo Lungma region. The Menkatun Formation has yielded a small fauna that is characterized by the presence of *Rhynchonella* aff. *R. paucicosta* and represents one of the Late Jurassic brachiopod assemblages in China. The second assemblage from the middle part of the Nynyxiongla Formation contains *Nyalamurhynchia mirifica, Rhactorhynchia lauta,* and other forms, which suggests a Bajocian-Bathonian age. The third assemblage, of Pliensbachian age, is

TABLE 6. SUCCESSION OF PERMIAN BRACHIOPOD ASSEMBLAGES IN CHINA

Stage	Southern Province	Central Province	Northern Province
Changhsingian		*Peltichia zigzag-Spinomarginifera chengyaoyengensis* Assemblage	
Lopingian		*Tyloplecta yangtzeensis-Squamularia grandis* Assemblage *Edriosteges poyangensis-Alatoproductus truncatus* Assemblage	
Maokouan	*Rugivistiva* *Urushtenia* *Comuquia* *Marginifera*	*Neoplicatifera huangi* Assemblage *Cryptospirifer striatus* Assemblage	*Transennatia* *Urushtenia* *Rugivistiva* *Pemundaria*
Chihsian	*Chonetella* Assemblage *Calliomarginatia* (*Taeniothaerus*) Assemblage	*Chaoina reticulata* Assemblage *Tyloplecta richthofeni-Orthotichia chekiangensis* Assemblage	*Paramarginifera* *Strophalosiella* *Yakovlevia* *Reticulatia* *Orthotichia*

marked by abundance of *Cirpa himalaica* and *Homoeorhynchia.*

The Jurassic brachiopods from the Yenshiping Group of Mount Tanggula and the Liuwan Formation of Mount Hengtuan are closely related to those from the Namyau series of the Shan States, Burma. Their age was traditionally referred to as Bathonian. Generally, this fauna is dominated by *Burmirhynchia* in the lower horizons and *Holothyris* in the upper. But in some places an assemblage consisting of *Kutchithyris degenenensis* and *Thurmanella rotunda* appears in a higher horizon and seems to be of Callovian age.

In addition to the foregoing faunas, a small fauna mainly composed of *Thurmannella* from a formation in northeastern Hilongjiang has been described (Sun, in press), representing another Late Jurassic brachiopod assemblage of China.

Cretaceous and Tertiary

Up until now, Cretaceous and Tertiary brachiopods have been collected from a few places only. An Upper Cretaceous brachiopod fauna containing *Orbirhynchia, Lunpolaia, Trochifera,* and *Yuezuella* was reported from the Langshan Formation in Bagon County, northern Xizang (Ye and Yang, 1979). Its age is believed to be Aptian to Cenomanian. In addition, a new brachiopod genus, *Xenothyris,* was described from the Upper Cretaceous of Yadong County, southern Xizang (Jin and others, 1976).

A small fauna from the Riukiu Limestone of Gaoxiong, Taiwan, was recorded by Hayasaka (1932). It includes *Picthyris, Hemithyris, Gryphus,* and others, indicating Pliocene age. Another small fauna was found in the Zhongpu Formation in Yadong County, southern Xizang, consisting of two new species of *Gryphus.* Its age is suggested to be Paleocene (Jin and others, 1976).

GENERAL REMARKS

1. The phases of the evolutionary history of Brachiopoda may be recognized at different scales on the basis of major morphologic and taxonomic changes. This fossil group seems to obey its own laws, with evolution proceeding at various rates through time. So the taxa of Brachiopoda are unlikely to have lived and died out in phase with other groups. The debate on the position of the Lower/Middle Ordovician boundary in South China may be cited as an example. A major change in the brachiopod succession occurs between the Dawan Formation and the Gunuetan Formation. About 80% of the total number of genera became extinct at

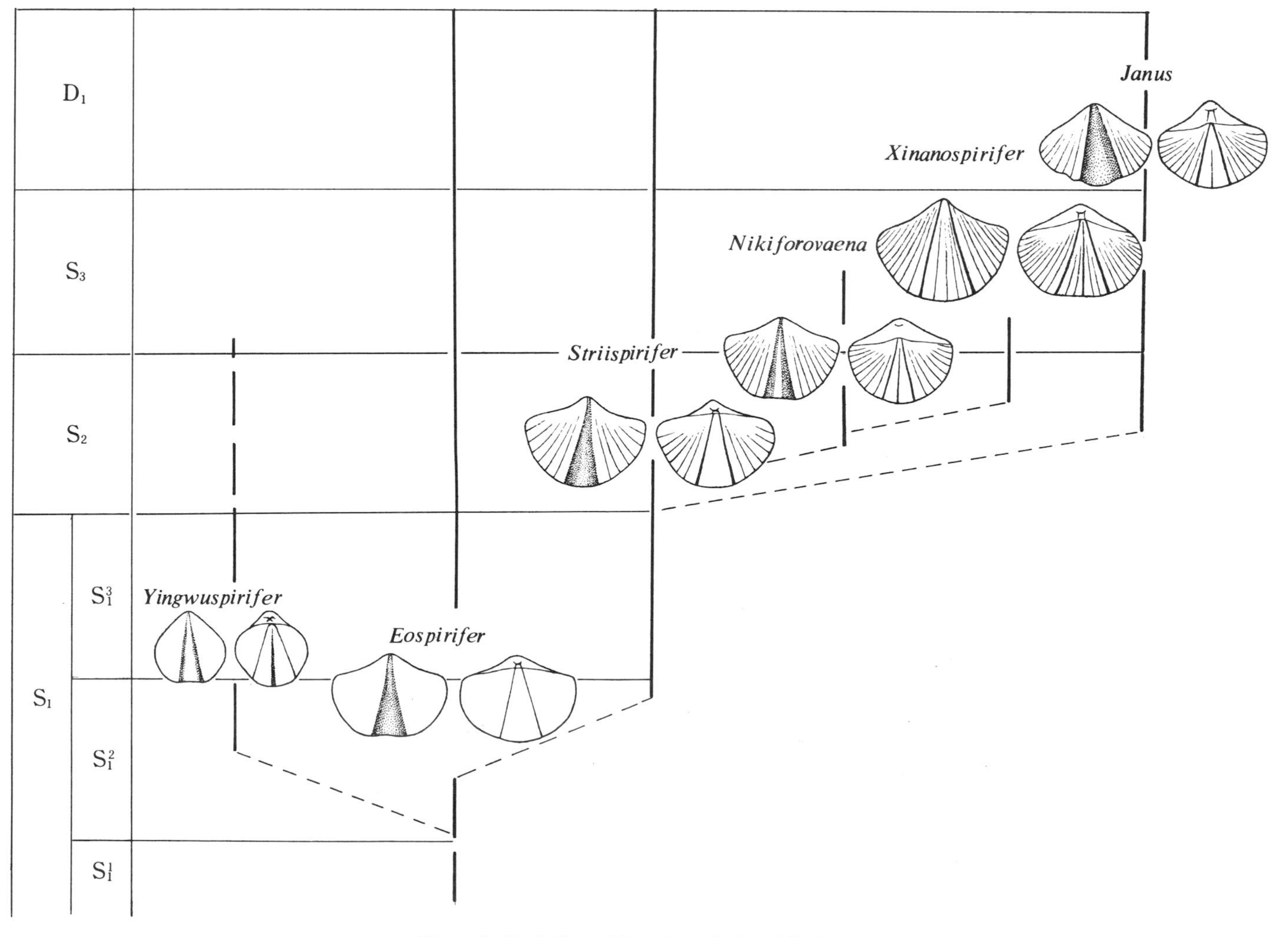

Figure 3. Evolution of the subfamily Eospiriferinae.

the end of the Dawan Stage. The brachiopod fauna of this formation includes small strophomenids with simple cardinal processes, typical orthids with ridgelike cardinal processes, and the endemic genus *Yangtzeella*. On the other hand, the Gunuetan Formation is characterized by the dominance of relatively large strophomenids with bilobed cardinal processes and by the rarity of the genus *Yangtzeella*. According to the evolutionary phase of Brachiopoda, the Lower/Middle Ordovician boundary is best drawn at the base of the Gunuetan Formation. It is equivalent to the boundary of Llanvirnian and Arenigian. As a result of correlation with graptolite zones, the Lower/Middle Ordovician boundary is placed at the top of the Gunuetan Formation, obviously in contradiction to the conclusion based on brachiopod data.

Though it has been suggested that evolutionary change of any segment of a fauna should have drastic effects in cohabiting groups and that brachiopod zones should conform with zones characterized by other groups in the same region, we should not overlook the fact that the evolutionary tempo of each group is commonly different in various degrees from that of the associated faunas. The frequent discovery so-called Permian brachiopods in basal Triassic rock is a striking example in illustration of this situation. Judging from the materials available from more than 15 localities in 11 provinces, the important elements of this fauna such as *Crurithyris* were collected from rocks of late Scythian, even from Anisian beds. The wide distribution and good preservation of these faunas indicate that the existence of some of the Permian-dominated brachiopods in basal Triassic rocks is a normal phenomenon. In a section near Wuxing County, Zhejiang, it can be seen that the diversity of brachiopods decreases progressively from the Changhsing limestone through a siliceous shale member and a calcareous shale bed into the basal Triassic. The number of brachiopod species decreases across the Permian/Triassic boundary from more than ten, to five to six, three to four, and finally only two, belonging to only two genera, *Crurithyris* and *Waagenites*. The other groups, such as fusulinids and Rugosa, did not appear above the limestone, whereas the bivalves common in Lower Triassic rocks made their first appearance in the calcareous shale bed.

2. Some ubiquitous brachiopod species show no detectable changes with time, but the brachiopods that do show change are very useful in correlation. The most useful groups represent primitive lineages, such as the subfamily Eospiriferinae, which represents one of the primitive lineages of Spiriferida. In the light of the Chinese materials, the phylogeny of the Eospiriferinae may be traced in terms of ornamental changes (Fig. 3) as well as the development of the cardinal process. Most of the Early Silurian (Llandoverian) *Eospirifer, Yingwuspirifer,* and *Striispirifer,* possess a smooth cardinal process. In the latest Early Silurian *Strii-*

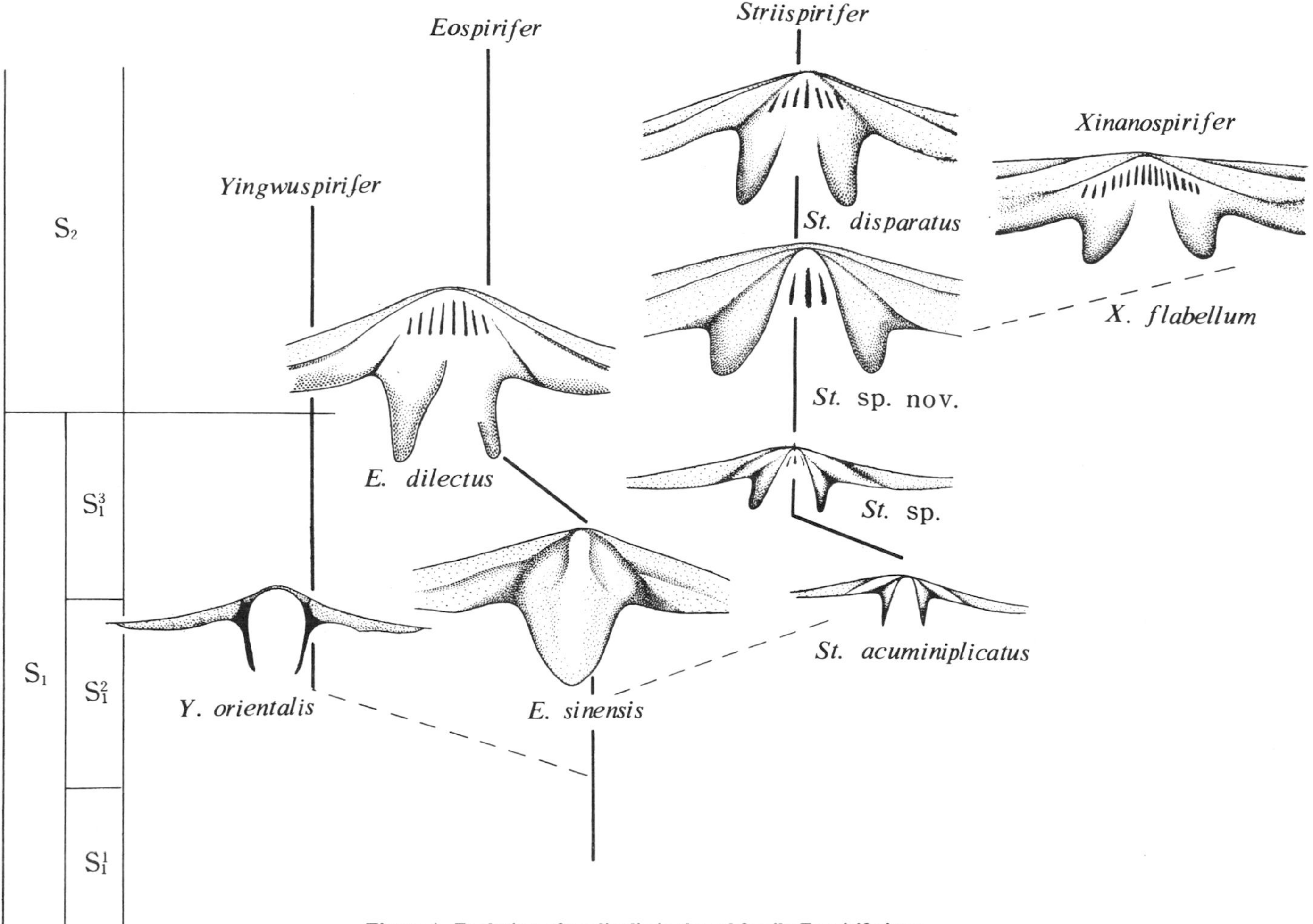

Figure 4. Evolution of cardinalia in the subfamily Eospiriferinae.

spirifer, a node or a few ridges appear. The Middle Silurian (Wenlockian) species of these genera are always characterized by striated cardinal processes (Fig. 4).

Apart from the primitive lineages, the endemic brachiopod groups also show a more rapid rate of evolution. For example, the phyletic production rate of the highly provincial family Lochengidae appears to be rather high during early Carboniferous and early Permian.

The early epochs of some periods are commonly characterized by the appearance of numerous primitive lineages and high provincialism. Consequently, these early epochs, for example, Early Ordovician, Early Silurian, and Early Devonian, provide more numerous brachiopod assemblages than the rest.

3. According to the distribution of brachiopod taxa, three biogeographic provinces in China may be distinguished in Ordovician and Devonian to Permian times, each possessing, as a rule, a number of endemic taxa, and being distinct in assemblage composition. On either side of the provincial boundary, different sequences of brachiopod assemblages may be found. An example will make this clear. The Llanvirnian, Visean, and Sakmarian assemblages in the central province are named *Hesperorthis, Gigantoproductus,* and *Tyloplecta* assemblages, respectively, of which the coeval assemblages in the southern province are the *Aportophyla, Syringothyris,* and *Stepanoviella* assemblages. It is worth noticing that the *Stepanoviella* assemblage extends northward through the collision zone running approximately along the Yalu Tsangpo River to the Nymqingtanglha-Gangdesi Mountains; no barrier ever prevented this Gondwana fauna from penetrating this collision zone.

4. It is important to recognize the environmental restrictions of brachiopod assemblages, because such knowledge is needed to establish more precise biostratigraphic correlations. For instance, three Early Silurian brachiopod assemblages have been recognized in southwestern China: the *Paraconchidium, Pentamerus,* and *Stricklandia* assemblages, which broadly represent three communities corresponding to the *Eocoelia, Pentamerus, Stricklandia,* and *Clorinda* communities of Ziegler (1965). As a matter of fact, these assemblages were not regarded as contemporaneous faunas until a few years ago.

Though very similar brachiopod assemblages occur sometimes in different places at different times in a similar environment, differences nevertheless exist. Generally speaking, it is found that the evolution of brachiopod taxa reflects genetic change through time within similar habitats. For example, the late Early Devonian brachiopods of the Tangxiang Formation and the Middle Devonian brachiopods of the Lofu Formation are all included in the Nandan type. Both are characterized by forms having small, thin shells with smooth surfaces and by paucity of individuals, but their composition is entirely different at the generic and species levels. Moreover, the brachiopod sequence of the Tanxiang Formation may be subdivided into three stages on the basis of taxonomic changes.

Conclusion: The foregoing analysis has shown that the Brachiopoda evolves at different rates at different times. Their distribution is controlled by ecological factors and each biogeographic province possesses a characteristic sequence of assemblages and community patterns.

ACKNOWLEDGMENTS

The data on the succession of brachiopod assemblages of each age were furnished by Liu Di-yong (Cambrian, Ordovician), Rong Jia-yu and Yang Xue-chang (Silurian), Wang Yu and Xu Han-kui (Devonian), Jin Yu-gan and Liao Zhuo-ting (Carboniferous, Permian), and Sun Dong-li (Mesozoic, Cenozoic). Wang Yu and Jin Yu-gan compiled the data and prepared other parts of this paper. We thank Zhang Wu-cong and Zhou Qi-yi who drew the figures.

REFERENCES CITED

Boucot, A. J., Johnson, J. G., and Talent, J. A., 1969, Early Devonian brachiopod zoogeography: Geological Society of America Special Paper 119, 106 p.

Feng Ru-ling, 1977, Discovery of *Syringothyris* from southern Gueizhou and its significance: Acta Palaeontologica Sinica, v. 16, no. 1, p. 53–57, Pl. 1 (in Chinese with English abstract).

Hayasaka, H., 1932, On the occurrence of *Pictothyris* in the so-called Riukiu Limestone of Takao, Taiwan (Formosa): The Venus, v. 3, no. 5, p. 249–251.

Hou Hong-fei, 1959, [Khou Khun-fey, Spiriferidy nizhnedevonskikh i eyfelskikh otlozheniy yuzhnoy chasti Guansi]: Acta Palaeontologica Sinica, v. 7, no. 6, p. 450–480, 2 pls. (in Chinese and Russian).

Hou Hong-fei, 1965, Early Carboniferous brachiopods from the Mengkungao Formation of Gieling, Central Hunan: Beijing, Professional Papers of Academy of Geological Science, Ministry of Geology, sec. 13, p. 111–146, Pls, 1–5 (in Chinese with English abstract).

Hou Hung-fei and Xian Shi-yuan, 1975, [Lower and Middle Devonian brachiopods from Kwangsi and Kweichow]: Beijing, Geological Publishing House, Professional Paper of Stratigraphy and Palaeontology, no. 1, p. 1–85, Pls. 1–32 (in Chinese).

Jin Yu-gan [Ching Yu-kan], 1963, *Urushtenia* from the Lower Permian of China: Acta Palaeontologica Sinica, v. 11, no. 1, p. 1–32, 2 pls. (in Chinese with English abstract).

Jin Yu-gan and Fan Bing-xing, 1976, [Upper Triassic brachiopod fauna from the area in the east of Hengduan Mountain, western Yunnan, *in* Mesozic fossils from Yunnan, Fasc. 2] Beijing, Science Press, p. 39–71, 6 pls. (in Chinese).

Jin Yu-gan, Sun Dong-li, and Rong Jia-yu, 1976, [Mesozoic and Cenozoic brachiopods from Mount Jolmolungma region, *in* A report of scientific expedition in the Mt. Jolmolungma region, 1966–1968, (Palaeontology), Fasc. II]: Beijing, Science Press, p. 271–357, 12 pls. (in Chinese).

Jin Yu-gan and Hu Shi-zhong, 1978, Brachiopods of the Kuhfeng Formation in southern Anhui and Nanjing Hills: Acta Palaeontologica Sinica, v. 17, no. 2, p. 101–127, 4 pls. (in Chinese with English abstract).

Jin Yu-gan, Liang Xi-luo [Ling Hsi-lo], and Wen Shih-hsuan, 1977, Additional material of animal fossils from the Permian deposition on the northern slope of Mount Jolmolungma Feng: Scientia Geologica Sinica, no. 3, p. 236–249, 1 pl. (in Chinese with English abstract).

Jin Yu-gan, Ye Song-ling, Xu Han-kui, and Song Dong-li, 1979, [Brachiopoda, *in* Fossil atlas of northwestern China, Qinghai section I]: Beijing, Geological Publishing House, p. 60–210, 39 pls. (in Chinese).

Liao Zhuo-ting, 1979, [Brachiopod assemblage zone of Changhsing stage and brachiopods from Permian-Triassic boundary beds in China]: Acta Stratigraphica Sinica, v. 3, no. 3, p. 200–208, 1 pl. (in Chinese).

——1979, Uppermost Carboniferous brachiopods from western Gueizhou:

Acta Palaeontologica Sinica, v. 18, no. 6, p. 527–542, 4 pls. (in Chinese with English abstract).

——1980, Brachiopod assemblages from the Upper Permian and Permian-Triassic boundary beds, South China: Canadian Journal of Earth Science, v. 17, p. 289–295.

Liu Di-yong, 1976, [Ordovician brachiopods from Mount Jolmolungma region, *in* A report of Scientific expedition to the Mt. Jolmolungma region, 1966–1968, (Palaeontology), Fasc. II]: Beijing, Science Press, p. 138–155, 2 pls. (in Chinese).

——1979, Earliest Cambrian brachiopods from Southwest China: Acta Palaeontologica Sinica, v. 18, no. 5, p. 506–511, 2 pls. (in Chinese with English abstract).

Rong Jia-yu, 1979, [The *Hirnantia* fauna of China with comments on the Ordovician-Silurian boundary]: Acta Stratigraphica Sinica, v. 3, no. 1, p. 1–29, 2 pls. (in Chinese).

Rong Jia-yu and Yang Xue-chang, 1977, On the *Pleurodium* and its relative genera: Acta Palaeontologica Sinica, v. 16, no. 1, p. 73–80 (in Chinese with English abstract).

——1978, Silurian spiriferoids from southwestern China with special reference to their stratigraphic significance: Acta Palaeontologica Sinica, v. 17, no. 4, p. 357–386, 3 pls. (in Chinese with English abstract).

——1980, Brachiopods of the Miaokao Formation (Uper Silurian) of Qujing, eastern Yunnan: Acta Palaeontologica Sinica, v. 19, no. 4, p. 263–288, 4 pls. (in Chinese with English abstract).

——1981, Middle and late Early Silurian brachiopod faunas and ecological communities: Memoirs of Nanjing Institute of Geology and Palaeontology, Academia Sinica, no. 13, p. 163–270, 26 pls. (in Chinese with English abstract).

Su Yang-zheng, Li Li, and Gu Feng, 1976, [Brachiopoda, *in* Fossil atlas of North China, Neimongol section II]: Beijing, Geological Publishing House, p. 155–306, 109 pls. (in Chinese).

Sun Don-li, 1981, New rhynchonellid brachiopods from Upper Jurassic of Huling County, Heilongjiang Province (in Chinese with English abstract) (in press).

Tong Zheng-xiang, Xu Zing-jian, and Wang Zheng-guan, 1979, [Brachiopoda, *in* Fossil atlas of western South China, Sichuan section II]: Beijing, Geological Publishin House, p. 210–315, 27 pls. (in Chinese).

Wang Yu, Jin Yu-gan, and Fang Da-wei, 1964, [Fossil Brachiopoda of China]: Beijing, Science Press, 777 p., 136 pls. (in Chinese).

Wang Yu, Fang Bing-xiang, Jin Yu-gan, Liao Zhuo-ting, Liu Di-yong, Rong Jia-yu, Sun Dong-li, Wu Qi, Xu Han-kui, and Yang Xue-chang, 1974, [Brachiopoda, *in* Handbook of Palaeontology and stratigraphy of southwestern China]: Beijing, Science Press, p. 113, 144–154, 195–208, 240–247, 275–283, 308–314, 351–353, Pls. 44, 64–66, 92–96, 121–126, 142–147, 162–165, 184 (in Chinese).

Wang Yu, Yu Chang-min and Wu Qi, 1974 [Advances in the Devonian biostratigraphy of South China]: Beijing, Science Press, Memoirs of Nanjing Institute of Geology and Palaeontology, Academic Sinica, no. 6, p. 1–72, 17 pls. (in Chinese).

Xu Han-kui, 1979, Brachiopoda from the Tangxiang Formation (Devonian) in Nandan of Guangxi: Acta Palaeontologica Sinica, v. 18, no. 4 p. 362–380, 2 pls. (in Chinese with English abstract).

Yang De-li, Ni Si-zhao, Chang Mei-li, and Zhao Ru-zuan, 1977, [Brachiopods: *in* Fossil atlas of central south China, II]: Beijing, Geological Publishing House, p. 303–470, 59 pls. (in Chinese).

Yang Shi-pu, 1964, [On the brachiopods from Middle and Lower Carboniferous deposits on northern slope of Borohoro Mountain, Xingjian, and their stratigraphic significance]: Beijing, Science Press, p. 1–179, 23 pls. (in Chinese with Russian abstract).

——1964, [Tournaisian brachiopods from southern Guizhou]: Acta Palaeontologica Sinica, v. 12, no. 1, p. 82–109, 3 pls. (in Chinese with Russian abstract).

Yang Zun-yi, Ting Pei-zhen, Yin Hon-fu, Chang Shou-xin, and Fang Jia-song, 1964 [The Carboniferous, Permian and Triassic Brachiopods from Qilian Shan Region, *in* Contribution to the geology of the Mt. Qilianshan]: Beijing, Science Press, v. 4, no. 4, p. 1–126, 68 pls. (in Chinese).

Yang Tsun-yi [Yang Zun-yi] and Xu Gui-rong [Xu Gueiyong], 1966 Triassic Brachiopoda of Central Guizhou Province: Zhonggua Gongye Chubanshe [China Industry Press]. Beijing, 124 p., 12 pls. (in Chinese with English summary.)

Ye Song-ling and Yang Shen-qiu, 1979, Brachiopods from the Bagon Lake Series, northern Tibet: Acta Palaeontologica Sinica, v. 18, no. 1, p. 64–71, 1 pl. (in Chinese with English abstract).

Zheng Qin-luan, 1976 [Brachiopoda, *in* Fossil atlas of central South China, I]: Beijing, Geological Publishing House, p. 27–70, 14 pls. (in Chinese).

Zhang Shou-xin and Jin Yu-gan, 1976, [Upper Palaeozoic brachiopoda from the Mount Jolmolungma region, *in* A report of scientific expedition in the Mount Jolmolungma region, 1966–1968, (Palaeontology), Fasc. II]: Beijing, Science Press, p. 159–242, 19 pls. (in Chinese).

Zhong Hua, 1977, Preliminary study on the ancient fauna of South China and its stratigraphic significance: Scientia Geologica Sinica, no. 2, p. 118–128 (in Chinese with English abstract).

Ziegler, A. M., 1965, Silurian marine communities and their environmental significance: Nature, v. 207, p. 270–272.

MANUSCRIPT ACCEPTED BY THE SOCIETY APRIL 22, 1981

Printed in U.S.A.

Geological Society of America
Special Paper 187
1981

The genus *Jiangxiella* and the origin of Pseudocardiniids (Bivalvia)

Liu Lu*
Nanjing Institute of Geology and Palaeontology, Academia Sinica, Nanjing 210008, People's Republic of China and Department of Geological Sciences, University of Rochester, Rochester, New York 14627, U.S.A.

Chen Jin-hua
Nanjing Institute of Geology and Palaeontology, Academia Sinica, Nanjing 210008, People's Republic of China

ABSTRACT

Recent study of the palaeoheterodont genus *Jiangxiella* and the related genera and subgenera *Guangdongella, Lilingella (Lilingella), L. (Xinyuella), L. (Hunanella), L. (Apseudocardinia),* and *Hamiconcha* gen. nov., has demonstrated that the Middle Jurassic freshwater genus *Pseudocardinia,* which is widely distributed in Asia, was derived from the genus *Jiangxiella,* which occurs in brackish-water, Upper Triassic facies of South China. The recognition of an evolutionary series extending from *Jiangxiella* through *Lilingella* to *Pseudocardinia* sheds new light on the origin of freshwater heterodont bivalves in Asia. Emended diagnoses of the family Pseudocardiniidae Martinson, 1961, and the genus *Jiangxiella* Liu (in Gu and others, 1976), are presented, and a new genus, *Hamiconcha,* is proposed. Extensive descriptions and discussions are given of the four known genera and subgenera of this evolutionary series. Two new species of *Jiangxiella (J. cuneata* and *J. xinyuensis)* and three species of *Guangdongella (G. equisita, G. longimorpha, G. brevicula)* are described.

INTRODUCTION

Abundant small bivalves occurring in the upper Triassic coal-bearing layers in Hunan and Jiangxi Provinces of South China have formerly been identified as species of *Cardinioides, Trigonodus,* and *Unionites* or, earlier, as Jurassic *Pseudocardinia, Sibireconcha,* and *Tutuella.* In 1967, however, Liu demonstrated that the hinge structures of these forms are distinctly different from those of the genera to which they had been formerly assigned. He proposed a new genus, *Jiangxiella,* for them and described and illustrated it in an unpublished fossil atlas of Late Triassic and Early Jurassic faunas and floras in South China. The name *Jiangxiella,* however, was widely adopted in Triassic biostratigraphic work in the southern provinces; *Jiangxiella* was formally published in 1976 in *Fossil Lamellibranchs of China* (Gu and others, 1976, p. 56). Similarly, Chen and Liu (in Liu, 1967) recognized the new Early Jurassic nonmarine genus *Lilingella* from South China on the basis of its peculiar configuration and hinge. This genus was published in 1977 (in Zhang and others, 1977). Subsequent investigations of Late Triassic to Middle Jurassic biostratigraphy in Guangdong, Hunan, and Jiangxi Provinces resulted in the discovery of many other new forms closely related to and intermediate between *Jiangxiella* and *Lilingella;* for example, the Late Triassic genus *Guangdongella* and the Early Jurassic genera *Hunanella* and *Xinyuella.* Recently, Liu Xie-zhang and Zhu Xian-zhu (in Cai and Liu, 1978) proposed a new genus *Apseudocardinia* for an Early to (?)Middle Jurassic bivalve from the Sichuan basin. These genera further bear a close relationship to the well-known genus *Pseudocardinia* Martinson, 1959, which is widely distributed in the Middle Jurassic deposits of Asia.

For many years we have been impressed with similarities in the hinge structures and general morphology of the shell in these diverse genera, as well as the continuous stratigraphic record and apparent successive migrations of these forms from marine to continental environments.

This study concerns the origin and dispersal of the pseudocardiniids in Asia from the Late Triassic to the Middle Jurassic, and their radiation into freshwater habitats. The related genera and subgenera that are presently recognized are shown in Table 1.

These taxa are fully described in the systematic section of this paper, but some general remarks on their unique morphologic features are necessary in relation to the succeeding discussions on evolutionary trends. The above taxa are generally characterized by (1) similar adductor and pedal muscle insertion areas, (2) one posterior lamellar tooth in the left valve and two in the right valve, and (3) fine transverse grooves in all pseudocardinal and lamellar teeth. Shells with similar outlines and external features appear repeatedly in the Upper Triassic, Lower Jurassic, and Middle

*Present address: Lu Liu, Amoco Orient Petroleum Company, P.O. Box 4381, Houston, Texas 77210.

TABLE 1. GENERA AND SUBGENERA OF ASIAN PSEUDOCARDINIIDAE

Taxon	Habitat and age
Genus *Pseudocardinia* Martinson, 1959	Nonmarine; Middle Jurassic
Genus *Jiangxiella* Liu, 1976 (in Gu and others, 1976)	Brackish water; Late Triassic
Genus *Guangdongella*, Li and Li, 1977 (in Zhang and others, 1977)	Brackish water; Late Triassic
Genus *Lilingella* Chen and Liu, 1977 (in Zhang and others, 1977)	Nonmarine; Early Jurassic
Subgenus *Lilingella* Chen and Liu, 1977	Nonmarine; Early Jurassic
Subgenus *Xinyuella* Chen and Xu, 1980	Nonmarine; Early Jurassic
Subgenus *Hunanella* Xiong and Wang, 1979	Nonmarine; Early Jurassic
Subgenus *Apseudocardinia* Liu and Zhu, 1978 (in Cai and Liu, 1978)	Nonmarine; Early Jurassic
Genus *Hamiconcha* Huang, Wei, and Chen (gen. nov., herein)	Nonmarine; Middle Jurassic

Jurassic, but they are homeomorphs and differ in their hingement. The strong pseudocardinal teeth in early forms became weakened and disappeared in later forms; or the strong pseudocardinals, after being weakened in early stages, ultimately reversed the trend and again became strong and gibbous in the youngest forms. The extensions at the anterior ends of pseudocardinals in older forms developed into initially small and successively elongated anterior lamellar teeth. In the above forms, therefore, the nature of the musculature, the posterior lamellar teeth, and the transverse grooves on the teeth are relatively stable features, whereas the pseudocardinals and the anterior lamellars underwent considerable evolution.

EVOLUTIONARY TRENDS

Discussing the evolution of hinge structures in Mesozoic heterodonts, Casey (1952) recognized that the connection or separation of the cardinals and the anterior laterals, and the strengthening or weakening of the cardinals were the major evolutionary trends of the heterodont hinge. Although the present study is confined to the paleoheterodonts, the evolution of their hinge also is restricted to pseudocardinals and anterior lamellars. The similarities of these genera to the heterodonts, particularly in the evolution of teeth, is interpreted as the result of close phylogenetic relationships. As a result of the present study, four evolutionary stages, or patterns, of hinge development have been recognized in the phylogeny of the lineage.

Early Jiangxielloid Stage.

This stage is represented by *Jiangxiella* and *Guangdongella* of Late Triassic brackish-water habitats. The hinge evolution of this stage may be subdivided into three phases (Fig. 1a–1c) which may be described as follows: (1) In the most primitive forms of the lineage, tooth 1 is acutely angular in shape; teeth 2a and 2b meet at an angle of less than 90°; teeth 3b and 4b run obliquely to the hinge margin; and anterior lamellar teeth are absent [Fig. 1a; Pl. 1(1–3); Pl. 3 (10, 12)]. (2) In later stages of evolution, pseudocardinal tooth 1 attained the shape of a right-angled triangle with the two basal corners protracting transversely; teeth 2a and 2b meet at an angle of approximating 90° [Fig. 1b; Pl. 1(4, 8, 13); Pl. 13(17, 18)]. (3) In the latest evolutionary phase of the early jiangxielloid hinge, pseudocardinal tooth 1 became laterally elongated, and either the apex of tooth 1 or the contact angle of teeth 2a and 2b became obtuse [Fig. 1c; Pl. 1(6–12); Pl. 3(16))].

Late Jiangxielloid Stage

This evolutionary stage is also represented by Late Triassic brackish-water species of *Jiangxiella* and *Guangdongella* and includes two successive phases of development. d and e in Figure 1. In phase d, pseudocardinal tooth 1 is extended to the upper side of the anterior adductor insertion area (Fig. 1d); and teeth 2a and 3a begin to extend anteriorly and to become lamellar in shape (Fig. 1d). The anterior prolongations of these pseudocardinal teeth represent the rudimentary elements of anterior lamellar teeth AI, AII, and AIII [Fig. 1d; Pl. 2(5–13); Pl. 3(8)].

In evolutionary phase e, lamellar teeth AI, AII, and AIII were more protracted anteriorly and curved along the anterodorsal margin (Fig. 1e); pseudocardinal teeth 1, 2a, and 3a remained connected with teeth AI, AII, and AIII, but were narrower; tooth 2a was entirely connected with 2b, the angle between them being wider, and tooth 2b appears to be the arched posterior portion of tooth 2a; teeth 3b and 4b are smaller, narrower, and oriented parallel to the shell margin; the anterior lamellar teeth are extended over the anterior adductor insertion area [Fig. 1e; Pl. 2(1–4); Pl. 3(1–3, 20)].

Lilingelloid Stage

This stage is represented by the Lower Jurassic nonmarine subgenera *Lilingella (Lilingella), L. (Hunanella), L. (Xinyuella),* and *L. (Apseudocardinia).*

Evolutionary stages leading to development of the lilingelloid hinge pattern include degeneration of the pseudocardinals of the late jiangxielloid hinge pattern and anterior protraction of the lamellar teeth. Pseudocardinal tooth 1 became laterally compressed, changed shape from obtusely triangular to a short lamellar form, and developed into the thick, curved posterior part of AI (Fig. 1f). At the same time, tooth 2a merged with the posterior

Figure 1. Evolutionary trends of pseudocardiniid hinge.
Early jiangxielloid stage. Upper Triassic. Three pseudocardinal teeth in each valve and two in right valve; no anterior lamellar teeth.
a. First phase: Tooth 1 is acutely angular; teeth 2a and 2b meet at an angle less than 90°.
b. Second phase: Tooth 1 is right-angled triangle; teeth 2a and 2b meet at an angle approximately 90°.
c. Third phase: Teeth 1, 2a, and 3a laterally elongated; either the apex of tooth 1 or the contact angle of teeth 2a and 2b become obtuse.
Late jiangxielloid stage. Upper Triassic. Appearance of anterior lamellar teeth resulted by the anterior protraction of pseudocardinal teeth 1, 2a, and 3a.
d. Early phase: Tooth 1 extends to upper side of anterior adductor insertion area, and teeth 2a and 3a begin to protract anteriorly, thus forming short anterior lamellar teeth AI, AII, and AIII.
e. Later phase: Lamellar teeth AI, AII, and AIII protracted to anterior side of anterior adductor insertion area; tooth 1 became weak.
Lilingelloid stage. Lower Jurassic.
f. Two anterior and two posterior lamellar teeth in right valves; one anterior and one posterior lamellar teeth in left valve.
Pseudocardinioid-hamiconchoid stage. Middle Jurassic.
g. Stage of pseudocardinioid hinge: Pseudocardinal teeth absent in both valves; two anterior and two posterior lamellar teeth in right valves; one anterior and one posterior lamellar teeth in left valve.
h. Stage of hamiconchoid hinge: Triangular tooth 1 and nodelike tooth 3a appeared again in right valve; teeth 2a and 2b redifferentiated in left valve; thus, recall the hinge pattern in phase d and e of late jiangxielloid stage.

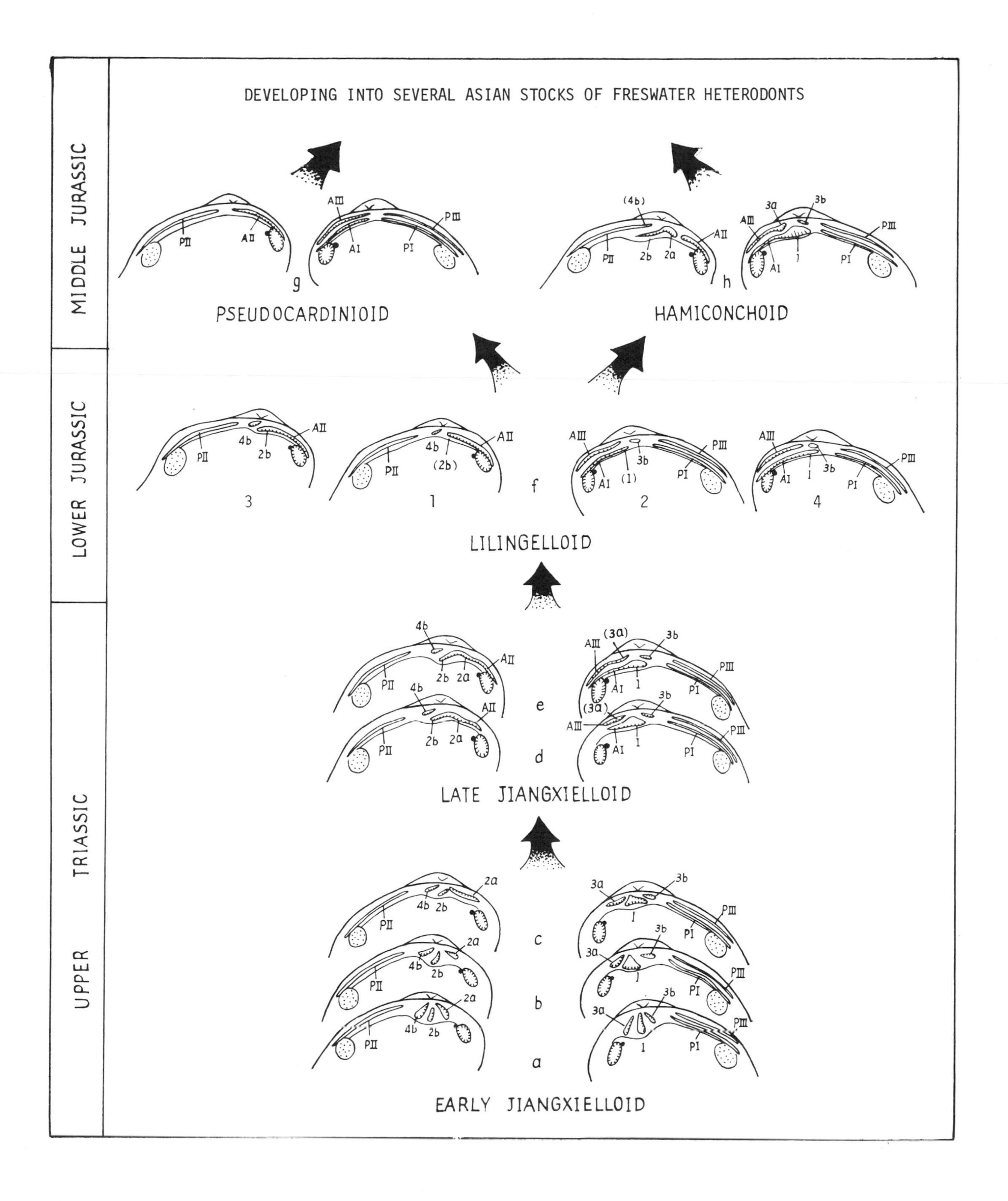

DEVELOPING INTO SEVERAL ASIAN STOCKS OF FRESWATER HETERODONTS
MIDDLE JURASSIC
LOWER JURASSIC
UPPER TRIASSIC
PSEUDOCARDINIOID
HAMICONCHOID
LILINGELLOID
LATE JIANGXIELLOID
EARLY JIANGXIELLOID
g
h
f
e
d
c
b
a
3
1
2
4
PII
AII
AIII
AI
PI
PIII
(4b)
2b
2a
3a
3b
4b
(2b)
(1)
(3a)
1

part of tooth AII. After junction of teeth 2a and 2b at their upper ends, tooth 2b became thick and merged with the curved poterior part of tooth AII. Teeth 3b and 4b became weakened, but are still distinguishable [Fig. 1f; Fig. 11e, f; Pl. 4; Pl. 4; Pl. 6; Pl. 7(11)]. In the lilingelloid hinge, the morphology of pseudocardinal teeth 1, 2b, 3b, and 4b vary according to the shell outline. In elongated shells, these teeth were always lamellar, whereas in shorter forms they become short lamellar or nodelike.

Pseudocardinioid-Hamiconchoid Stage

This stage is represented by the freshwater Middle Jurassic *Pseudocardinia* and *Hamiconcha*. During this stage, the Lower Jurassic lilingelloid hinge evolved along two contrasting lines: (1) the hinge was further simplified, as in *Pseudocardinia*, or (2) it became again complicated, as in *Hamiconcha*.

In the first case, pseudocardinal teeth 1, 2a, and 3a were fused to the anterior lamellar teeth AI, AII, and AIII; yet these anterior lamellars evolved from those pseudocardinals. The posterior lamellar tooth PII extended forward and became fused with tooth 4b [Fig. 1g; Pl. 7(4–7). Owing to the degeneration of the pseudocardinal teeth during this evolution, the hinge structures became simplified and, with few exceptions, were composed of only lamellar teeth. Although pseudocardinals appear in some species of *Pseudocardinia*, they are merely very weakly developed, nodelike processes.

In contrast, in the evolution of the second lineage, the weak pseudocardinals of the lilingelloid stage (for example, in *Apseudocardinia)*, became prominent again. Central pseudocardinal tooth 1 has a triangular outline and is similar in strength to the hinge of the jiangxielloid stage; teeth 2a and 2b became differentiated again; tooth 3a remained connected to tooth AIII, became nodelike, and was stronger than 3a of the early jiangxielloid stage [Fig. 1h; Pl. 7(1–3, 10)]. Moreoever, tooth 6b is developed in some individuals, a condition never observed in the preceding stages. With restrengthening in this lineage, the anterior and posterior lamellars of *Hamiconcha* remain very well developed. The uniqueness of the hinge in this form lies in the fact that the pseudocardinals regained a level of prominence found in the ancestral jiangxielloid pattern, whereas the anterior lamellar teeth retained their advanced character previously derived from the lilingelloid hinge. This indicates that the hinge pattern of *Hamiconcha* is not a reversion to the primitive jiangxielloid hinge, but a more advanced level of evolution that represented an adaptive advantage in response to physical environmental factors similar to those present during the jiangxielloid stage. The differentiation of pseudocardinals and lamellars in *Hamiconcha* is also noteworthy; the hinge characteristics of *Hamiconcha* are similar to those of primitive heterodonts. This evolutionary series probably signals the appearance of a principal nonmarine heterodont lineage.

PHYLOGENETIC RELATIONSHIPS

At the time he erected the genus, Martinson (1959) considered that *Pseudocardinia* was derived from marine Early Jurassic *Cardinia*. Later, Kolesnykov (1974) reaffirmed this assumption, regarding *Cardinia* as the ancestor of *Pseudocardinia*. After a careful study of the hinge and muscle characteristics of *Pseudocardinia* and *Cardinia*, we do not believe that *Pseudocardinia* can be derived from *Cardinia*. The two genera are not related, and there is no real morphological linkage between them. In *Pseudocardinia*, both the anterior and posterior lamellar teeth are quite narrow and long, extending from near the umbo to the anterolateral and posterolateral margins, respectively, of the anterior and posterior adductor insertion areas [Fig. 3g, 3h; Pl. 7(6, 7); Gu and others, 1976, Pl. 98, figs. 2, 10; Ma and others, 1976, Pl. 12, fig. 22]. The lamellar teeth are transversely grooved. The lateral teeth in *Cardinia*, however, are all nodelike or very short lamellar processes that are weak relative to the wide hinge plate of the genus, and none of the teeth in *Cardinia* shows any trace of grooves or denticles. Furthermore, the two abductor muscle insertion areas in *Pseudocardinia* are quite different in their relative development; the anterior area is more prominent and is marked by irregular transverse grooves; there is a pedal retractor insertion area near the dorsoposterior edge of the adductor insertion area. In contrast, the posterior one is usually weak and smooth. In *Cardinia*, however, the anterior and posterior adductor insertion areas are nearly equal in strength and lack transverse grooves (Hayami, 1958, Pl. 11).

In *Cardinia*, the hinge is characterized by a typical heterodont dental pattern, whereas in *Pseudocardinia*, a paleoheterodont type of hinge prevails. It is generally accepted in the study of bivalves that the paleoheterodont hinge is ancestral to the heterodont hinge. Martinson (1959, 1961) and Kolesnykov (1974), in their study of the genus *Pseudocardinia*, suggested the opposite view. Furthermore, *Cardinia s. s.* has been found only in fully marine deposits. There are no stratigraphic or paleoecologic data to support the concept that *Cardinia* was euryhaline. The widespread extinction of this genus after the Middle Jurassic may reflect, in part, the inability of *Cardinia* to adapt more variable, regressive environments of the Late Jurassic, including widespread development of brackish and freshwater facies. The stenohaline nature of *Cardinia* thus prevented some members of the genus from finding refuge in less saline facies as marine habitats diminished. If stenohline marine *Cardinia* is the ancestor of the freshwater *Pseudocardinia*, intermediate forms should be found in intermediate environments through time. This is not the case in the preserved stratigraphic record.

The discovery of abundant *Jiangxiella, Guangdongella, Lilingella*, and *Hunanella* in the Upper Triassic and Lower Jurassic of Hunan, Jiangxi, and Guangdong Provinces of South China, has provided important clues for the interpretation of the phylogeny of *Pseudocardinia* and related genera and for tracing of their origin. These genera exhibit close affinities with *Pseudocardinia* in shell morphology, muscle characteristics, and hinge structures, and a successive evolutionary pattern of hinge development has been detected among them (see previous discussion).

In Guangdong, Hunan, and Jiangxi Provinces of South China, the Upper Triassic coal-bearing deposits yielding *Jiangxiella* and *Guangdongella*, in association with *Bakevelloides* and *Waagenoperna*, represents a brackish estuarine facies; the Lower Jurassic rocks that yield *Lilingella (Lilingella), L. (Xinyuella)*, and *L. (Hunanella)* are brackish delta front deposits, or at least, strongly brackish estuarine facies. In the Middle Jurassic (and questionably the late Early Jurassic), most areas of China, including

Guangdong, Hunan, and Jiangxi Provinces, were areas of continental deposition; freshwater lacustrine facies in these regions contain abundant *Pseudocardina, Apseudocardinia,* and *Hamiconcha* (Si and Zhou, 1962; Zhao and others, 1962; Gu, 1962; Fan, 1963; Gu and others, 1976; Li, 1977; Huang and Chen, 1980; Chen and others, 1980).

From Late Triassic to Middle Jurassic, important changes from marine to continental environments occurred in South China. The discovery of prolific jiangxielloid bivalve assemblages in deposits of the Late Triassic brackish estuarine facies indicates that the ancestors of *Jiangxiella* may have been marine paleoheterodonts that became adapted to brackish water environments. The appearance of lilingelloid forms exclusively in the more brackish facies of the Lower Jurassic demonstrates broad euryhaline adaptations in jiangxielloid forms. The geologic and geographic distributions of the jiangxielloid to pseudocardinioid stages are shown in Table 2, and the geographic migration and distribution of these genera are illustrated in Figure 2.

The recognition of this evolutionary series sheds new light on the origins and dispersal of the widely distributed late Mesozoic freshwater heterodonts in Asia. We believe that the pseudocardinioid and hamiconchoid patterns of hinge development have very close affinities with the patterns of certain heterodonts of the late Mesozoic, such as some corbiculids and pisidiids. In Pisidiidae, the number of anterior and posterior laterals is identical with those of the latest stage in the pseudocardiniid evolutionary series, whereas the cardinals of Pisidiidae are weakened. It seems probable that primitive forms of the Pisidiidae might have been derived from forms with a pseudocardinioid type of hinge. In corbiculids, however, the identical number of laterals and the better development of the cardinals suggest their probable derivation from pseudocardiniids with a hamiconchoid type of hinge. This will be further discussed by us in future publications.

We are aware of the probable polygenesis of freshwater heterodonts known throughout eastern Asia. In Japan, the widespread brackish-water deposits during Late Triassic and Early Jurassic times, indicate environmental conditions similar to those in South China. Certain endemic Japanese elements such as *Cardinioides* (Kobayashi and Ichikawa, 1952) and *Crenotrapezium* (Hayami, 1958) occur in these facies (Hayami, 1961, 1975). Some of these forms might be ancestral to certain late Mesozoic freshwater heterodonts in Japan and northeastern China, such as *Tetoria* (Kobayashi and Suzuki, 1937), *Paracorbicula* (Kobayashi and Suszuki, 1939), and *Fulpioides* (Gu and others, 1976), representing an independent line of nonmarine heterodont evolution.

TABLE 2. DISTRIBUTION OF PSEUDOCARDINIIDAE THROUGH TIME AND SPACE

Genera and Subgenera	Geological Age	Geographical Distribution	Environment
Pseudocardinia	Middle	South and Northwest China, Mongolia, Siberia, and Central Asia	Freshwater
Hamiconcha	Jurassic		
Apseudocardinia			
Lilingella			
Hunanella	Early	South China	Nonmarine
	Jurassic		
Xinyuella			
Apseudocardinia			
Jiangxiella	Late	Jiangxi, Hunan and Guangdong, South China	Brackish
Guangdongella	Triassic		

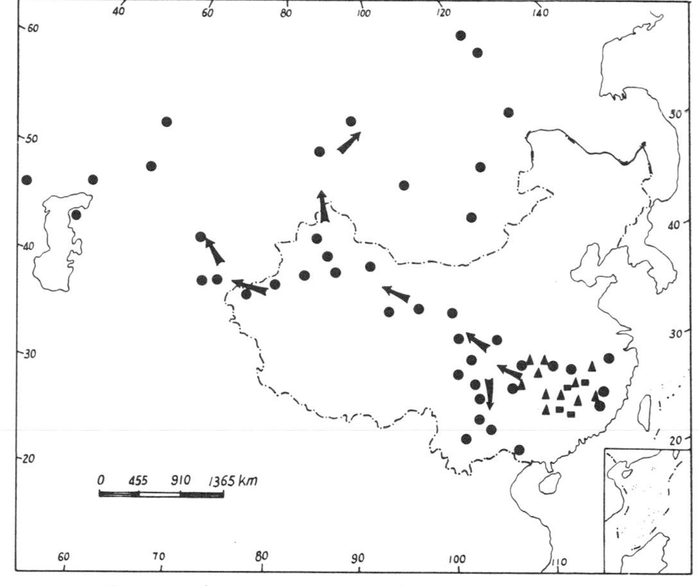

• *Pseudocardinia, Hamiconcha, Apseudocardinia*
▲ *Lilingella, Xinyuella, Hunanella*
■ *Jiangxiella, Guangdongella*

Figure 2. Geographic migration and dispersal of Pseudocardiniidae. Circles: Known fossil localities of freshwater Middle Jurassic *Pseudocardinia* and *Hamiconcha* and freshwater upper Lower Jurassic *Lilingella (Apseudocardinia).* Triangles: Known fossil localities of nonmarine Lower Jurassic *Lilingella (Lilingella), L. (Xinyuella),* and *L. (Hunanella).* Quadrates: Known fossil localities of brackish-water Upper Triassic *Jiangxiella* and *Guangdongella.* Arrows: Major directions of migration and dispersal of pseudocardiniids from Late Triassic to Middle Jurassic. (Data outside China after Kolesnykov, 1974.)

SYSTEMATIC DESCRIPTIONS

Subclass PALAEOHETERODONTA Newell, 1965
Order UNIONOIDA Stoliczka, 1871
Superfamily PSEUDOCARDINIACEA Martinson, 1961
Family PSEUDOCARDINIIDAE Martinson, 1961

The diagnosis of this family is herewith emended as follows: Shell of small to medium size, oval to cuneiform, ophisthodetic. Paleoheterodont hinge, all teeth transversely grooved; pseudocardinal and anterior lamellar teeth variable: primitive forms with two or three developed pseudocardinal teeth in each valve but without anterior lamellar teeth; advanced forms with well-developed anterior lamellar teeth, one in left valve and two in right valve. Posterior lamellar teeth always one in left valve and two in right valve. Anisomyarian: anterior adductor scar deep, irregularly grooved; distinct isolated pedal scar located at upper rear side of anterior adductor insertion area; posterior adductor scar weak and smooth; pallial line entire.

Discussion: When Martinson (1961) erected the family Pseudocardiniidae, he placed two genera, *Pseudocardinia* and *Sibirecon-*

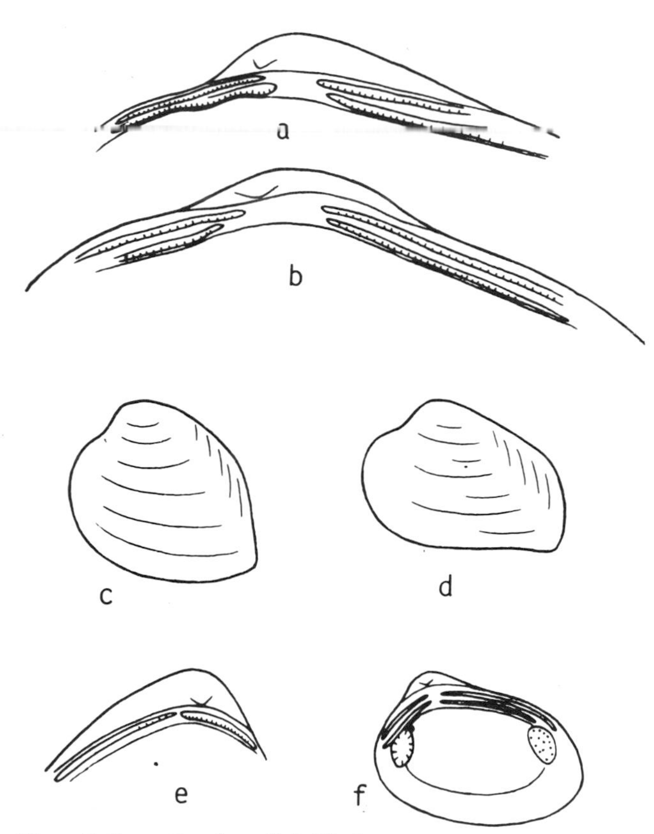

Figure 3. Genus *Pseudocardinia* Martinson.
a, b. *Pseudocardinia lanceolatus* (Chernyshev), Middle Jurassic. Hinge of right valves, $\times 3$, showing phase g of pseudocardinioid hinge.
c. *Pseudocardinia carinata* (Martinson), Middle Jurassic. Left valve exterior, $\times 2$, showing general outline of species.
d. *Pseudocardinia submagma* (Martinson), Middle Jurassic. Schematic diagrams showing internal structure of the left valve (after Martinson, 1959), Middle Jurassic. Left valve exterior, $\times 2$, showing general outline of species.
e, f. *Pseudocardinia kweichouensis* (Grabau), Middle Jurassic. Left and right valve interiors, $\times 2$, showing phase g of pseudocardinioid hinge.

cha, into the family and assigned it to the suborder Preheterodonta. Vokes (1967) followed Martinson's opinion and placed the Pseudocardiniidae in the superfamily Unionacea. On the other hand, Cox (*in* Cox and others, 1969, p. *N*410) assigned the family doubtfully to the superfamily Anthracosiacea. Both authors, however, agreed that the Pseudocardiniidae is an independent family and belongs to the subclass Palaeoheterodonta.

In *Fossil Lamellibranchs of China* (Gu and others, 1976), the taxon Pseudocardiniidae was not accepted, and the genera *Pseudocardinia* and *Jiangxiella* were placed in the family Pachycardiidae, whereas Zhu (1976) continued to consider the Pseudocardiniidae to be a valid family.

Recently, the scope of Pseudocardiniidae has been enlarged by the addition of *Guangdongella, Lilingella, Hunanella, Xinyuella,* and *Apseudocardinia,* and the new genus *Hamiconcha* established in this paper. The anisomyarian adductor, the location of the anterior pedal scar, the identical number of posterior lamellar teeth, and the transverse grooves on the teeth reveal close relations between *Pseudocardinia* and *Jiangxiella.* It is evident that the hingement and the musculature of *Jiangxiella* and related genera differ from those of the Pachycardiidae; in the latter, two posterior lamellar teeth occur in the left valve and one in the right valve; teeth are rarely grooved, and two pedal scars are located above the anterior and posterior adductor insertion areas, respectively. These facts show that the Pachycardiidae is closely related to the Unionidae and that the Pseudocardiniidae and Pachycardiidae probably represent two parallel evolutionary series within the order Unionoida.

Kolesnykov (1977) revised the diagnosis of the family Pseudocardiniidae, raised it to superfamily rank, and considered that because of the presence of weak nodelike "cardinals" this family should be placed in the Heterodonta, whereas we believe that these nodelike "cardinals" are diminished pseudocardinal teeth 3b and 4b of the lilingelloid hinge pattern. Some Chinese paleontologists assigned *Lilingella, Hunanella,* and *Apseudocardinia,* respectively, to the heterodont families Arcticidae, Neomiodontidae, and Corbiculidae (Zhang and others, 1977; Xiong and Wang, 1979; Cai and Liu, 1979). But in these heterodont families, the anterior lateral teeth are well differentiated from the cardinal teeth, the adductors are homomyrian, and the anterior pedal scars are absent.

Distribution and age: Asia, Late Triassic to Middle Jurassic.

Genus *Pseudocardinia* Martinson, 1959
Type species: *Pseudocardinia submagna* Martinson, 1959

Shell of medium size, oval, subcircular to elliptical; umbonal carina present or absent; surface with concentric sculptures only. Left valve with one anterior and one posterior lamellar teeth, weakened towards the umbo. Right valves with two anterior and two posterior lamellar teeth; all teeth with transverse grooves and ridges. Anisomyrian: anterior adductor scar deep, oval, with irregular grooves; posterior adductor scar weak, oval; pallial line entire.

Hinge formula: $\dfrac{\text{AIII} \quad \text{AI} \quad \text{PI} \quad \text{PIII}}{\text{AII} \qquad \text{PII}}$

Discussion: The illustrations of the hingement given by Martinson (1959, p. 34, Figs. 1–3) indicate that the two valves of *Pseudocardinia* do not fit into each other along the hinge. There have been no adequate illustrations of *Pseudocardinia's* hinge in later Russian publications.

Martinson's diagnosis of the genus was critically revised by Gu in *Fossil Lamellibranchs of China* (Gu and others, 1976, p. 346). This revision was based upon the hinge characteristics of Chinese specimens from Hami in Xinjiang, but the hinge of the Hami form is very different from that of Martinson's original, and it is inappropriate to revise the generic diagnosis on the basis of materials collected hundreds of miles from the type locality of the type species. In our opinion, the Hami form represents a new taxon, *Hamiconcha,* as proposed by Huang, Wei, and Chen (written comm). On the other hand, forms that are similar to the type species of *Pseudocardinia* Martinson do occur in the Middle Jurassic of China, for example, *Pseudocardinia kweichouensis* (Grabau) from Hupei, (Gu and others, 1976) and *P. carinata* Martinson from Yunnan (Ma and others, 1976).

Distribution and age: Asia; Middle Jurassic.

Genus *Jiangxiella* Liu, 1976 emend. nov.
Type species: *Jiangxiella subovata* Liu, 1976 (in Gu and others, 1976)

Shell oval, triangular or elliptical in outline. Three pseudocardinal teeth in each valve: tooth one of right valve acutely to obtusely triangular; other teeth nodelike or short lamellar. Anterior lamellar teeth not developed in primitive forms, while in advanced forms one anterior lamellar tooth present in left valve and two in right valve. Posterior lamellar teeth: one in left valve and two in right valve, beginning from below external ligament and extending to upper side of posterior adductor insertion areas; all teeth transversely grooved. Anterior adductor scar deep, oval, with irregular grooves; distinct pedal scar inserted at its posterior upper side. Posterior adductor scar weak, oval, pallial line entire.

Hinge formula:

Early jiangxielloid type

3a	1	3b	PI	PIIII
2a	2b	4b	PII	

Late jiangxielloid type

AIII	AI	3a	1	3b	PI	PIII
AII		2a	2b	4b	PII	

Discussion: This genus was first proposed by Liu (1967), but not formally established until 1976 (in Gu and others, 1976). In the original description of the genus, the triangular pseudocardinal tooth was incorrectly described as being in the left valve, and two posterior lamellar teeth were mentioned in the left valve (Gu and others, 1976, p. 56), although, in fact, the triangular pseudocardinal tooth is in the right valve, and there is in most cases only one posterior lamellar tooth in the left valve.

On the basis of the morphologic variations of the pseudocardinal tooth one and the gradual appearance of anterior lamellar teeth in both valves, forms that are at present placed in *Jiangxiella* can be divided into two major types according to their hinge development; namely, those without anterior lamellar teeth in both valves (early jiangxielloid stage) and those with anterior lamellar teeth in both valves (Late jiangxielloid stage).

1. The early jiangxielloid stage includes the following known species: *Jiangxiella subovata* Liu, 1976 (in Gu and others, 1976, p. 57, Pl. 23, figs. 2, 5–7, *non* Fig. 1); *J. datianensis* Liu, 1976 (in Gu and others, 1976, p. 57, Pl. 23, figs. 8–12); *J. plana* Liu, 1976 (in Gu and others, 1976, p. 57, Pl. 23, figs. 13, 14); *J. elliptica* Liu, 1976 (in Gu and others, 1976, p. 57, Pl. 23, figs 15, 16); *J. ?oblonga* Z. Li, 1977 (Li, 1977, p. 6–7); and *J. cuneata* Liu and J. Chen, sp. nov. Three relevant phases of pseudocardinal teeth development are recognized at this stage: (1) The apical angle of pseudocardinal

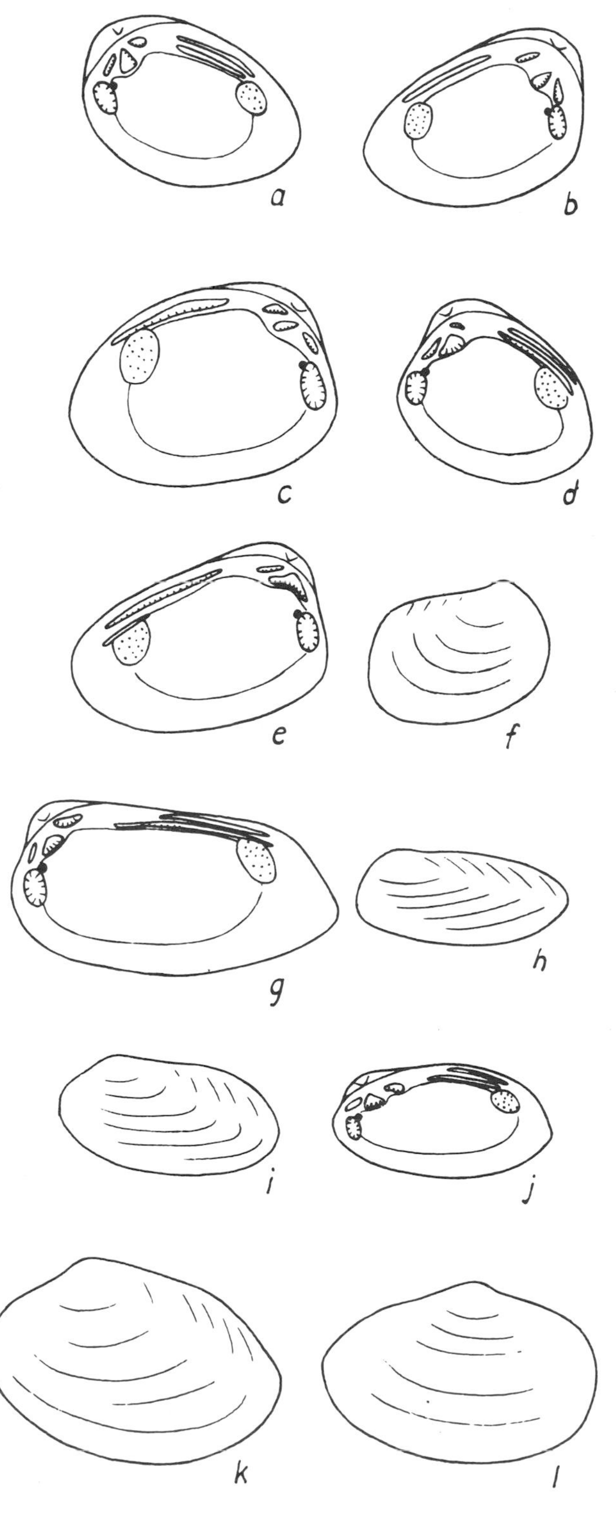

Figure 4. Genus *Jiangxiella* Liu

a, b. *Jiangxiella cuneata* Liu and Chen, sp. nov., Upper Triassic. Right left valve interiors, ×2, showing phase a of early jiangxielloid hinge.

c–f. *Jiangexiella subovata* Liu, Upper Triassic.
- **c. Left valve interior, ×2, showing phase c of early jiangxielloid hinge.**
- **d. Right valve interior, ×2, showing phase b of early jiangxielloid hinge.**
- **e. Left valve interior, ×2, showing phase c of early jiangxielloid hinge.**
- **f. Right valve exterior, ×2, showing general outline of species.**

g, h. *Jiangxiella datianensis* Liu, Upper Triassic.
- **g. Right valve interior, ×2, showing phase c of early jiangxielloid hinge.**
- **h. Left valve exterior, ×1, showing general outline of species.**

i, j. *Jiangxiella elliptica* Liu, Upper Triassic.
- **i. Left valve exterior, ×1, showing general outline of species.**
- **j. Right valve exterior, ×1, showing phase c of early jiangxielloid hinge.**

k, l. *Jiangxiella plana* Liu, Upper Triassic.
Left and right valve exteriors, ×2, ×1, respectively, showing general outline of species.

tooth 1, and the angle of convergence of 2a and 2b are acute [Fig. 1a; Fig. 4a, 4b; Pl. 1(1–3)].(2) These two form right angles or near right angles [Fig. 1b; Fig. 4d; Pl. 1(13)]. (3) They are obtusely angular [Fig. 1c; Fig. 4e, 4g, 4j; Pl. 1(6, 14)]. Obviously, recognition of the above three phases of hinge development affords us the elementary clue for interpreting the evolutionary relationship between Early and late jiangxielloid hinge.

2. The late jiangxielloid stage includes the following species: *Jiangxiella ninyuanensis* Chen and Xu, 1980, and *J. xinyuensis* Liu and Chen, sp. nov. Besides, some individuals of *Jiangxiella subovata* and *J. datianensis* also reveal the late jiangxielloid hinge pattern [Pl. 2,(5, 11)]. The hinge characteristics of this stage are the following:

In the right valve, there is an obtusely angular pseudocardinal tooth 1, its anterior end protracting anteriorly and developing into an anterior lamellar tooth (AI), 3a also extends anteriorly and becomes the second anterior lamellar tooth (AII); 3b is short and lamellar and there are two posterior lamellar teeth. In left valves, an obtusely angular socket is present under the umbo; the two pseudocardinal teeth, 2a and 2b, connect at their apical ends; the anterior ends of 2a protract anteriorly and become the anterior lamellar tooth of the left valve; 2b is short; 4b is short, lamellar, and nearly parallel with the dorsal margin; there is one posterior lamellar tooth.

Two important phases can be distinguished in this late jiangxielloid stage. In the first phase, the anterior lamellar teeth are still rather short; they reach only the upper side of the anterior adductor insertion area [Fig. 1e; Fig. 5a, 5b, 5e; Pl. 2(1)]. This phase is morphologically very close to the last phase of the Early Jiangxielloid hinge, except for the appearance of anterior lamellar teeth. In the second phase of this stage, the anterior lamellar teeth extend to the anterior side of the anterior adductor insertion area and curve slighty downward parallel to the anterodorsal margin.

The differences between the Early and Late Jiangxielloid hinge may provide criteria for separating them as two subgenera. But, for reasons discussed elsewhere, we do not intend to make this separation at present (see discussion under *Guangdongella*).

Distribution and age: South China; Late Triassic.

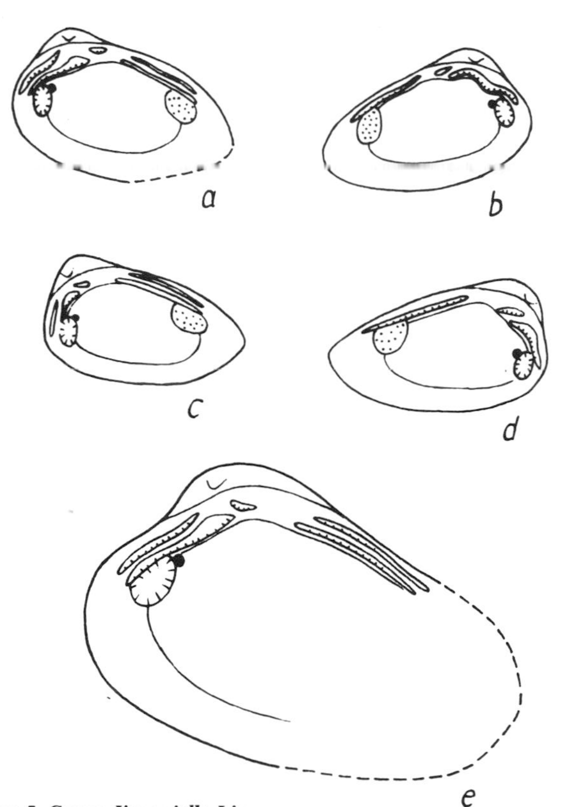

Figure 5. Genus *Jiangxiella Liu*
***a–e. Jiangxiella xinyuensis* Liu and Chen, sp. nov., Upper Triassic.**
a, b. Right and left valve interiors, ×2, showing phase e of late jiangxielloid hinge.
c, d. Right and left valve interiors, ×2, showing phase d of late jiangxielloid hinge.
e. Right valve interior, ×4, showing phase e of late jiangxielloid hinge.

Jiangxiella xinyuensis Liu and Chen, sp. nov.
Figure 5a–5e; Plate 2(1–4)

1976 *Jiangxiella subovata* Liu (in Gu and others, 1976, p. 57, Pl. 23, fig. 1; *non* Figs. 2, 5–7)

1977 *Jiangxella subovata* Liu; Zhang and others, 1977, p. 24, Pl. 1, Fig. 28; *non* Figs. 26, 27, 29

Shell obliquely oval, moderately gibbous; umbos anteriorly located; umbonal carina wide, obscure; surface with growth lines only. Right valve with an obtusely triangular pseudocardinal tooth 1, connected anteriorly with the anterior lamellar tooth AI; pseudocardinal teeth 3a and 3b short, lamellar; two anterior lamellar teeth extend to anterior side of anterior adductor insertion area; two posterior lamellar teeth. Left valve with three pseudocardinal teeth, one anterior and one posterior lamellar tooth; 2a connected with AII; 2b and 4b short, lamellar; all teeth transversely grooved. Anterior adductor scar oval, deep, with irregular grooves, with small, round pedal scar at its upper posterior side; posterior adductor scar suboval, smooth; pallial line entire.

Comparison: This new species differs from *Jiangxiella subovata* Liu in the appearance of the anterior lamellar teeth in both left and right valves.

Locality and horizon: Jiangxi, Hunan; Late Triassic; Anyuan Group.

Jiangxiella cuneata Liu and J. Chen, sp. nov.
Figure 4a, 4b; Plate 1(1–4)

Shell short, cuneiform, of medium size, usually 15 to 20 mm long; umbos high, inflated, located anteriorly one-sixth of the shell length. Anterior dorsal margin steep, slightly curved; posterior dorsal margin long, oblique; ventral margin widely arched; posterior end slightly narrower than anterior one. Surface with fine growth lines or smooth; umbonal craina absent. Each valve with three pseudocardinal teeth; apical angle of tooth 1 and angle between 2a and 2b varies from acute [Pl. 1(1, 2, 3)] to right-angle [Pl. 1(4)]; two posterior lamellar teeth in right valve, one in left valve. Anterior adductor scar strong, oval, irregularly grooved; posterior adductor scar relatively weak, suboval, smooth; pallial line entire.

Comparison: The hingement of this species is similar to *Guangdongella exquisita* Z. Li and Y. Li [Pl. 3(10–14)], whereas externally the present form lacks the acute posterior carina and concentric rugae characteristic of the genus *Guangdongella,* and it is much larger.

Locality and horizon: Xinyu, Jiangxi; Sanjiachong Formation of the Anyuan Group; Upper Triassic.

Genus *Guangdongella* Li and Li, 1977 (in Zhang and others, 1977)
Type species: *Guangdongella exquisita* Li and Li, 1977 (in Zhang and others, 1977); designated herein

Shell small, less than 10 mm long; posterior carina strong, acute; surface with regular concentric rugae in front of the carina, but smooth behind it; hingement and musculature same as in *Jiangxiella.*

Discussion: In the original publication of the genus, three species were described but none was selected as type. According to the suggestion of the original authors (1979, oral comm.), we here designate *Guangdongella exquisita* Z. Li and Y. Li as the type species of the genus.

The genus *Guangdongella* has undergone the same hinge development as *Jiangxiella.* Criteria for distinguishing *Guangdongella* are the size of the shell, the strong and acute posterior carina, and the concentric surface rugae.

As mentioned in the discussion of the genus *Jiangxiella,* the two major hingement patterns in *Jiangxiella* provide criteria for dividing the genus into two separate genera or, at least, subgenera, but the problem is complicated because the same evolutionary stages of the hinge, comparable to the Early and Late Jiangxielloid hinge stages, have also been recognized in the genus *Guangdongella.* The question is which criteria to use for generic and subgeneric distinctions. Some authors have insisted that differences in hinge characteristics should be considered as essential for distinguishing genera and differences in external morphological features for subgenera. According to this opinion, species of *Jiangxiella* and *Guangdongella* with an early jiangxielloid hinge type should be placed in one genus, and those with a late jiangxielloid type of hinge in another; each genus can further be subdivided into two subgenera on the basis of absence or presence of a sharp carina and concentric rugae. Such a classification would, however, cause much confusion even in generic identification, especially when the hinge structures are not exposed.

We classify the two genera by combining the hinge characteristics with external morphological features; that is, the hinge in both groups went through the same evolutionary stages, while the external morphologic features differ. Then, according to whether they have early or late jiangxielloid hinges, each of the two genera can be further divided into two subgenera. The advantage of this classification is apparent not only in the recognition of the parallelism in hinge development in morphologically different groups, but also in the generic identification of specimens in which the hingement is not exposed. The reason we do not formally erect the four subgenera in this paper is that we are still trying to find out some more feasible criteria in addition to the appearance of anterior lamellar teeth to distinguish these subgenera.

Distribution and age: South China; Late Triassic.

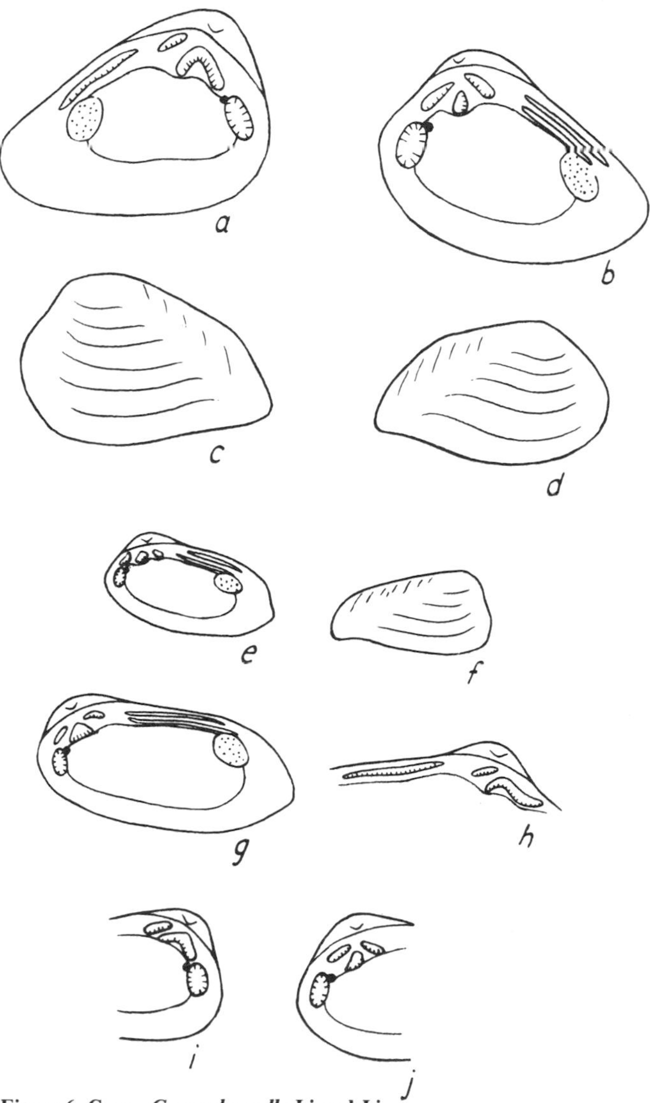

Figure 6. Genus *Guangdongella* Li and Li
a–d. *Guangdongella exquisita* Li and Li, Upper Triassic.
a, b. Left and right valve interiors, ×5, showing phase a of early jiangxielloid hinge.
c, d. Left and right valve exteriors, ×5, showing general outline of species.
e–j. *Guangdongella longimorpha* Li and Li, Upper Triassic.
e, g. Right valve interiors, ×2.5, showing phase b of early jiangxielloid hinge.
f. Right valve exterior, ×2.5, showing general outline of species.
h. Hinge of left valve, ×2.5, showing phase c of early jiangxielloid hinge.
i, j. Hinges of left and right valves, ×5, showing phase b of early jiangxielloid hinge.

***Guangdongella exquisita* Li and Li, 1977**
Figure 6a–6d; Plate 3(10–14)

1977 *Guangdongella exquisita* Li and Li (in Zhang and others, 1977, p. 26, Pl. 2, figs. 1, 2, 4, *non* fig. 3)

Shell small, obliquely triangular, gibbous, umbos located anteriorly; anterior end rounded; posterior end narrow; carina strong, acute. Surface with 10 more-or-less regular concentric rugae in front of, and growth lines only behind, the carina. Left valve with three pseudocardinal teeth; 2a and 2b cuneiform, forming an angle of about 80°; 4b short, lamellar; one posterior lamellar tooth. Right valve with three pseudocardinal teeth; tooth 1 forming an acute triangle; 3a and 3b short, lamellar; two posterior lamellar teeth; all teeth transversely grooved. Anterior adductor scar deep, oval, with irregular grooves; pedal scar located at its upper posterior side; posterior adductor scar relatively weak.

Discussion: In the original illustrations of the species, one internal mold of a right valve (Zhang and others, 1977, Pl. 2, fig. 3) shows the appearance of anterior lamellar teeth. Thus, the species is close to *Guangdongella brevicula* Li and Li [Pl. 3(8)].

Since no holotype was designated in the original publication, we here designate the original of Plate 2, figure 4 (NG 304) as the holotype of the species.

Locality and horizon: Xiaoshui, Guangdong; Hongweikeng Formation; Upper Triassic.

Guangdongella longimorpha Li and Li, 1977
Figure 5e–5j; Plate 3(16–18)

1977 *Guangdongella longimorpha* Li and Li (in Zhang and others, 1977, p. 26, Pl. 2, figs. 1, 2, 4, *non* fig. 3)

Shell small, transversely elliptical; carina acute; surface with eight to ten concentric rugae in front of carina. Right valve with three pseudocardinal and two posterior lamellar teeth; tooth 1 forms right-angled to obtusely angled triangle; 3a and 3b short, lamellar; left valve with three pseudocardinal and one posterior lamellar teeth; angle of convergence between 2a and 2b about 85° to 110°; 4b short, lamellar. Anterior adductor scar strong, oval irregularly grooved, pedal scar inserted at its upper posterior side.

The hinge pattern of the present species belongs to the early jiangxielloid stage.

Locality and horizon: Xiaoshui, Guangdong; Hongweikeng Formation; Upper Triassic.

Guangdongella brevicula Li and Li, 1977
Figure 7a–7d; Plate 3(1–9)

1977 *Guangdongella brevicula* Li and Li (in Zhang and others, 1977, p. 26, Pl. 2, figs. 5–7)

1977 *Guangdongella exquisita* Li and Li (in Zhang and others, 1977, p. 26, Pl. 2, fig. 3, *non* figs. 1, 2, 4)

Shell small, subtriangular; umbos prominent, located anteriorly. Carina strong; posterior ventral end acute, somewhat protracted; anterior end round; ventral margin incurved near posterior ventral angle. Late Jiangxielloid pattern of hinge; pallial line entire.

Locality and horizon: Xiaoshui, Guangdong; Hongweikeng Formation; Upper Triassic.

Genus _Lilingella_ Chen and Liu, 1977 (in Zhang and others, 1977)
Type species: _Lilingella simplex_ Chen and Liu, 1977 (in Zhang and others, 1977, p. 42, Pl. 2, figs. 26, 27); here designated

Shell of medium size; triangular, subcircular, elliptical, or cuneiform; umbonal cavity deep. Each valve with two pseudocardinal teeth; 3b and 4b short, lamellar, nearly parallel with posterior dorsal margin; 1 and 2b connected, respectively, with AI and AII. Right valve with anterior and two posterior lamellar teeth; all teeth transversely grooved. Anisomyarian; anterior adductor scar deep, oval, with irregular grooves; posterior adductor scar weak, subcircular; pallial line entire.

Hinge formula: $\dfrac{\text{AIII}\quad\text{AI}}{\text{AII}} - \dfrac{(1)\quad\text{3b}\quad\text{PI}\quad\text{PIII}}{(2\text{b})\,(4\text{b})\quad\text{PII}}$

Discussion: This genus was proposed by Chen and Liu in 1967 (in a fossil atlas printed for internal circulation by the Nanjing Institute of Geology and Palaeontology) (see Liu, 1967) mainly on the basis of its peculiar shape—the cuneiform outline, the strong and sharp carina, and the acute and protracted posteroventral end. It was validated through publication by Zhang in Zhang and others (1977). Based on his observation on materials of *Lilingella (Lilingella) cuneata* from the Lower Jurassic at Lingling, Hunan Province, Zhang added to the original diagnosis the description of the hingement. He mentioned that there were three cardinal teeth in each valve, two anterior and two posterior lateral teeth in the right valve and one anterior and one posterior lateral teeth in the left valve; thus, he placed the genus into the heterodont family

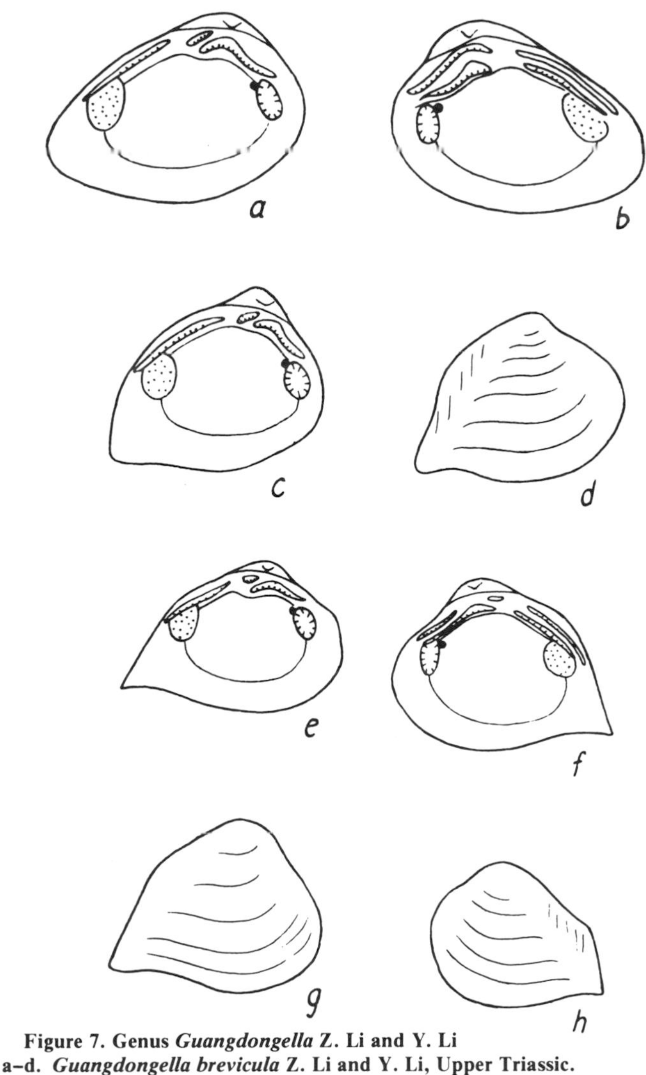

Figure 7. Genus _Guangdongella_ Z. Li and Y. Li
a–d. _Guangdongella brevicula_ Z. Li and Y. Li, Upper Triassic.
a, b. Left and right valve interiors, ×5, showing phase e of late jiangxielloid hinge.
c. Left valve interior, ×5, showing phase e of late jiangxielloid hinge.
d. Right valve exterior, ×5, showing general outline of species.
e–h. _Guandongella? elegans_ J. Chen and Xu, Upper Triassic.
e, f. Left and right valve interiors, ×5, showing phase e of late jiangxielloid hinge.
g, h. Right and left valve exteriors, ×5, showing general outline of species.

Arcticidae. Recent study of abundant materials of *Lilingella (Lilingella) cuneata,* collected by the junior author from the same locality and horizons, shows that there are only two pseudocardinal teeth in each valve, and the pseudocardinal teeth 1 and 2b are not entirely differentiated from the anterior lamellar teeth. Therefore, this is clearly a paleoheterodont hinge.

In the original manuscript, Chen and Liu (in Liu, 1967) designated *Lilingella simplex* as the type species of the genus, but this was not mentioned by Zhang when he published the genus. Thus, we here reaffirm the original designation of the *Lilingella simplex* as the type species of the genus.

At present, this genus includes four subgenera: *Lilingella (Lilingella)* Chen and Liu, 1977 (in Zhang and others, 1977); *L. (Hunanella)* Xiong and Wang, 1979; *L. (Xinyuella)* Chen and Xu, 1980; and *L. (Apseudocardinia)* Liu and Zhu, 1978 (in Cai and X. Liu, 1978).

It should be mentioned that these four subgenera were proposed by different authors within a short period of time. Some of them seem very closely related, and their validity remains to be reviewed.

Distribution and age: South China; Early to Middle Jurassic.

Subgenus *Lilingella (Lilingella)* Chen and Liu, 1977 (in Zhang and others, 1977)

Shell of medium size, triangular, trapezoid or cuneiform. A strong, acute carina extends from the umbo to the more-or-less protracted posteroventral end, the ventral margin just in front of this is slightly incurved; surface in front of carina with regular concentric rugae, smooth behind carina, 1 and 2b weak, greatly diminished in forms with strongly elongated shells; 3b and 4b thin, lamellar.

This subgenus includes three known species: (1) *Lilingella (Lilingella) simplex* Chen and Liu, 1977 (in Zhang and others, 1977, p. 42, Pl. 2, figs. 26, 27); Figure 8c; Plate 4(1); (2) *L. (L.) cuneata* Wu, 1977 (in Zhang and others, 1977, p. 42, Pl. 2, figs. 30, 31); Figure 8a,8b; Plate 4(2–9); and (3) *L. (L.) robusta* Zhou, 1977 (in Zhang and others, 1977, p. 42, Pl. 2, figs. 28, 29); Figure 8d.

Distribution and age: Guangxi, Hunan and Jiangxi Provinces; Early Jurassic.

Subgenus *Lilingella (Xinyuella)* Chen and Xu, 1980
Type species: *Trigonodus? liuyangensis* Gu and Liu, 1976 (in Gu and others, 1976, p. 56, Pl. 22, figs. 21, 22)

Similar to *Lilingella (Lilingella),* but differs in following aspects: (1) teeth 1 and 2b diminished, 3b and 4b very weak; (2) carina obtusely angled in transverse section; (3) posteroventral end not protracted backward, and ventral margin just in front of posteroventral end not incurved; and (4) concentric rugae extend to surface behind carina.

This subgenus contains the following species: *Lilingella (Xinyuella) liuyangensis* (Gu and Liu, 1976); *(Trigonodus? liuyangensis* Gu and Liu, 1976 in Gu and others, 1976, p. 56, Pl. 22, figs. 21, 22); Figure 9c; Plate 7(8); and *L. (X.) glota* Chen and Xu, 1980 (p. 407, Pl. 2, figs. 1–3); Figure 9a,9b; Plate 7(11).

Distribution and age: Jiangxi, Hunan and Guangdong Provinces; Early Jurassic.

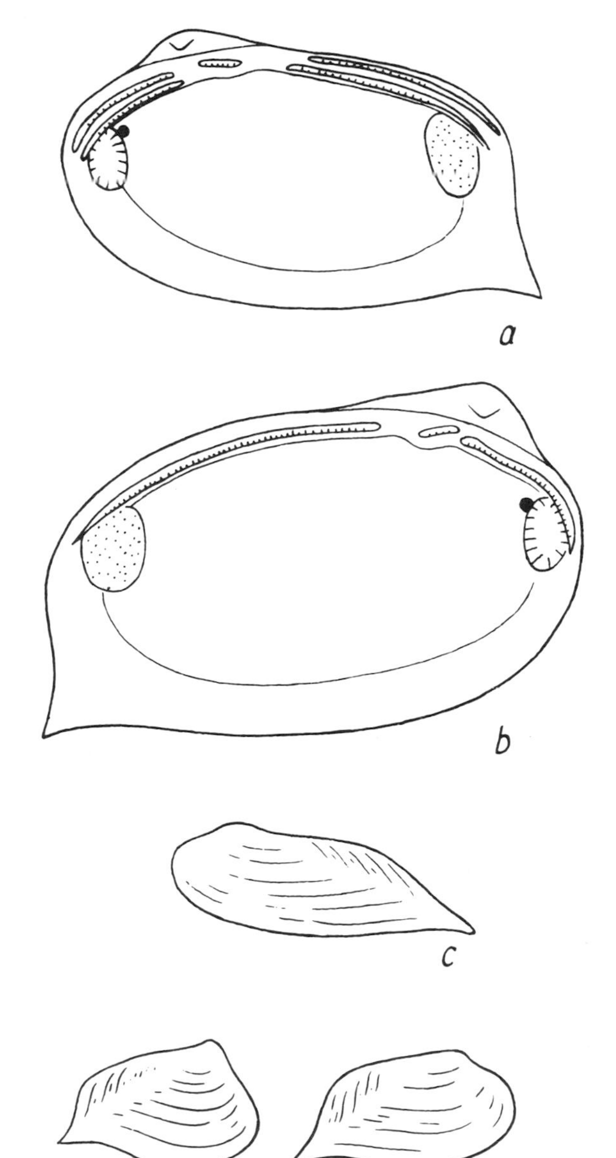

Figure 8. Subgenus *Lilingella (Lilingella)* Chen and Liu
a,b,e. *Lilingella (Lilingella) cuneata,* Wu. Lower Jurassic.
a, b. Right and left valve interiors, ×5, showing phase f of lilingelloid hinge.
e. Right valve exterior, ×1.5, showing general outline of species.
c. *Lilingella (Lilingella) simplex* Chen and Liu, Lower Jurassic. Left valve exterior, ×1, showing general outline of species.
d. *Lilingella (Lilingella) robusta* Zhou, Lower Jurassic. Right valve exterior, ×1, showing general outline of species.

Subgenus *Lilingella (Hunanella)* Xiong and Wang, 1979
Type species: *Hunanella guangyintanensis* Xiong and Wang, 1979

This subgenus is characterized by oval to elliptical outline, weak and rounded carina, absence of concentric rugae, usually nodelike pseudocardinal teeth 1 and 2b, and prominent short lamellar 3b and 4b.

Known species of the subgenus are *Lilingella (Hunanella) guanyintanensis* Xiong and Wang, 1979; Figure 10a–10c; Plate 5(1–

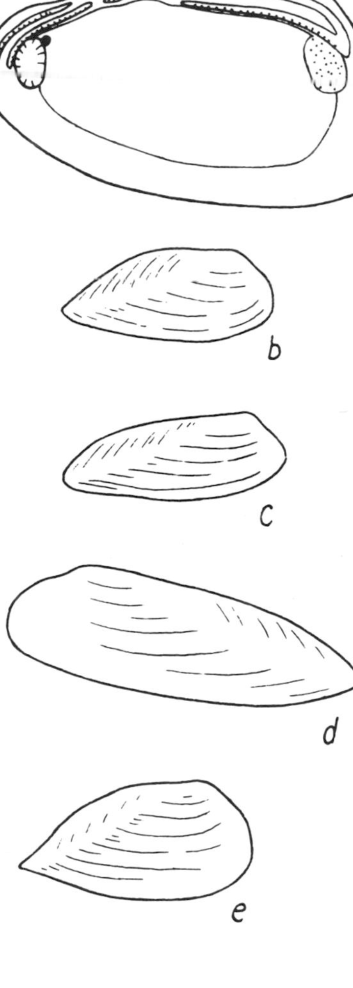

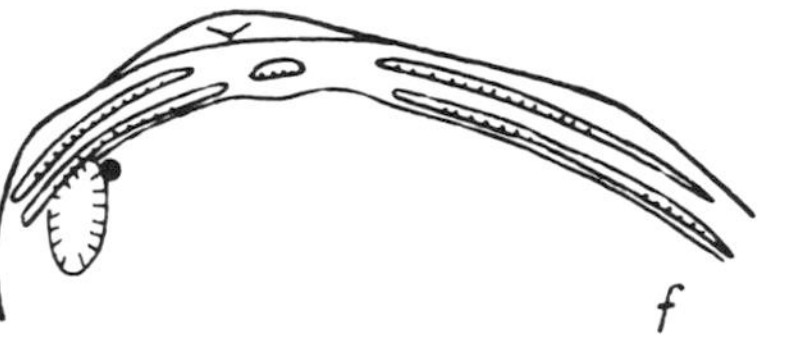

Figure 9. Subgenus *Lilingella (Xinyuella)* J. Chen and Xu.
a,b. *Lilingella (Xinyuella) glota* J. Chen and Xu, Lower Jurassic.
a. Right valve interior, ×5, showing phase f of lilingelloid hinge.
b. Right valve exterior, ×2, showing general outline of species.
c, d. *Lilingella (Xinyuella) liuyangensis* (Gu and Liu), Lower Jurassic. Right and left valve exteriors, ×1, showing general outline of species.
e,f. *Lilingella (Xinyuella) pinglingensis* J. Chen, Lower Jurassic.
e. Right valve exterior, ×2, showing general outline of species.
f. Right valve interior, ×4, showing phase f of lilingelloid hinge.

6); *L. (H.) ovata* Xiong and Wang, 1979; Figure 10d, 10e; *L. (H.) oblongiformis* Xiong and Wang, 1979; *L. (H.) elliptica* Chen and Xu, 1980 (p. 406, Pl. 3, figs. 13–16); Figure 10h, 10i; Plate 5(7–10); and *L. (H.) eilotes* Chen and Xu, 1980 (p. 406, Pl. 3, figs. 1, 2): Figure 10f, 10g.

Distribution and age: South China; Early Jurassic.

Subgenus *Lilingella (Apseudocardinia)* Liu and Zhu, 1978 (in Cai and Liu, 1978)
Type species: *Apseudocardinia sichuanensis* Liu and Zhu, 1978 (Fig. 11a–d,g,h)

This subgenus differs slightly from *Hunanella* in the presence of obtuse-triangular pseudocardinal tooth 1 and in the generally large size. It obviously differs from *Pseudocardinia* Martinson, 1959, in the presence of pseudocardinal teeth.

More than 10 species of *Apseudocardinia* have been recognized,

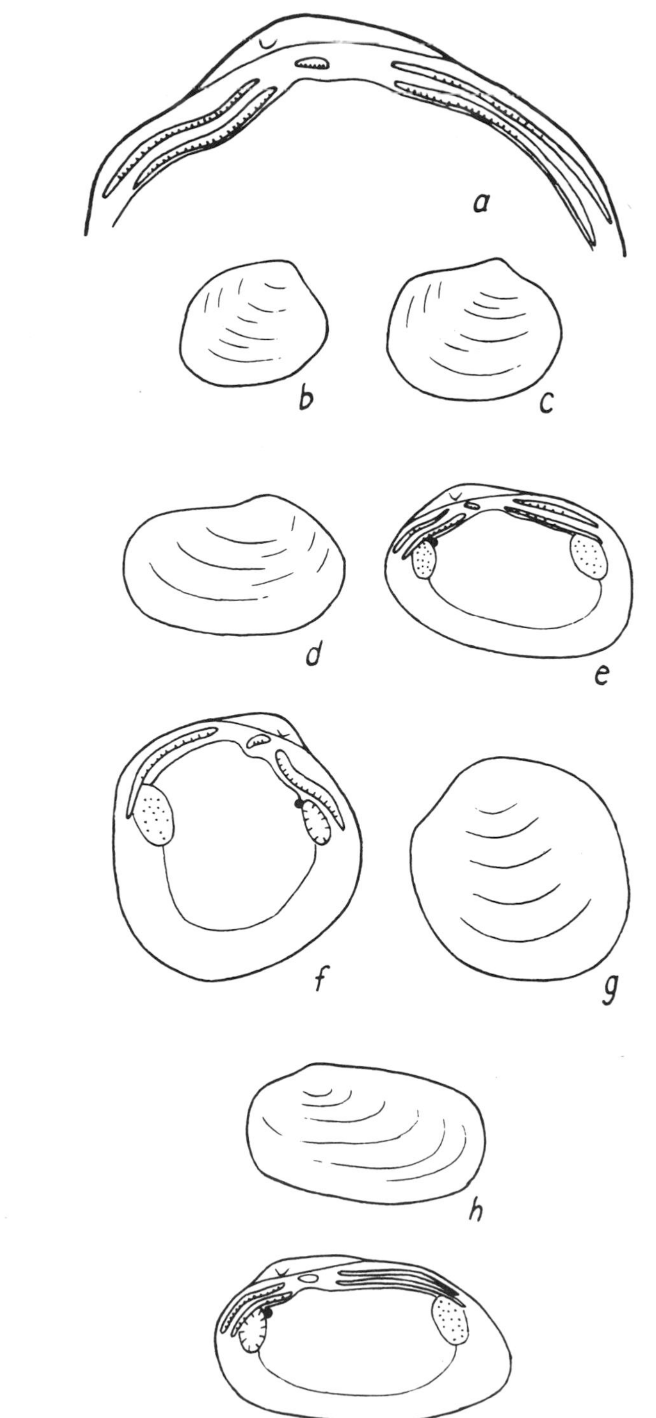

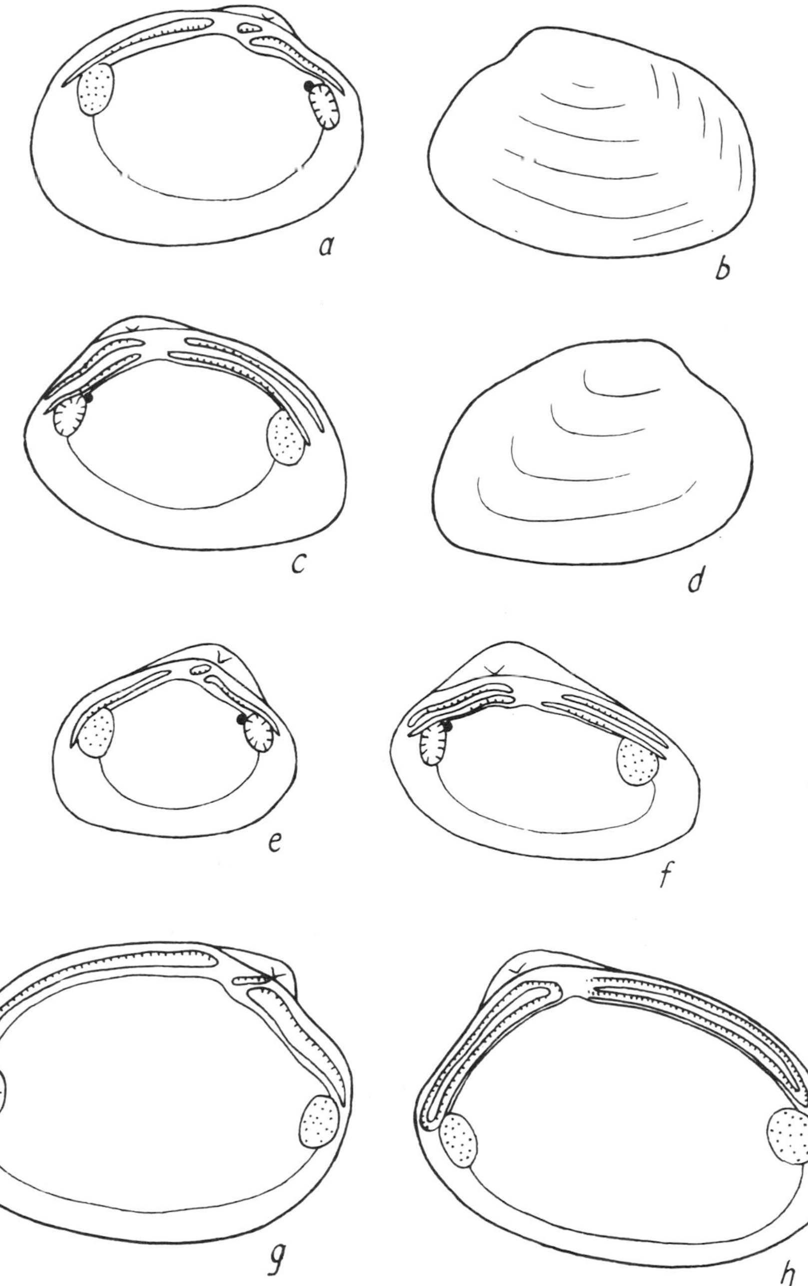

Figure 11. Subgenus *Lilingella (Apseudocardinia)* X. Liu and Zhu

a–d. *Lilingella (Apseudocardinia) sichuanensis* X. Liu and Zhu, Lower Jurassic.

a,c. Left and right valve interiors, ×2, showing phase f of lilingelloid hinge.

b,d. Left and right valve exteriors, ×2, showing general outline of species.

e. *Lilingella (Apseudocardinia) longiovalis* X. Liu and Zhu, Lower Jurassic.

Left valve interior, ×2, showing phase f of lilingelloid hinge.

f. *Lilingella (Apseudocardinia) weiyuanensis* X. Liu, Lower Jurassic.

Right valve interior, ×2, showing phase f of lilingelloid hinge.

g,h. *Lilingella (Apseudocardinia) sichuanensis* X. Liu and Zhu, Lower Jurrasic.

Schematic diagrams of both valve interiors (after Cai and X. Liu, 1978). But the close at ends of lamellar teeth in right valve are not true.

Figure 10. Subgenus *Lilingella (Hunanella)* Xiong and Wang

a–c. *Lilingella (Hunanella) guanyintanensis* Xiong and Wang, Lower Jurassic.

a. Right valve interior, ×5, showing phase f of lilingelloid hinge.

b,c. Right valve exteriors, ×1.5, showing general outline of species.

d,e. *Lilingella (Hunanella) ovata* Xiong and Wang, Lower Jurassic.

d. Right valve exterior, ×2, showing general outline of species.

e. Right valve interior, ×2, showing phase f of lilingelloid hinge.

f,g. *Lilingella (Hunanella) eilotes* J. Chen and Xu, Lower Jurassic.

f. Left valve interior, ×2, showing phase f of lilingelloid hinge.

g. Left valve exterior, ×2, showing general outline of species.

h,i. *Lilingella (Hunanella) elliptica* J. Chen and Xu, Lower Jurassic.

h. Left valve exterior, ×3, showing general outline of species.

i. Right valve interior, ×3, showing phase f of lilingelloid hinge.

some of them are no doubt comparable with several Chinese species formally identified as *Pseudocardinia*. Further work concerning the reassignment of these taxa remains to be carried out.

Distribution and age: South China; Early to Middle Jurassic.

Genus *Hamiconcha* Huang, Wei, and Chen, gen. nov. (herein)
Type species: *Hamiconcha xinjiangensis* Huang, Wei, and Chen, gen. et sp. nov.

This genus was established by Huang Bao-yu, Wei Jing-ming, and Chen Jin-hua during their recent study on some Mesozoic freshwater bivalves of Xinjiang.

Shell medium to large, oval to triangular, inflated; umbos wide, projected above the hinge line, anterior to mid-length; beaks prosogyrate; lunule and escutcheon developed. Umbonal carina present; external ligament; opisthodetic. Surface concentrically sculptured; right valve with three pseudocardinal teeth, two anterior and two posterior lamellar teeth; tooth 1 strong, obtusely triangular; nodelike 3a and short lamellar 3b fused at their apices; posterior ends of AI and AIII fused, respectively, with anterior ends of 1 and 3a. Both anterior and posterior lamellar teeth extend to external sides of anterior and posterior adductor insertion areas. Left valves with one anterior and one posterior lamellar tooth, three pseudocardinal teeth; 2a and 2b fused at apex, occasionally 2b shows a tendency to protract backward; in some cases 4b extends posteriorly and connects with anterior end of PII; all teeth have transverse grooves. Anterior adductor scar strong, kidney-shaped, irregularly grooved, small pedal scar inserted at its posterior upper side; posterior adductor scar relatively weak, oval.

Comparison and discussion: The holotype of the type species of *Hamiconcha* was collected from the Middle Jurassic of Hami County, Xinjiang, and was identified as *Pseudocardinia cardinia carinata* Martinson (Gu and others, 1976, p. 346, Fig. 15, Pl. 96, fig. 34). The specimen clearly shows that there are three pseudocardinal teeth developed below the umbo, an obtusely triangular tooth 1, a nodelike tooth 3a, and a short lamellar tooth 3b; the anterior ends of 1 and 3a are connected with AI and AIII, respectively. It is evident that the hingement of this Hami form is quite different from that of *Pseudocardina* Martinson, although neither Martinson (1959) nor Kolesnykov (1977) mentioned the existence of pseudocardinal teeth in the type material of *Pseudocardinia*. Therefore, it is not possible to revise the diagnosis of *Pseudocardinia* Martinson to agree with the hinge characters shown in the Xinjiang material (Gu, in Gu and others, 1976, p. 346).

The hinge characters of the new genus closely resemble those of the Late Jiangxielloid pattern in *Jiangxiella* and *Guangdongella*, but the present genus has a stronger, nodelike 3a and more strongly developed anterior lamellar teeth.

Distribution and age: Asia; Middle Jurassic.

***Hamiconcha xinjiangensis* Huang, Wei and J. Chen, sp. nov. (herein)**
Figure 12a,12b; Plate 7(9–10)

1976 *Pseudocardinia carinata*, Gu in Gu and others, 1976, p. 347, Fig. 15, Pl. 96, fig. 34

Shell roundly triangular in outline; umbos highly projected; umbonal carina strong, somewhat acute; posteroventral angle prominent; surface with irregular concentric sculptures.

Locality and horizon: Hami, Xinjiang; Qitai Formation; Middle Jurassic.

***Hamiconcha* sp.**
Figure 12e–12g; Plate 7(1–3)

Shell suboval in outline, umbonal carina weak; 2b shows tendency to protract backward.

Locality and horizon: Wuqia (Ulugqat) County, Tarim Basin; Middle Jurassic; Yangyeer Formation.

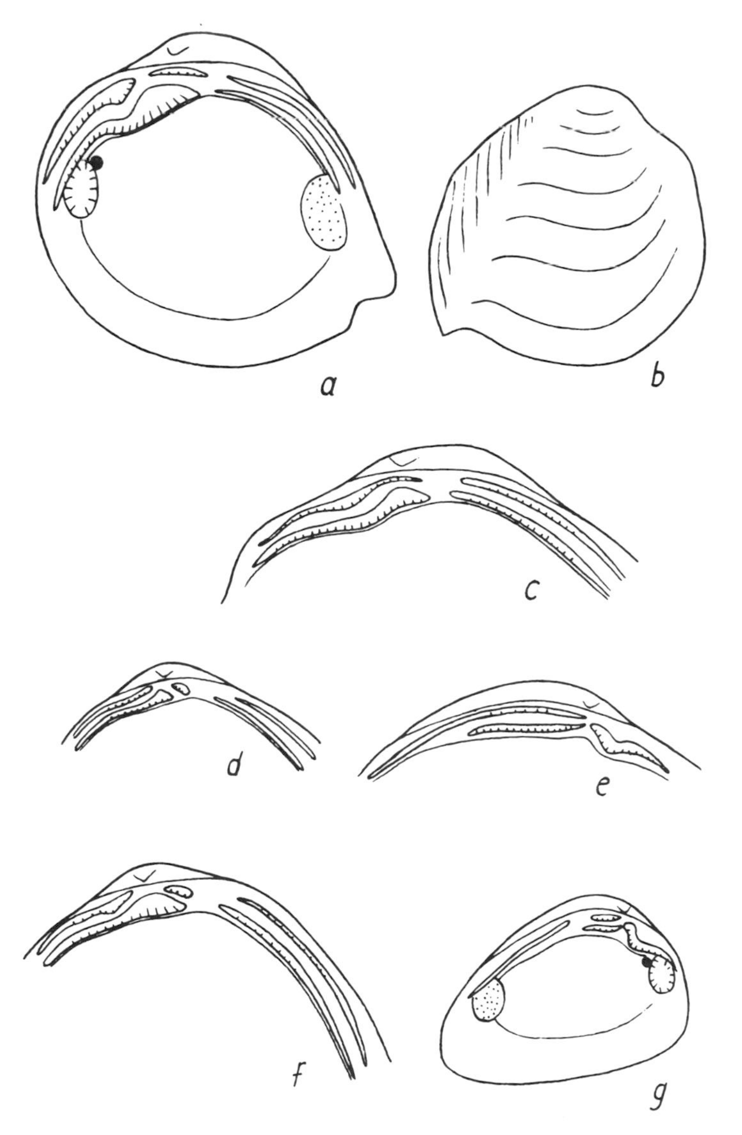

Figure 12. Genus *Hamiconcha* Huang, Wei and J. Chen (gen. nov.)
a,b. *Hamiconcha xinjiangensis* Huang, Wei and J. Chen (gen. et sp. nov.), Middle Jurassic.
a. Right valve interior, ×2, showing phase h of hamiconchoid hinge.
b. Right valve exterior, ×18, showing general outline of species.
c–g. *Hamiconcha* sp., Middle Jurassic.
All figures showing phase h of hamiconchoid hinge.
c,d,f. Right valve interiors, ×2, ×1.5, ×2, respectively.
e,g. Left valve interiors, ×1.5, ×2, respectively.

ACKNOWLEDGMENTS

We are much indebted to the following paleontologists and technicians in the preparation of this manuscript: Professor Wang Yu (Nanjing Institute of Geology and Palaeontology) for reading the manuscript; Professor James Conkin (University of Louisville) for reading the manuscript and correcting the language; Dr. Erle G. Kauffman (Smithsonian Institution) for reviewing the manuscript; Professor Curt Teichert (University of Rochester) for the final reading and editing of the manuscript; Huang Bao-yu (Nanjing Institute of Geology and Palaeontology) and Wei Jing-ming (Xinjiang Administrative Bureau of Petroleum) for their permission to publish the new genus *Hamiconcha* Huang, Wei and J. Chen; Liu Xie-zhang (Chengdu Institute of Geology and Mineral Resources) and Li Zi-shun and Li Yun-tong (Institute of Geology, Ministry of Geology) for providing photos of type specimens of *Apseudocardinia* Liu and Zhu, and *Guangdongella* Li and Li; Yang Yong-qing and Song Zi-yao for their assistance in preparing drawings and photographs at our Institute; and Zhu Zhi-kang of our Institute for typing the manuscript.

SELECTED REFERENCES

Cai Shao-yin and Liu Xie-zhang, 1978, [Non-marine Lamellibranchia, *in* Palaeontological atlas of Southwest China, Sichuan, no. 2]: Beijing, Geological Press, p. 365–403, Pls. 114–130 (in Chinese).

Casey, R., 1952, Some genera and subgenera, mainly new, of Mesozoic heterodont lamellibranchs: Proceedings of the Malacological Society of London, v. 29, pt. 4, p. 121–176, 3 pls.

Chen Jin-hua and Xu Yu-ming, 1980, New materials of bivalve fossils from the Mesozoic coal series of Southwest Hunan: Acta Palaeontologica Sinica, v. 19 (in Chinese with English abstract).

Chen Jin-hua, Zhou Zhi-yan, Pan Hua-zhang, Cao Mai-zheng, Lin Qi-bing, Sang Yu-ke, and Xu Yu-ming, 1980, [On the Mesozoic coal series of Southwest Hunan and its faunas and floras]: Bulletin of the Nanjing Institute of Geology and Palaeontology, Academia Sinica, v. 1 (in Chinese).

Cox, L. R., and others, 1969, Mollusca 6, Bivalvia, *in* Moore, R. C., ed., Treatise on invertebrate paleontology, Part N, Volume 1: Boulder, Colorado, Geological Society of America (and University of Kansas Press), 519 p.

Fan Jia-song, 1963, On Lower Liassic Lamellibranchiata from Guangdong (Kuangtung): Acta Palaeontologica Sinica, v. 11, no. 4, p. 508–542, 5 pls. (in Chinese with English abstract).

Gu Zhi-wei, 1962, [The Jurassic and Cretaceous stratigraphy of China, *in* Proceedings of National Stratigraphical Conference 1959]: Beijing, Science Press, 84 p. (in Chinese).

Gu Zhi-wei, Huang Bao-yu, Chen Chu-zhen, Wen Shi-xuan, Ma Qi-hong, La Xiu, Xu Jun-tao, Liu Lu, Wang Shu-mei, Wang De-you, Qiu Ran-zhong, Huang Zao-qi, Zhang Zao-ming, Chen Jin-hua, and We Pei-li, 1976, [Fossil lamellibranchs of China]: Beijing, Science Press, 522 p., 150 pls. (in Chinese).

Hayami, I., 1958, Taxonomic notes on *Cardinia* with description of a new species from the Lias of western Japan: Journal of the Faculty of Science, Imperial University of Tokyo, sec. 2, v. 11, pt. 2, p. 115–130, Pl. 9.

——1961, On the Jurassic pelecypod faunas in Japan: Journal of the Faculty of Science, Imperial University of Tokyo, sec. 2, v. 13, pt. 2, p. 243–343, Pl. 14.

——1975, A systematic survey of the Mesozoic bivalvia from Japan: University of Tokyo Press, 228 p., 10 pls.

Huang Bao-yu and Chen Jin-hua, 1980, [Late Triassic and Middle Jurassic nonmarine lamellibranchs from Zhejiang and South Anhui, *in* Divisions and comparisons of the Mesozoic volcanic depositions of Zhejiang and Anhui]: Beijing, Science Press, p. 69–96, 8 pls. (in Chinese).

Kobayashi, T., and Ichikawa, K., 1952, The Triassic fauna of the Heki Formation in the Province of Tamba (Kyoto Prefecture), Japan: Japanese Journal of Geology and Geography, v. 22, p. 55–84, 3 pls.

Kobayashi, T., and Suzuki, K., 1937, Non-marine shells of the Jurassic Tetori series in Japan: Japanese Journal of Geology and Geography, v. 14, p. 33–51, 2 pls.

——1939, The brackish Wealden Fauna of the Yoshimo beds in Prov. Nagato, Japan: Japanese Journal of Geology and Geography, v. 16, p. 213–224, 2 pls.

Kolesnykov Ch. M., 1974, Mikrostruktura rakovin urskogo limnicheskogo rod Pseudocardinia (Bivalvia): Paleontologicheskiy Zhurnal, no. 1, p. 66–71 (in Russian).

——1977, Sistema i proishozdenie limnicheskikh dvustuorok Mezozoya: Paleontologicheskiy Zhurnal, no. 3, p. 42–54 (in Russian).

Li Zi-sun, 1977, [On the stratigraphical problems of the Mesozoic coal series in northern Guangdong Province; *in* Professional papers on coal-bearing series of South China]: Beijing, Coal Industrial Press, p. 1–17 (in Chinese).

Liu Lu, 1967, [Late Triassic and Early Jurassic lamellibranchs from Hunan and Jiangxi, *in* Late Triassic and Early Jurassic faunas and floras from Hunan and Jiangxi]: Nanjing Institute of Geology and Palaeontology, p. 1–115, 47 pls. (in Chinese; printed for internal circulation).

Ma Qi-hong, Chen Jin-hua, Lan Xiu, Gu Zhi-wei, Chen Chu-zhen, and Lin Ming-ji, 1976, [The Mesozoic lamellibranch fossils of Yunnan, *in* Mesozoic fossils of Yunnan]: Beijing, Science Press, p. 161–386, 42 pls. (in Chinese).

Martinson, G. G., 1959, O novom rode urskikh plastinchatozabernykh Pseudocardinia: Paleontologicheskiy Zhurnal, no. 3, p. 33–40, Pl. 3 (in Russian).

——1961, Mezozoyski i kaynozoyskie molluski kontinentalnykh otlozheniy Sibirskoy platformy, Zabaykalya i Mongolii: Moskva, Izdatelstvo Akademiy Nauk SSSR, 332 p., 26 pls. (in Russian).

Si Xing-jian and Zhou Zhi-yan, 1962, [The Mesozoic continental stratigraphy of China, *in* Proceedings of National Stratigraphical Conference, 1959]: Beijing, Science Press, 180 p. (in Chinese).

Vokes, H. E., 1967, Genera of bivalvia. A systematic and bibliographic catalogue: Bulletins of American Paleontology, v. 51, no. 232, p. 105–394.

Xiong Cun-wei and Wang Sai-yi, 1979, [Some Lower Jurassic bivalve fossils of Hunan and Jiangxi with their stratigraphic significances]: Coal-Geology and Exploration, no. 3, p. 1–12, 2 pls. (in Chinese).

Zhang Ren-jie, Wang De-you, and Zhou Zhu-ren, 1977, [Bivalvia, *in* Palaeontological atlas of central South China, Part e]: Beijing, Geological Press, p. 4–65, Pls. 1–24 (in Chinese).

Zhao Jin-ke, Chen Chu-zhen, and Lian Xi-luo, 1962, [The Triassic stratigraphy of China, *in* Proceedings of National Stratigraphical Conference, 1959]: Beijing, Science Press, 130 p. (in Chinese).

Zhu Guo-xing, 1976, [Lamellibranchia, *in* Palaeontological Atlas of Northeast China, Inner Mongolia, no. 2]: Beijing, Geological Press, p. 17–35, Pls. 1–12 (in Chinese).

MANUSCRIPT ACCEPTED BY THE SOCIETY APRIL 22, 1981

PLATE 1

All specimens from the Sanjiachong Formation of the Anyuan Group, Upper Triassic; Xinyu County, Jiangxi Province. NIGP: Nanjing Institute of Geology and Palaeontology, Academica Sinica.

Figure

1–4. ***Jiangxiella cuneata* Liu and Chen, sp. nov.**
1. **Right internal mold, holotype (NIGP 64401), ×4, showing phase a of early jiangxielloid hinge.**
2. **Left internal mold, paratype (NIGP 64402), ×6.**
3. **Plaster cast of specimen shown in figure 1, ×4.**
4. **Left internal mold, paratype (NIGP 64404), ×3.**

5–12. ***Jiangxiella subovata* Liu**

5. **Left internal mold, (NIGP 64405), ×6.**
6. **Right internal mold (NIGP 24779), ×4, showing phase c of early jiangxielloid hinge.**
7. **Left internal mold (NIGP 64417), ×6.**
8. **Left internal mold (NIGP 64406), ×2.**
9. **Plaster cast of figure 11, ×6.**
10. **Left internal mold (NIGP 64406), ×6.**
11. **Right internal mold (NIGP 64408), ×6.**
12. **Left internal mold, holotype (NIGP 24780), ×6.**

13, 14. ***Jiangxiella datianensis* Liu**

13. **Right internal mold (NIGP 64409), ×4, showing phase b of early jiangxielloid hinge.**
14. **Right internal mold, holotype (NIGP 24785), ×4, showing phase c of early jiangxielloid hinge.**

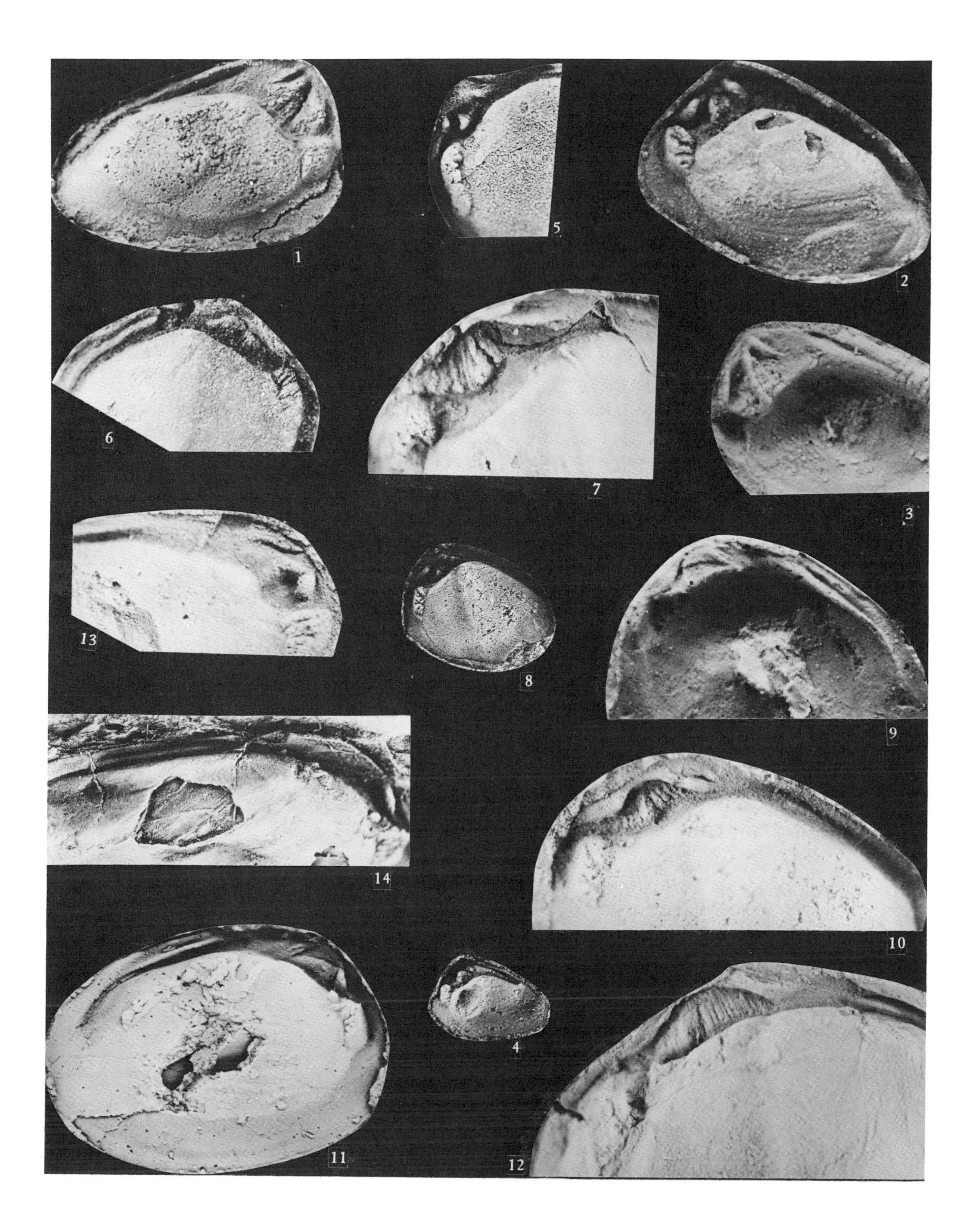
1
5
2
6
7
3
13
8
9
14
10
11
4
12

PLATE 2

Figure

1–4. ***Jiangxiella xinyuensis* Liu and Chen, sp. nov.**

1. **Right internal mold, paratype (NIGP 24778), ×6, showing phase e of late jiangxielloid hinge. Locality: Anyuan, Jiangxi, Sanjiachong Formation of Anyuan Group; Upper Triassic.**
2. **Right internal mold, paratype (NIGP 64410), ×6. Locality: Xinyu, Jiangxi, Sanjiachong Formation of Anyuan Group; Upper Triassic.**
3. **Right internal mold, holotype (NIGP 64411), ×4. Locality: same as figure 2.**
4. **Plaster cast of figure 2, ×8.**

5–10. ***Jiangxiella subovata* Liu**

5. **Plaster cast from a left internal mold (NIGP 64413), ×6, showing phase d of late jiangxielloid hinge. Locality: same as figure 2.**
6. **Plaster cast from a left internal mold (NIGP 64414), ×4. Locality: same as figure 2.**
7. **Right internal mold (NIGP 45840), ×6. Locality: same as figure 2.**
8. **Right internal mold (NIGP 45841), ×4. Locality: Liuyang, Hunan; Sanjiachong Formation of Anyuan Group; Upper Triassic.**
9. **Gypsum cast of a left internal mold (NIGP 64415), ×4. Locality: same as figure 2.**
10. **Right internal mold (NIGP 64416), ×3. Locality: same as figure 2.**

11. ***Jiangxiella datianensis* Liu**

Right internal mold (NIGP 45824), ×6, showing phase d of late jiangxielloid hinge. Locality: Anyuan, Jiangxi, Sanjiachong Formation of Anyuan Group; Upper Triassic.

12, 13. ***Jiangxiella ninyuanensis* Chen and Xu**

12. **Right internal mold (NIGP 47683), ×10, showing phase d of late jiangxielloid hinge. Locality: Ninyuang, Hunan; Yanbaichong Formation; Upper Triassic.**
13. **Bivalved internal mold, holotype (NIGP 47674), ×10. Locality: same as figure 12.**

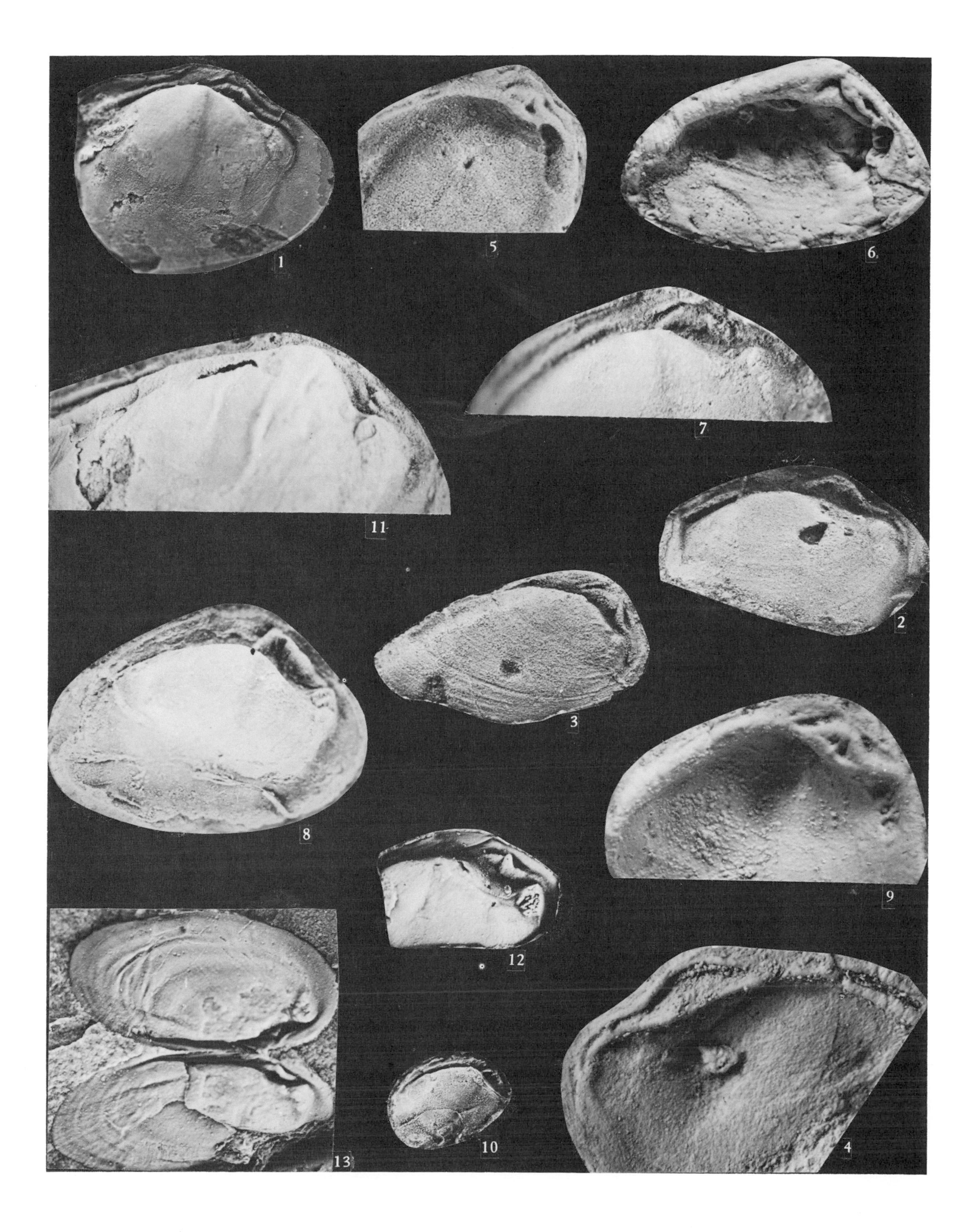
1
5
6
7
11
2
3
8
9
12
13
10
4

PLATE 3

Because the authors were not able to examine the types of the species of *Guangdongella,* figures 1–18 were prepared from prints made from the original negatives kindly loaned by Z. Li and Y. Li.

Figure

1–9. *Guangdongella brevicula* Li and Li
Locality: Xiaoshui, Guangdong; Hongweikeng Formation; Upper Triassic.
1. **Right internal mold (NG 310), ×5.**
2. **Right internal mold, holotype (NG 300), ×5.**
3. **Left internal mold (NG 303), ×5.**
4. **Left external mold (NG 328), ×5.**
5. **Left internal mold (NG 329), ×5.**
6. **Right internal mold (NG 327), ×5.**
7. **Left internal mold (NG 339b), ×5, showing phase d of late jiangxielloid hinge.**
8. **Right internal mold (NG 3032), ×10, showing phase d of late jiangxielloid hinge.**
9. **Bivalved internal mold (NG 330), ×5.**

10–14. *Guangdongella exquisita* Z. Li and Y. Li
Locality: same as figure 1–9.
10. **Left internal mold (NG 306), ×5, showing phase a of early jiangxielloid hinge.**
11. **Left internal mold (NG 307), ×5.**
12. **Left internal mold, holotype (NG 304), ×5, showing phase a of early jiangxielloid hinge.**
13. **Right valve (NG 315), ×8.**
14. **Left internal mold (NG 309), ×5.**

15. *Guangdongella* sp.
Left internal mold (NG 305), ×5. Locality: same as figures 1–9.

16–18. *Guangdongella longimorpha* Li and Li
Locality: same as figures 1–9.
16. **Right internal mold (NG 802), ×2.5.**
17. **Left internal mold, holotype (NG 336), ×2.5.**
18. **Right internal mold (NG 334), ×5.**

19, 20. *Guangdongella? elegans* Chen and Xu
Locality: Liuyang, Hunan; Sanjiachong Formation of Anyuan Group; Upper Triassic.
19. **Right internal mold (NIGP 47689), ×6.**
20. **Right internal mold, holotype (NIGP 46788), ×6.**

1
2
15
3
4
10
5
11
16
17
6
18
7
12
19
9
20
13
8
14

PLATE 4.

Figure

1. ***Lilingella (Lilingella) simplex*** **Chen and Liu**
Bivalved internal mold (NIGP 47634), ×2. Locality: Hengnan, Hunan; Tabakou Member of Guanyintan Formation; Lower Jurassic.

2–9. ***Lilingella (Lilingella) cuneata*** **Wu**
Locality: Lingling, Hunan; Tabakou Member of Guanyintan Formation: Lower Jurassic.

2. **Right internal mold (NIGP 47629) ×10, showing phase f of lilingelloid hinge.**
3. **Left internal mold (NIGP 47625), ×10.**
4. **Right valve (NIGP 47639), ×3.**
5. **Left internal mold (NIGP 47627), ×6.**
6. **Left internal mold (NIGP 47628), ×10.**
7. **Left internal mold (NIGP 47628), ×6.**
8. **Right internal mold (NIGP 47626), ×8.**
9. **Right internal mold (NIGP 47623), ×6.**

1
2
3
4
5
6
7
8
9

PLATE 5

Figure

1–6. *Lilingella (Hunanella) guanyintanensis* Xiong and Wang

1. Right valve interior (NIGP 47639), ×10, showing phase f of lilingelloid hinge. Locality: Lanshan, Hunan, Tabakou Member of Guanyintan Formation: Lower Jurassic.
2. Right valve interior (NIGP 47637), ×10. Locality: same as figure 1.
3. Left internal mold (NIGP 47644), ×10. Locality: Lingling, Hunan; Tabakou Member of Guanyintan Formation; Lower Jurassic.
4. Right internal mold (NIGP 47645), ×10. Locality: same as figure 3.
5. Right internal mold (NIGP 47611), ×10. Locality: same as figure 3.
6. Right internal mold (NIGP 47646), ×7.5. Locality: same as figure 3.

7–10. *Lilingella (Hunanella) elliptica* Chen and Xu

7. Left valve interior (NIGP 47368), ×10. Locality: Lanshan, Hunan; Tabakou Member of Guanyintan Formation; Lower Jurassic.
8. Right internal mold (NIGP 47641), ×6. Locality: Lingling, Hunan; Tabakou Member of Guanyintan Formation; Lower Jurassic.
9. Right valve interior (NIGP 47640), ×7.5. Locality: same as figure 7.
10. Right valve interior (NIGP 47636), ×7.5. Locality: same as figure 7.

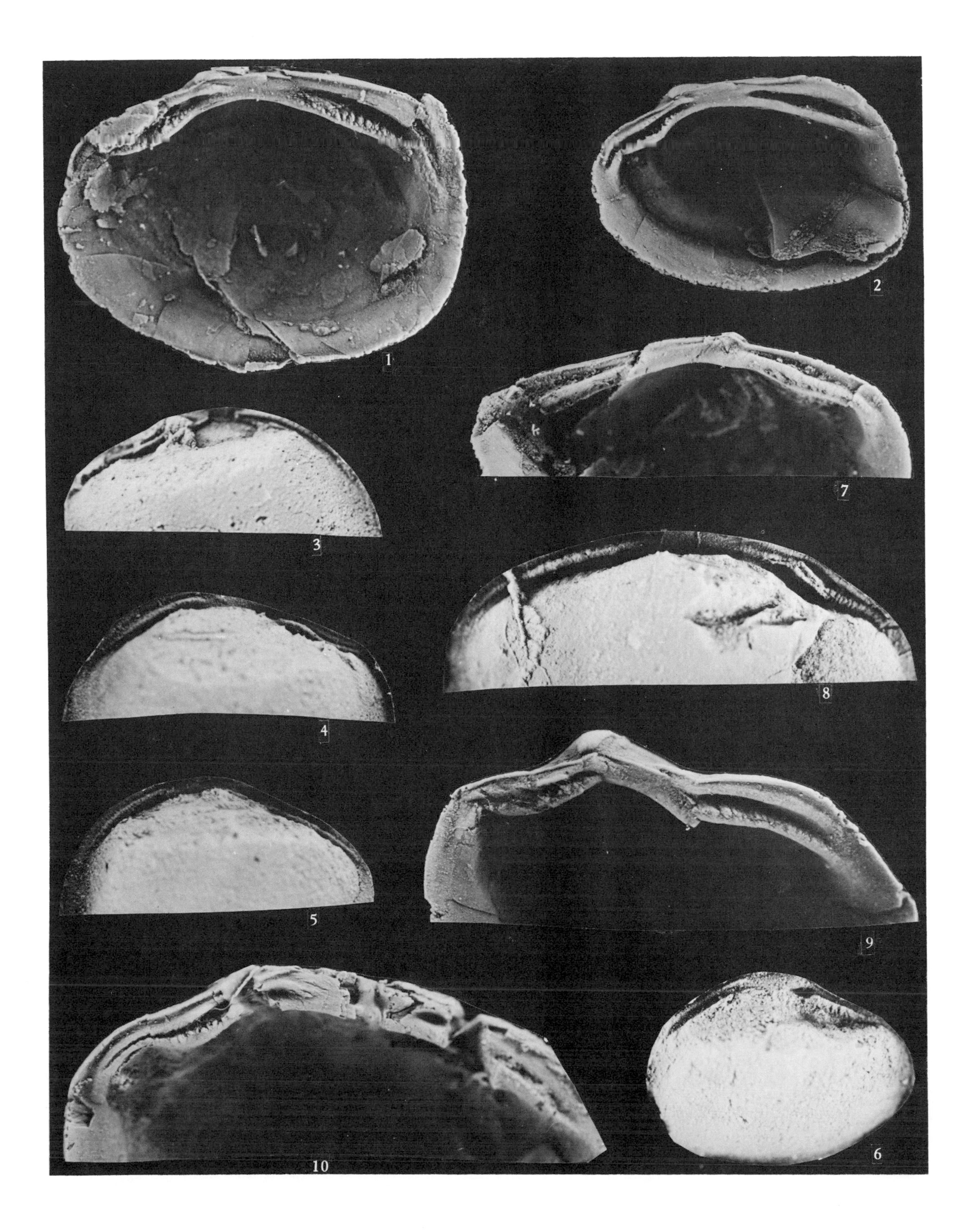
1
2
7
3
8
4
5
9
10
6

PLATE 6

All specimens, except figure 21, are from the Dongyuemiao Member of the Ziliujin Formation; Lower Jurassic; Qianwei, Sichuan.

Figure

1–3. ***Lilingella (Apseudocardinia) convexa*** **Liu**
- **1. Right valve interior (Sbi 1001), ×2.**
- **2. Right valve exterior (Sbi 1002), ×2.**
- **3. Left valve interior (Sbi 390), ×2.**

4–7. ***Lilingella (Apseudocardinia) subovata*** **Liu and Zhu**
- **4. Right valve interior (Sbi 1003), ×2.**
- **5. Right valve interior (Sbi 1004), ×2.**
- **6. Left valve interior (Sbi 1005), ×2.**
- **7. Right valve interior (Sbi 1006), ×2.**

8, 9. ***Lilingella (Apseudocardinia) elegans*** **Liu and Zhu**
- **8. Right valve interior (Sbi 441), ×2.**
- **9. Left valve interior (Sbi 1007), ×2.**

10–12. ***Lilingella (Apseudocardinia) sichuanensis*** **Liu and Zhu**
- **10. Right valve interior (Sbi 1008), ×2.**
- **11. Left valve interior (Sbi 1009), ×2.**
- **12. Left valve interior (Sbi 1010), ×2.**

13,14. ***Lilingella (Apseudocardinia) cuneata*** **Liu and Zhu**
- **13. Right valve interior (Sbi 1011), ×2.**
- **14. Left valve interior (Sbi 1012), ×2.**

15–17. ***Lilingella (Apseudocardinia) qianweiensis*** **Liu and Zhu**
- **15. Left valve interior (Sbi 407), ×2.**
- **16. Left valve exterior (Sbi 407), ×2.**
- **17. Left valve interior (Sbi 408), ×2.**

18. ***Lilingella (Apseudocardinia) subcentralis*** **Liu and Zhu**
- **Left valve interior (Sbi 397), ×2.**

19–21. ***Lilingella (Apseudocardinia) planula*** **Liu**
- **19. Left valve exterior (Sbi 409), ×2.**
- **20. Left valve interior (Sbi 409), ×2.**
- **21. Left internal mold (NIGP 47642), ×4. Locality: Lingling, Hunan; Tabakou Member of Guanyintan Formation; Lower Jurassic.**

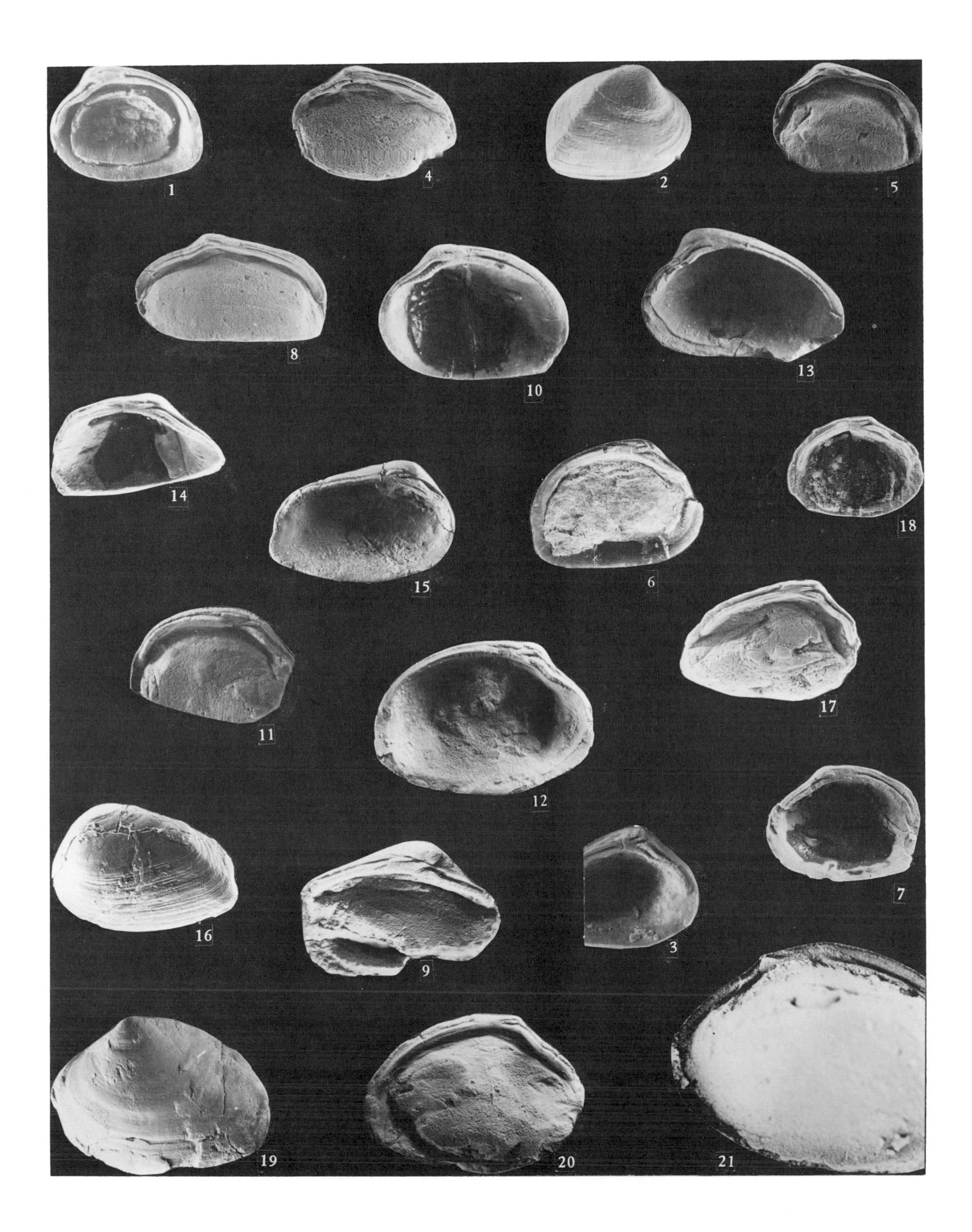
1
4
2
5
8
10
13
14
18
15
6
11
17
12
16
9
3
7
19
20
21

PLATE 7

Figure

1–3. *Hamiconcha* sp.
(Prints of negatives prepared by M. M. Wei.)
Locality: Wucha, Xinjiang, Yangyel Formation; Middle Jurassic.
1. Left valve interior (XBB 0288), ×4, showing phase h of hamiconchoid hinge.
2. Right valve interior (XBB 0288a), ×4.
3. Right valve interior (XBB 0290), ×2.

4,5. *Pseudocardinia lanceolata* (Chernyshev)
(Prints of negatives prepared by Z. W. Gu *in* Gu and others, 1976, Pl. 96, figs. 29, 30.) Locality: Fergana Basin, Central Asia; Middle Jurassic.
4. Right valve interior (NIGP 64403), ×3, showing phase g of pseudocardinioid hinge.
5. Right valve interior (NIGP 64407), ×3.

6,7. *Pseudocardinia kweichowensis* (Grabau)
(Reproductions of Pl. 96, figs. 14 and 16, *in* Gu and others, 1976.)
Locality: Yunnan; Hepingxian Formation; Middle Jurassic.
6. Left valve interior (NIGP 25302), ×3.
7. Left valve interior (NIGP 25301), ×3.

8. *Lilingella (Xinyuella) liuyangensis* (Gu and Liu)
Left valve interior (NIGP 47658), ×1.5, showing phase f of lilingelloid hinge. Locality: Xinyu, Jiangxi; Shikang Formation; Lower Jurassic.

9,10. *Hamiconcha xinjiangensis* Huang, Wei and Chen, gen. et sp. nov.
(Reproductions of Pl. 96, figs. 34 and 36, *in* Gu and others, 1976.)
9. Right valve exterior, paratype (NIGP 25315), ×2. Locality: Wucha, Xinjiang; Talke Formation; Middle Jurassic.
10. Right valve interior, holotype (NIGP 25313), ×3. Locality: Hami, Xinjiang; Qiketai Formation; Middle Jurassic.

11. *Lilingella (Xinyuella) glota* Chen and Xu
Right valve interior (NIGP 47654), ×10. Locality: Lanshan, Hunan; Tabakou Member of Guanyintan Formation; Lower Jurassic.

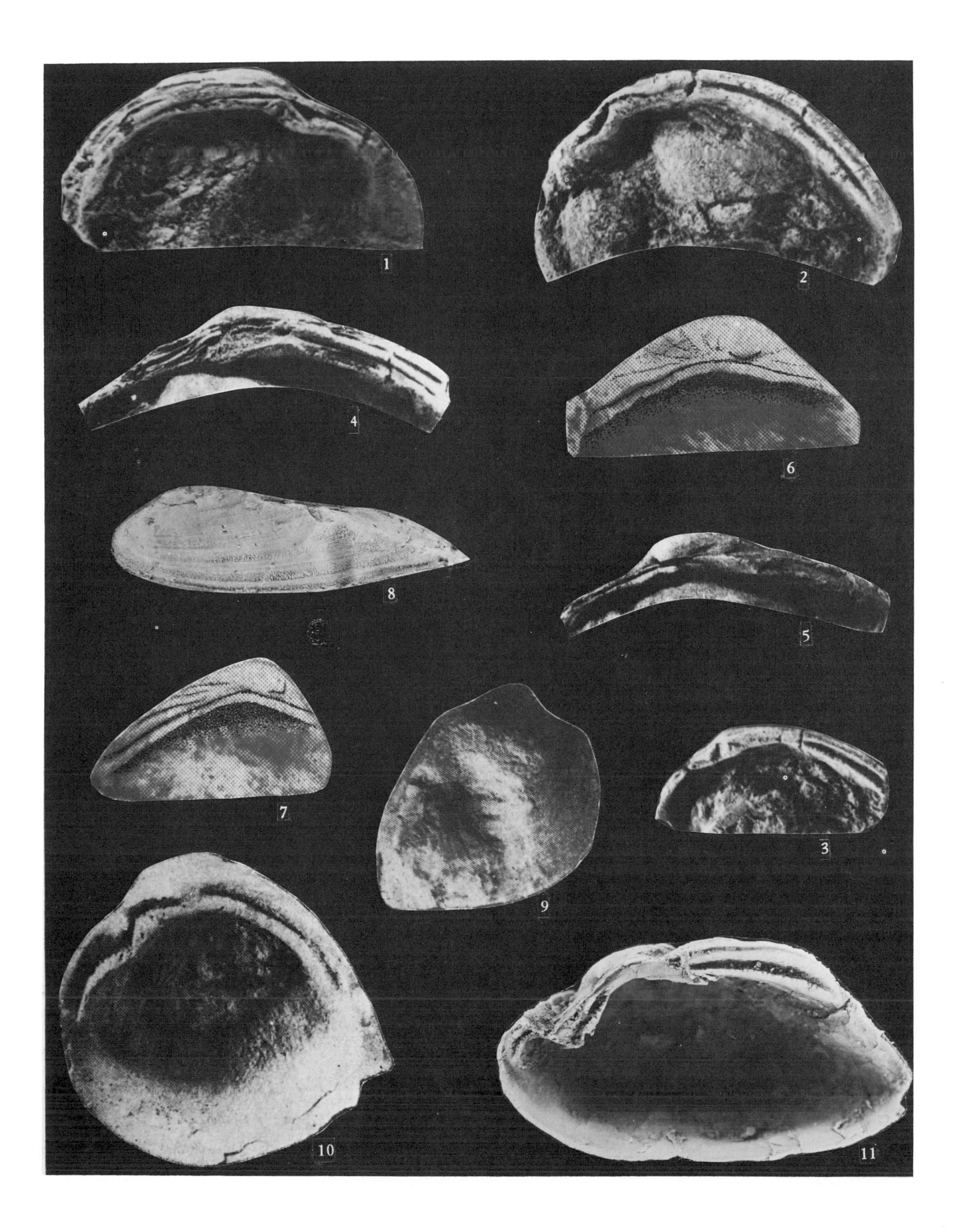
1
2
4
6
8
5
7
9
3
10
11

Printed in U.S.A.

Geological Society of America
Special Paper 187
1981

Upper Cambrian cephalopods from western Zhejiang

Chen Jun-yuan
Nanjing Institute of Geology and Palaeontology, Academia Sinica, Nanjing 210008, People's Republic of China

Qi Dun-lun
Regional Geological Survey of Anhui Province, Geological Bureau of Anhui, Chaoxian, Anhui, People's Republic of China

ABSTRACT

The purpose of this paper is to describe the primitive cephalopods occurring in the upper Cambrian Siyangshan Formation of Zhejiang. Six species belonging to four genera are recognized. Among them one new genus, *Zhuibianoceras,* supposedly an ancestral form of the Oncocerida, is proposed. A short disucssion of their stratigraphic significiance is presented.

INTRODUCTION

Although Upper Cambrian rocks containing rich trilobite faunas are widespread in western Zhejiang, no cephalopods were described before. However, new finds have been made during recent years at some localities such as Jiangshan, Changshan, and Xiaoshan in Zhejiang Province.

This report is a record of the cephalopods from the Upper Cambrian Siyangshan Formation, 8 km north of Jiangshan County in western Zhejiang (Fig. 1). Here, the Siyangshan Formation is overlain conformably by the Lower Ordovician Yichupu Formation containing *Staurograptus, Anisograptus,* and other graptolite genera, and is composed of limestone interbedded with shale, attaining a thickness of more than 50 m. According to a recent study, Lu Yan-hao and Lin Huan-ling (1980) subdivided the Siyangshan Formation into two zones: the upper *Lotagnostus hedini* Zone and the lower *Lotagnostus punctatus-Hedinaspis* Zone.

Characteristic trilobites of the *Lotagnostus hedini* Zone are the following:

Lotagnostus hedina (Troedsson)
Plicatolina spp.
Pseudagnostus spp.

Characteristic trilobites of the *Lotagnostus punctatus-Hedinaspis* Zone are the following:

Lotagnostus punctatus Lu
L. asiaticus Troedsson
Wujiajiania spp.
Hedinaspis spp.
Charchaqia spp.
Olenus sinensis Lu
Micragnostus spp.
Proceratopyge (Sinoproceratopyge) spp.

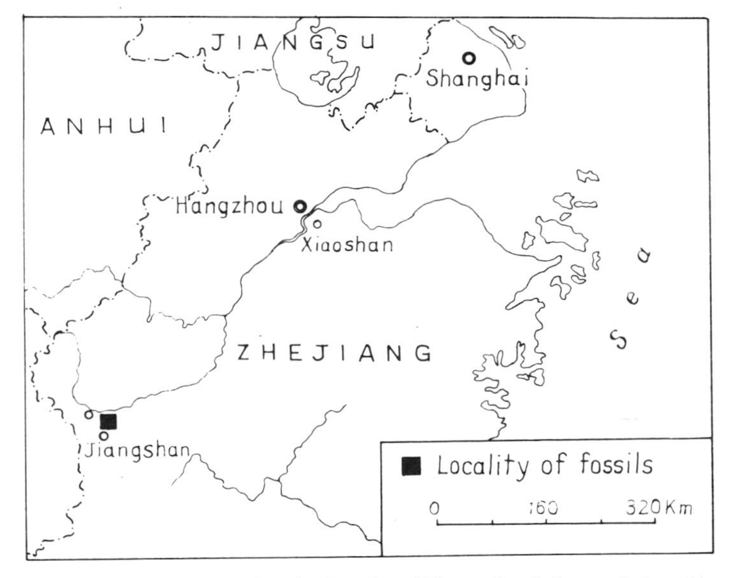

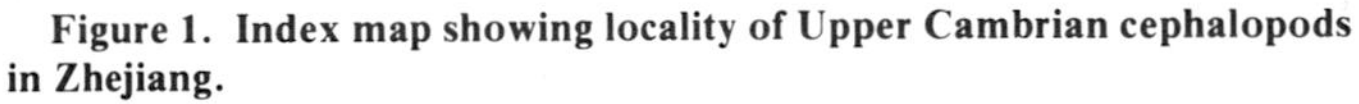

Figure 1. Index map showing locality of Upper Cambrian cephalopods in Zhejiang.

The cephalopods described below were collected from the cephalopod-bearing bed 2 to 3 m below the *Lotagnostus hedini* Zone and a few metres above the *Lotagnostus punctatus-Hedinaspis* Zone. The cephalopod-bearing bed is a black limestone, 0.2 to 0.3 m in thickness. The collection under study consists of the following forms:

Ectenolites primus Flower

E. penecilin Flower
Acaroceras endogastrum Chen, Qi, and T. E. Chen
A. elongatum Chen and Qi (sp. nov.)
Zhuibianoceras conicum Chen and Qi (gen. et sp. nov.)
Huaiheceras hanjiaense Zou and T. E. Chen

The discovery of the cephalopods, such as *Acaroceras endogastrum, Huaiheceras hanjiaense,* and *Ectenolites primus* is of important significance in the correlation of the Siyangshan Formation in South China with the Fengshan Formation in North China and the Trempealeauan San Saba Formation in North America (Table 1).

It should be mentioned that the Siyanghsan Formation has yielded exogastric forms such as *Huaiheceras hanjiaense* and *Zhuibianoceras conicum* assigned here. Of particular interest is the occurrence of the latter, *Zhuibianoceras,* of which the phragmocone is more rapidly expanded orad than that of any other Cambrian exogastric forms, such as *Huaiheceras* and *Balkoceras.* It is supposed that *Zhuibianoceras* is morphologically similar to oncoceroids, and may be well regarded as an ancestral type of the Oncocerida, or at least very close to it.

TABLE 1. CORRELATION OF CEPHALOPOD-BEARING FORMATIONS OF LATEST CAMBRIAN AGE

	Zhejiang Province	North China
Lower Ordovician	Yinchupu Formation:	Yehli Formation:
	Staurograptus, Anisograptus, Hysterolenus at its bottom	Staurograptus, Anisograptus, Leiostegium (Alloleiostegium), Onychopyge at its bottom
Upper Cambrian	Siyangshan Formation:	Fengshan Formation:
	Lotagnostus hedini Zone	Mictosaukia Zone
	Acaroceras Zone	Acaroceras-Aburoceras Zone
		or Sinoeremoceras Zone
	Lotagnostus punctatus -	Quadraticephalus Zone
	Hedinaspis Zone	Ptychaspis-Tsinania Zone
	Huayansi Formation	Changshan Formation

SYSTEMATIC DESCRIPTION

All the specimens described here came from the Upper Cambrian Siyangshan Formation, near Zhuibian Town, 8 km north of Jiangshan County in western Zhejiang Province.

Family ELLESMEROCERATIDAE Kobayashi, 1934
Genus *Ectenolites* Ulrich and Foerste, 1936
Ectenolites primus, Flower, 1964
Plate 1(6–7, 11a, 11b); Figure 2

1964 Ectenolites primus Flower, p. 54, Pl. 5, figs., 3, 5–6

Our collection includes two specimens that can be assigned to this species. The longer one, which was cut into two pieces by carelessness, is well preserved. It is a straight, rapidly tapering phragmocone, slightly compressed in cross section. The siphuncle is small, ventral in position, with a diameter of 0.8 mm at the base, where the conch diameter is 5.6 mm. Corresponding measure-

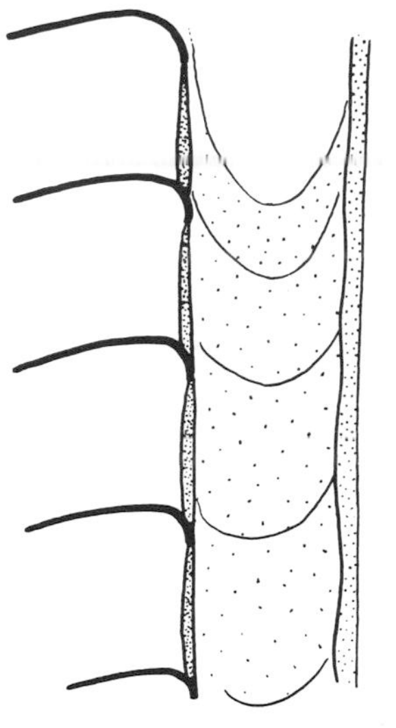

Figure 2. *Ectenolites primus* Flower. Adapical part of siphuncle showing diaphragms traversing the siphuncle, ×16. Cat. No. 63871.

ments near the adoral part of the specimen are about 1.8 and 10.5 mm, respectively. Septal necks are short, slightly sloping inward; connecting ring is moderately thick. The apical end of the siphuncle is traversed by concave diaphragms and filled with calcite deposited in the spaces between them.

Five to nine camerae occupy a distance equal to the diameter of the conch. The concavity of the septa is equal to the height of one or two camerae. Camerae are very short, 1 mm long on the average.

The smaller specimen is 24 mm long and expands rapidly from a diameter of 6 mm apically to 9 mm adorally in a length of 18 mm.

Horizon and locality: Upper part of the Upper Cambrian Siyangshan Formation, near Zhuibian Town, 8 km north of Jiangshan County in western Zhejiang.

Cat. Nos. 63871, 63874.

Ectenolites penecilin Flower, 1964
Plate 1(10)

1964 *Ectenolites penecilin* Flower, p. 55, Pl. 7, figs. 10–12

The specimen described here represents a short fragment of a living chamber and a considerable portion of a phragmocone. It is essentially straight, slowly expanding, compressed in cross section. The siphuncle is small and tubular, ventral in position, having a maximum diameter equal to one-seventh of the conch diameter. Septal necks are short and slightly sloping inward. Connecting rings are straight, moderately thick. The siphuncle is occupied by calcareous deposits. Camerae are short, about 1 mm on an average, closed in arrangement. Eight camerae together have a length equal to the diameter of the conch.

Horizon and locality: Same as the preceding form.

Cat. No. 63873.

Family ACAROCERATIDAE Chen, Qi, and T. E. Chen, 1979

This family was erected for ellesmerocerids which differ from the Ellesmeroceratidae in that their connecting rings are poorly calcified, whereas the ring of the Ellesmeroceratidae is well calcified, primitively showing layering structure, and is of some appreciable thickness.

Genus *Acaroceras* Chen, Qi, and T. E. Chen, 1979
Type species: *Acaroceras endogastrum* Chen, Qi, and T. E. Chen, 1979

Conch is small, slightly curved endogastrically, and laterally compressed in cross section. Living chamber is rather long. The siphuncle is small, ventral in position, and tubular in shape. Septal necks are short and straight. Connecting rings are thin, slightly concave, and without calcareous deposits.

This genus resembles both *Ellesmeroceras* and *Ectenolites* in general aspects, but differs from these two genera in its thin, poorly calcified rings and absence of calcareous deposits between the diaphragms.

Occurrence: All known representatives of *Acaroceras* come from the Siyangshan Formation of Zhejiang, and the Fenghsan Formation of Anhui, Shandong, Liaoning, and Shanxi. Both formations are of Late Cambrian age.

Acaroceras endogastrum Chen, Qi, and T. E. Chen, 1979
Plate 1(1, 2); Figure 3

1979 *Acaroceras endogastrum* Chen, Zou, T. E. Chen, and Qi, p. 115, Pl. III, figs. 3, 4, 17; Pl. IV, figs. 11, 12; Text-fig. 9

Only a single specimen in our collection can be assigned to this species. It is about 32 mm long, slightly compressed in cross section. The conch is curved endogastrically, moderately expanded, a portion 28 mm long increasing in width from 4 to 8 mm. The siphuncle is small, ventral in position, having a maximum diameter equal to one-eighth of the diameter of the conch. Septal necks are short, equal to one-fourth of the camerae in length. Connecting rings are straight and thin. The apical part of the siphuncle shows numerous and regular diaphragms. There are no organic calcareous deposits in the spaces between the diaphragms.

Four to eight camerae occupy a distance equal to the dorsoventral diameter. Camerae are short and closed in arrangement, 0.9 to 1 mm in length. Septa are shallowly concave.

Horizon and locality: Upper part of the Upper Cambrian Syiangshan Formation, near Zhuibian Town, 8 im north of Jiangshan County in western Zhejiang.

Cat. No. 63868.

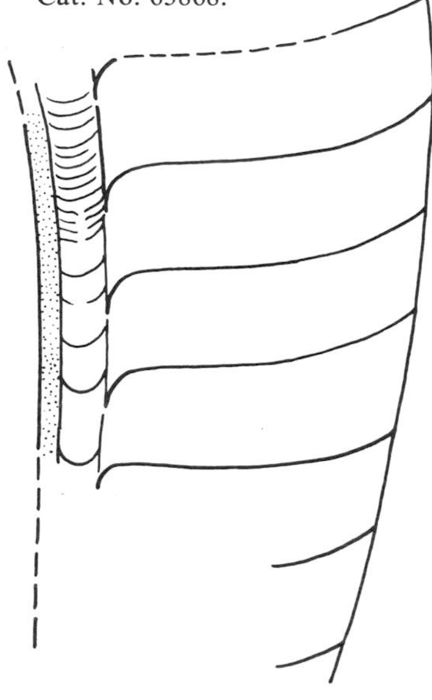

Figure 3. *Acaroceras endogastrum* Chen, Qi, and T. E. Chen. Adapical part of phragmocone showing diaphragms traversing the siphuncle, ×11. Cat. No. 63868.

Acaroceras jianghsanense Chen and Qi (sp. nov.)
Plate 1(5)

The single known representative of this species is about 32 mm long. The conch is straight, slowly expanding orad from 7.5 to 10 mm in a length of 25 mm. The cross section is compressed laterally, about 7.6 mm high and 5.8 mm wide at the adapical end. The siphuncle is small, ventral in position, its diameter about one-seventh of the conch diameter. Septal necks are short, slightly sloping inward. Connecting rings not preserved.

Camerae are short, very closed in arrangement, 0.6 mm in length. Fourteen camerae together have a length equal to the diameter of the conch.

Remarks: The most distinctive features of this species seem to be its large size, slender conch, and very short camerae.

Horizon and locality: Same as the preceding form.

Cat. No. 63870 (holotype)

Family HUAIHECERATIDAE Zou and T. E. Chen, 1979

This family is erected for ellesmerocerids which differ from the endogastric Ellesmeroceratidae and Acaroceratidae in the development of exogastric curvature. They are very similar in external feature to the Balkoceratidae, but differ from the latter in having straight and well-calcified rings.

Genus *Huaiheceras* Zou and T. E. Chen, 1979
Type species: *Huaiheceras hanjiaense* Zou and T. E. Chen, 1979

Conch is large, slowly expanding, compressed in cross section, slightly curved exogastrically. Siphuncle is small, ventral in position; septal necks are short, orthochoanitic; connecting rings are thick and straight.

Occurrence: All known representatives of this genus are of Late Cambrian age. They come from the Siyangshan Formation of Zhejiang and the Fengshan Formation of Anhui and Shandong.

Huaiheceras hanjiaense Zou and T. E. Chen, 1979
Plate 1(3, 4)

1979b *Huaiheceras hanjiaense* Chen, Zou, T. E. Chen, and Qi, p. 117, Pl. II, figs. 3, 4; Text-fig. 11

The only known specimen is about 40 mm long and represents much of the living chamber and a small portion of the phragmocone. The ventral outline is somewhat convex and the dorsal outline straight or slightly concave. The conch appears to be slightly compressed laterally in cross section. The diameter of the phragmocone is moderately expanding orad, but that of the living chamber is nearly constant.

The siphuncle is small, close to the ventral wall of the conch. Its diameter measures about 2 mm where the conch is 12 mm wide. The septal necks are short, slightly sloping inward. The connecting rings are moderately thick. The camerae are short; the septa are very closely spaced. There are 11 camerae in a distance of 9 mm.

Horizon and locality: Upper part of the Upper Cambrian Siyangshan Formation, near Zhuibian Town, 8 km north of Jiangshan County in western Zhejiang.

Cat. No. 63869.

Genus *Zhuibianoceras* Chen and Qi (gen. nov.)
Type species: *Zhuibianoceras conicum* Chen and Qi (gen. et sp. nov.)

Conch is rather small, slightly annulated, curved exogastrically, and laterally compressed in cross section. Living chamber is short, slowly expanding. Phragmocone is breviconic, expanding rapidly. Siphuncle is tubular and small, ventral in position. Septal necks are short, slightly sloping inward. Connecting ring is thin. Slightly concave diaphragms are present in early state. Camerae are short, closely spaced; septa are shallowly concave.

Remarks: The sinphuncle of this genus is closely similar to that of *Huaiheceras;* however, it is more rapidly expanding orad in phragmocone.

Occurrence: The single known representative of this genus, *Zhuibianoceras conicum,* comes from the Upper Cambrian Siyangshan Formation of western Zhejiang Province.

Zhuibianoceras conicum Chen and Qi (gen. et sp. nov.)
Plate 1(8); Figure 4

The single known representative of this species is well preserved, about 32 mm long. The conch is small, slightly annulated, very slightly curved exogastrically, with its dorsal profile straight and its ventral convex. The phragmocone is rapidly expanding orad, a portion 14 mm long increasing in width from 4.8 to 8.8 mm. The cross section is oval, compressed laterally, more narrowly rounded dorsally than ventrally. At the apical end the longer diameter of the conch is about 4.8 mm, the shorter 3.8 mm.

The living chamber is nearly complete, about 18 mm long, about twice the conch diameter. It is gradually expanding orad, from 8.8 to 10 mm in a length of 16 mm.

The siphuncle is small, ventral in position, having a maximum diameter equal to one-eighth of the conch diameter. The septal necks are short, slightly sloping inward, about one-third of a camera in length. The connecting ring is straight, moderately thick. The apical part of the siphuncle is traversed by two diaphragms, with calcite deposited in the space between them.

The camerae are short, very closely spaced, about 0.5 mm in average height. Twelve to fourteen camerae occupy a distance equal to the dorsoventral diameter of the conch. Septa are shallowly concave.

Horizon and locality: Upper part of the Upper Cambrian Siyangshan Formation, near Zhuibian Town, 8 km north of Jiangshan County in western Zhejiang.

Cat. No. 63872 (holotype).

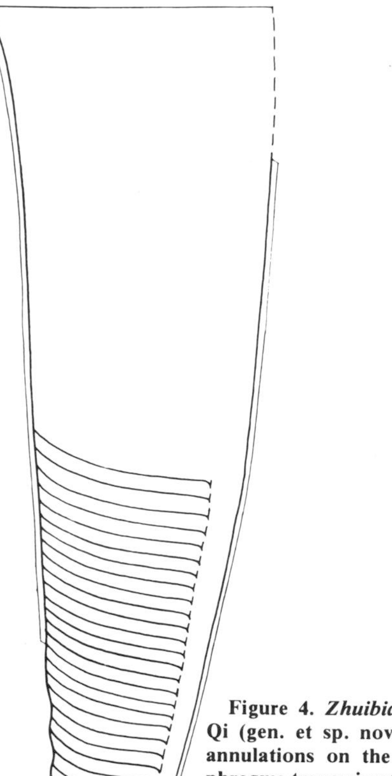

Figure 4. *Zhuibianoceras conicum* Chen and Qi (gen. et sp. nov.). Vertical section showing annulations on the surface and traces of diaphragms traversing the adapical part of the siphuncle, holotype, ×3. Cat. No. 63872.

ACKNOWLEDGMENTS

We are greatly indebted to Professor C. Teichert for his courtesy and kindness in giving us numerous valuable suggestions and ciriticisms. We also tender our deep appreciation to Xu Bao-rei and Zhang Fu-tian for their help in drawing and photographic work.

REFERENCES CITED

Chen Jun-yuan, Tsou Si-ping, Chen Ting-en, and Qi Dun-luan, 1979a, Late Cambrian cephalopods of North China—Plectronocerida, Protactinocerida (ord. nov.) and Yanhecerida (ord. nov.): Acta Palaeontologica Sinica, v. 18, no. 1, p. 1–24, Pls. 1–4 (in Chinese with English abstract).

Chen Jun-yuan, Zou Si-ping, Chen Ting-en, and Qi Dun-luan, 1979b, Late Cambrian Ellesmerocerida (Cephalopoda) of North China: Acta Palaeontological Sinica, v. 18, no. 2, p. 103–124, Pls. 1–4 (in Chinese with English abstract).

Flower, R. H., 1954, Cambrian cephalopods: New Mexico Institute of Mining and Technology, Bulletin 40, p. 1–50, Pls. 1–3.

——1964, The nautiloid order Ellesmeroceratidae (Cephalopoda): New Mexico Institute of Mining & Technology, Memoir 12, p. 1–234, Pls. 1–32.

Lu Yan-hao [Lu Yen-hao], and Lin Huan-ling, 1980, Cambro-Ordovician boundary in western Zhejiang and the trilobites contained therein: Acta Palaeontologica Sinica, v. 19, no. 2, p. 118–135, Pls. 1–3 (in Chinese with English abstract).

MANUSCRIPT ACCEPTED BY THE SOCIETY APRIL 22, 1981

PLATE 1

All specimens illustrated here came from the Upper Cambrian Siyangshan Formation, near Zhuibian Town, 8 km north of Jiangshan County in western Zhejiang Province. They are stored in Nanjing Institute of Geology and Palaeontology, Academia Sinica.

Figures

1–2. ***Acaroceras endogastrum*** **Chen, Qi, and T. E. Chen, 1979. 1, ×3, vertical section; 2, ×5, enlargement showing details of siphuncle of the same specimen near the apical part. Cat. No. 63868.**

3–4. ***Huaiheceras hanjiaense*** **Zou and T. E. Chen, 1979. 3, ×1, vertical section; 4, ×3, enlargement of the same specimen. Cat. No. 63869.**

5. ***Acaroceras jiangshanense*** **Chen and Qi, (sp. nov.). Holotype, ×3, vertical section. Cat. No. 63870.**

6–7. ***Ectenolites primus*** **Flower, 1964. 6, ×2; 7, ×5, enlargement of the same specimen. Cat. No. 63871.**

8–9. ***Zhuibianoceras conicum*** **Chen and Qi, (gen. et sp. nov.). 8, ×3; ×2, vertical section of holotype. Cat. No. 63872.**

10. ***Ectenolites penecilin*** **Flower, 1964. Vertical section, ×2, Cat. No. 63873.**

11. ***Ectenolites primus*** **Flower, 1964. 11a, ×3, spical part; 11b, ×3, anterior part of same. Cat. No. 63874.**

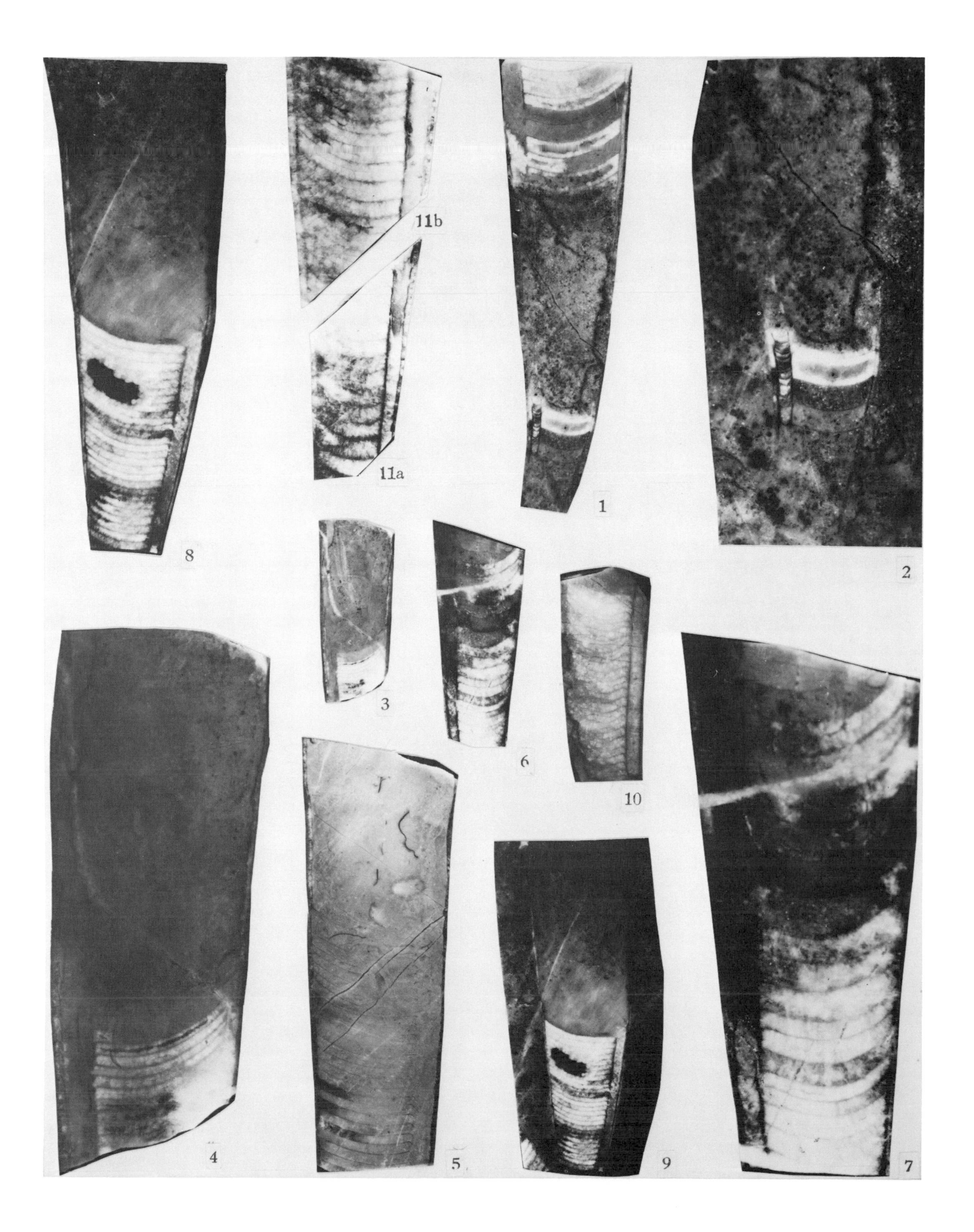
11b
11a
1
8
2
3
6
10
4
5
9
7

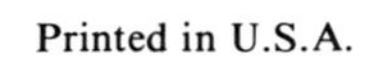
Printed in U.S.A.

Geological Society of America
Special Paper 187
1981

Provincialism, dispersal, development, and phylogeny of trilobites

Lu Yan-hao
Nanjing Institute of Geology and Palaeontology, Academia Sinica, Nanjing, 210008 People's Republic of China

ABSTRACT

Four topics are discussed in this paper: (1) provincialism, (2) dispersal, (3) larval developments, and (4) evolutionary trends of trilobites. The emphasis is placed on the discussion of relevant problems in appearance, migration, evolution, and extinction of trilobites.

INTRODUCTION

Before the 1950s, the study of organic evolution was principally based on observation of morphological changes of organs and structures of organisms and their sequence in geological history. During the 1950s, paleontologists began to use the results of modern studies in provincialism and faunal dispersion as well as diversity and various patterns of organic evolution that had been suggested and discussed. With the help of the electronic mircoscope the microstructures of fossils can be clearly observed; hence, the taxonomic position of some taxa, previously wrongly identified or regarded as *incertae sedis,* has been settled or amended, and the principles of classification have also been improved to a certain degree.

Trilobites are among the earliest Metazoa to appear on earth, and they represent the first burst of the explosive development of animals. Trilobites lived in various ecological conditions and had different living habits, such as pelagic and benthic. Their remains are found in different kinds of sedimentary rocks; they help us to understand the processes of appearance, distribution, and extinction of other kinds of life also. The study of the Trilobita is, therefore, a "typical" example of the study of fossil faunas in general.

PROVINCIALISM OF TRILOBITES

According to abundant data accumulated during the past 50 years in China and recent achievements elsewhere, the universal provincialism of trilobites during the Cambrian may be summarized as follows (Lu and others, 1974) (Fig. 1):

Oriental Realm:
- I. North China Province
  - A. North China-southern East China Subprovince
  - B. Yangzi-Qinghai-Sichuan-Tibet Subprovince
- II. Southeast China Province
  - A. Jiangnan-Northwest China Subprovince
  - B. Zhu-Jian Subprovince
- III. Transitional mixture of North China and Southeast China Provinces

Occidental Realm:
- I. North American Province (or "Pacific" Province)
- II. Atlantic Province
- III. Transitional mixture of North American Province and Atlantic Province

Transitional or mixed faunas
- Intermediate areas between Oriental and Occidental faunas

The factors controlling provincialism and geographic distribution are as follows: (1) Trilobites have two different places of origin. During the latest Cryptozoic and earliest Phanerozoic, the Oriental and Occidental faunas or realms evolved and began to differentiate in the shallow seas of both sides of Pangaea (Fig. 2). (2) Trilobites of the North China Province of the Oriental Realm and those of the North American Province of the Occidental Realm on the one hand, and trilobites of Southeast China Province of the Oriental Realm as well as those of the Atlantic Province of the Occidental Realm on the other hand underwent a parallel development in morphology, heredity, ecology, and adaptation. The former two provinces cover mainly the shelf and epeiric seas, characterized by marginal and shallow-marine conditions suitable for benthic life; the latter two provinces occupied enclosed deep-water troughs or basins and were characterized by floating and swimming forms (Figs. 3, 4). Of the two factors mentioned above, the first one (the place of origin), is the most important and controls the geographical distribution and the transitional (or mixed) characteristics of two distinct faunas. For example, most representatives of the Oriental Realm, such as redlichiids, drepanyopygids, palaeonenids, damesellids, tsinaniids, and kaolishaniids, are distributed in Asia, Oceania, and Antarctica (Daily, 1957; Öpik, 1961, 1963, 1967; Palmer and Gatehouse,

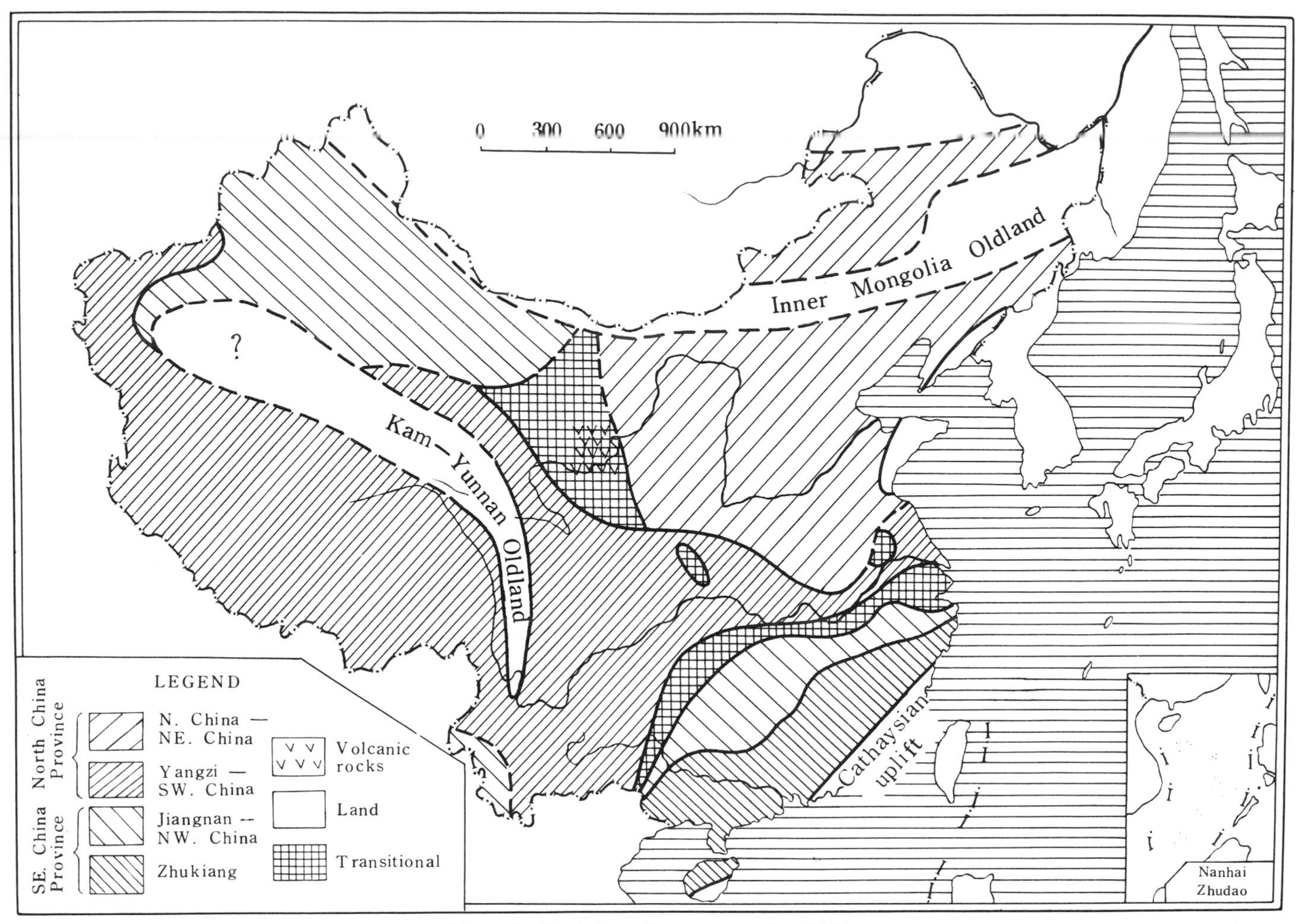

Figure 1. Cambrian paleozoogeography of China.

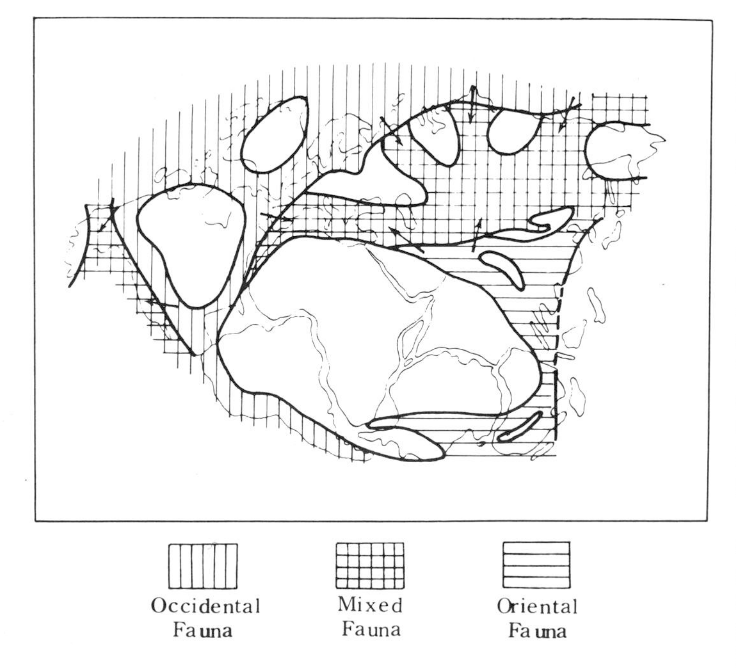

Figure 2. Distribution of Cambrian faunas on Pangea.

1972). Representatives of the Occidental Realm, such as olenellids and paradoxidids, occupied an area extending from western Europe, eastern North America, northwestern Europe to South America (V. Poulsen, 1958; C. Poulsen, 1960). In the transitional areas such as Siberia (Repina, 1972, Ivshin and Pokrovskaya, 1968), Tethys (Dean, 1971; Wolfart, 1970; Linan 1978), North Africa (Hupé, 1952, 1960), and western North America (Cook and Taylor, 1975; Fritz, 1972; Palmer, 1968, 1977; Taylor, 1975), elements of both the Oriental and Occidental Realms, such as olenellids, redlichiids, *Fallotaspis,* paradoxidids, dorypygids, anomocarids, and *Bailiella,* intermingled with each other (Fig. 5). The second factor (parallelism) was controlled by geographical environment, and its progressive development resulted in distinctive provinciality; that is, both the Oriental and Occidental Realms had differentiated into shelf or epeiric seas and deep-water troughs or basins, respectively, as shown in (Fig. 1).

In the shelf area are found mainly benthic or nektobenthic elements, such as redlichiids, olenellids, protolenids, inouyiids, damesellids, tychaspidids, tsinaniids, kaolishaniids, saukiids, and changshaniids. In the deep-water troughs lived abundant floating forms, such as eodiscids, agnostids, and olenids. The latter two kinds of trilobites are cosmopolitan; besides, there were paradoxidids (*Xystridura* and *Galahetes* of the Oriental Realm) (Zhu and

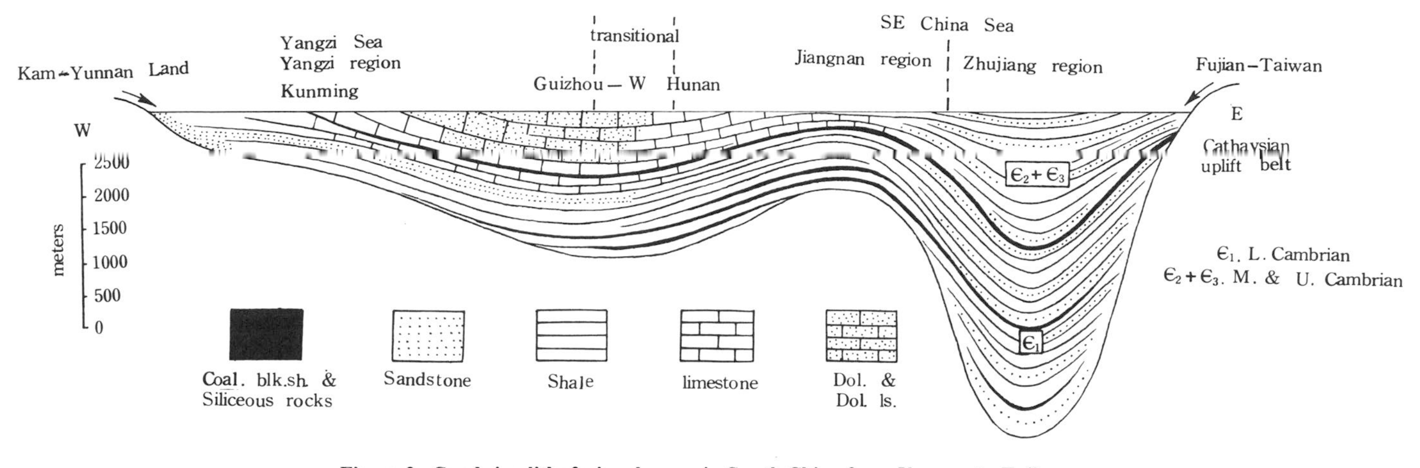

Figure 3. Cambrian lithofacies changes in South China from Yunnan to Fujian.

Lin, 1978) or *Paradoxides* of the Occidental Realm as well as *Centropleura* (Liu and Zhang, 1979), which is an element common to the Oriental and Occidental Realms. The genera of Cambrian olenids had undergone parallel development in the Oriental and Occidental deep-water troughs (the Jiangnan, Northwest China Subprovince and the Atlantic Province). Their representatives on the genus-group level did not become common in both provinces until Late Cambrian and Orodovician time (*Parabolinella, Parabolina,* and others). After the Ordovician, the marine interconnections having been established for a long time, the characteristics of the Oriental and Occidental faunas disappeared progressively. The faunas of the whole globe became intermingled, and the important controlling provincialism determined the ecological adaptations; that is, trilobites spread to various geographical environments. Shelf or shallow water, deep-water troughs, and the transitional areas in between determined the major patterns of provinciality. For example, the North China Province and the North American Province were fundamentally characterized by shallow-water shelves and epeiric seas, where heavy-shelled cephalopods, such as actinoceroids, gastropods, and nektobenthonic trilobites, such as *Hystricurus,* as well as benthonic dendroid graptolites were found. The Southeast China Province and the Atlantic Province occupied deep-water troughs, that contained the swimming or floating trilobites *Hysterolenus, Ceratopyge,* and *Parabolinella,* and floating graptolites (Lu and others, 1976). Until the Ordovician, the "Yangzi Province" was differentiated from the North China Province as an independent province that communicated southward with Australia and westward with the Tethys and France. In this province the trilobites *Dactylocephalus, Hanchungolithus, Taihungshania,* and others occur (Dean, 1971; Webby, 1973).

In addition to the two important factors mentioned above, other subordinate factors such as temperature, salinity, alkalinity, turbidity, and submarine relief also had some influence upon the distribution of the faunas.

The factors controlling provincialism had a profound influence on the dispersal, rate of evolution, differentiation, and distinction of trilobites.

DISPERSAL OF TRILOBITES THROUGH GEOLOGIC AGES

Judging from the provincialism of Cambrian trilobites, the presence of a so-called Pangaea in the Cambrian Period is beyond any doubt. During that time, the Atlantic, or proto-Atlantic, had not yet come into existence; North America, eastern South America, Europe, and western Africa, joined eastern Africa and India to form Gondwanaland, and the latter was attached to southern and southeastern Asia. The assumed presence of a Pangaea is very

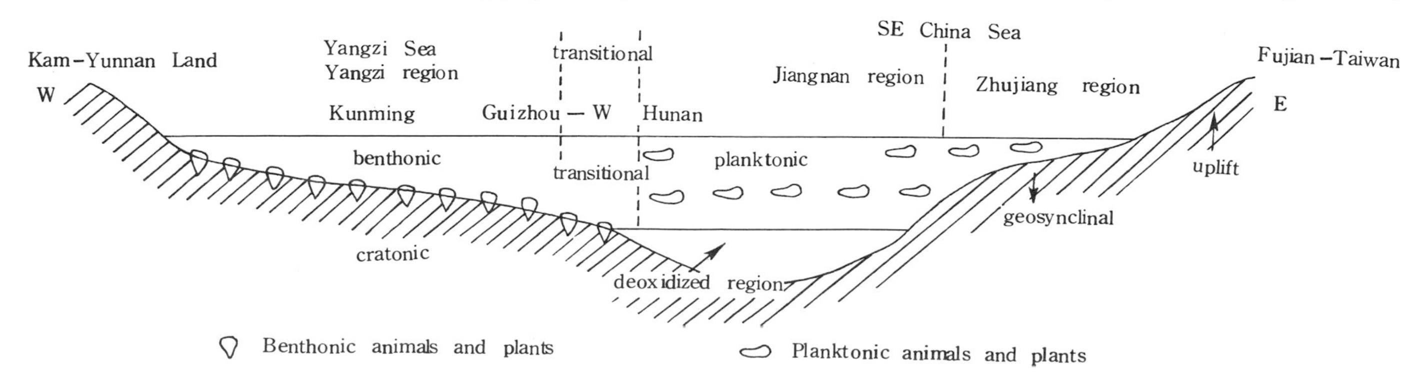

Figure 4. Ecology of Cambrian seas in South China.

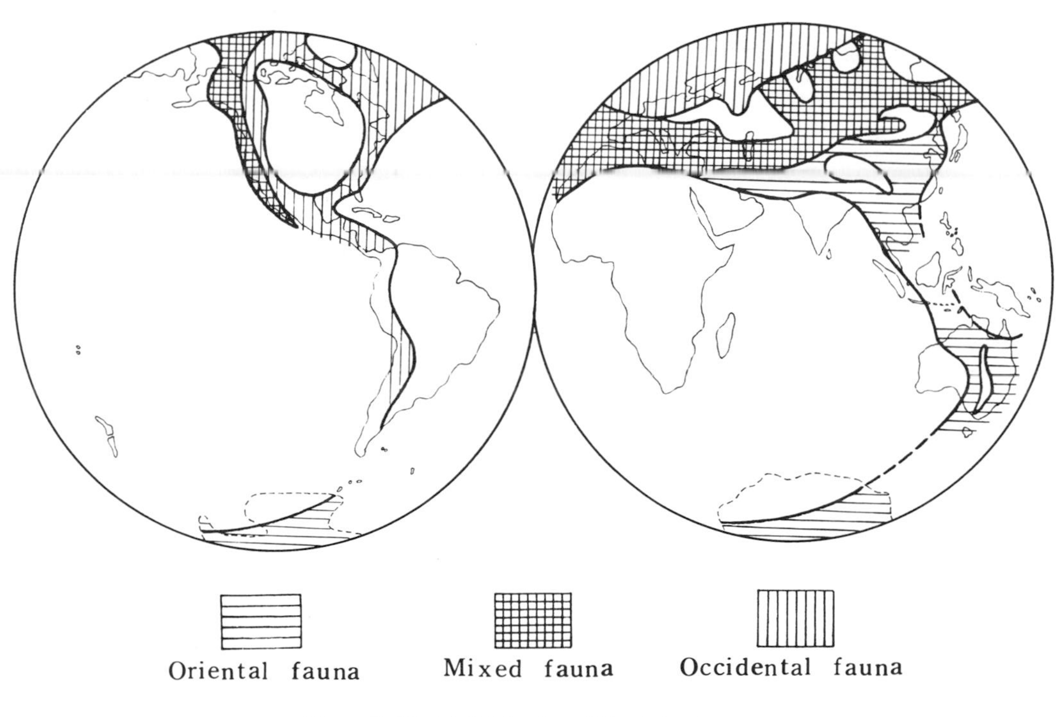

Figure 5. World distribution of Cambrian faunas.

useful in explaining the dispersion of trilobites. I consider that North and South America, especially the former, were the places of origin of the Occidental fauna, and South China, Southeast Asia, and Oceania were the places of origin of the Oriental fauna. The direction of dispersal of the Occidental fauna (olenellids, paradoxidids, olenids, and others) was from west to east, whereas that of the Oriental fauna (redlichiids, damesellids, dorypygids, and others) was from east to west. In the northwestern angle of North Africa, Tethys, southern Europe, Siberia, and western North America, these two faunas met and mixed with each other to form an intermediate fauna (Fig. 2). For example, redlichiids of the Oriental fauna and olenellids and *Fallotaspis* of the Occidental fauna coexisted in Morocco in North America (Hupé, 1952, 1960), in Spain (Lotze and Sdzuy, 1961; Linan, 1978), in Sardinia (Nicosia and Rasetti, 1970; Rasetti, 1972), and in Siberia where, in addition, *Paradoxides* occurred (Repina, 1972; Ivshin and Pokrovskaya, 1968). In western North America, except for the olenellids and *Fallotaspis,* the *Hedinaspis* assemblage of the Late Cambrian Oriental fauna also mixed with the Occidental fauna (Taylor, 1975; Cook and Taylor, 1975; Palmer, 1968, 1977). In Asia, the direction of dispersion of trilobites from east to west was very conspicuous not only in the Cambrian but also in the Ordovician; for instance, the genus *Hammatocnemis* in southwestern China had migrated westward since Early Orodvician time and disappeared from Central China in the Late Ordovicin (Ashgillian); at this time it migrated from Zinjiang through Kazakhstan to Uzbekistan, and flourished in Poland as late as the Ashgillian (Lu and Zhou, 1979) (Fig. 6). Trilobites originating in the Occidental fauna and then migrating eastward also are not very rare; for example, *Dalmanitina* is found in the Middle Ordovician of the Occidental fauna, but in China, it flourished in Late Ordovician and Early Silurian times (Lu and Wu, 1979).

LARVAL DEVELOPMENT OF TRILOBITES

If two evolutionary lineages of organisms are distinctively different, they are assumed to have different ancestors. This is a general law of heredity. In determining whether the lineages are the same or not, the recapitulation theory of Haeckel probably is most important. The earliest representatives of the Oriental and Occidental faunas are Early Cambrian redlichiids and olenellids. Both lineages are quite different in morphology of various larval development stages, including protaspis, meraspis, and holaspis. Many genera and species of redlichiids in the Oriental fauna are extremely similar in their protaspid and meraspid stages, as may be seen in *Eoredlichia, Redlichia (Redlichia), R. (Pteroredlichia)* of the Redlichiidae, and *Yinites* of Yinitidae (Pl. 1) (Lu, 1940, 1945; Kobayashi and Kato, 1951; Zhang and others, 1980). Likewise, the protaspis and meraspis stages of many genera of olenellids, for example, *Olenellus, Elliptocephala,* and *Paedeumias,* are quite similar (Fig. 7). Thus, trilobites from different biogeographical provinces exhibit quite different morphologies, although they are contemporaneous. Judging from the view of larval development, they should have differnt ancestors. From these facts it may be inferred that the Oriental and Occidental faunas developed in geographically separated areas. The evolution of trilobites of both Early and Middle Cambrian age was controlled by this law. For instance, the larvae and adults of the Middle Cambrian Paradoxididae of the Occidental fauna are obviously more closely related to those of the olenellids and have nothing to do with the redlichiids (Lu, 1940; Whittington, 1957). After the Late Cambrian, the Oriental and Occidental faunas intermingled with each other again and again, the characteristic features of the original faunas disappeared progressively, and geographical provinciality became more and more apparent, along with the results of

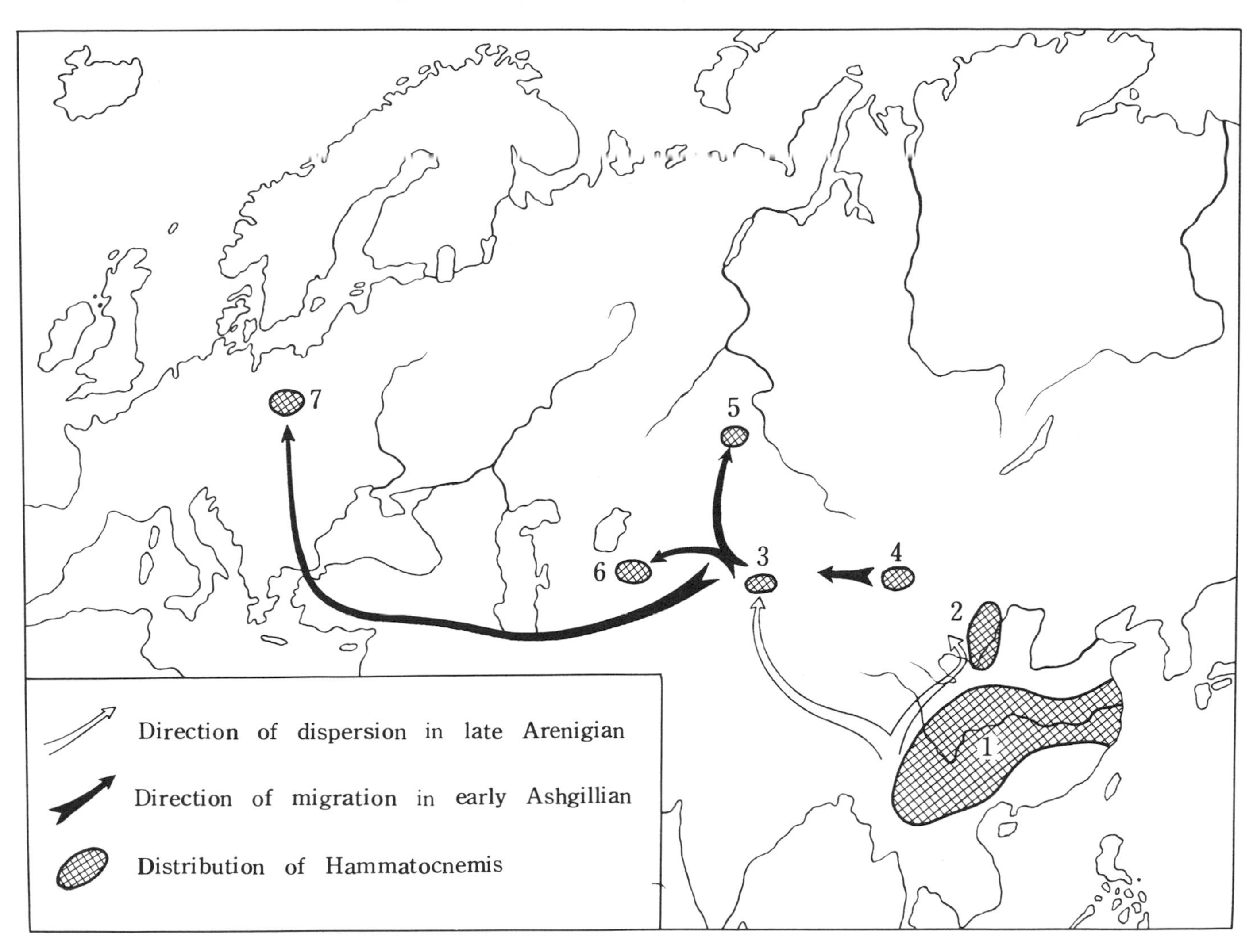

Figure 6. Distribution, dispersal, and migration of *Hammatocnemis* 1, South China and Northern Shan States, Burma; 2, border region of Shaanxi, Gansu, Ningxia, and Nei Mongol; 3, Kalping in Xinjiang; 4, Beishan in Gansu; 5, Kazakhstan; 6, Uzbekistan; 7, Poland.

parallel evolution, so that many families and genera then were common to both the North China and North American Provinces. Only the shelf faunas retained their own regional features. In Europe and Asia, provincial similarity was also shown in the ontogeny and phylogeny of the Early and Late Ordovician trinucleids as well as the Late Ordovician and the Early Silurian dalmanitinids. The characteristics of larvae and adults of early *Hanchungolithus* are similar to those of larvae of *Reedolithus,*

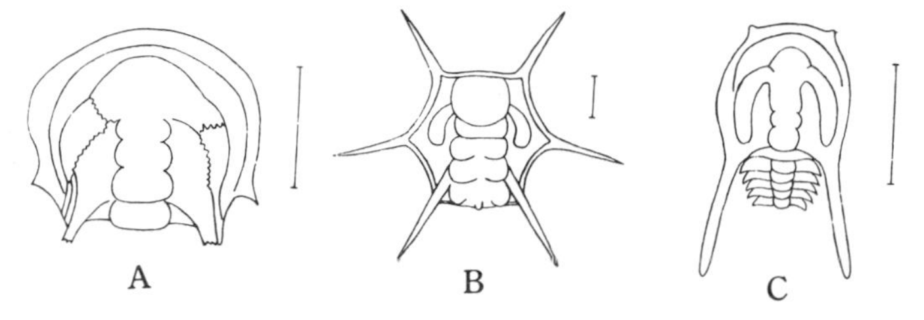

Figure 7. Protaspis and meraspis stages of olenellids. A, *Elliptocephala* (after Whittington, 1957); B, *Olenellus* (after McNamara, 1978); C, *Paedeumias* (after Walcott, 1910). Length of bars 1 mm.

Nankinolithus, and *Onnia,* all of which belong to the trinucleids of Late Ordovician age of Asia and Europe; the latter forms recapitulate the development of earlier stages of the former (Störmer, 1930; Lu, 1964; Hughes and others, 1975). Likewise, the morphology of the protaspis and meraspis stages of European *Dalmanitina socialis* (Barrande, 1852) and *D. olini* (Temple, 1952) in the Late Ordovician are extremely similar to those of *Dalmanitina nanchengensis* Lu in the Early Silurian of China. *Dalmanitina* probably migrated to Asia from Europe through Siberia, and judging from its ontogeny, we may infer that it should belong to one faunal province in a strict taxonomical sense.

EVOLUTION OF TRILOBITES

In discussing trilobite evolution, beginning with the Cambrian, first, the Oriental fauna represented by redlichiids and the Occidental fauna represented by olenellids should be distinguished; second, the differences in evolutionary rate operating in shelf and deep-water trough or basin environments and the differences in abundances of species, genera, and individuals should be noted. The evolution of the redlichiids proceeded mainly in shelf areas and is summarized in Figure 8.

The main evolutionary trend of redlichiids is from the *Eoredlichia* group through the *Redlichia takooensis* group to the *R. chinensis* group. This trend has the following characteristics: (1) Structures developed from simple to complicated; for example, Early Cambrian *Eoredlichia* does not have a row of pits in the border furrow of the cranidium nor a row of papillate processes on the ventral side of the doublural border, whereas such pits and processes appeared in the middle Early Cambrian *R. takooensis* and the late Early Cambrian *R. chinensis* [Pl. 1, (4, 16, 18, 19)]. (2) From *Eoredlichia* to *R. chinensis,* the palpebral lobe became increasingly longer, the included angle between the anterior section of the facial sutures and the axial line of the glabella became larger (from 45° to 60° to 90°)[Pl. 1(4, 16, 18–20)]. (3) The shape of the glabella changed from rounded to sharp conical [Pl. 1(4, 16–20)] (Lu, 1940; Kobayashi and Kato, 1951; Zhang and others, 1980). The tendency of these three morphological changes is contrary to the general trends of trilobite evolution and may be regarded as an example of orthogenesis; however, from a viewpoint of ontogeny, this is a phenomenon of neoteny or paedomorphosis. This is one of the evolutionary trends of redlichiids.

The second evolutionary trend observed in redlichiids is from the *Eoredlichia* group to the *R. takooensis* group to the *R. nobillis* group, and represents an "conservative" trend of trilobite evolution. From the middle Early Cambrian *R. takooensis* group to the late Early Cambrian *R. nobilis* group, the length of the palpebral lobe and the angle between the anterior section of the facial sutures and the axial line of the glabella remained essentially unchanged [Pl. 1(4, 16, 21)].

The third trend is from the *Eoredlichia* group [Pl. 1(1–4)] to the *Yinites* group [Pl. 1(5–9)], in which the length of the palpebral lobe, the shape of the glabella, and the anterior furrow in front of the glabella did not become specialized. The long, postero-lateral pygidial spines developed in the Yinitidae are an important feature derived from the *Eoredlichia* group (Zhang and others, 1980).

The shallow-water shelf is an area in which the environmental change is most active. Dramatic changes may even be caused by minor oscillations of sea level. Organisms have to modify their morphology and structure to adapt to changing conditions; otherwise, they would be doomed to extinction. Such situations cause rapid diversification of organisms, and the emergence of new genera and new species as well as the extinction of old ones occur rapidly. This is the reason why many trilobite zones occur in the North China and North American Provinces. In this sense the shelf may be called a "cradle" of new organisms (or even life) (Lu, 1980).

In the deep-water troughs or basins, benthic organisms could not live. In such areas most trilobites are planktonic or nektonic, such as the Cambrian agnostids, eodiscids, olenids, and *Centropleura* (Liu and Chang, 1979), the Ordovician cyclopygids, *Carolinites,* and telephinids, as well as some Devonian trilobites of the Nandan type (Zhang, 1974). The rate of evolution in these trilobites is rather slow, for the oscillations of sea level have not influenced them very much. Therefore, much more detailed work has to be done if we want to use trilobites to establish fossil zones for stratigraphic purposes. For instance, trilobite zones in the North

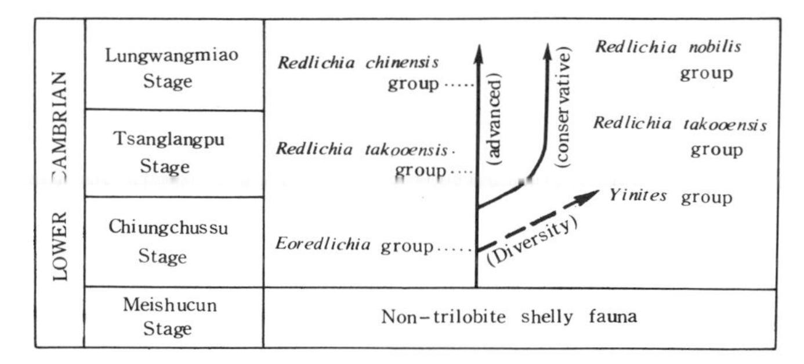

Figure 8. Evolution of redlichiid tiolobites in the Early Cambrian.

China Province can be easily recognized in the field, while in the Jiangnan area it is rather difficult to distinguish fossil zones because the trilobites in this region are nektonic and planktonic types with lower evolutionary rates, and we often have to use species rather than genera to establish biostratigraphic zones.

SELECTED REFERENCES

Barrande, J., 1852, Systême Silurien du Centre de la Boheme 1. Recherches Paléontologiques, Volume 1, Trilobites: Paris, Praha, 935 p., 51 pls.

Chang Wen-Tang [Zhang Wen-tang], 1966, On the classification of Redlichiacea, with description of new families and new genera: Acta Palaeontologica Sinica, v. 14, no. 2, p. 135–184, Pls. 1–4.

Cook, H. E., and Taylor, M. E., 1975, Early Palaeozoic continental margin sedimentation, trilobite biofacies, and the thermocline, Western United States: Geology, v. 3, no. 10, p. 559–562.

Daily, R., 1957, The Cambrian in South Australia; Australia Bureau of Mineral Resources, Geology and Geophysics Bulletin no. 49, p. 91–147.

Dean, W. T., 1966, The Lower Ordovician stratigraphy and trilobites of the Landeyran Valley and the neighboring district of The Montague Noire, South-Western France: Bulletin of the British Museum (Natural History), Geology, v. 12. no. 6. p. 245–353, 21 pls.

——1971, The lower Palaeozoic stratigraphy and faunas of the Taurus Mountains near Beysehir, Turkey, II. The trilobites of the Sydisehir Formation (Ordovician): Bulletin of the British Museum (Natural History), Geology, v. 20, no. 1. p., 1–24, 5 pls.

Fritz, H. W., 1972, Lower Cambrian trilobites from the Sekwi Formation, type section, Mackenzie Mountains, Northwestern Canada: Geological Survey of Canada Bulletin 212, p. 1–58, Pls. 1–20.

Henningsmoen, G., 1957, The trilobite family Olenidae: Skrifter utgitt av det Norske Videnskaps-Akademi i Oslo, I. Mat.-Naturv. Klasse, no. I, 303 p., 31 pls.

Henningsmoen, G., 1958, The Upper Cambrian faunas of Norway: Norsk Geologisk Tidsskrift, bd. 38, H. 2, p. 179–196, 7 pls.

Hughes, C. P., Ingham, G. K., and Addison R., 1975, The morphology, classification and evolution of the trinucleidae (Trilobita): Philosphical Transactions of the Royal Society of London, v. 272 (920), p. 537–607, Pls. 1–10.

Hupé, P. R., 1952, Contribution a l'étude du Cambrien supérieur et du Précambrien III de l'Anti-Atlas marocain: Service géologque Maroc (Rabat), Notes et Mémoirs, no. 103, 402 p., 24 pls.

——1960, Sur le Cambrien inférieur du Maroc: Norden, International Geological Congress, 21st, sec. 8, pt. 8, p. 75–85.

Ivshin, N. K., and Pokrovskaya, N. V., 1968, Stage and zonal subdivision of the Upper Cambrian: International Geological Contress 24th, Montreal, sec. 9. p. 97–108.

Kobayashi, T., 1971, The Cambro-Ordovicin faunal provinces and the interprovincial correlation, discussed with special reference to the trilobites in eastern Asia: Journal of the Faculty of Science, University of Tokyo, sec. 2, v. 18, no. 1, p. 129–299.

—— 1972, Three faunal provinces in the Early Cambrian Period: Proceedings of Japanese Academy, v. 48, no. 4, p. 242–247.

Kobayashi, T., and Hamada, T., 1961, On a new Malayan species of *Dalmanitina:* Japanese Journal of Geology and Geography, v. 35, nos. 2–4, p. 101–113, Pl. 4.

Linan, E. G., 1978, Bioestratigrafia de la Sierra de Cordaba: Departemento de Palaeontologia Facultad Ciencias, Universidad Granada, 212 p., 17 pls.

Liu Yiren and Zhang Tairong, 1979, [Discovery of Middle Cambrian Centropleura (Trilobita) in China]: Geological Review, v. 25, no. 2, p. 1–6, 1 pl. (in Chinese).

Lochman, C., 1956, Stratigraphy, paleontology and paleogeography of the *Elliptocephala asaphoides* strata in Cambridge and Hoosick quadrangles, New York: Geological Society of America, Bulletin, v. 67, p. 1331–1396.

—— 1970, Upper Cambrian faunal pattern on the craton: Geological Society of America Bulletin, v. 81, p. 3197–3224.

—— 1972, Cambrian paleoenvironments on the craton of the United States: Proceedings International Paleontological Union, International Geological Congress, 23rd, Prague, p. 267–282.

Lotze, F., and Sdzuy, K., 1961, Das Kambrium Spaniens: Abhandlungen der Mathematisch-naturwissenschaftlichen Klasse, Akademie der Wissenschaften und der Literatur, Mainz, Jahrgang 1961. Teil I, Stratigraphie, by F. Lotze; no. 6, 285–498. Teil II, Trilobiten, by K. Sdzuy, nos. 7–8, 499–693, Pls. 1–34.

Lu Yan-hao and Wu Hong-ji, 1979, [Ontogeny of *Dalmanitina nanchengensis* Lu and *Platycoryphe sinensis* (Lu)]: Papers contributed to the 50th Anniversary Meeting of the Palaeontological Society of China [Abs.], p. 62, (in Chinese).

Lu Yan-hao and Zhou Zhi-yi, 1979, Systematic position and phylogeny of *Hammatoonemis* (Trilobites): Acta Palaeontologica Sinica, v. 18, no. 5, p. 415–434, 4 pls.

Lu Yen-hao [Lu Yan-hao], 1940, On the ontogeny and phylogeny of *Redlichia intermedia* Lu (sp. nov.): Bulletin of the Geological Society of China v. 20, nos. 3–4, p. 333–342.

—— 1950, [On the genus *Redlichia* (Trilobita) with descriptions of some new species]: Geological Review, 15, nos. 4–6, p. 157–170 (in Chinese).

—— 1964, Ontogeny of *Hanchungolithus multiseriatus* (Endo) and *Ningkianolithus welleri* (Endo) (Trilobita): Scientia Sinica, v. 13, no. 2, p. 291–308, 3 pls.

—— 1980: Ontogeny and evolutionary trends of trilobites: Papers submitted to the International Geological Congress, 26th, Paris, Abstracts, v. 1, p. 174.

Lu Yen-hao [Lu Yan-hao], Chu Chao-ling [Zhu Zhou-ling], Chien Yi-yuan [Zian Yi-yuan], Lin Huan-ling, Chow Tse-yi [Zhou Zhi-yi], and Yuan Ke-xing, 1974, [Bio-environmental control hypothesis and its application to the Cambrian biostratigraphy and palaeozoogeography: Nanjing Institute of Geology and Palaeontology, Academia Sinica, no. 5, p. 27–112, 4 pls. (in Chinese).

Lu Yen-hao [Lu Yan-hao], Chu Chao-ling [Zhu Zhao-ling], Chien Yi-yuan [Qian Yi-yuan], Zhou Zhi-yi, Chen Jun-yuan, Liu Geng-wu, Yu Wen, Chen Xu, and Xu Han-kui, 1976, [Ordovician biostratigraphy and palaeozoogeography of China]: Nanjing Institute of Geology and Palaeontology, Academia sinica, Memoirs, no. 7, p. 1–83, 14 pls. (in Chinese).

McNamara, K. J., 1978, Paedomorphosis in Scottisch Olenellid trilobites (Early Cambrian): Palaeontology, v. 21, pt. 3, p. 635–655, pl. 71.

Nicosia, M. L., and Rassetti, F., 1970, Revisioni dei trilobiti del Cambriano dell' Iglesiente (Sardegna) descritti da Meneghini: Atti della Accademia Nazionale dei Lincei, Memorie Classe di Scienze fisiche, matematiche e naturali, Serie VIII, 10, sec. II, no. 1, p. 1–20. Pls. 1–3.

Öpik, A. A., 1961, Cambrian geology and palaeontology of the headwaters of the Burke River, Queensland: Australia of Bureau of Mineral Resources, Geology and Geophysics, Bulletin 53, p. 1–249, Pls.
1962, Early Upper Cambrian fossils from Queensland: Australia Bureau of Mineral Resources, Geology and Geophysics, Bulletin 64, p. 1–133, Pls. 1–9.

—— 1967, The Mindyallan fauna of North-Western Queensland: Australia Bureau of Mineral Resources, Geology, and Geophysics, Bulletin 74, v. 1, p. 1–404, v. 2, p. 1–167, Pls. 1–67.

—— 1968, Early Ordovician at Claravale in the Ferguson River Area, Northern Territory: Australia Bureau of Mineral Resources, Geology and Geophysics, Bulletin 80, p. 161–165.

Öpik, A. A., and others, 1957, The Cambrian geology of Australia: Australia Bureau of Mineral Resources, Geology and Geophysics, Bulletin 49, p. 1–284.

Palmer, A. R., 1957, Ontogenic development of two olenellid trilobites: Journal of Palaeontology, v. 31, p. 105–128.

—— 1968, Cambrian trilobites of East-Central Alaska: U.S. Geological Survey Professional Paper 559-B, p. 1–115, Pls. 1–15.

—— 1977, Biostratigraphy of the Cambrian System—A progress report: Annual Reviews of Earth and Planetary Sciences, no. 5, p. 13–33.

Palmer, A. R., and Gatehouse, C. G., 1972, Early and Middle Cambrian trilobites from Antarctica: U.S. Geological Survey Professional Paper 456–D, p. D1–D37, 6 pls.

Palmer, A. R., and Stewart, J. H., 1968, A paradoxidid trilobite from Nevada: Journal of Paleontology, v. 42, no. 1, p. 177–179.

Poulsen, C., 1960, Fossils from the late Middle Cambrian *Bolaspidella* zone of Mendoza, Argentina: Matematisk-Fysiske Meddelelser udgivet af Det Kongelige Danske Videnskabernes Selskab, Bind 32, Nr. 11, p. 1–42, 3 pls.

Poulsen, V., 1958, Contributions to the Middle Cambrian palaeontology and stratigraphy of Argentina: Matematisk-Fysiske Meddelelser udgivet af Det Kongelige Danske Videnskabernes Selskab, Bind 32, Nr. 11, p. 1–42, 3 pls.

Rasetti, F., 1972, Cambrian trilobite faunas of Sardinia: Academia der Memorie Scienze fisiche, ser. 8, v. 11, Sez. II, Fasc. 1, p. 1–100, 19 pls.

Repina L. N., 1972, Biogeography of Early Cambrian of Siberia according to trilobites: Proceedings International Palaeontological Union, International Geological Congress 23rd, Prague, p. 289–300.

Rong Jia-yu, 1979, The *Hirnantia* fauna of China with comments on the Ordovician-Silurian boundary: Acta Stratigraphica Sinica, v. 3, no. 1, p. 1–29.

Störmer, L., 1930, Scandinavian Trinucleidae with special reference to Norwegian species and varieties: Skrifter utgitt av det Norske Videnskaps Akademi i Oslo 1, Mat.-Naturv. Klasse, no. 4, p. 1–111, 14 pls.

Taylor, M. E., 1975: Late Cambrian of western North America: Trilobite biofacies, environmental significance, and biostratigraphic implications, *in* Kauffman, E. G., and Hazel, J. E., eds., Concepts and methods in biostratigraphy: Stroudsburg, Pennsylvania, Dowden, Hutchison & Ross, Inc., p. 397–425.

—— 1976, Indigenous and redeposited trilobites from Late Cambrian basinal environments of central Nevada: Journal of Paleontology, v. 50, no. 4, p. 668–700, 2 pls.

Temple, J. T., 1952, The ontogeny of the trilobite *Dalmanitina olini:* Geological Magazine, v. 89, no. 4, p. 251–262, Pls. 9–10.

Webby, B. D., 1973, *Remopleurides* and other Upper Ordovician trilobites from New South Wales: Palaeontology, v. 16, no. 3, p. 445–475, Pls. 51–55.

Westergaard, A. H., 1946, Agnostidea of the Middle Cambrian of Sweden: Sveriges Geologiska Undersökning, ser. C, no. 477, 140 p., 16 pls.

Whittington, H. B., 1957, The ontogeny of *Elliptocephala, Paradoxides,*

Sao, *Blaina* and *Triarthurs* (Trilobita): Journal of Paleontology, v. 31, p. 934–946, Pls. 115–116.

—— 1957, The ontogeny of trilobites: Biological Reviews, v. 32, p. 421–469.

Whittington, H. B., and Hughes, C. P., 1972, Ordovician geography and faunal provinces deduced from trilobite distribution: Philosophical Transactions, Royal Society of London, B, v. 263, no. 850, p. 235–278.

Wolfart, R., 1970, The age of the Tremadocian and of the Saukia Zone and the boundary between Cambrian and Ordovician: Newsletters in Stratigraphy, v. 1, p. 10–18.

Zhang, Wentang, 1974: [Devonian trilobites], *in* [A handbook of palaeontology and stratigraphy of Southwestern China]: Science Press, p. 235–237, Pl. 117 (in Chinese).

Zhang Wentang [Chang Wen-tang], Lu Yanhao [Lu Yen-hao], Zhu Zhaoling [Chu Chao-ling], Qian Yiyan [Chien Yi-yuan], Lin Huanling, Zhou Zhiyi, Zhang Sengui, and Yuan Jinliang, 1980, Cambrian trilobite faunas of southwestern China, Palaeontologica Sinica, whole no. 159, new ser. B, No. 16, p. 1–500, 134 pls.

Zhu Zhao ling and Lin Tian rui, 1978, Some Middle Cambrian trilobites from Yaxian, Hainan Island: Acta Paleontologica Sinica, v. 17, no. 4, p. 439–443, 1 pl.

MANUSCRIPT ACCEPTED BY THE SOCIETY 22, 1981

PLATE 1

Repository of all specimens is the Nanjing Institute of Geology and Palaeontology.

Figure

1–4. *Eoredlichia intermedia* (Lu)
- **1. Anameraspid cephalon, ×20, Cat. No. 37687.**
- **2. Meraspis, Degree 4, ×15, Cat. No. 37688.**
- **3. Meraspis, Degree 7, ×15 (Lu, 1940, Pl. 1, Fig. 4).**
- **4. Holaspid cranidium, ×3, (Lu, 1940, Pl. 1, Fig. 10).**

Horizon and locality: Zone of *Eoredlichia intermedia,* lower Lower Cambrian Chiungchussu Formation, Chungsintsun of Kunming, eastern Yunnan.

5–9. *Yinites wanshougongensis* Zhang and Lin
- **5. Protaspis, ×20, Cat. No. 37918.**
- **6. Anameraspid cranidium, ×20, Cat. No. 37920.**
- **7. Meraspis, Degree 3, ×20, Cat. No. 37921.**
- **8. Meraspis, cranidium with 2 thoracic segments attached, ×20, Cat. No. 37922.**
- **9. Holaspid oranidium, ×1, Cat. No. 37930.**

Horizon and locality: Middle Lower Cambrian Minghsingssu Formation, Wanshougon, south of Meitan city, northern Guizhou.

10–16. *Redlichia takooensis* Lu
- **10. Anaprotaspis, ×20, Cat. No. 37586.**
- **11. Metaprotaspis, ×20, Cat. No. 37587.**
- **12. Metaprotaspis, ×20, Cat. No. 37589.**
- **13. Anameraspid cranidium, ×20, Cat. No. 37590.**
- **14. Anameraspid cranidium, ×20, Cat. No. 37591.**
- **15. Metameraspid cranidium, ×20, Cat. No. 37592.**
- **16. Holaspid cranidium, ×3, Cat. No. 37596.**

Horizon and locality: Upper part of Lower Cambrian, South of Takoo, northeast of Guiyang, Guizhou.

17–18. *Redlichia murakaui* Resser and Endo
- **17. Metameraspis, ×10, Cat. No. 37611.**

Horizon and locality: Upper Lower Cambrian Lungwangmiao Formation, Zhenguanying Yiliang, eastern Yunnan.
- **18. Holaspid cranidium, ×5, Cat. No. 37613.**

Horizon and locality: Upper Lower Cambrian Chingshutung Formation, Jingpingshui, Meitan, Guizhou.

19. *Redlichia chinensis* Walcott

Holaspid cranidium, ×3, Cat. No. 37610.

Horizon and locality: Upper part of Lower Cambrian, Guiyang, Guizhou

20. *Redlichia guizhouensis* Zhou

Holaspid cranidium, ×3, Cat. No. 21473.

Horizon and locality: Top of upper Lower Cambrian Chingshutung Formation, Yankung of Jinsha, Guizhou.

21. *Redlichia nobilis* Walcott

Cranidium, ×3, Cat. No. 37546.

Horizon and locality: Top of upper Lower Cambrian Lungwangmiao Formation, Shuanglungtang, Quzhing, eastern Yunnan.

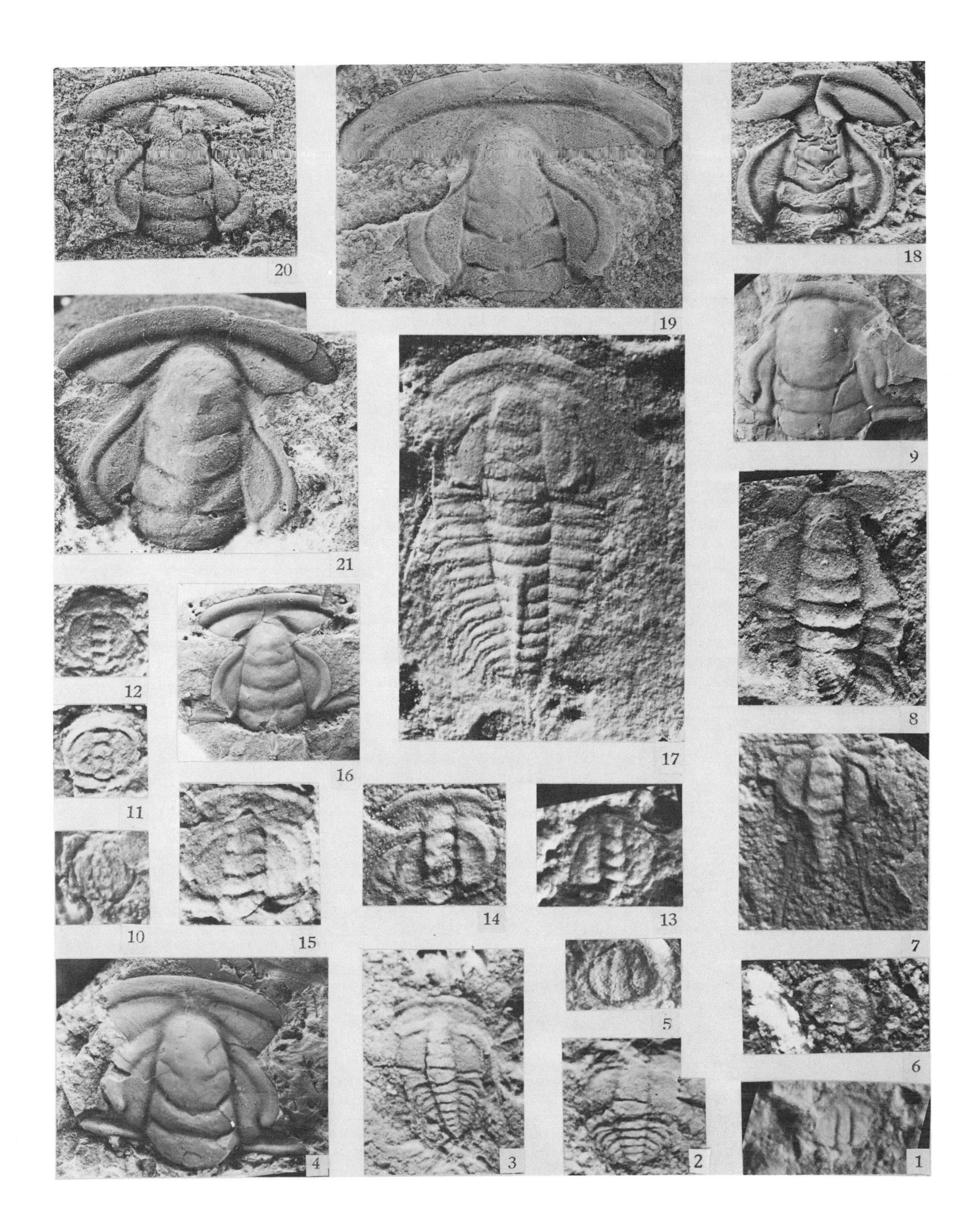

Printed in U.S.A.

Geological Society of America
Special Paper 187
1981

An Early Cambrian trilobite faunule from Yeshan, Luhe District, Jiangsu

Zhu Zhao-ling
Nanjing Institute of Geology and Palaeontology, Academia Sinica, Nanjing 210008, People's Republic of China

Jiang Li-fu
Regional Geological Survey, Geological Bureau of Anhui, Chaoxian, Anhui, People's Republic of China

ABSTRACT

In this paper two new genera and four new species of middle Early Cambrian trilobites from Yeshan of the Luhe District, Jiangsu Province, are described. They are *Redlichia (Redlichia) nanjiangensis* Zhang and Lin, *Yeshanaspis subconicus* (gen. et sp. nov.), *Paraichangia jiangsuensis* (sp. nov.), *P. spinosus* (sp. nov.), and *Paraprotolenella subcylindrica* (gen. et sp. nov.). A brief account of their paleozoographical significance is given.

INTRODUCTION

The age of the carbonate rocks exposed in the Yeshan region of Luhe District, about 47 km north of Nanjing (Fig. 1), remained in doubt for a long time after Liu Ji-chen and Zhao Ru-jin made the first geologic reconnaissance to this region in 1924 (Liu Chi Chen and Chao Ju Chün, 1924). During detailed geologic mapping work in 1974, Jiang Li-fu and his colleagues collected some trilobites, found with inarticulate brachiopods and algae, from the upper part of this carbonate sequence. Additional fossil collections were made by us with Qian Yi-yuan, Zhang Sen-gui, Yuan Jin-liang, and Zhao Wen-jie. A detailed stratigraphic section measured by Jiang Li-fu and his colleagues is given in Figure 2.

Upper Sinian: Töngying Formation

The trilobite faunule from Bed No. 7 of the Mufushan Formation comprises forms that occur in middle Lower Cambrian rocks in several localities of Southwest China. *Redlichia (Redlichia) nanjiangensis* Zhang and Lin has been found in the Yanwangbian Formation of Nanjiang and the Yingzuiyan Formation of Chengkou, both in Sichuan Province. *Paraichangia* is a characteristic form of the Yingzuiyan Formation of the Zhenping District of South Shaanxi. The new genus *Paraprotolenella* is very similar to *Protolenella* from the Yingzuiyan Formation of Chengkou and the Bianmachong Formation of Kaili, Southeast Guizhou. The chief features of *Yeshanaspis* (nov. gen) are comparable to those of *Pseudoichangia* which has a wide distribution in Southwest China. All strata containing *Paraichangia, Protolenella, Pseudoichangia,* and *Redlichai (R.) nanjiangensis,* that is, the Yanwangbian, the Yingzuiyan, and the Bianmachong Formations of Southwest China, are assigned to the middle Early Cambrian.

The lithological characters of the rocks containing the Yeshan faunule are the same as those of the lower part of the Mufushan Formation at the southern slope of Mount Mufushan, about 10 km north of Nanjing city (Yu Jian-hui and others, 1962). In the upper part of this formation at Mufushan, *Paokannia* and *Redlichia* have been collected and studied by Lin Tian-rei (1965), and their age is also considered to be middle Early Cambrian. The correlation of the above-cited formations is shown in Table 1.

It has been pointed out by Lu and others (1974) that the Cambrian trilobite faunas of China may be divided into a North China Faunal Province, a Southeast China Faunal Province, and a Transitional Province. The North China Faunal Province may again be divided into two subprovinces according to environments; that is, the North China–Southern Northeast China and the Yangzi-Qinghai-Sichuan-Xizhang [Tibet] subprovinces. The trilobite assemblage of the Yeshan faunule strongly suggests a Yangzi-Qinghai-Sichuan-Xizang affinity, with analogous faunas from Tangquan of Jiangpu District of Jiangsu Province, Bantang of Chaoxian, Central Anhui, and Mufushan of Nanjing, whereas in the Cambrian faunule found in Chuxian District of East Anhui, the trilobites appcar to be of transitional type, with *Hunanocephalus* and *Eodiscus* as the leading forms. Figure 1 shows the provinciality of the faunule of Yeshan and its neighboring regions; the provincial boundary runs from Yeshan in the north southwestward to Bantang. East of this line, the Cambrian trilobites belong

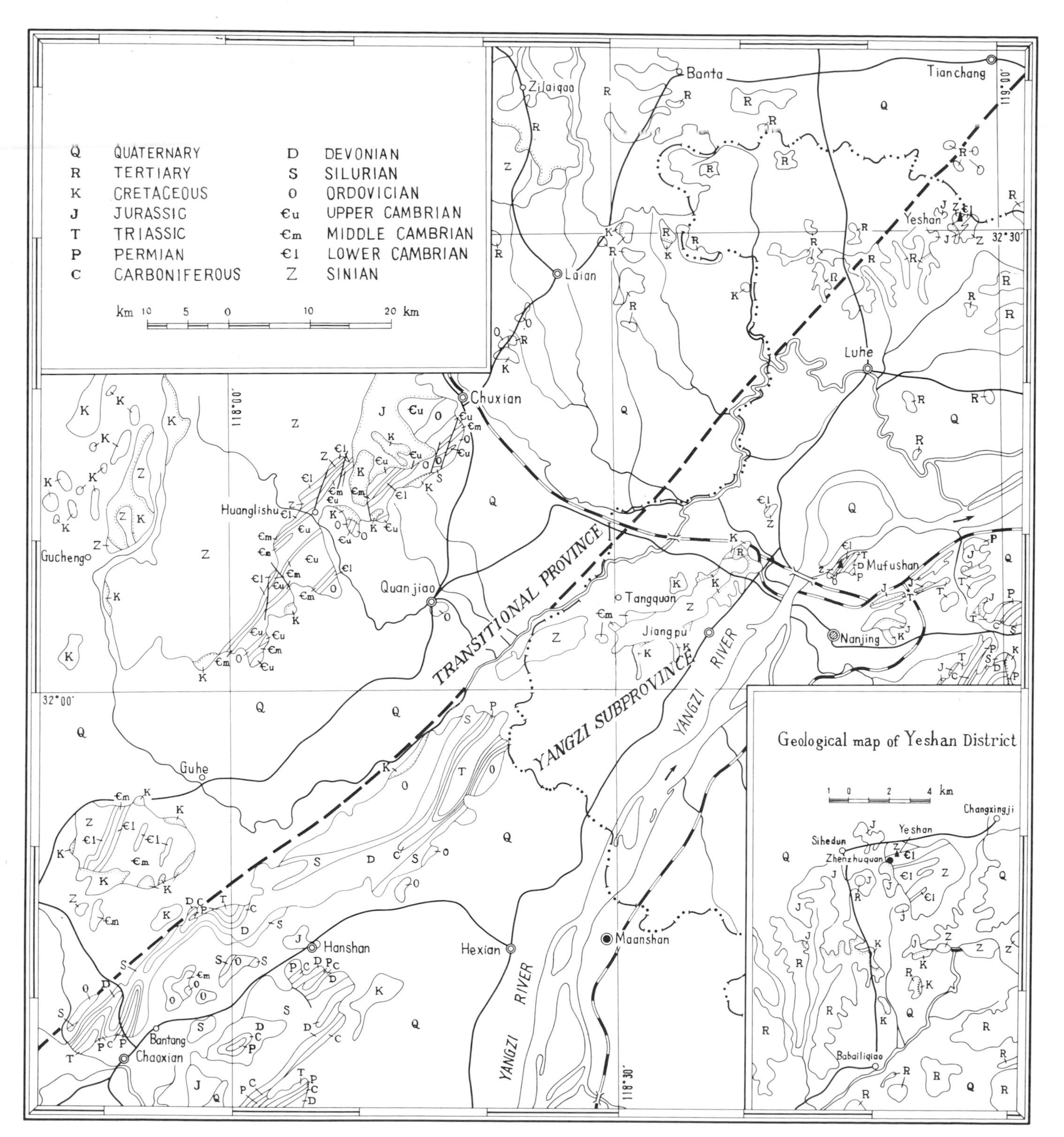

Figure 1. Provinciality of Cambrian trilobite faunule in Luhe and its neighboring district.

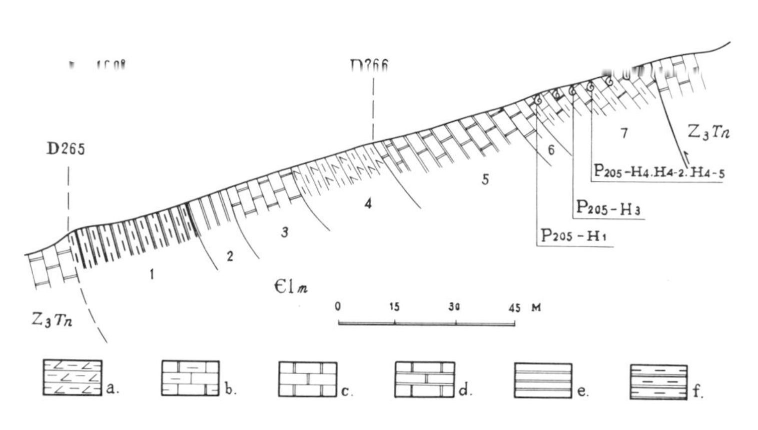

	Thickness (m)
Lower Cambrian: Mufushan Formation	148.42
7. Dark gray, thin-bedded, argillaceous, dolomitic limestone, containing *Redlichia nanjiangensis* Zhang and Lin (P205-H4-2), *Paraichangia jiangsuensis* sp. nov., *P. spinosus* sp. nov. (P205-H3-2,H4-5), *Paraprotolenella subcylindrica* gen. and sp. nov. (P205-Hi), *Yeshanaspis subconica* gen. and sp. nov. (Zhenzhuquan-H1), *Obolus* sp. (P205-H1)	28.45
6. Gray, medium- to thick-bedded dolomite	12.88
5. Light gray, thick-bedded, siliceous dolomite	29.48
4. Grayish black and black, siliceous shale	21.78
3. Light gray, medium- to thick-bedded dolomite	15.76
2. Grayish black, light gray and grayish brown, siliceous rock	9.78
1. Grayish black and black carbonaceous slate	30.29

Figure 2. Section of Lower Cambrian Mufushan Formation at Yeshan of Luhe, Jiangsu. Z_3T_n = Tongying Formation, Sinian: a = siliceous shale; b = argillaceous dolomitic limestone; c = dolomite; d = siliceous dolomite; e = siliceous rock; and f = carbonaceous slate.

to the Yangzi subprovince, whereas those west of this line are a part of the Transitional Province.

SYSTEMATIC DESCRIPTIONS

Family REDLICHIIDAE Poulsen, 1927
Genus *Redlichia* Cossman, 1902
Subgenus *Redlichia (Redlichia)* Cossmann, 1902
Redlichia (Redlichia) nanjiangensis Zhang and Lin, 1978
Plate 1(1–2)

1978 *Redlichia (Redlichia) nanjiangensis* Zhang and Ling, Atlas of the palaeontology in Southwest China, Sichuan Part 1, p. 180; Pl. 89, figs. 11–13.

Description: Glabella convex, large, elongate conical, rounded in front, reaching to marginal furrow, with three pairs of glabellar furrows; posterior pair long, extending from the dorsal furrows obliquely backward and inward, connecting with each other in the middle, arcuate backward; second pair shallow, extending in the same direction as the posterior one; anterior pair faintly marked. Occipital furrow well-defined, curving backward; occipital ring wide in middle, extending backward into a long and strong occipital spine. Anterior border narrow and up-turned; preglabellar field absent; frontal margin slightly arching forward; marginal furrow shallow in middle, deepening toward both sides. Preocular limb small, subtriangular in shape. Palpebral lobe long, arcuate, its posterior end reaching the level of the base of the glabella; ocular ridge short, extending obliquely from the dorsal furrow opposite the antero-lateral sides of the glabella just in front of the anterior glabellar furrows. Fixed cheeks evenly convex, narrow. Anterior facial sutures short, running divergent from the palpebral lobes forwards and cutting the anterior border to form α angle (axial angle formed by the axial line and the anterior facial sutures) about 40°.

TABLE 1. LOWER CAMBRIAN FORMATIONS OF SOUTHERN CHINA

Nanjiang, Sichuan	Chengkou, Sichuan	Mufushan, Nanjing	Yeshan, Luhe
Yanwangbian Formation	Yingzuiyan Formation	Mufushan Formation	Mufushan Formation
	Palaeolenus		
Paokannia	Paokannia	Paokannia	
Redlichia (R.) nanjiangensis		Redlichia	
	Protolenella		Paraprotolenella
	Chengkouia		Redlichia (R.) nanjiangensis
Mayiella			
	Metaredlichioides		
Qiaotingaspis			
	Redlichia (R.) nanjiangensis		Paraichangia
			Yeshanaspis
Shatania			
Xiannudong	Liangshuijing Formation		

Surface ornamented with numerous fine pustules.

Remarks: The chief features of the cranidium are the same as those of *Redlichia (R.) nanjiangensis* from the Yanwangbian Formation of Nanjiang, North Sichuan Province, but our specimen which is preserved in limestone is more convex than the shale specimens of Nanjiang.

Family PROTOLENIDAE Richter and Richter, 1948
Subfamily *Ichangiinae* Zhu, 1980
Genus *Yeshanaspis* Zhu and Jiang, gen. nov.
Type species: *Yeshanaspis subconicus* Zhu and Jiang, sp. nov.

Description: Glabella narrow and long, subconical, strongly tapering anteriorly and contracted at the middle, with three pairs of short glabellar furrows; dorsal furrows deep at the sides, shallowing in front of the glabella. Occipital furrow well-defined, deep and narrow; occipital ring gently convex, slightly expanded towards the axial line, narrowing laterally. Preglabellar field slighty convex, with a short longitudinal ridge in the middle portion; anterior border of uniform width (tr.), marginal furrow narrow and deep; ocular ridge prominent; palpebral lobe of medium size. Fixed cheeks gently convex; anterior facial sutures diverging forward, posterior ones short.

Remarks: This new genus is similar to *Pseudoichangia* Chu and Zhou in some respects, but it differs in the shape of the glabella which is narrow in the middle and strongly tapering anteriorly, in the distinct marginal furrow, in the absence of an occipital spine, and in the more divergent anterior facial sutures.

Yeshanaspis subconicus Zhu and Jiang, sp. nov.
Plate 1(3–5)

Description: Cranidium moderately convex, subquadrate in outline. Glabella convex, narrow and long, subconical in shape, front portion tapering rapidly toward the pointed front and contracted at the middle, with three pairs of short, faint, oblique glabellar furrows, dorsal furrows deep and wide at the sides, shallowing in front of the glabella; occipital furrow deep and narrow, nearly straight; occipital ring slightly expanded toward the axial line, narrowing laterally. Preglabellar field slightly convex, with a short longitudinal furrow extending from the front of the glabella to marginal furrow; anterior border slightly upturned, uniform in width (tr.), marginal furrow distinct, narrow and deep; frontal margin rounded, arching forward. Ocular ridge well-defined, extending slightly oblique backward and outward; palpebral lobe of medium size, situated posteriorly to the midlength of glabella. Fixed cheeks gently convex, less than their width at the base of the glabella; posterior marginal furrow deep and wide. Anterior facial sutures diverge from eyes forward; posterior ones short, running obliquely outward and backward to cut the posterior marginal border.

Genus _Paraichangia_ Lee, 1978
Paraichangia jiangsuensis Zhu and Jiang, sp. nov.
Plate 1(6–9)

Description: Cranidium moderately convex, subrectangular in outline. Glabella convex, gently contracted at the middle and slightly expanded in front, roundly angulate at frontal end. Three pairs of glabellar furrows discernible: posterior pair longer, deep on sides, gradually shallowing inward, not meeting with each other at the middle of the glabella; second pair shorter, parallel to the posterior one; anterior pair weakly defined, extending slightly oblique forward; dorsal furrows deep at the sides, shallowing forward. Preglabellar field gently convex, a little longer (sag.) than the anterior border which is slightly convex, uniform throughout, and gently arched forwards; marginal furrow shallow. Occipital furrow deep and wide, curving back slightly; occipital ring convex from side to side, wide in middle portion, gradually narrowing laterally, arched rearward. Ocular ridge distinct, extending slightly oblique backward from the dorsal furrow opposed to the first glabellar furrow; palpebral lobe longer, situated behind the midlength of the glabella. Fixed cheek wider, about two-thirds the width of the base of glabella; posterior marginal furrow narrow, gradually widening toward the sides. Anterior facial sutures diverging anteriorly, cutting the anterior border in a rounded angle; posterior sutures very short, passing directly outward and then bending strongly backward to cut the posterior marginal border.

Remarks: Two cranidia are referred to this species. In all features this species is closely comparable to the type species, *P. shensiensis* (Lee, 1978), which, however, differs in the shape of glabella, in the slender palpebral lobe and ocular ridge, in the slightly convex anterior border, and in the absence of an occipital apine.

P. jiangsuensis is also somewhat similar to *Hsuaspis sinensis* (Chang), but differs in having a shorter and narrower glabella, in the comparatively flat anterior border, in the wider fixed cheek, in the slender palpebral lobe, and in the narrower (sag.) occipital ring.

Paraichangia spinosa Zhu and Jiang, sp. nov.
Plate 1(10–11)

Description: Cranidium relatively broad, subrectangular in outline. Glabella convex, gradually expanding forward, slightly pointed in frontal end, with three pairs of glabellar furrows; posterior pair deep on sides, gradually shallowing inward, second pair parallel to the posterior one, anterior pair weak, extending slightly oblique forward. Preglabellar field gently convex, as long (sag.) as the anterior border, which is convex, and gently arched forward; marginal furrow distinct. Occipital furrow deep and wide, slightly curving backward; occipital ring wide in middle portion, with a short occipital spine. Ocular ridge well-defined, extending from dorsal furrow opposite to the first glabellar furrows; palpebral lobe long, situated posterior to midlength of glabella. Fixed cheek wide, a little narrower than the width of the base of glabella. Posterior marginal furrow deep and wide. Anterior facial sutures extending slightly divergent from eyes forward, cutting the anterior border in a rounded angle; posterior sutures very short, running directly outward and then bending strongly backward to cut the posterior marginal border.

Comparison: This species is closely similar to *P. jiangsuensis,* but differs in the presence of an occipital spine, in the comparatively narrower preglabellar field, and in the wider anterior border.

Subfamily PROTOLENINAE Richter and Richter, 1949
Genus _Paraprotolenella_ Zhu and Jiang, gen. nov.
Type species: _Paraprotolenella subcylindrica_ Zhu and Jiang, sp. nov.

Description: Glabella wide, subcylindrical, slightly contracted at the middle, well-rounded in front, with three pairs of glabellar furrows; occipital furrow well-defined, narrow and straight; occipital ring wide in middle portion, narrowing laterally. Preglabellar field slightly convex, marginal furrow distinct, anterior border raised. Palpebral lobe and ocular ridge continuous, long, extending from the anterio-lateral corners of the glabella almost to the posterior marginal furrow. Fixed cheeks evenly convex; facial sutures diverging anteriorly and oblique posteriorly.

Remarks: This new genus is closely allied to *Protolenella* Chien and Yao in the general shape of the cranidium and the evenly convex fixed cheeks, but differs in having a broad, subcylindrical glabella, in the gently convex anterior border, in the comparatively shallower dorsal furrows; furthermore, the ocular ridge is stronger and the palpebral lobe is longer in *Protolenella* than in *Paraprotolenella.*

Paraprotolenella may also be compared with *Palaeolenus* Mansuy, but differs from the latter in the broader glabella, in the stronger ocular ridge, in the large palpebral lobe, in the more divergent anterior and shorter posterior facial sutures.

Paraprotolenella subcylindrica **Zhu and Jiang, sp. nov.**
Plate 1(12–15)

Description: Cranidium broad, moderately convex, subrectangular in outline; glabella wide, subcylindrical, well-rounded in front, slightly contracted in the middle, with three pairs of glabellar furrows; posterior pair long and deep at the sides, gradually shallowing inward and also extending obliquely backward and inward; second pair shorter, running almost straight; anterior pair very weak. Dorsal furrows deep along the sides of the glabella; shallowing forward; occipital furrow deep, narrow and straight; occipital ring wide in middle portion, narrowing laterally. Preglabellar field gently convex, marginal furrow well-defined; anterior border convex, a little shorter than the preglabellar field; frontal margin slightly arched forward in the middle. Ocular ridge stout, extending outward and backward from the antero-lateral corners of the glabella to the palpebral lobe. Palpebral lobe medium-sized, situated behind the midlength of the glabella. Fixed cheeks evenly convex, less than two-thirds the width of the glabella at the base. Posterior marginal border well-marked from posterior limb by a deep and narrow posterior marginal furrow, narrowing slightly toward the occipital ring. Anterior facial sutures diverging from anterior end of palpebral lobe forward to cut the anterior border in an even curve; posterior sutures bending outward and backward to cut the posterior marginal border.

ACKNOWLEDGMENTS

The specimens described in this paper are in the Nanjing Institute of Geology and Palaeontology, Academia Sinica. Photography by Mao Ji-liang, Hu Shang-qing, and Zhang Fu-tian.

SELECTED REFERENCES

Chang Wen Tang [Zhang Wen-tang], 1966, On the classification of Redlichiacea, with description of new families and new genera: Acta Palaeontologica Sinica, v. 14, no. 2, p. 135–184, Pls. 1–6 (in Chinese with English summary).

Kobayashi, T., 1944, The Cambrian Formation in the middle Yangtze Valley and some trilobites contained therein: Japanese Journal of Geology and Geography, v. 19, p. 89–138, Pls. 1–11.

Lee Shanji, 1978, [Trilobita, *in* Chengdu Institute of Geology and Mineral Resources, ed., Atlas of fossils of Southwest China, Sichuan, Part 1: p. 179–283, Pls. 89–114 (in Chinese).

Lin Tian-rui, 1965, A Cambrian trilobite fauna from Mufushan, Nanking: Acta Palaeontologica Sinica, v. 13, no. 3, p. 552–559, Pl. 1 (in Chinese with English summary).

Liu Chui Chen [Liu Ji-chen] and Chao Ju Chün [Zhao Ru-jun], 1924, A preliminary report on the geology and the mineral resources of Kiangsu: Memoirs of the Geological Survey of China, ser. A, no. 4, 34 p. (in English), 82 p. (in Chinese).

Lu Yan-hao [Lu Yen-hao], Zhang Wen-tang [Chang Wen Tang], Zhu Zhao-ling [Chu Chao-ling], Qian Yi-yuan [Chien Yi-yuan], and Xiang Li-wen [Hsiang Lee-wen], 1965, [Trilobita (Chinese fossils of all groups)]: Beijing, Science Press, v. 1–2, 766 p., 135 pls. (in Chinese).

Lu Yan-hao, Zhu Zhao-ling, Qian Yi-yuan, Lin Huan-ling, Zhou Zhi-yi, and Yuan Ke-xing, 1974, [Bio-environmental control hypothesis and its application to the Cambrian biostratigraphy and Palaeozoogeography]: Memoirs of Nanjing Institute of Geology and Palaeontology, Academia Sinica, no. 5, p. 28–110, Pls. 1–4 (in Chinese).

Moore, R. C., ed., 1959, Treatise on invertebrate palaeontology, Part O, Arthropoda 1: New York, Geological Society of America and University of Kansas Press, 560 p. (reprinted 1968).

Richter, R., and Richter, E., 1941, Studien im Palaeozoikum der Mittelmeer-Länder. 6. Die Fauna des Unter-Kambriums von Cala in Andalusien: Senckenbergiana Lethaea, Abhandlungen 455, p. 4–83, Pls. 1–4.

Saito, K., 1934, Older Cambrian trilobites and conchostraca from northwestern Korea: Japanese Journal of Geology and Geography, v. 11, nos. 3–4, p. 211–237, Pls. 25–27.

Suvorova, N. P., 1956, Trilobity Kembriya vostoka Sibirskoy platformy: Trudy Paleontologicheskogo Instituta Akad. Nauk SSSR, v. 63, 158 p., 12 pls.

Walcott, C. D., 1913, The Cambrian faunas of China, *in* Research in China, Volume 3: Carnegie Institution, Washington, Publication 54, p. 3–276, Pls. 1–24.

Yu Jian-hua, Chen Min-juan, and Zhang Qian-shen, 1962, [The discovery of Cambrian strata from Mufushan, Nanjing]: Journal of Nanjing University, Geological part, no. 1, p. 55–59 (in Chinese).

Zhang Wen-tang and Zhu Zhao-ling, 1979, Notes on some trilobites from the Lower Cambrian Houjiashan Formation in southern and southwestern parts of North China: Acta Palaeontologica Sinica, v. 18, no. 6, p. 513–525, Pls. 1–3. (in Chinese with English abstract).

MANUSCRIPT RECEIVED BY THE SOCIETY APRIL 22, 1981

PLATE 1

Figure

1–2. *Redlichia (Redlichia) nanjiangensis* Zhang and Lin, 1978.
1. Cranidium, ×5.
2. As above, ×2, Cat. No. 17088.

3–5. *Yeshanaspis subconica* Zhu and Jiang (Gen. et sp. nov.).
3. Cranidium, ×5, holotype.
4. As above, ×8.
5. Cranidium, ×8; Cat. Nos. 17089–17090.

6–9. *Paraichangia jiangsuensis* Zhu and Jiang (sp. nov.).
6. Cranidium, ×4, holotype.
7. As above, ×6.
8. Cranidium, ×6.
9. As above, ×6; Cat. Nos. 17091, 17092.

10–11. *Paraichangia spinosa* Zhu and Jiang (sp. nov.).
10. Cranidium, ×4, holotype.
11. As above, ×6; Cat. No. 17093.

12–15. *Paraprotolenella subcylindrica* Zhu and Jiang (gen. et sp. nov.).
12. Cranidium, ×6, holotype.
13. As above, ×6.
14. Cranidium, ×4.
15. As above, ×6; Cat. Nos. 17094, 17095.

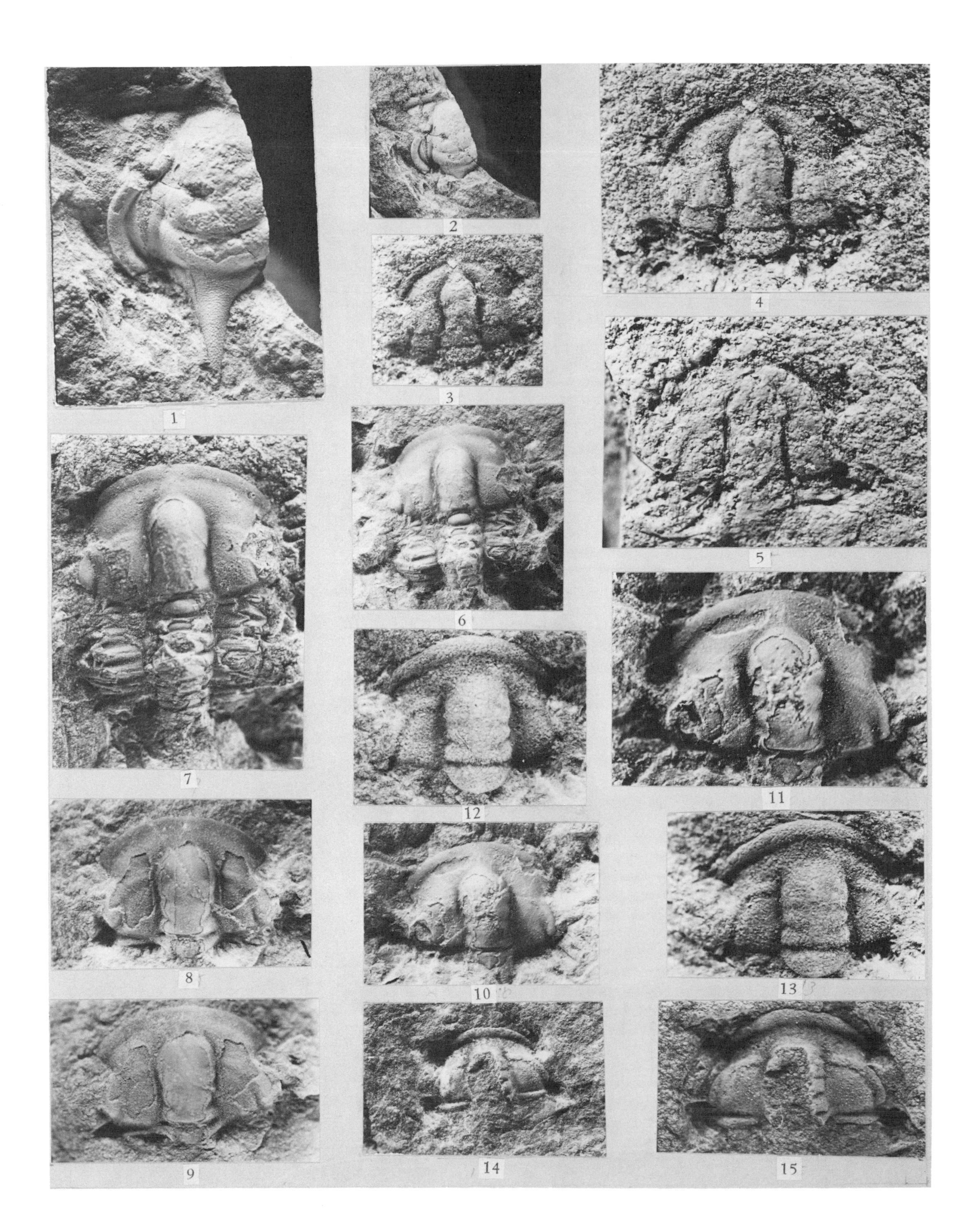

Printed in U.S.A.

Geological Society of America
Special Paper 187
1981

Trilobites from the Hsuchuang Formation (lower Middle Cambrian) in western marginal parts of the North China platform

Zhang Wen-tang
Yuan Jin-liang
Nanjing Institute of Geology and Palaeontology, Academia Sinica, Nanjing 210008, People's Republic of China

ABSTRACT

The Hsuchuang Formation in western marginal parts of the North China platform has by far the most richly fossiliferous and fully developed lower Middle Cambrian succession in North China. The purpose of this paper is both to describe some new trilobites, including 18 new species and 8 new genera, and to present briefly the new trilobite zones of this formation.

INTRODUCTION

Cambrian rocks crop out in the eastern (central Anhui, northern Jiangsu, Shantung, eastern Hebei, Liaoning), middle (western Henan, Shanxi), and western marginal parts (Shaanxi, Ningxia, Inner Mongolia) of the North China platform. They consist chiefly of neritic limestone, oolitic limestone, edgewise conglomerate, marl, and shale, and have a thickness of about 500 to 600 m. The Cambrian of the North China platform disconformably overlies the nonmetamorphosed late Precambrian sediments, (western, southern, and eastern marginal parts, Yenshan area) or unconformably overlies Archaean metamorphics (central Shantung). It is conformably overlain by Orodovician limestone. The Middle and Upper Cambrian are well developed in North China. The lower part of the Lower Cambrian is missing in North China.

In China, the stratigraphic stages of the Lower Cambrian are based on the biostratigraphic sequence of the Southwest China platform, and the stages of the Middle and Upper Cambrian are based on the biostratigraphic sequence of the North China platform. In order to reach a better understanding of the stratigraphic relationships of the Middle Cambrian Hsuchuang Formation, the Chinese biostratigraphic units in current usage are briefly tabulated in descending order as follows:

Late Cambrian: Fengshan Stage, Changshan Stage, Kushan Stage.

Middle Cambrian: Changshia Stage, Hsuchuang Stage, Maochuang Stage.

Early Cambrian: Lungwangmiao Stage, Tsanglangpu Stage, Chiungchussu Stage, Meishucun Stage.

The Hsuchuang Formation in the western marginal parts of the North China platform, which conformably overlies the Maochuang Formation and is conformably overlain by the Changshia Formation, consists chiefly of gray limestone, oolitic limestone, and green or purple shale, with a thickness of 150 to 206 m.

As a result of the study on the Cambrian biostratigraphy and trilobite fauna in the western marginal parts of the North China platform since 1977, the Hsuchuang Formation has by far the most richly fossiliferous and fully developed early Middle Cambrian succession in North China. So far as we are aware, trilobite faunas from the Hsuchuang Formation in this area consist of more than 300 species belonging to 90 genera and 24 families. It should be noted that the trilobite faunas found in this area are all benthic and endemic Ptychoparida, with the exception of a very few pandemic Corynexochida, Eodiscina, and Agnostina.

The purpose of this paper is both to describe some new trilobites (18 new species and 8 new genera) and to present briefly the new trilobite zones of this formation. The full description of all the trilobite faunas and the stratigraphic sections and discussion of the classification and evolution of the early Middle Cambrian trilobites in North China will be published in a monograph *Early Middle Cambrian Trilobite Faunas of Western Marginal Parts of North China Platform.*

On the basis of the succession of the trilobite faunas in the Hsuchuang Formation in this area, the nine fossil zones, from which trilobites (cited in the parentheses) are described in this paper, are briefly listed in descending order as follows:

9. *Bailiella* Zone *(Proasaphiscina taosigouensis* n. sp., *Bailiella lantenoisi* [Mansuy])

8. *Poriagraulos* Zone *(Poriagraulos dactylogrammacus* n. sp.)

7. *Inouyops* Zone *(Inouyops latilimbatus* n. sp.)

6. *Metagraulos* Zone (*Zhuozishania typica* n. gen., n. sp., *Metagraulos truncatus* n. sp.)

5. *Sunaspidella* Zone (*Sunaspidella rara* n. gen., n. sp., *Wuhushania cylindrica* n. gen., n. sp.)

4. *Sunaspis* Zone (*Sunaspis triangularis* Lin and Wu, *Wuania venusta* n. sp., *Taitzuina lubrica* n. gen., n. sp., *Wuanoides situla* n. gen., n. sp.)

3. *Pagetia jinnanensis* Zone (*P. [Sinopagetia] neimengguensis* n. subgen., n. sp., *Parainouvia prompta* n. gen., n. sp.)

2. *Ruichengaspis* Zone (*Wuhanina lubrica* n. gen., n. sp., *Ruichengaspis neimengguensis,* n. sp.)

1. *Kochaspis-Ruichengella* Zone (*Luaspides brevica* n. sp., *Zhongtiaoshanaspis similis* n. sp., *Catinauyia typica* n. gen., n. sp.)

SYSTEMATIC DESCRIPTIONS

Order AGNOSTIDA Kobayashi, 1935
Suborder EODISCINA Kobayashi, 1939
Family PAGETIIDAE Kobayashi, 1935
Genus *Pagetia* Walcott, 1916
Subgenus *Sinopagetia* n. subgen.
Type species: *Sinopagetia neimengguensis* n. sp.

Diagnosis: Pagetid trilobite with broad and tapered glabella; frontal lobe of glabella well rounded; preglabellar depression broad and prominent. Occipital furrow very slightly impressed at sides; occipital ring without occipital spine. Axis of pygidium broad, well defined, and furrowed; terminal section without spine.

Discussion: *Sinopagetia* closely resembles *Pagetides* Rasetti (1945, p. 311) and *Pagetia* Walcott (1916, p. 407). It can be easily distinguished from the former by the shape of its glabella, its semicircular occipital ring, its anterior border without median inbend, and its pygidial axis which is wider and has fewer axial segments. It differs from the latter in that it has longer palpebral lobes, its anterior border has less uniform width, it has no occipital spine or axial spine of the pygidium, it has no preglabellar field, and the posterior border of its cranidium has obtuse geniculations. It can be also compared with *Pagetia latilimbata* Chien (1961, p. 94, Pl. 1, figs. 5–7), *Pagetia howardi* Jell (1975, p. 43, Pl. 4, figs. 1–12), *Pagetia jinnanensis* Lin and Wu (in press), *Pagetia gaotaiensis* S. G. Zhang (1980, Pl. 12, figs. 20–22; Pl. 13, figs. 14–19). *P. latilimbata* has a triangular occipital ring and a narrower axis of the pygidium. *P. howardi* has an occipital spine, distinct brim, and a moderately impressed interpleural furrow of the pygidium. *P. jinnanensis* possesses a triangular occipital ring, faint glabellar furrow, and obsolete pleural and interpleural furrows of the pygidium. *P. gaotaiensis* has a triangular occipital ring, distinct eye ridges, and transverse cranidium. In addition to *P. howardi,* all the species mentioned above may belong to this subgenus.

Age and distribution: Early Middle Cambrian, China.

Sinopagetia neimengguensis n. sp.
Plate 2(1–2)

Description: Small proparian trilobites with cephalon and pygidium of subequal size. Glabella convex, expanded posteriorly, then tapering forward, moderately rounded in front, glabellar furrows effaced; occipital furrow faintly impressed at the sides, occipital ring elevated, semicircular; axial furrow wide and deep; fixigenal wider than the glabella, attaining their maximum relief mesially; palpebral lobe of moderate length; eye ridges very narrow and obscure; anterior border furrow narrow and distinct at the sides, wide and shallow in the middle; anterior border broad mesially and narrowing laterally, marked with a row of more or less distinct radial grooves; posterior border furrow deep, with obtuse geniculations. Anterior sections of facial suture perpendicular to the lateral margin of the cephalon; posterior sections obliquely backward and outward; pygidium elliptical and convex. Axis moderately wide, strongly convex about pleural region, almost reaching to posterior margin, with five to six segments; pleural lobes smooth or with faint interpleural furrows; marginal furrow deep and narrow; border convex and extremely narrow.

Occurrence: *Sinopagetia jinnanensis* Zone of the Hsuchuang Formation, Gangdeershan, Haibowan Shi, Nei mongol Zizhiqu (NZ 02) (Inner Mongolia Autonomous Region).

Order PTYCHOPARIIDA Swinnerton, 1915
Suborder PTYCHOPARIINA Richter, 1933
Family PTYCOPARIIDAE Matthew, 1887
Genus *Wuhaina* n. gen.
Type species: *Wuhaina lubrica* n. sp.

Diagnosis: Opisthoparian ptychopariid with a subtrapezoidal cranidium. Axial furrow very shallow; glabella convex, truncato-conical, broadly rounded in front, usually unfurrowed; occipital furrow absent or very faint; occipital ring extending backward into a large occipital spine; eye ridges slightly oblique; palpebral lobe of medium size. Fixigenae gently convex; preglabellar field broad, gently downsloping toward border furrow; anterior border furrow shallow and slightly curved forward; anterior border narrow and convex; posterior lateral limb subtriangular. Posterior border furrow distinct; posterior border gently convex, about four-fifths the width of the glabella near the base. Anterior sections of facial suture divergent from palpebral lobe to border furrow; posterior sections slanting outward and backward. Surface of test smooth.

Discussion: This genus is characterized by its trunacto-conical glabella, absence of glabellar furrows, broad brim, large occipital spine, and slightly oblique eye ridges. It is somewhat similar to *Probowmaniella* Chang (1963, p. 447), *Shantungaspis* Chang (1957, p. 24), *Stoecklinia* Wolfart (1974, p. 40) and *Alanisia* Hupé (1953, p. 147). It differs from *Probowmaniella* in having occipital spines, an obsolete axial furrow, and a shallower anterior border furrow; from *Shantungaspis* in its unfurrowed glabella, shallower axial furrow, and anterior border furrow; from *Stoecklinia* in the broad brim, narrower anterior border, glabella gradually tapering forward, shallower axial furrow, and horizontal eye ridges; and from *Alanisia* by its longer posterior lateral limb and shorter palpebral lobes.

Age and distribution: Early Middle Cambrian, China and South Asia.

***Wuhaina lubrica* n. sp.**
Plate 1(7–9)

Description: Glabella slightly convex, truncato-conical, without glabellar furrows, and defined by shallow axial furrow; occipital furrow obscure, occipital ring extended backward into long spine. Preglabellar field broad, anterior border convex and narrow, defined by shallow border furrow; anterior sections of facial suture slightly divergent; eye ridges distinct, slanted slightly backward.

Occurrence: *Ruichengaspis* Zone of the Hsuchuang Formation, Gangdeershan, Haibowan Shi, Nei Mongol Zizhiqu (NZ 5) (Inner Mongolia Autonomous Region).

Family LUASPIDIDAE n. fam.

Opisthoparian, micropygous ptychopariids with elongately ovate exoskeleton. Cephalon semielliptical; glabella large, truncato-conical, reaching to anterior border furrow, broadly rounded in front, with three pairs of lateral glabellar furrows. Occipital furrow deep at the sides, shallow in the middle; occipital ring elevated, with or without occipital spine; eye ridges present; palpebral lobes small. Anterior border furrow slightly curved backward in the middle, preglabellar field absent; anterior border narrow, elevated, horizontal or slightly curved backward in the middle; anterior sections of facial suture usually divergent from palpebral lobe to the anterior border furrow; posterior sections oblique outward and backward. Thorax with 15 to 22 segments; pygidium moderately small to tiny, semi-elliptical, having 2 to 3 axial rings.

Discussion: *Luaspides* has been assigned by Guo and Duan (1978, p. 447) to the Antagminae, but trilobites of this subfamily, such as *Eoptychoparia, Antagmus, Poulsenia,* and *Periomma,* have a relatively shorter glabella, more or less well-developed preglabellar field, anterior margin of cranidium curved forward, and not more than 15 thoracic segments. Because the taxonomic assignment of *Luaspides* is doubtful, the new family Luaspididae is proposed to include *Luaspides* Duan, 1965 and *Beldirella* Pokrovskaya, 1960.

Age and distribution: Early Middle Cambrian, China and Siberia.

Genus *Luaspides* Duan, 1965

1965 *Luaspides* Duan, Symposium of Changchun Geological Institute, no. 4, p. 144

1978 *Luaspides,* Guo and Duan, Acta Palaeontologica Sinica, no. 4, p. 447

Type species: *Luaspides lingyuanensis* Duan, 1965
***Luaspides brevica* n. sp.**
Plate 1(1–2)

Description: Carapace elongate-oval, relative proportions of cephalon, thorax, and pygidium about 10:23:3. Cephalon semicircular; cranidium subquadrate except posterior lateral limb; glabella convex, truncato-conical, broadly rounded in front, with four pairs of lateral glabellar furrows; the first two pairs very short and faint, directed inward and forward; the posterior two pairs longer and deeper, oblique inward and backward. Occipital furrow deep at the sides, rapidly shallowing inward, slightly arched forward; occipital ring convex, broad in the middle, narrowing outward, with a short occipital spine. Fixigenae gently convex, about one-third the width of glabella at its base. Eye ridges broad and elevated, straight-slanting outward and backward, divided into two segments; palpebral lobe small, palpebral furrow narrow, located opposite the midlength of glabella. Anterior border furrow very deep and wide, slightly curved forward at the sides, mesially curved backward and connected with axial furrow; anterior border narrow and convex, slightly curved backward in the middle; posterior lateral limb subtriangular, extending outward. Anterior sections of facial suture strongly divergent from palpebral lobes to the border furrow; posterior sections slanted outward and backward. Free cheek gently convex, with broad border furrow, narrow and convex border; genal spine broad-based, rapidly tapering and extending backward to the level of the sixth thoracic segments. Thorax of 15 segments, convex; maximum width of axis in the anterior part, gradually tapering posteriorly, each axial ring has a small node. Pleural lobes gently convex; pleural spines long and directed obliquely outward and backward. Pygidium small, semicircular; axis broad and short, with two rings and terminal piece showing a pair of swellings; pleural region flat, with two pairs of pleural furrows; border narrow and flat.

Comparison: This species is similar to *Luaspides lingyuanensis* Duan, type species of this genus (Guo and Duan, 1978, p. 447, Pl. 1, figs. 3–6), but differs in having shorter and wider glabella, smaller palpebral lobes, narrower palpebral area of fixigenae, and 15 thoracic segments.

Occurrence: *Kochaspis-Ruichengella* Zone of the Hsuchuang Formation, Gangdeershan, Haibowan Shi, Nei Mongol Zizhiqu (NZ2a).

Family INOUYIIDAE Chang 1963
Genus *Parainouyia* Lin and Wu, in Zhang and others (1980)
Type species: *Parainouyia lata* Lin and Wu
***Parainouyia prompta* n. sp.**
Plate 2(3)

Description: Cranidium subtrapezoidal, moderately rounded anteriorly; glabella rather convex, tapering forward, somewhat truncated in front, with a faint longitudinal keel; glabellar furrows effaced, impressions of glabellar furrows on expoliated specimens; occipital furrow faint, slightly curved backward; occipital ring elevated, extending backward into short spine. Fixigenae wide, about 1.2 times the width of glabella at the base, rising rapidly from the dorsal furrow to the palpebral lobes, steep-sloping downward to the posterior lateral limb; palpebral lobes small, about one-fourth the length of glabella, situated at a level somewhat in advance of glabella. Eye ridges extend outward and forward from antero-lateral corner of glabella, with a double ridge structure (that is, the ocular striga of Öpik, 1961). Order furrow very faint, slightly arched forward; anterior border flat and very narrow, somewhat downsloping outward; preglabellar field gently convex, with a median swelling. Posterior area of fixigenae very wide and long, extending laterally beyond the palpebral lobes; posterior border furrow wide and deep, distally curved downward; anterior sections of facial suture strongly convergent from palpebral lobes to the border furrow; posterior sections long, extended outward and backward.

Comparison: The present species is characterized by its eye ridges extending outward and forward, its palpebral lobes located opposite the level in advance of glabella. It differs from the type species *P. lata* Lin and Wu (in press) in having narrower palpebral area of fixigenae, in its eye ridges extending obliquely outward and forward, and in more advanced location of palpebral lobes.

Occurrence: *Sinopagetia jinnanensis* Zone of the Hsuchuang

Formation, Gangdeershan, Haibowan Shi, Nei Mongol Zizhiqu (NZ 02) (Inner Mongolia Autonomous Region).

Genus *Catinouyia* n. gen.
Type species: *C. typica* n. sp.

Diagnosis. Cranidium of Ptychoparid pattern. Glabella convex, truncato-conical. Preglabellar field and palpebral area of fixigenae broad, faint circular boss in front of the glabella; anterior border narrow and convex; border furrow narrow and deep.

Discussion: This genus differs from *Inouyia* Walcott (1911, p. 80) chiefly in having a narrow and convex anterior border, a subrectangular cranidium, and in a very narrow and deep border furrow. The differences between *Catinouyia* and *Eoinouyia* Lo (Lo, 1974, p. 640) from the early Middle Cambrian of Southeast Yunnan area as follows: (1) Anterior border furrow regularly curved forward in *Eoinouyia;* (2) palpebral lobes of *Eoinouyia* smaller than those of *Catinouyia;* (3) posterior area of fixigenae of *Eoinouyia* wider than that of *Catinouyia.*

Age and distribution: early Middle Cambrian, North China.

***Catinouyia typica* n. sp.**
Plate 1(5–6)

Description: Cranidium subrectangular in outline; glabella small and short, tapering forward, truncated in front, with three pairs of faintly lateral glabellar furrows; occipital furrow narrow and shallow, slightly arched backward; occipital ring elevated, semi-elliptical; fixigenae gently convex, wider than width of glabella at the base; eye ridges elevated, nearly horizontal, with a double ridge structure; palpebral lobes median in size, situated at the level of glabellar midlength; anterior border furrow narrow and deep; preglabellar field broad, rather convex, with a large ill-defined median boss. Anterior border narrow and gently convex; posterior border furrow distinct, broadening distally; posterior border gently convex, extending outward and slightly backward, with obtuse geniculation. Anterior sections of facial suture slightly divergent from palpebral lobes to anterior border furrow; posterior sections extending outward and backward.

Occurrence: *Kochaspis-Ruichengella* Zone of the Hsuchuang Formation, Gangdeershan, Haibowan Shi, Nei Mongol Zizhiqu (NZ 3) (Inner Mongolia Autonomous Region).

Family *Wuaniidae* n. fam.

Cranidium moderately convex, subtrapezoidal or subquadrate; glabella convex, tapering forward, subquadrate or subelliptical in outine, with three or four pairs of faint lateral glabellar furrows; anterior border furrow wide and shallow running into axial furrow in front of glabella. No preglabellar field; anterior border wide and elevated; anterior sections of facial suture divergent; posterior sections extending outward and backward. Pygidium subtriangular or subquadrate; axis highly elevated, tapering backward, reaching to posterior margin, with a narrow and flat border.

Discussion: This family includes *Wuania* Chang (1963, p. 485), *Ruichengaspis* Zhang and Yuan, *Wuhushania* (n. gen.), and *Wuanoides* (n. gen.). It differs from the Inouyiidae in having the anterior border furrow divided into three sections, in the absence of preglabellar field, in its elevated anterior border, in its narrower fixigenae, and in its longer palpebral lobes.

Age and distribution: Early Middle Cambrian, North China.

Genus *Ruichengaspis* Zhang and Yuan, in Zhang Wen-tang and others, (1980)
Type species: *R. mirabilis* Zhang and Yuan
***Ruichengaspis neimengguensis* n. sp.**
Plate 1(10–12)

Description: Cranidium moderately convex, subtrapezoidal or subelliptical, frontal margin strongly arched forward; axial furrow very broad and deep; glabella strongly convex, tapering forward, truncated in front, with four pairs of faint glabellar furrows, of which the first two pairs are short, horizontal, and situated at about one-third the length of glabella from its frontal end; the third pair long and oblique; the fourth pair longer and more oblique and sometimes bifurcate. Occipital furrow with medium width and depth, curving backward; occipital ring elevated, lunate, with a small node. Fixigenae convex, as wide as the average width of glabella; eye ridges convex, extended outward and slightly backward; palpebral lobes of medium size, about one-half the length of glabella; palpebral furrow shallow and narrow. Anterior border furrow very wide and shallow, running into the preglabellar furrow in front of glabella; anterior border very wide, swollen mesially, rapidly narrowing outward; posterior area of fixigenae narrow and short. Posterior border furrow relatively narrow and deep; posterior border convex and very narrow, its distal end pointing outward and forward. Anterior sections of facial sutures nearly parallel or slightly convergent from palpebral lobes to anterior border furrow; posterior sections short, slanting outward and backward.

Comparison: The present species resembles the type species of this genus *Ruichengaspis mirabilis* Zhang and Yuan (Zhang and others, Pl. 6, figs. 8–9, in press) in glabella, frontal border, axial and border furrow, relatively narrow and deep; posterior border convex and very narrow, its distal end pointing outward and forward. Anterior sections of facial sutures nearly parallel or slighty convergent from palpebral lobes to anterior border furrow; posterior sections short, slanting outward and backward.

Comparison: The present species resembles the type species of this genus *Ruichengaspis mirabilis* Zhang and Yuan (Zhang and others, Pl. 6, figs. 8–9, in press) in glabella, frontal border, axial and border furrow, but differs chiefly in the transversely wider cranidium, nearly parallel anterior sections of facial sutures, and lunate occipital ring.

Occurrence: *Ruichengaspis* Zone of the Hsuchuang Formation, Gangdeershan, Haibowan Shi, Nei Mongol Zizhiqu (NZ 7, NZ 8) (Inner Mongolia Autonomous Region).

Genus *Wuania* Chang, 1963

1963 *Wuania,* Chang, Acta Palaeontologica Sinica, v. 11, no. 4, p. 485

1965 *Wuania,* Lu and others, Fossils of China, Chinese Trilobites, v. 1, p. 249

1976 *Wuania,* Schrank, Zeitschrift für Geologisene Wissenschaften, Berlin, v. 4, p. 896

Type species: *Inouyia fongfongensis* Chang, 1959
***Wuania venusta* n. sp.**
Plate 2(6–7)

Description: Cranidium semielliptical or subquadrate in outline, moderately convex, strongly arched anteriorly. Axial furrow relatively narrow; glabella well defined, moderately convex, subquadrate in outline, slightly contracted in the middle, broadly rounded anteriorly, with a distinct longitudinal keel; four pairs of lateral glabellar furrows, the first pair very short, extending inward and forward, situated at proximal end of eye ridges; the second pair longer and parallel to the first pair, situated at about one-fourth the length of glabella from its frontal end; the posterior two pairs longer, extended inward and backward, bifurcating not far from axial furrow. Occipital ring moderately convex, broad in the middle; occipital furrow broad and shallow, straight, not connected distally with distinct axial furrow. Fixigenae gently convex, about one-half the width of glabella; eye ridges convex and narrow, extending outward and slightly backward from its proximal ends; palpebral lobes of medium size. Anterior border furrow wide and deep, running into the preglabellar furrow in front of glabella; anterior border wide, swollen in the middle, narrowing outward. Posterior area of fixigenae narrow and short; posterior border furrow wide and deep; posterior border gently convex and very narrow. Anterior sections of facial sutures divergent from palpebral lobes to the anterior border furrow; posterior sections slanting outward and backward. Pygidium triangular, with straight-slanting lateral margin; axis wide, prominent, tapering backward, reaching to the posterior margin, composed of four or five axial rings and a terminal piece; pleural regions slightly convex, with five pairs of pleural furrows; border narrow and flat.

Comparison: The present species differs from *Wuania fongfongensis* (Chang) (Chang, 1963, p. 472, Pl. 2, fig. 20), the type species of this genus, in the subquadrate glabella, in the narrower fixigenae, in the strongly curved anterior border, and in the strongly divergent anterior facial suture.

Occurrence: *Sunaspis* Zone of the Hsuchuang Formation, Gangdeershan, Haibowan Shi, Nei Mongol Zizhiqu (Inner Mongolia Autonomous Region).

Genus *Wuanoides* n. gen.
Type species: *Wuanoides situla* n. sp.

Diagnosis: Glabella convex, broad and urn-shaped, with four pairs of glabellar furrows; axial furrow broad and deep. Fixigenae relatively narrow and convex; palpebral lobe of medium size, palpebral furrow shallow and broad. Anterior border thickened in the middle, with a transverse lenticular boss in front of the glabella.

Discussion: This genus resembles closely *Wuania* Chang (1963, p. 485). It can be easily distinguished from *Wuania* by its broad and urn-shaped glabella, a large transverse swelling in front of glabella, and relatively narrow fixigenae.

Age and distribution: Early Middle Cambrian, North China.

***Wuanoides situla* n. sp.**
Plate 2(10)

Description: Cranidium convex, quadrate in outline, glabella strongly convex, urn-shaped, about one-half of the length of cranidium, with four pairs of lateral glabellar furrows. Occipital furrow wide and deep; occipital ring gently convex; axial furrow wide and deep. Fixigenae convex and narrow; palpebral lobes medium in size; eye ridges present, strongly oblique and backward. Anterior border furrow wide and deep, running into the preglabellar furrow in front of glabella; anterior border thickened in the middle, with a large transverse swelling in front of glabella, narrowing rapidly outward. Posterior area of fixigenae narrow and short; posterior border furrow wide and deep. Anterior sections of facial sutures divergent from palpebral lobe to anterior border; posterior sections extending outward and backward.

Occurrence: *Sunaspis* Zone of the Hsuchuang Formation, Helanshan, Ningxia Huizu Zizhiqu (NH 12) (Ningsia Hui Autonomous Region).

Genus *Wuhushania* n. gen.
Type species: *Wuhushania cylindrica* n. sp.

Diagnosis: Opisthoparian, isopygous or macropygous trilobites; glabella strongly convex, slightly tapering forward, unfurrowed, bluntly rounded in front; occipital furrow shallow; occipital ring convex, broad in the middle. Fixigenae upsloping from dorsal furrow to palpebral lobes; eye ridges present; palpebral lobes prominent, situated opposite midlength of glabella; posterior lateral limb very narrow. Anterior border furrow broad and deep, anterior border convex, thickened in the middle. Preglabellar field narrow, downsloping to the border furrow. Pygidium subquadrate; axis long and strongly convex, club-shaped or cylindrical, composed of 9 axial rings and a lunate terminal piece; pleural regions gently convex, crossed by six pairs of deep and relatively broad pleural furrows; border flat, widening backward laterally; posterior margin broadly rounded.

Age and distribution: Early Middle Cambrian, North China.

***Wuhushania cylindrica* n. sp.**
Plate 2(13–14); Plate 3)11)

Description: Cranidium semi-elliptical, anterior margin strongly arched forward; glabella well defined, slightly tapering forward, broadly rounded in front, unfurrowed; occipital furrow shallow and slightly bent backward; occipital ring convex, broad in the middle. Fixigenae upsloping from axial furrow to palpebral lobes; eye ridges present, slanting backward from its proximal end; palpebral lobes upturned, of medium size; posterior lateral limb extremely narrow. Anterior border furrow moderately wide and deep; anterior border convex, thickened in the middle. Preglabellar field present, narrow, downsloping forward. Anterior sections of facial sutures convergent from palpebral lobes to anterior border; posterior sections extending outward and backward. Pygidium large, convex, subquadrate; axis strongly convex, cylindrical, reaching to posterior margin, with 9 axial rings and a lunate terminal piece. Pleural regions gently convex, with six pairs of broad and deep pleural furrows; border flat, widening posterior laterally.

Occurrence: *Sunaspidella* Zone of the Hsuchuang Formation, Wuhushan, Wuda Shi, Nei Mongol Zizhiqu (Q5-VIII—14–40) (Inner Mongolia Autonomous Region).

Family ORDOSSIDAE Lu, 1954
Genus *Taitzuina* n. gen.
Type Species: *Taitzuina lubrica* n. sp.

Diagnosis: Cranidium convex, quadrate, anterior margin rounded forward; axial furrow broad, of medium depth; glabella broad, convex, subquadrate, and slightly tapering forward, with four pairs of faint glabellar furrows. Occipital furrow deep and broad, slightly curved backward; occipital ring convex, and very narrow, with a small median node. Anterior border broad and convex, thickened in the middle, with a pair of triangular plectra pointing backward; border furrow broad and deep, composed of three sections. Pygidium semi-elliptical, both axial lobe and pleural regions segmented; border flat of moderate width.

Discussion: This genus closely resembles *Eotaitzuia* Zhang and Yuan, *Taitzuia* Resser and Endo, 1935, and *Poshania* Chang, 1957, in the characteristics of the anterior border furrow and anterior border, and in the shape of the pygidium. It differs from *Eotaitzuia* in the less-tapered glabella, in the shape of the quadrate cranidium, in the longer palpebral lobes, and in the presence of a pair of triangular plectra in the posterior margin of the anterior border; from *Poshania* and *Taitzuia* in the broad and convex anterior border with a pair of triangular plectra.

Age and distribution: Early Middle Cambrian, North China.

Taitzuina lubrica n. sp.
Plate 2(8–9)

Description: Cranidium convex, anterior margin strongly rounded anteriorly; axial furrow broad, of medium depth; glabella convex, broad, subquadrate, slightly tapering forward, with four pairs of faint glabellar furrows; the anterior two pairs of lateral glabellar furrows faint and very short, the posterior two pairs long and bifurcate. Occipital furrow deep and broad, slightly curved backward; occipital ring convex, very narrow, with a small median node. Anterior border broad and convex, thickened in the middle, with a pair of triangular plectrum pointing backward; anterior border furrow broad and deep, composed of three sections, running into the preglabellar furrow in front of glabella. Fixigenae convex, slightly downsloping to the axial and posterior border furrows; eye ridges present, extending outward and backward; palpebral lobes of medium size, situated opposite posterior two-thirds of length of glabella. Posterior area of fixigenae extremely narrow; anterior sections of facial sutures divergent from palpebral lobes to anterior border furrow; posterior sections extending transversely outward and backward. Pygidium semi-elliptical; axis convex, tapering backward, rounded posteriorly, with 6 or 7 axial rings and a terminal piece; pleural region gently convex, with five pairs of distinct pleural furrows and faint interpleural furrows; border flat.

Occurrence: *Sunaspis* Zone of the Hsuchuang Formation, Hejin Xian, Shanxi and Helanshan, Ningxia Huizu Zizhiqu (NH 48) (Southern Shanxi and Ningxia Hui Autonomous Region).

Family AGRANULIDAE Raymond, 1913
Genus _Metagraulos_ Kobayashi, 1935
Type species: _Agraulos nitida_ Walcott
Metagraulos truncatus n. sp.
Plate 3(3–5)

Description: Cranidium convex, subquadrate; glabella strongly elevated, slightly tapering forward, truncato-conical; glabellar furrows obscure; occipital furrow faint; occipital ring with a stout occipital spine. Fixigenae gently upsloping from shallow axial furrow to the palpebral lobes; palpebral lobes of medium size; palpebral furrow obscure; eye ridges faint, slightly slanting outward and backward from its proximal end. Anterior border furrow obsolete; anterior border slightly downsloping forward; anterior sections of facial sutures parallel or slightly convergent from palpebral lobes to anterior border; posterior sections short, extending outward and backward from palpebral lobe to posterior border; posterior area of fixigenae very short. Pygidium small, fusiform in outline; axis convex, wide, tapering backward, composed of 4 axial rings and a terminal piece. Pleural region narrower (tr.) than axis, with three pairs of shallow pleural furrows; border furrow faint; border narrow; when the exoskeletal crust of the cranidium is exfoliated, the border furrow, axial furrow, eye ridges, and glabellar furrows are more or less visible.

Comparison: The present species resembles closely *Metagraulos dolon* (Walcott) (Walcott, 1913, p. 156, Pl. 15, fig. 6) in the shape of cranidium. It differs from *M. dolon* in having visible eye ridges, in the truncato-conical glabella, and in the relatively longer cranidium.

Occurrence: *Metagraulos* Zone of the Hsuchuang Formation, Gangdeershan. Haibowan Shi, Nei Mongol Zizhiqu (NZ 20, NZ 23) (Inner Mongolia Autonomous Region).

Genus _Poriagraulos_ Chang, 1963
Type species: _Poriagraulos perforatus_ Chang
Poriagraulos dactylogrammacus n. sp.
Plate 3(7)

Description: Cranidium semi-elliptical, convex; glabella strongly convex, conical, narrowly rounded in front, unfurrowed, ornamented with concentric striations; occipital furrow very shallow; occipital ring semi-elliptical, strongly convex, with a small node, ornamented with more or less vertical striations on each side of the occipital node. Fixigenae gently convex, about one-half the width of the glabella at the base; eye ridges very faint; palpebral lobe small. Anterior border wide and gently convex; anterior border furrow shallow at the sides, very shallow in front of the glabella; preglabellar field gently downsloping forward; posterior area of fixigenae subtriangular; posterior lateral furrow shallow; posterior border of cranidium flat, broadened gradually outward. Anterior sections of facial sutures parallel or convergent from palpebral lobes to anterior border; posterior sections extending obliquely outward and backward.

Comparison: The present species is very similar to *Poriagraulos nanus* (Dames) (Schrank, 1975, p. 859, Pl. 1, figs. 1–6), but differs chiefly in its conical or triangular glabella, in the shape of occipital ring, and it is ornamented with striations of dactylogram on the surface of glabella and more or less vertical striations on the occipital ring.

Occurrence: *Poriagraulos* Zone of the Hsuchuang Formation, Niuxinshan, Longxian, western Shaanxi. (L23).

Family LORENZELLIDAE Chang, 1963
Genus _Inouyops_ Resser, 1942
Type species: _Ptychoparia titiana_ Walcott
Inouyops latilimbatus n. sp.
Plate 3(6)

Description: Cranidium moderately convex, quadrate in outline, anterior margin arched forward; glabella convex, slightly tapering forward, truncato-conical, with four pairs of lateral glabellar furrows, the posterior pair bifurcate. Occipital furrow broad and deep; occipital ring elevated and extending backward into a wide and long occipital spine. Fixigenae sloping from axial furrow to the palpebral lobes; eye ridges extending obliquely outward and backward, about one-half the length of glabella; palpebral furrow distinct. Anterior border furrow arched forward, deep and narrow; preglabellar field strongly elevated in front of glabella, ornamented with radiating striations; anterior border gently convex, broad or with moderate width. Posterior area of fixigenae very narrow; posterior border furrow of cranidium very deep and wide. Anterior sections of facial sutures divergent from palpebral lobes to the anterior border furrow; posterior sections extended outward and backward.

Comparison: The present species can be easily distinguished from *Inouyops titiana* (Walcott) (Walcott, 1913, p. 155, Pl. 14, fig. 9) by its quadrate cranidium, longer palpebral lobes, wider anterior border, and longer and wider occipital spine.

Occurrence: *Inouyops* Zone of the Hsuchuang Formation, Gangdeershan, Haibowan Shi, Nei Mongol Zizhiqu (NZ 21) (Inner Mongolia Autonomous Region).

Family PROASAPHISCIDAE Chang, 1963
Genus *Zhongtiaoshanagia* Zhang and Yuan, in Zhang Wen-tang and others (1980)
Type species: *Zhongtiaoshanaspis ruichengensis* Zhang and Yuan
***Zhongtiaoshanaspis similis* n. sp.**
Plate 1(3, 4)

Description: Cranidium convex, quadrate in outline; glabella strongly convex, tapering forward, rounded in front, with four pairs of lateral glabellar furrows; the first pair short, shallow, slightly oblique forward, situated opposite the proximal end of eye ridge; second pair longer, nearly horizontal, situated at about one-fourth the length of glabella from its frontal end; third pair longer, slightly oblique backward and inward; the posterior pair the longest, wider and bifurcating. Occipital furrow narrow and deep, bending backward; occipital ring convex, slightly thickened in the middle; axial furrow narrow and deep. Fixigenae gently convex, slightly downsloping toward the axial furrow and posterior border furrow of cranidium; maximum width of the palpebral area of fixigenae about two-thirds the width of the glabella at the base; eye ridges convex, slightly oblique outward and backward, with a double ridge structure; palpebral lobe long, arcuate outward; palpebral furrow narrow and prominent. Posterior area of fixigenae very narrow, posterior border furrow wide and deep; posterior border narrow and convex. Anterior border furrow of moderate depth; preglabellar field wide, gently convex, slightly downsloping toward the border furrow, ornamented with anastomosing striations; anterior border gently convex or slightly upturned, slightly thickened in the middle, rapidly narrowing outward. Anterior sections of facial sutures divergent from palpebral lobes, curving inward and forward, then crossing anterior border furrow; posterior sections transversely extending outward and backward.

Comparison: The present species looks like *Zhongtiaoshanaspis rara* Zhang and Yuan (Zhang and others, Pl. 7, figs. 13–15, in press), but it differs from *Z. rara* in possessing a broader cranidium and fixigenae, deeper anterior border furrow and occipital furrow, and more divergent anterior facial sutures. At first sight it is closely similar to *Alokistocarella* aff. *mexicana* Lochman-Balk (Repina and others, 1975, p. 166, Pl. 26, figs. 9–10) from the early Middle Cambrian of the Northern Submountain region of Turkestan-Alaj Ridge, but it is distinguished from *Alokistocarella* aff. *mexicana* in having longer palpebral lobes, narrower anterior border, and eye ridges with double ridge structure.

Occurrence: *Kochaspis-Ruichengella* Zone of the Hsuchuang Formation, Helanshan, Ningxia Huizhu Zizhiqu (NH 32).

Genus *Zhuozishania* n. gen.
Type species: *Zhuozishania typica* n. sp.

Diagnosis: Cranidium quadrate, flatly convex, glabella medium-sized, tapering forward, broadly rounded in front, with faint glabellar furrows; eye ridge oblique, prominent; palpebral lobes relatively long, strongly arcuate outward; palpebral furrow narrow and shallow. Preglabellar field broad; anterior border broad and flat. Pygidium large, semi-elliptical or subtriangular in outline. Anterior lateral margin with four pairs of short spines; posterior margin rounded.

Discussion: This new genus closely resembles *Proasaphiscus* Resser and Endo (in Kobayashi, 1935), *Szeaspis* Chang, 1959, and *Zhongtiaoshanaspis* Zhang and Yuan, in the shape of cranidium, but differs from these genera in the characteristic pygidium with four pairs of lateral pygidial spines.

Age and distribution: Early Middle Cambrian, North China.

***Zhuozishania typica* n. sp.**
Plate 3(1–2)

Description: Cranidium quadrate, exclusive of posterior area of fixigenae; glabella short, tapering forward, broadly rounded in front, about one-half the length of cranidium; four pairs of faint lateral glabellar furrows; occipital furrow straight and shallow, occipital ring gently convex, broad in the middle, with a small node. Fixigenae wide, gently upsloping from dorsal furrow to the palpebral lobes. Eye ridge convex, extending obliquely outward and backward from its proximal end; palpebral lobe long, strongly arcuate outward. Posterior area of fixigenae very narrow; posterior border furrow distinct; posterior border of cranidium gently convex and very narrow; anterior border broad and flat; anterior border furrow moderately wide and deep, regularly curved forward; preglabellar field wide, gently convex, slightly downsloping to the border furrow; anterior sections of facial sutures divergent from palpebral lobes to the anterior border furrow; posterior sections transversely extending outward and backward. Pygidium large, semi-elliptical or subtriangular. Axis convex, tapering backward, bluntly rounded at the rear end, with 4 or 5 axial rings and a terminal piece; pleural region gently convex, subtriangular, crossed by pleural furrows, interpleural furrow absent. Border broad. Anterior lateral margin with four pairs of short spines; posterior margin rounded.

Occurrence: *Metagraulos* Zone of the Hsuchuang Formation, Gangdeershan, Haibowan Shi, Nei Mongol Zizhiqu (NZ 20) (Inner Mongolia Autonomous Region).

Genus *Proasaphiscina* Lin and Wu, in Zhang Wen-tang and others (1980)
Type species: *Proasaphiscina quadrata* Lin and Wu
***Proasaphiscina taosigouensis* n. sp.**
Plate 3(8–9)

Description: Cranidium moderately convex, subquadrate in outline; glabella convex, slightly tapering forward, flatly rounded in front, with three or four pairs of faint lateral glabellar furrows;

occipital furrow shallow, slightly curved backward; occipital ring convex, thickened in the middle; axial furrow distinct. Fixigenae gently convex, about one-half the width of glabella at the basal part; eye ridges distinct, short, extending obliquely outward and backward from its proximal end; palpebral lobes long, narrow and arcuate; posterior area of fixigenae very narrow; anterior border furrow distinct. Preglabellar field flat and narrow; anterior border convex, without plectrum, narrowing at the sides. Pygidium ovate or semi-elliptical in outline, medium in size. Axis convex, tapered backward, with 4 axial rings and a narrow and long terminal piece. Pleural region gently convex, subtriangular in outline, with five pairs of pleural furrows, its distal part extending to the border which is wide and slightly depressed.

Comparison: This species is similar to *Proasaphiscina quadrata* Lin and Wu (type species of this genus) (Zhang and others, 1980) in the general outline of the cranidium, but differs in the longer cranidium, slightly tapered glabella, in the relatively narrower eye ridge and palpebral lobe, and in the relatively broad and depressed pygidial border.

Occurrence: *Bailiella* Zone of the Hsuchuang Formation, Hulusitai of Helanshan, Ningxia Huizhu Zizhiqu (NH 55) (Ningxia Hui Autonomous Region).

Family CONOCORYPHIDAE Angelin, 1854
Genus *Bailiella* Matthew, 1885
Type species: *Conocephalites baileyi* Hartt, 1868

Discussion: *Bailiella* is the only known genus of Conocoryphidae in China. This genus is widespread in North China and is confined to the top part of the Hsuchuang Formation.

***Bailiella lantenoisi* (Mansuy)**

1916 *Conocoryphe lantenoisi* Mansuy, Faunes Cambriennes Mémoires du Service Géologique de l'Indochine, v. 5, fasc. 1, p. 30, Pl. 4, figs. 6a–g; Pl. 5, fig. 3
1924 *Conocoryphe lantenoisi,* Hayasaka, Brief note on the Cambrian fossils from Chin-chia Cheng-Tzu, Huhsien and Liaotung, South Manchuria. Journal of Geography, Tokyo, v. 35, no. 412, p. 209
1935 *Conocoryphe lantenoisi,* Kobayashi, The Cambro-Ordovician formations and faunas of South Chosen. Journal of the Faculty of Science, Imperial University of Tokyo, sec. 2, v. 4, pt. 2, p. 218, Pl. 23, figs. 13–14
1937 *Bailiella ulrichi,* Resser and Endo, Manchurian Science Museum Bulletin 1, p. 193, Pl. 41, figs. 5–8; Pl. 42, Pl. 59, fig. 21
1957 *Bailiella lantenoisi,* Lu [Index fossils of China, Invertebrate fossils pt. 3], Geological Press, Peking, p. 265, Pl. 141, figs. 10–12 (in Chinese)
1965 *Bailiella lantenoisi,* Lu and others, [Fossils of China, Chinese trilobites, v. 1], Peking, Science Press, p. 150, Pl. 24, figs. 15–17 (in Chinese)
1973 *Bailiella lantenoisi,* Korobov, p. 119

Comparison: The present cranidium from Longxian in Shaanxi is very similar to the cranidium of *Bailiella lantenoisi* from Yunnan (Mansuy, 1916, p. 30, Pl. 4, figs. 6a–g, fig. 7; Pl. 55, fig. 3), but has a small occipital node, deeper anterior border furrow, and narrower convex anterior border.

Occurrence: *Bailiella* Zone of the Hsuchuang Formation, Niuxinshan of Longxian, Shaanxi (L 25).

Family TSINANIIDAE Kobayashi, 1933

This family is composed of *Tsinania* Walcott, 1914; *Tsinania (Dictyites)* Kobayashi, 1936; *Sunaspis* Lu, 1953; *Leiaspis* Wu and Lin (in Zhang and others, this volume) and *Sunaspidella* (Zhang and Yuan, gen. nov.). *Tsinania* and *Dictyites* are of Late Cambrian Fengshanian age while the remaining three are early Middle Cambrian Hsuchuangian. *Sunaspis* and *Sunaspidella* differ from *Tsinania* and from *Dictyites* in the relatively smaller palpebral lobe, in the more or less distinct glabellar and occipital furrows on the exfoliated surface of the cranidium, in the absence of a pygidial border on either testaceous or exfoliated pygidia, and in the more or less horizontal extensions of the pleural furrows on the pygidium. *Leiaspis* is very similar to *Tsinania,* but the former has a distinct narrow and convex anterior border. Considering the stratigraphic occurrence of these genera, the writers are inclined to believe that either *Leiaspis* or *Sunaspis* may be the ancestor of *Tsinania.*

Age and distribution: Early Middle Cambrian and Late Cambrian, China, South Asia, Australia.

Genus *Sunaspis* Lu, 1953

1953 *Sunaspis* Lu, Acta Geologica Sinica, v. 32, no. 3, p. 171
1957 *Sunaspis,* in Ku C. W., Yang Tsun-yi, Hsü C., Yin Tsan-hsun, Yü C. C., Chao K. K., Lu Yen-hao, Hou T. Y., Index fossils of China, Invertebrate, v. 3, p. 267
1965 *Sunaspis,* in Lu and others [Fossils of China. Chinese Trilobites, v. 1, p. 343] Peking, Science Press (in Chinese).

Type species: *Sunaspis laevis* Lu

Diagnosis: Opisthoparian and isopygous trilobites; cranidium gently convex, semicircular or semi-elliptical, characterized by its effaced surface; glabella rectangular, with four pairs of faint lateral glabellar furrows on exfoliated specimens. Palpebral lobes medium in size; eye ridges present. Preglabellar field gently convex, slightly downsloping forward; anterior border wide and flat; anterior border furrow faint. Fixigenae about one-half the width of glabella at its base. Pygidium semicircular; axis slightly convex, tapering backward, extending to the posterior margin, axial ring very weak. Pleural region triangular; pleural furrow weak, nearly horizontal, without border and border furrow.

Discussion: This genus is very similar to *Tsinania* Walcott (1914) from Upper Cambrian in North China, but differs from the latter in the rectangular glabella with four pairs of faint lateral glabellar furrows, in the well-defined occipital furrow and narrow occipital ring, in the smaller palpebral lobe, in the absence of pygidial border, and in the horizontal extension of the faint pleural furrows on the pygidium.

Age and distribution: Early Middle Cambrian, North China.

***Sunaspis triangularis* Lin and Wu, in Zhang and others (1980)**
Plate 2(4–5)

Description: Cranidium gently convex, subtriangular, anterior margin strongly curved forward, axial furrow faint and shallow; glabella convex, rectangular, flatly rounded in front; three or four pairs of faint lateral glabellar furrows. Occipital furrow distinct; occipital ring very narrow, with a small node. Fixigenae gently convex, about one-half the width of the glabella at its base; eye ridges slightly oblique outward and backward; palpebral lobe small. Anterior border furrow very shallow and narrow; Preglabellar field gently convex, downsloping to the anterior border furrow; anterior border flat or slightly upturned, broad in the middle, frontal margin strongly arched forward, posterior area of fixigenae subtriangular, posterior border of cranidium gently convex and narrow. Anterior sections of facial sutures slightly divergent from palpebral lobe to the anterior border furrow; posterior sections extending outward and backward. Pygidium semielliptical or subtriangular; axis gently convex, tapering backward, consisting of 12 or 13 axial rings on exfoliated specimens; pleural region triangular; pleural furrows shallow and nearly horizontal.

Occurrence: *Sunaspis* Zone of the Hsuchuang Formation, Niuxinshan, Longxian, Shaanxi (L 13).

Genus *Sunaspidella* n. gen.
Type species: *Sunaspidella rara* n. sp.

Diagnosis: Cranidium convex and subtrapezoidal, glabella convex, long and subquadrate, with four pairs of glabellar depressions; occipital furrow wide and shallow, occipital ring convex and very narrow, with a small node. Fixigenae narrow and gently downsloping outward; eye ridges present; palpebral lobe small; posterior area of fixigenae triangular. Anterior border furrow obsolete; preglabellar field absent; anterior border narrow and slightly downsloping forward. Pygidium semicircular; axis convex and wide, tapering backward, consisting of 6 to 8 axial rings; pleural region gently convex; pleural furrow shallow, nearly horizontal, without border.

Discussion: This genus can be easily distinguished from *Sunaspis* Lu, 1953, by its larger and broader glabella, absence of a preglabellar field, more oblique eye ridges, and the narrower fixigenae.

Age and distribution: Early Middle Cambrian, North China.

***Sunaspidella rara,* n. sp.**
Plate 2(11–12)

Description: Cranidium strongly convex, subtrapezoidal; glabella large, long and subquadrate in outline, with four pairs of lateral glabellar depressions; first two pairs narrow and weak, extending inward and slighty forward, third pair broad and oblique inward and backward, posterior pair wide and bifurcate. Occipital furrow wide and shallow; occipital ring convex and very narrow, with a small node. Fixigenae gently downsloping outward, about one-third the width of the glabella at the base; eye ridges narrow, strongly oblique outward and backward; palpebral lobes small; posterior area of fixigenae triangular. Posterior border furrow very shallow; posterior border very narrow and gently convex. Anterior border furrow obsolete; preglabellar field absent; anterior border narrow and slightly downsloping forward; anterior sections of facial sutures almost parallel, then curved inward; posterior sections extending outward and backward. Pygidium semicircular; axis wide, strongly convex, tapering backward, reaching to posterior margin; axial ring furrows obsolete; pleural region subtriangular; pleural furrows shallow and extending horizontally.

Occurrence: *Sunaspidella* Zone of the Hsuchuang Formation, Gangdeershan, Haibowan Shi, Nei Mongol Zizhiqu (76-302-F 38) (Inner Mongolia Autonomous Region).

ACKNOWLEDGMENTS

All trilobites described in this paper are deposited in the Museum of the Nanjing Institute of Geology and Palaeontology, Academia Sinica.

REFERENCES CITED

Chang Wen Tang (Zhang Wen-tang), 1957, Preliminary note on the Lower and Middle Cambrian stratigraphy of Poshan, central Shantung: Acta Palaeontologica Sinica, v. 5, no. 1, p. 13–32, 1 pl. (in Chinese with English abstract).

——1959, New trilobites from the Middle Cambrian of North China: Acta Palaeontologica Sinica, v. 7, no. 3, p. 193–236, 4 pls. (in Chinese with English abstract).

——1963, A classification of the Lower and Middle Cambrian trilobites from the North and Northeastern China, with description of new families and new genera: Acta Palaeontologica Sinica, v. 11, no. 4, p. 447–491, 2 pls. (in Chinese with English summary).

Chien Yi-yuan, 1961, Cambrian trilobites from Sandu and Duyun, southern Kweichow: Acta Palaeontologica Sinica, v. 9, no. 2, p. 91–140 (in Chinese and English).

Courtessole, R., 1973, Le Cambrien Moyen de la Montagne Noire (Biostratigraphie; Paléontologie): Laboratoire de Géologie Cearn de La Faculté des Sciences de Toulouse, 248 p., 27 pls.

Duan Ji-ye, 1965, [Cambrian stratigraphy and trilobites from Northeastern Hebei and western Liaoning]: Symposium of Changchun Geological Institute, v. 4, p. 137–152, 2 pls. (in Chinese).

Fritz, W. H., 1968, Lower and early Middle Cambrian trilobites from the Pioche shale, east-central Nevada, U.S.A.: Palaeontology, v. 11, no. 2, p. 183–235, Pls. 36–43.

Gu Zhi-wei [Ku Chih Wei], Yang Zun-yi [Yang Tsun-yi], Xu Jie [Hsü Singwu C.], Yin Zan-xun [Yin Tsan Hsun], Yu Jian-zhang [Yü Chien Chang], Zhao Jin-ki [Chao Kingkoo], Lu Yan-hao [Lu Yen-hao], and Hou You-tang, 1957, [Index fossils of China, Invertebrate, Part 3]: Beijing, Geological Publishing House (in Chinese).

Guo Hong-jun [Kuo Hung-tsun] and Duan Ji-ye, 1978, Cambrian and Ordovician trilobites from Northeastern Hebei and western Liaoning: Acta Palaeontologica Sinica, v. 17, no. 4, p. 445–458, 3 pls. (in Chinese with English abstract).

Harrington, H. J., and othes, 1959, Arthropoda 1, *in* Moore, R. C., ed., Treatise on invertebrate palaeontology, Part O: Geological Society of America and University of Kansas Press, 560 p.

Hupé, 1953, Contribution a l'étude du Cambrien inférieur et du Précambrien III de l'Anti-Atlas marocain: Service géologique du Maroc (Rabat), Notes et Mémoires, no. 103, p. 1–402.

Jell, A. P., 1975, Australian Middle Cambrian eodiscids with a review of the superfamily: Palaeontographica, Abstract A, Band 150, 97 p., 29 pls.

Kobayashi, Teiichi, 1935, The Cambro-Ordovician formations and faunas of South Chosen: Paleontology; Part III, Cambrian faunas of South Chosen with a special study on the Cambrian trilobite genera and families: Journal of the Faculty of Science, Imperial University of Tokyo, sec. 2, v. 4, pt. 2, p. 49–344, 24 pls.

——1961, The Cambro-Ordovician formations and faunas of South Korea, Part 8, Palaeontology 7, Cambrian faunas of the Mun'gyong (Bunkei) district and the Samposan Formation of the Yongwal (Neietsu) district: Journal of the Faculty of Science, Imperial University of Tokyo, sec. 2, v. 13, pt. 2, p. 181–241, Pls. 9–13.

——1962, The Cambro-Ordovician formations and faunas of South Korea, Part. 9, Palaeontology 8, The Machari fauna: Journal of the Faculty of Science, Imperial University of Tokyo, sec. 2, v. 14, pt. 1, 152 p., 8 pls.

——1967, The Cambro-Ordovician formations and faunas of South Korea, Part 10, Section C, The Cambrian of eastern Asia and other parts of the continent: Journal of the Faculty of Science, Imperial University of Tokyo, sec. 2, v. 16, pt. 3, p. 391–534.

Korobov, M. N., 1973, Trilobity semeystva Conocoryphidae i ikh znachenie dlya stratigrafii Kembriyskikh otlozheniy: Moskva, Akademiya Nauk SSSR, Geologicheskiy Institut, Trudy, v. 211, 161 p., Izdatel'stvo "Nauka."

Lo, H. L., 1974, [Cambrian trilobites, *in* Atlas of fossils from Yunnan Province]: Geological Bureau of Yunnan, Ministry of Geology, ed., Kunming, Yunna, People's Press, p. 1–864, 296 pls.

Lochman, Christina, 1948, New Cambrian trilobites genera from northwest Sonora, Mexico: Journal of Paleontology (Tulsa) v. 22, p. 451–464, Pls. 69, 70.

Lu Yan-hao [Lu yen-hao] and Dong Nan-ting, 1953, [Revision of the Cambrian type sections in Shantung]: Acta Geologica Sinica, v. 32, no. 3, p. 164–201 (in Chinese).

Lu Yan-hao [Lu Yen-hao], Zhang Wen-tang [Chang Wen Tang], Zhu Zhao-ling [Chu Chao-ling], Qian Yi-yuan [Chien Yi-yuan], and Xiang Le-wen [Hsiang Lee-wen], 1965, [Fossils of China. Chinese Trilobites, Volume 1], 362 p., 66 pls., Peking Science Press (in Chinese).

Mansuy, H., 1916, Faunes Cambriennes de l'extrême orient meridional: Mémoires du Service Géologique de l'Indochine, v. 5, fascicule 1, p. 1–48, Pls. 1–7.

Palmer, A. R., and Gatehouse, C. G., 1972, Early and Middle Cambrian trilobites from Antarctica: U.S. Geological Survey Professional Paper 456-D, 66 p., 6 pls.

Poprovskaya, N. V., 1960, *in* Khalfin, L. L., ed., Biostratigrafiya Paleozoya Sayano-Altayskoy gornoy oblasti, Volume 1, Nizhniy Paleozoy: Sibirskiy Nauchno-Issled. Institut Geologii, Geofiziki i Mineralnogo Syrya (SNIIGGIMS), Trudy, v. 19, p. 1–498.

Qian Yi-yuan [Chien Yi Yuan], 1961, Cambrian trilobites from Sandu and Duyun, Southern Kweichou [Quizhou]: Acta Palaeontologica Sinica, v. 9, no. 2, p. 110–129, Pls. 1–5 (in Chinese and English).

Rasetti, Franco, 1945, Fossiliferous horizons in the "Sillery Formation" near Levis, Quebec: American Journal of Science, v. 243, p. 305–319.

——1951, Middle Cambrian stratigraphy and faunas of the Canadian Rocky Mountains: Smithsonian Miscellaneous Collections, v. 116, no. 5, p. 1–270, Pls. 34.

——1963, Middle Cambrian ptychoparioid trilobites from the conglomerates of Quebec: Journal of Paleontology, v. 37, no. 3, p. 575–594, Pls. 66–70.

——1966, Revision of the North American species of the Cambrian trilobite genus *Pagetia:* Journal of Paleontology, v. 40, no. 3, p. 502–511, Pls. 59–60.

Reed, F.R.C., 1910, The Cambrian fossils of Spiti: Palaeontologia Indica, ser. 15, v. 7, Memoir No. 1, p. 1–70, Pls. 1–6.

Repina, L. N., and others, 1975, Stratigrafiya i fauna nizhnego Paleozoya severnykh predgoriy Turkestanskogo i Alayskogo Khrebtov: Isdatel'stvo "Nauka," Sibirskoe Otdelenie, Novosibirsk, p. 1–248, 48 pls.

Resser, C. E., 1942, Fifth contribution to nomenclature of Cambrian fossils: Smithsonian Miscellaneous Collections, v. 101, no. 15, p. 1–58.

Resser, C. E., and Endo, R., 1937, Description of the fossils, *in* Endo, R., and Resser, C. E., The Sinian and Cambrian formations and fossils of southern Manchoukuo: Mukden, Manchurian Science Museum Bulletin 1, p. 103–301, Pls. 18–73.

Savitskiy, V. E., Yevtushenko, V. M., Yegorova, L. I., Kontorovich, A. E., and Shabanov, U. Ya, 1972, Kembriy Sibirskoy platformy Yudomo-Olenekskii tip razreza. Kuonamskii kompleks otlozheni: Trudy Sibirskogo Nauchno-issledovatel'skogo Instituta Geologii, Gefiziki i Mineral'nogo Syr'ya (SNIIGGIMS), v. 130, p. 1–200, 23 pls.

Schrank, Ernst, 1976, Kambrische Trilobiten der China—Kollektion v. Richthofen, Teil 3, Mittelkambrische Faunen von Taling: Berlin, Zeitschrift für Geologische Wissenschaften, Jahrgang 4, Heft 6, p. 891–919, Pls. 1–6.

Walcott, C. D., 1911, Cambrian geology and paleontology II—Cambrian fauna of China: Smithsonian Miscellaneous Collections, v. 57, no. 4, p. 69–108, Pls. 14–17.

——1913, The Cambrian faunas of China: Research in China, Carnegie Institution of Washington, Publication 54, p. 1–276, 24 pls.

——1916, Cambrian geology and palaeontology III, no. 5, Cambrian trilobites: Smithsonian Miscellaneous Collections, v. 64, no. 5, p. 301–456.

Wolfart, Reinhard, 1974, Die Fauna (Brachiopoda, Molluska, Trilobita) aus dem Unter Kambrium von Kerman, Südost-Iran: Hannover, Geologisches Jahrbuch, Band 8, p. 5–70, Pls. 1–9.

Yegorova, L. I., 1960, *in* Khalfin, L. L., ed., Biostratigrafiya Paleozoya Sayano-Altayskoy gornoy oblasti: Nizhniy, Trudy Sibirskogo Nauchno-Issledovatel'skogo Instituta Geologii, Geofiziki i Mineral'nogo Syr'ya (SNIIGGIMS) v. 19, tome 1, p. 1–498.

Yegorova, L. I., and Savitskiy, V. E., 1969, Stratigrafiya i biofatsii kembriya Sibirskoy platformy, Zapadnoe Prianabar'ye: v. 43, 408 p.

Zhang Wen-tang [Chang Wen Tang], Lin Huan-ling, Wu Hong-ji, Yuan Jin-liang, 1980, Cambrian stratigraphy and trilobite fauna from Zhongtiao Mountains, Southern Shanxi: Memoirs of Nanjing Institute of Geology and Palaeontology, Academia Sinica, no. 16, p. 39–97, 11 pls. (in Chinese with English abstract).

Zhang Wen-tang [Chang Wen Tang], Lu Yan-hao, Zhu Zhao-ling, Qian Yi-yuan, Lin Huan-ling, Zhou Zhi-yi, Zhang Sen-gui, and Yuan Jin-liang, 1980, Cambrian trilobite faunas of southwestern China: Peking, Science Press, Palaeontologia Sinica, whole ser. no. 159, new ser. B, no. 16, 497 p., 134 pls. (in Chinese with English summary).

Zhang Wen-tang [Chang Wen Tang], Zhu Zhao-ling, Wu Hong-ji, Yuan Jin-liang, Shen Hou, Yao Bao-qi, Luo Kun-quan, and Wang Xu-ping, 1980, [Cambrian stratigraphy of the western and southern marginal parts of the Ordos Platform]: Peking, Science Press, Journal of Stratigraphy (Dicengxue Zashi), v. 4, no. 2, p. 106–109 (in Chinese).

MANUSCRIPT ACCEPTED BY THE SOCIETY APRIL 22, 1981

PLATE 1

Figure

1–2. ***Luaspides brevica*** **n. sp. Gangdeershan, Nei Mongol Zizhiqu (NZ 2a),** ***Kochaspis-Ruichengella*** **Zone of the Hsuchuang Formation.**
- **1. Complete exoskeleton, ×4, Cat. No. 62236.**
- **2. holotype, cranidium, ×3, Cat. No. 62237.**

3–4. ***Zhongtiaoshanaspis similis*** **n. sp. Hulusitai of Helanshan, Ningxia Huizu Zizhiqu (NH 32),** ***Kochaspis-Ruichengella*** **Zone of the Hsuchuang Formation.**
- **3. Cranidium, ×4, Cat. No. 62238.**
- **4. holotype, cranidium, ×4, Cat. No. 62239.**

5–6. ***Catinouyia typica*** **n. gen., n. sp. Gangdeershan, Haibowan Shi, Nei Mongol Zizhiqu (NZ 3),** ***Kochaspis-Ruichengella*** **Zone of the Hsuchuang Formation.**
- **5. Holotype, cranidium, ×4, Cat. No. 62240.**
- **6. Cranidium, ×5, Cat. No. 62241.**

7–9. ***Wuhaina lubrica*** **n. gen., n. sp. Gangdeershan, Haibowan Shi, Nei Mongol Zizhiqu (NZ 5),** ***Ruichengaspis*** **Zone of the Hsuchuang Formation.**
- **7. Cranidium, ×6, Cat. No. 62242.**
- **8. Holotype, cranidium, ×6, Cat. No. 62243.**
- **9. Cranidium, ×3, Cat. No. 62244.**

10–12. ***Ruichengaspis neimengguensis*** **n. sp. Gangdeershan, Haibowan Shi, Nei Mongol Zizhiqu (NZ 7, NZ 8),** ***Ruichengaspis*** **Zone of the Hsuchuang Formation.**
- **10. Cranidium, ×4, Cat. No. 62245.**
- **11. Cranidium, ×6, Cat. No. 62246.**
- **12. Holotype, cranidium, ×6, Cat. No. 62247.**

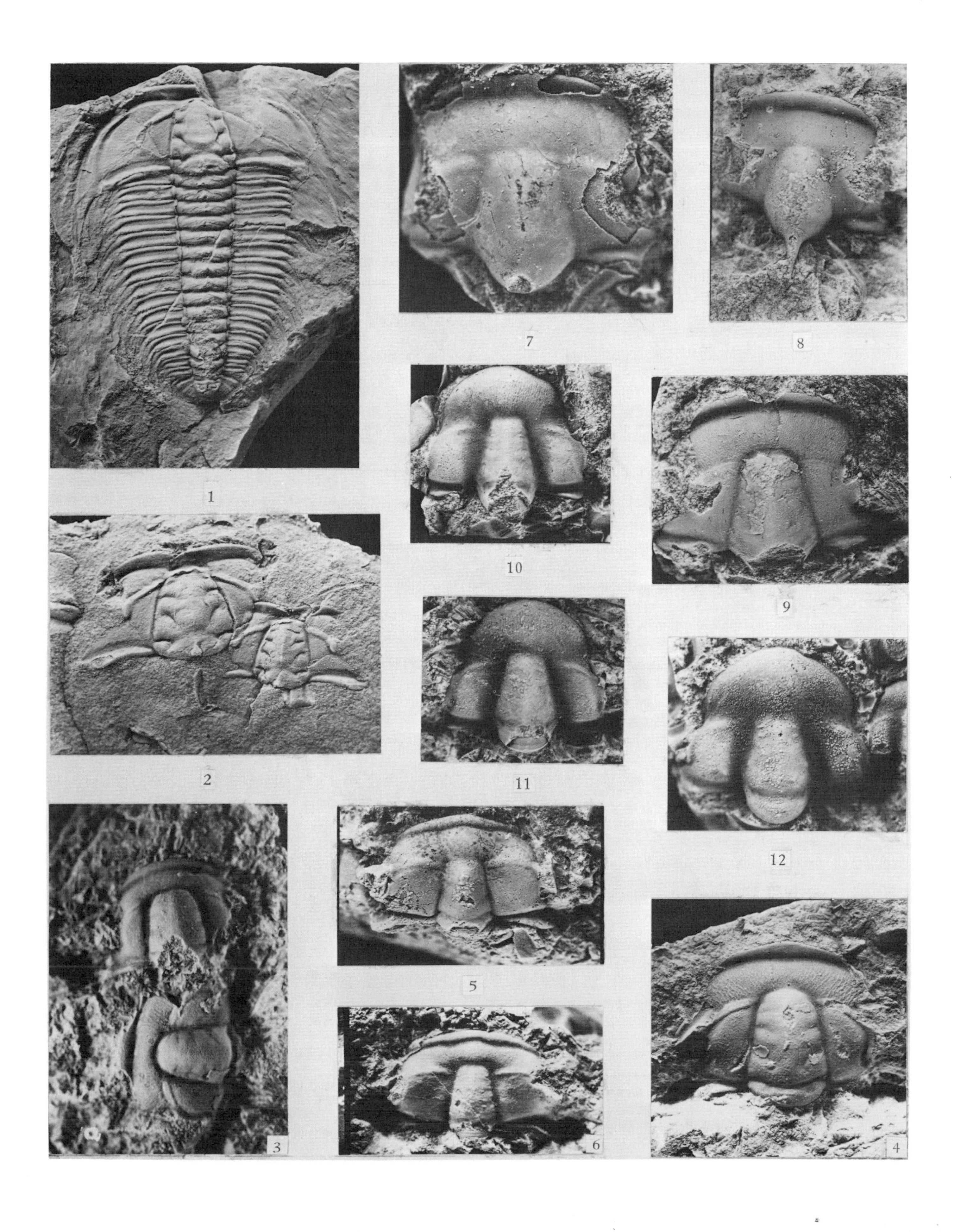
1
2
3
4
5
6
7
8
9
10
11
12

PLATE 2

Figure

1–2. *Pagetia (Sinopagetia) neimengguensis* n. subgen., n. sp. Gangdeershan, Haibowan Shi, Nei Mongol Zizhiqu (NZ 02), *Pagetia (Sinopagetia) jinnanensis* Zone of the Hsuchuang Formation.
1. Holotype, cranidium, ×20, Cat. No. 62248.
2. Pygidium, ×18, Cat. No. 62249.

3. *Parainouyia prompta* n. sp. Gangdeershan, Haibowan Shi, Nei Mongol Zizhiqu (NZ 02), *Pagetia (Sinopagetia) jinnanensis* Zone of the Hsuchuang Formation. Holotype, cranidium, ×5, Cat. No. 62250.

4–5. *Sunaspis triangularis* Lin and Wu. Niuxinshan, Longxian, Shaanxi Province (L 13), *Sunaspis* Zone of the Hsuchuang Formation.
4. Cranidium; ×5, Cat. No. 62251.
5. Pygidium, ×5, Cat. No. 62252.

6–7. *Wuania venusta* n. sp. Gangdeershan, Haibowan Shi, Nei Mongol Zizhiqu, *Sunaspis* Zone of the Hsuchuang Formation.
6. Holotype, cranidium, ×3, Cat. No. 62253.
7. Pygidium, ×4, Cat. No. 62254.

8–9. *Taitzuina lubrica* n. gen., n. sp. Hejin Xian, Shanxi Province (F 15'); Hulusitai of Helanshan, Ningxia Huizu Zizhiqu (NH 48), *Sunaspis* Zone of the Hsuchuang Formation.
8. Holotype, cranidium, ×8, Cat. No. 62255.
9. Pygidium, ×8, Cat. No. 62256.

10. *Wuanoides situla* n. gen., n. sp. Helanshan, Ningxia Huizu Zizhiqu (NH 12), *Sunaspis* Zone of the Hsuchuang Formation. Holotype, cranidium, ×3, Cat. No. 62257.

11–12. *Sunaspidella rara* n. gen., n. sp. Gangdeershan, Haibowan Shi, Nei Mongol Zizhiqu (76-302-F38), *Sunaspidella* Zone of the Hsuchuang Formation.
11. Holotype, cranidium, ×4, Cat. No. 62258.
12. Pygidium, ×4, Cat. No. 62259.

13–14. *Wuhushania cylindrica* n. gen., n. sp. Wuhushan, Nei Mongol Zizhiqu (Q 5-VIII-H 40), *Sunaspidella* Zone of the Hsuchuang Formation. Dorsal and lateral views of pygidium, holotype, ×4, Cat. No. 62260.

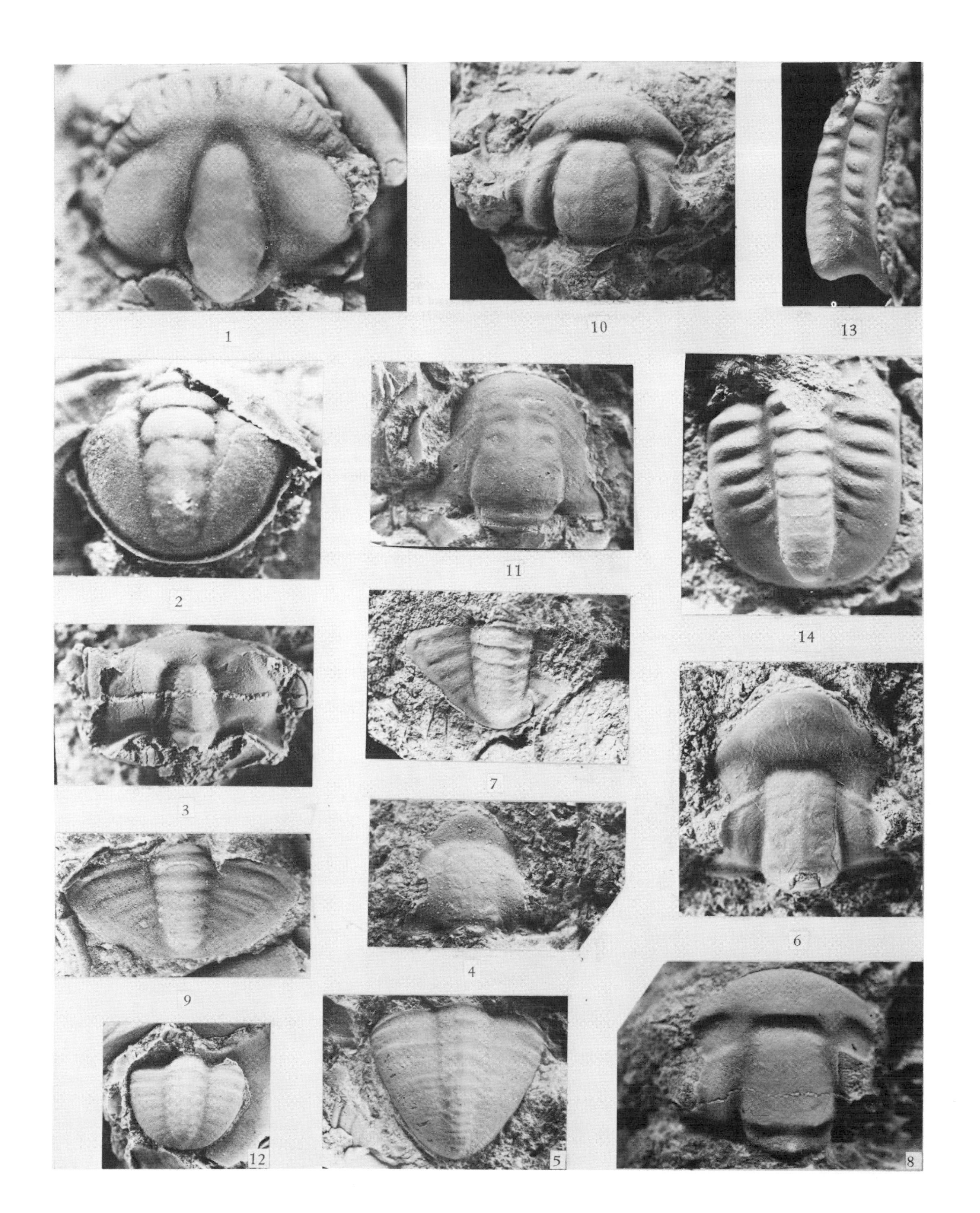

1
10
13
2
11
14
3
7
9
4
6
12
5
8

PLATE 3

Figure

1–2. *Zhuozishania typica* n. gen., n. sp. Gangdeershan, Haibowan Shi, Nei Mongol Zizhiqu (NZ 20), *Metagraulos* Zone of the Hsuchuang Formation.
   1. Holotype, cranidium, ×6, Cat. No. 62261.
   2. Pygidium, ×4, Cat. No. 62262.

3–5. *Metagraulos truncatus* n. sp. Gangdeershan, Haibowan Shi, Nei Mongol Zizhiqu (NZ 23, NZ 20, NZ 19), *Metagraulos* Zone of the Hsuchuang Formation.
   3. Pygidium, ×8, Cat. No. 62263.
   4. Holotype, cranidium, ×4, Cat. No. 62264.
   5. Cranidium, ×6, Cat. No. 62265.

6. *Inouyops latilimbatus* n. sp. Gangdeershan, Haibowan Shi, Nei Mongol Zizhiqu (NZ 21), *Inouyops* Zone of the Hsuchuang Formation. Holotype, cranidium, ×5, Cat. No. 62266.

7. *Poriagraulos dactylogrammacus* n. sp. Niuxinshan, Longxian, Shaanxi Province (L 23), *Poriagraulos* Zone of the Hsuchuang Formation. Holotype, cranidium, ×8, Cat. No. 62267.

8–9. *Proasaphiscina taosigouensis* n. sp. Hulusitai of Helanshan, Ningxia Huizu Zizhiqu (NH 55), *Bailiella* Zone of the Hsuchuang Formation.
   8. Holotype, cranidium, ×4, Cat. No. 62268.
   9. Pygidium, ×8, Cat. No. 62269.

10. *Bailiella lantenoisi* (Mansuy) Niuxinshan, Longxian, Shaanxi Province (L 25), *Bailiella* Zone of the Hsuchuang Formation. Cranidium, ×6, Cat. No. 62270.

11. *Wuhushania cylindrica* n. gen., n. sp. Wuhushan, Nei Mongol Zizhiqu (Q 6—VIII-H 40), *Sunaspidella* Zone of the Hsuchuang Formation. Cranidium, ×5, Cat. No. 62271.

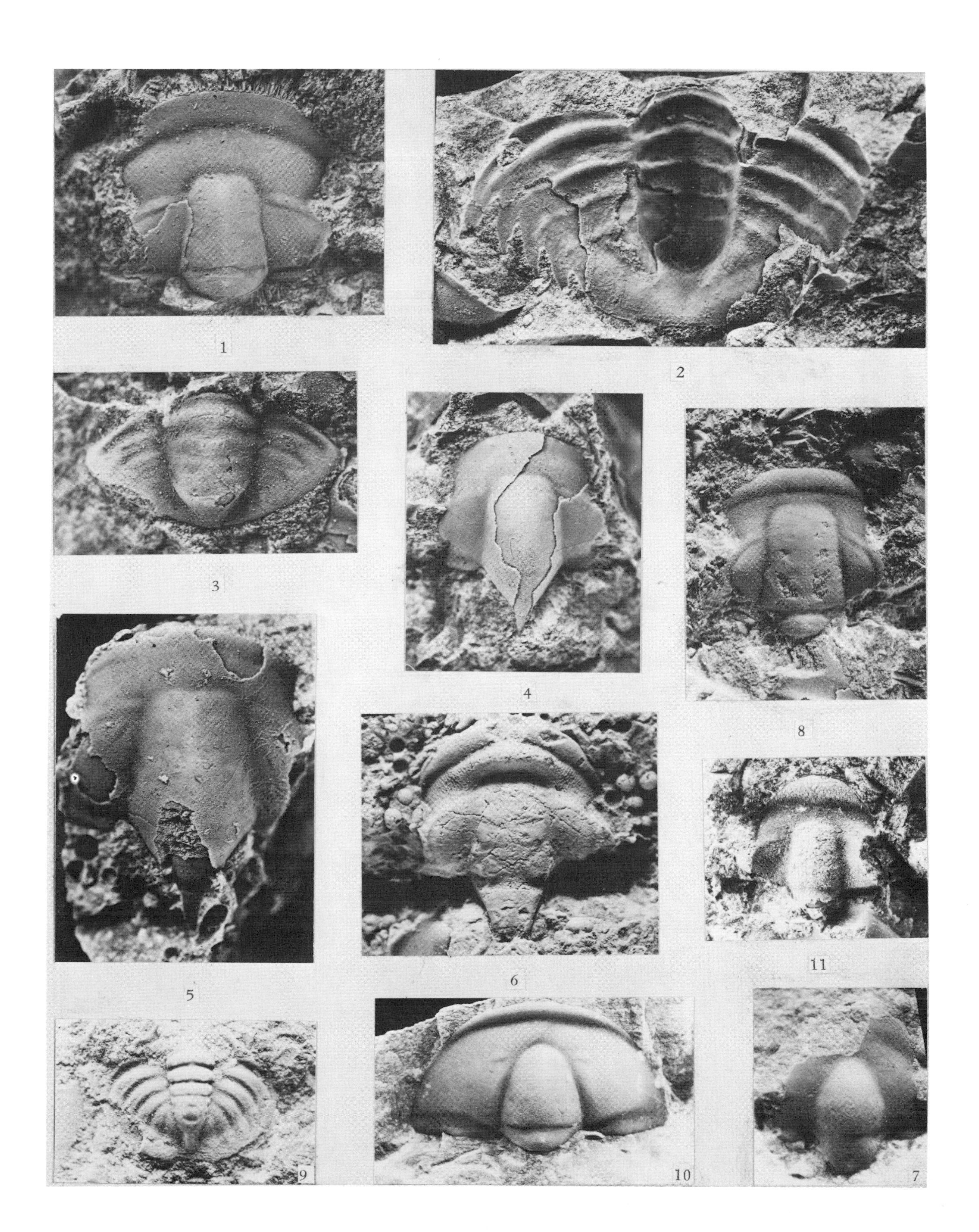
1
2
3
4
8
5
6
11
9
10
7

Printed in U.S.A.

Geological Society of America
Special Paper 187
1981

Systematic position of *Sarkia* Klouček (Trilobita)

Zhou Zhi-yi
Nanjing Institute of Geology and Palaeontology, Academia Sinica, Nanjing 210008, People's Republic of China

ABSTRACT

The occurrence of the submarginal suture in well-preserved specimens of *Sarkia superioris* Zhou from the Caradocian of North Jiangxi indicates that the genus *Sarkia* seems to be an aberrant conocoryphid. The family name Sarkiidae, therefore, ought to be abandoned.

INTRODUCTION

The specimens of the type species of *Sarkia, S. bohemica,* described by Klouček (1916) and redescribed and figured by Prantl and Přibyl (1948) from the Dobrotiva shales (dγ2b, Llandeilian) of Bohemia are somewhat distorted and too fragmentary to furnish convincible evidence for determining the taxonomic position of the genus. Klouček (1916) considered that this genus shows a certain relationship to the genus *Harpes* Goldfuss and Prantl and Přibyl (1948) placed it in *Familiae incertae* under the order Hypoparia. In 1953, Hupé ascribed it to the family Trinucleidae, and in 1955 he removed it from the Trinucleidae and erected Sarkaidae (corrected as Sarkiidae by Henningsmoen in 1959) as an independent new family. In Hupé's opinion, *Sarkia* bears some characteristics of the Richardsonellidae and may be an offshoot of it. Hupé's new family was accepted by Henningsmoen (1959) in *Treatise on Invertebrate Paleontology,* but the problem to which order or superfamily it should belong has not been settled. In 1960, Maksimova (1960, p. 179) placed *Sarkia* doubtfully in the Harpidae. Recent discovery of the cephalic suture and diagnostic shell structures of conocoryphid type in well-preserved specimens of *Sarkia superioris* Zhou from the Middle and Upper Ordovician of North Jiangxi, however, demonstrates conclusively Hupé's Sarkiidae to be an invalid family and that *Sarkia* may be more correctly interpreted as a member of the conocoryphids.

SYSTEMATIC DESCRIPTION

Superfamily CONOCORYPHACEA Angelin, 1854
Family CONOCORYPHIDAE Angelin, 1854
Genus *Sarkia* Klouček, 1916, emend. Prantl and Přibyl, 1948

Diagnosis: Cephalon broadly semicircular, with stout genal spines. Glabella slightly tapering forward, with 3 pairs of short but deeply incised lateral furrows. Occipital ring bearing a median node. Eyes absent. Eye ridges short, oblique. Cephalic suture submarginal, curving across genal angles on dorsal side. Cephalic border furrow pitted and developing a median inbend. Librigenae small, represented only by genal spines, fused with doublure which bears a remarkable marginal girder. Thorax consists of at least 10 segments; pygidium small, transverse. Axial lobe subconical, with 3 to 5 rings. Pleural lobes with 3 to 5 pairs of pleural furrows and very ill-defined interpleural furrows.

Remarks: *Sarkia* differs so markedly from all the genera of the Harpididae, the Trinucleidae, and the Richardsonellidae that it must be considered neither allied to nor identical with any of them. Making comparisons of *Sarkia* with various Cambrian and Ordovician trilobites, the writer comes to the conclusion that it does not fit in well with any family except the Conocoryphidae and it may be an aberrant form of the conocoryphids. There are many characteristics seen commonly in *Sarkia* and some typical conocoryphids, such as *Conocoryphe* Hawle and Corda, *Atops* Emmons, *Meneviella* Stubblefield, *Ctenocephalus* Hawle and Corda, and *Bailiella* Matthew. Among other features, the most significant are the submarginal cephalic suture, the broad semicircular cephalon with genal spines, the subsemielliptical and furrowed pygidium, the well-defined and tapering glabella with 3 pairs of short lateral furrows, and the absence of eyes. In addition, the oblique eye ridges, the presence of an occipital node, and the thorax that contains more than 10 segments are features in common with the aforementioned genera. *Sarkia,* however, is distinguished from them by the more convex cephalon with pitted border furrow, by the broader and less tapering glabella, by the deeper glabellar furrows, and by the heteropygous, instead of micropygous, pygidium. The cephalic doublure of *Sarkia* is broad (sag.) and entire, with a more developed marginal girder, whereas that of *Conocoryphe* (Barrande, 1852, Pl. 2B, fig. 24; Pl. 14, fig. 8; Rasetti, 1952, p. 891; Šnajdr, 1958, Pl. 34, figs. 1, 3) is narrow (sag.), with a rostral plate separated by connective sutures. The hypostomata of *Sarkia* (Prantl and Přibyl, 1948, Text-fig. 3) and *Conocoryphe* (Barrande, 1852, Pl. 2B, fig. 24; Pl. 14, figs. 10, 16, 17; Šnajdr, 1958, Pl. 34, figs. 4, 8, 9a) are also quite different. Summing up, it seems to the writer that the differences between these two genera are too small to be regarded as familial characters, rather than being of only generic importance, and that *Sarkia* may be referred to the family Conocoryphidae. The Sarkiidae established by Hupé should be regarded as a subjective synonym of the Conocoryphidae.

Sarkia is known from the Llandeilian of Bohemia and from the Caradocian of South China.

Sarkia superioris Zhou
Plate 1(1–5); Figures 1–5

1976 *Sarkia superioris* Zhou, in Lu and others, p. 62, Pl. X, figs. 1, 2

Description: Cephalon convex, broad semicircular in outline, almost straight in front. Glabella vaulted, longer than wide, with a width at the base slightly more than one-quarter the length of the cephalon, somewhat tapering forward, truncated in front, rounded in antero-lateral angle, bounded by broad and deep dorsal and preglabellar furrows. Three pairs of deeply incised lateral glabellar furrows present: posterior pair long, curving obliquely inward and backward; middle pair about half the length of the posterior one and extending parallel to it; anterior pair very short, pitted, close to the dorsal furrows. Occipital furrow shallow and curving forward in middle portion, deepening laterally and bending obliquely forward toward the dorsal furrows. Occipital ring with broadly rounded posterior margin and a median node near the occipital furrow, bending forward laterally. Cheeks convex, broad, confluent, sloping steeply down anteriorly and laterally. Eyes absent. Eye ridge visible, short, about one-third the width of the cheek, originating at the junction of the dorsal furrow and the anterior lateral glabellar furrow, running obliquely backward toward the genal angle. Cephalic suture submarginal, curving inward and backward opposite the middle pair of lateral glabellar furrows and passing across the genal angles on dorsal side. Cephalic border narrow, developing a median inbend, bearing a row of about 54 pits distributed at regular intervals. Cephalic border convex, narrowing inward. Posterior border furrows deep and broad. Genal spines stout, longer than two-thirds the length of cephalon, extending postero-laterally at an angle of about 130° to the posterior margin of the posterior border. Doublure broad (sag.), continuously extending to and linking up with the genal spines. Rostral plate and connective sutures absent. Marginal girder deep in dorsal view, with a row of granules internal to the girder on dorsal surface corresponding to the pits in border furrow.

Only three segments of thorax attached to the cephalon preserved. Axis strongly convex, delimited by deep dorsal furrows, about two-sevenths the total width of thorax, bearing a slightly elevated knot on each side of the axial ring. Pleurae horizontal for two-thirds of their width from the dorsal furrow, curving downward and backward laterally, obliquely truncated in an arc at the distal end. Pleural furrows deep and broad.

Pygidium small and broad, subsemielliptical in shape, with posterior margin somewhat bent forward, about one-third as long as wide. Axis elevated, narrow (trans.), well-defined by deep dorsal furrows, with a length about four-fifths the length of pygidium, composed of four rings and a rounded terminal ring. Pleural lobe subtriangular, with five pairs of transverse pleural furrows and three pairs of very shallow and inconspicuous interpleural furrows. Marginal border weakly defined. Doublure narrow, about one-quarter the length of pygidium, with 6 to 7 terrace lines.

Surface of cephalon ornamented densely with small granules.

Comparison: *S. superioris* is readily distinguished from the type species, *S. bohemica,* by the broader preglabellar field (sag.), by the narrower glabella with shorter frontal glabellar lobe, and by the genal angle forming a right rather than an acute angle. A closer comparison between the two species is impossible, because the specimens of the Bohemian species are too fragmentary and badly preserved.

Horizons and locality: Zone of *Xiushuilithus,* Yenwashan Formation, upper Middle Ordovician (middle Caradocian) and Zone of *Nankinolithus,* Huangnekang Formation, lower Upper Ordovician (upper Caradocian), Wuning district, North Jiangxi.

ACKNOWLEDGMENTS

The material on which this paper is based was collected by the writer and Zhang Yin, to whom the writer is grateful for ready help and collaboration in the field. The writer must record his gratitude to Mao Ji-liang who prepared the photographs and to Zhang Wu-chong who drew the text figures.

REFERENCES CITED

Barrande, J., 1852, Système Silurien du Centre de la Bohème, I, *in* Recherces Paléontologiques, Volume I, Trilobites: Paris, Praha, 935 p., 51 pls.

Henningsmoen, G., 1959, Family Sarkiidae, *in* Moore, R. C., ed., Treatise on invertebrate paleontology, Part O, Arthropoda 1: Geological Society of America and University of Kansas Press, p. 512.

Hupé, P., 1935, Classe des trilobites, *in* Piveteau, J.: Traité de Paléontologie, III, Masson, Paris, p. 44–246.

——1955, Classification des trilobites: Annales de Paléontologie, v. 41, p. 91–325.

Klouček, C., 1916, o vrstvách $d_1\gamma$, jich trilobitech a nalezištich. Rozpr. II, tř. Čes. akad., Volume 25, no. 39, Praha. (Über die $d_1\gamma$ Schichten und ihre Trilobiten-Fauna): Bulletin International de L'Académie des Sciences de la Bohème, Praha, p. 231–246, 1 pl.)

Lu Yen-hao [Lu Yan-ho], Chu Chao-ling [Zhu Zhao-ling], Chien Yi-yuan [Qian Yi-yuan], Zhou Zhi-yi, Chen Jun-yuan, Jin Geng-wu, Yü Wen, Chen Xu, and Xu Han-kui, 1976, [Ordovician biostratigraphy and palaeozoogeography of China]: Memoirs of Nanjing Institute of Geology and Palaeontology, Academia Sinica, no. 7, p. 1–83, 14 pls. (in Chinese).

Maksimova, Z. A., 1960, Nadsemeystvo Harpoidea, *in* Yu. A. Orlov, ed., Osnovy Paleontologi, Chlenistonogie, Trilobitoobraznye i Rakoo-

Figures 1–5. ***Sarkia superioris*** **Zhou.**

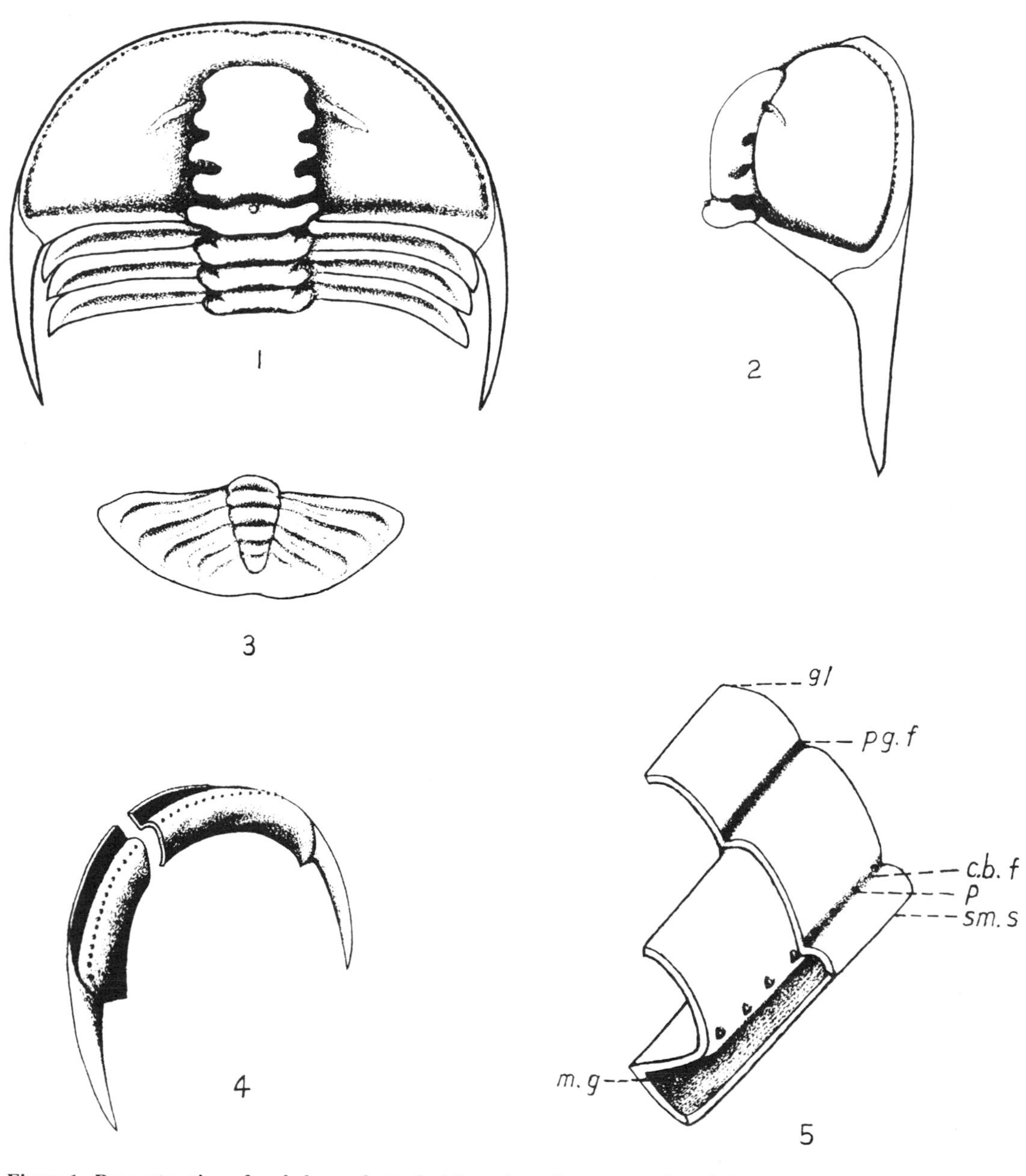

Figure 1. Reconstruction of cephalon and attached three thoracic segments, dorsal view, ×3.
Figure 2. Reconstruction of cephalon, lateral view, ×3.
Figure 3. Reconstruction of pygidium, ×3.
Figure 4. Restored doublure fused with genal spines, oblique view.
Figure 5. Restored portion of anterior area of cephalon; *gl* = glabella; *pg. f* = pregrabellar furrow; *P* = pit; *c.b.f* = cephalic border furrow; *sm.s* = submarginal suture; *m.g* = marginal girder.

braznye: Gosudarstvennoe Nauchno-Tekhnicheskoe. Izdatel'stvo, Moskva, p. 178–179.
Prantl, F., and Přibyl, A., 1948, Some new or imperfectly known Ordovician trilobites from Bohemia: Bulletin International de l'Académie Tcheque des Sciences, v. 49, no. 8, p. 1–23, 3 pls.
Rasetti, F., 1952, Ventral cephalic sutures in Cambrian trilobites: American Journal of Science, v. 250, p. 885–898.
Šnajdr, M., 1958, The trilobites of the Middle Cambrian of Bohemia: Rozpravy Ustředniho Ustavu Geologického, Svazek, v. 24, p. 1–279, 46 pls.

Manuscript Accepted by the Society April 22, 1981

PLATE I. ***Sarkia superioris*** **Zhou**

All the specimens were collected from a locality about 5 km northwest of Wuning City, North Jiangxi, and deposited in the Nanjing Institute of Geology and Palaeontology, Academia Sinica. Specimens shown in figures 1–11 are from the Huangnikang Formation (lower Upper Ordovician); figures 12–15 from the Yenwashan formation (upper Middle Ordovician).

Figures

1–3.	**Holotype, cephalon and three thoracic segments attached, dorsal, right lateral, and anterior views, ×3, Cat. No. 23912.**
4–5.	**Cephalon, dorsal and right lateral views, ×4, Cat. No. 63481.**
6.	**External mould of cranidium, ×6, Cat. No. 63482.**
7.	**Incomplete cranidium, ×3, Cat. No. 63483.**
8–9.	**Incomplete cranidium, dorsal and oblique views, ×2, Cat. No. 23913.**
10.	**Pygidium, ×4, Cat. No. 63484.**
11.	**Pygidium, ×3, Cat. No. 63485.**
12–14.	**Cephalon, dorsal, right lateral, and anterior views, ×6, Cat. No. 63486.**
15.	**Incomplete cephalon, oblique view, ×4, Cat. No. 63487.**

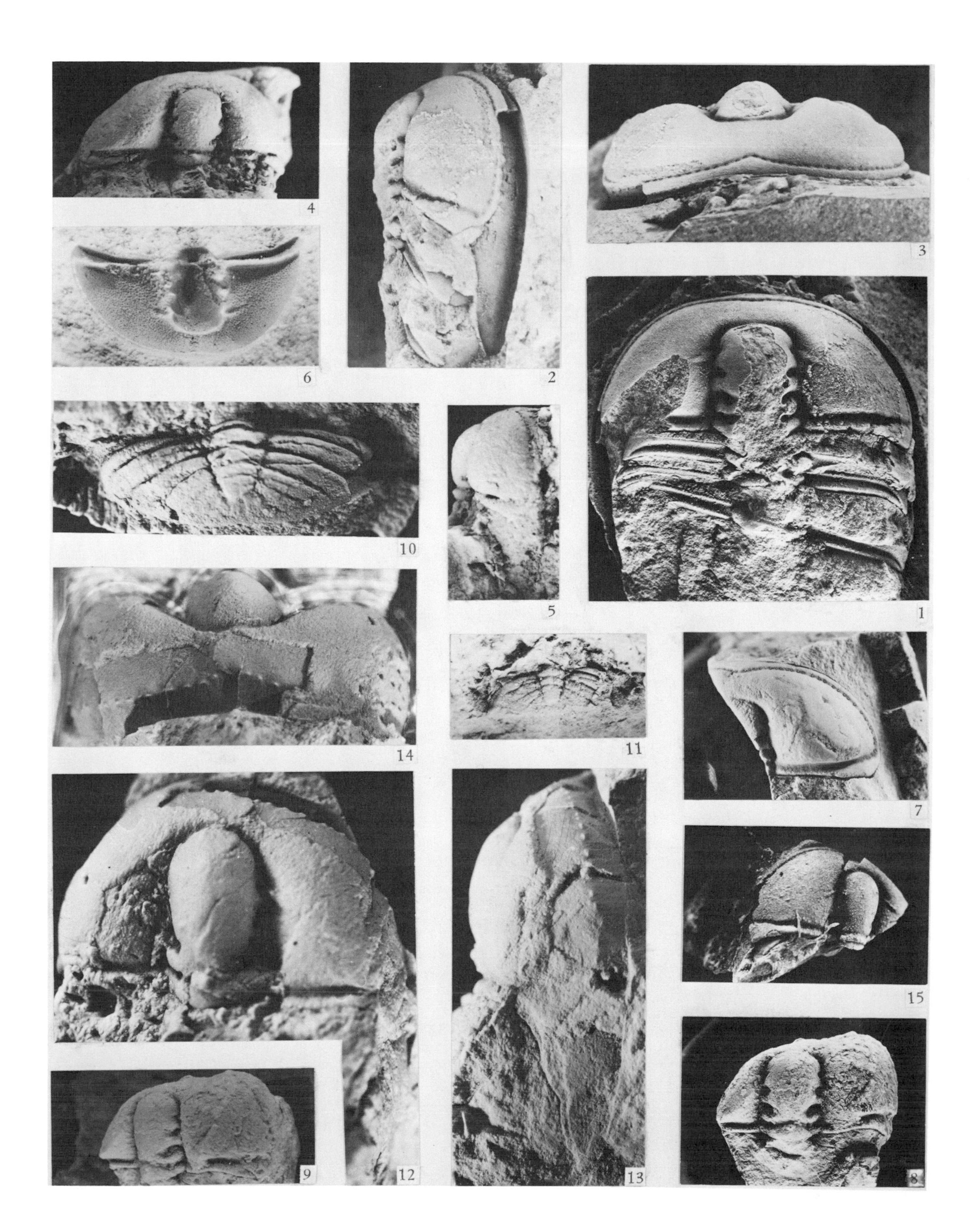

4
3
6
2
10
5
1
14
11
7
15
9
12
13
8

Printed in U.S.A.

Geological Society of America
Special Paper 187
1981

Some Late Devonian trilobites of China

Xiang Li-wen
Institute of Geology, Chinese Academy of Geological Sciences, Peking, People's Republic of China

ABSTRACT

Late Devonian trilobites are here reported from three localities in China: (1) Daihua, Changshun County, Guizhou province; (2) Mianduhe, Hulun Buir Meng, Nei Mongol [Inner Mongolia], central part of Da Hingan [Great Khingan]; and (3) Hoxtolgay, West Junggar, Xinjiang. Four genera, three species, and one subspecies are represented: *Typhloproetus laevis* sp. nov., *Chaunoproetus daihuaensis* sp. nov., *Dianops guizhouensis* sp. nov., *Phacops (Omegops) accipitrinus mobilis* subsp. nov. and a phacoid, gen. et sp. indet. These are associated with ammonoids. It the latest Devonian the ammonoid and trilobite faunas were cosmopolitan, so the trilobite-bearing horizons may be used for worldwide correlation.

INTRODUCTION

Since the discovery of *Cheiloceras-Sporadoceras* beds in Da Hingan in 1957 by the Da Hingan Ling regional geological surveying team and of the *Wocklumeria* fauna in Guizhou in 1959 by Li Jiecai (Chang An-zhi, 1958, 1960; Sun Yun-zhu and Shen Yao-ting, 1965), latest Devonian deposits (= Famennian of Europe) have been known to exist in China. The ammonoid horizon in Da Hingan is of early Famennian age and that in Guizhou late Famennian. A number of trilobites are associated with these ammonoids, and late Upper Devonian trilobites were also collected when the author carried out a stratigraphic survey in West Unggar in 1973. These three collection of trilobites form the basis of this study.

The Late Devonian (Famennian) trilobite fauna here reported and described from China for the first time has been little known throughout Asia. In China the only two Upper Devonian trilobite species described so far are *Dechenella? chengi* (Sun), collected by Sun Yunzhu from Chaling, Hunan Province, in 1937, and *Typhloproetus sinensis* Chang, established by Zhang Wen-tang from Lingchuan, Guangxi Province, in 1955, but these are not necessarily of Famennian age. In East Asia, trilobites of probably Famennian age are *Langgonbole vulgaris* Kobayashi and Hamada and *Waribole perlisensis* Kobayashi and Hamada, reported by Kobayashi and Hamada (1966, 1973) from the Langgon red beds of Malaysia. Only *Phacops accipitrinus* (Phillips) of Afghanistan (Durkoop and others, 1967; Pillet and Lapparent, 1969) is definitely Famennian in age.

The Late Devonian trilobites here described include 4 genera, 3 species, and 1 subspecies; that is, *Typhloproetus laevis* sp. nov., *Chaunoproetus daihuaensis* sp. nov., *Dianops guizhouensis* sp. nov., *Phacops (Omegops) accipitrinus mobilis* subsp. nov., and a phacopid, gen. et sp. indet.

STRATIGRAPHY

The trilobites occur in three localities: (1) Daihua Formation, Daihua of Changshun County, Guizhou Province; (2) Shangdamingshan (Upper Taminshan) Formation, Mianduhe of Hulun Buir Meng, Nei Mongol; and (3) Honggueleng Formation, Hoxtolgay of West Junggar, Xinjiang.

Daihua Formation

The Diahua Formation may be divided into two members: the lower member is composed of argillaceous banded limestone, 108.8 m thick, whereas the upper member is represented by gray dense limestone, 35.4 m thick. A brachiopod-rich horizon, characterized by *Eoperegrinella baschkirica* (Tschernychev), occurs at the base of the upper member, but abundant ammonoids and numerous trilobites are concentrated at the top. Ammonoids: *Wocklumeria sphaeroides* (Richter), *Parawocklumeria distorta* (Tietze), *P. paradoxa* (Wedekind), *Glatziella helenae* Renz, *Kosmoclymenia tabulata* Sun and Shen, *K. undulata* (Münster), *Cyrtoclymenia perinvoluta* Sun and Shen, *Cymaclymenia aurita* Sun and Shen, *C. parinvolvens* Ruan and Hu, *C. spiralia* Sun and Shen, *C. striata* (Münster), *Sporadoceras muensteri* (von Buch), *Imitoceras substriatum* (Münster). Trilobites: *Dianops guizhouensis* sp. nov., *Typhloproetus laevis* sp. nov., *Chaunoproetus daihuaensis* sp. nov. The Daihua Formation forms a continuous sedimentary sequence with the underlying Xiangshudong Formation (Frasnian), but a stratigraphic break is present between its top and the overlying Lower Carboniferous beds that yield *Pseuduralina.*

The trilobite-bearing horizon evidently corresponds to the Famennian *Wocklumeria*-Stufe (VI), for which a *Dianops-Typhloproetus* trilobite assemblage has been established. *Dianops* occurs in the Upper Famennian (IV–VI) all over the world, mainly in the *Clymenia*-Stufe and *Wocklumeria*-Stufe (V–VI). *Typhloproetus* and *Chaunoproetus* are also limited to the upper Famennian. When only trilobites are considered, it is safe to say that they are late Famennian in age; whereas, if the associated ammonoid fauna is also considered, a latest Famennian age may be assigned to the trilobites.

Shangdamingshan Formation

The lower part of the Shangdamingshan Formation is composed of grayish brown tuffaceous conglomerate; the middle part of purplish red, bioclastic limestone and tuff; the upper part of grayish green, andesinite-porphyrite lavas; the top part is incompletely exposed and more than 60 m thick. The purplish red bioclastic limestone yields many ammonoids associated with some trilobites and corals. They are named the *Cheiloceras-Sporadoceras* beds; in the lower part of the beds the fauna is represented by *Cheiloceras subpartitum* (Münster), *Sporadoceras biferum* (Phillips), *S. latilobatum* Schindewolf, *S. pompeckji* Wedekind, S. *rotundum* Wedekinds, and in the upper part by *Sporadoceras contiguum* (Münster), *S. muensteri* (Buch), *Platyclymenia (Platyclymenia) walcotti* Wedekind, *Tornoceras (Protornoceras) planidorsatum euryomphala* Wedekind, *T. (P.) planidorsatum planidorsatum* (Münster), *Sinotites* sp. Trilobites: Phacopid, gen. et sp. indet. Corals: *Nalivkinella* sp. According to the ammonoid fauna, the beds are considered to be early Late Devonian and correspond approximately to *Cheiloceras*-Stufe and *Platyclymenia*-Stufe (II-IV). The Shangdamingshan Formation and the underlying Xiadamingshan (Lower Taminshan) Formation (Frasnian) are comformable with each other.

Hongguleleng Formation

This formation, characterized by an alternation of marine and continental facies, consists of yellowish brown, purplish red, grayish green tuffaceous sandstone and conglomerate, siliceous siltstone, and a few limestone beds, which yielded abundant fossils such as corals (*Nalivkinella*), brachiopods (*Cyrtospirifer, Eoparaphorhynchus, Plicatifera*), trilobites (*Phacops (Omegops) accipitrinus mobilis* subsp. nov.), and plant remains (*Leptophloeum rhombicum* Dawson); in addtion, an ammonoid (*Cymaclymenia*) was discovered in the Hongguleleng Formation. *Phacops (Omegops) accipitrinus* is known from the uppermost Famennian of Germany, France, Belgium, England, Czechoslovakia, Soviet Union, Morocco, Afghanistan, and elsewhere; therefore, the Honggueleng Formation is of late Famennian age. This is supported by the associated brachiopods and the ammonoid. There is a sedimentary break or a conformity between the Hongguleleng Formation and Zhulumute Formation (Upper Devonian), but the former is continuous with the overlying Heishantou Formation (Lower Carboniferous).

PALEOECOLOGY AND PALEOBIOGEOGRAPHY

The Daihua trilobites are small with dismembered carapaces but are so well preserved that postmortem transportation must have been almost negligible. They occur in dense limestone, which reflects an environment of rather quiet and deeper water. The biota comprising only ammonoids and trilobites belongs, respectively, to the nekton and vagrant benthos. It is to be noted that the trilobites are all blind, because they were confined to the aphotic zone of rather deep water.

The Shangdamingshan trilobites which lived in a sedimentary environment of the active eugeosynclinal region have degenerated eyes. The biota is composed of abundant clymeniids, rare trilobites and corals. The trilobites may be taken as inhabitants of the elevated area in the deep-sea basin, where they lived, presumably, in the dysphotic zone.

Great numbers of Upper Devonian trilobites were found in West Junggar, but they are represented by only one subspecies, namely, *Phacops (Omegops) accipitrinus mobilis* subsp. nov., associated with many brachiopods and corals. Such an association indicates an environment of littoral sea, where oxygen was sufficiently rich to support various benthic biotas. *Phacops,* a trilobite with well-developed eyes, is adapted to an aerated photic habitat. *Cymaclymenia,* a nektonic form, may swim into the neritic zone.

During the latest Devonian, marine biotas in China may be differentiated into an ammonite-trilobite facies and a coral-brachiopod facies. The Daihua and Shangdamingshan trilobite faunas belong to the former, whereas in West Junggar, where trilobites and ammonoid coexist, the trilobites are benthic and belong to the coral-brachiopod biofacies. The Famennian ammonoid-trilobite facies is evidently marked by cosmopolitan biotas belonging to the same paleobiogeographical realm; hence the trilobite-bearing horizons may serve as means for worldwide correlation.

SYSTEMATIC DESCRIPTIONS

Genus *Typhloproetus* Richter, 1913
***Typhloproetus laevis* sp. nov.**
Plate 1(1–4)

Material: 2 cephala, 6 pygidia.

Holotype: Cephalon (T1257) figures in Plate (1).

Description: Cephalon semicircular, vaulted and distinctly sloping laterally. Glabella elongately bottle-shaped, extending forward beyond the anterior border furrow to reach the border; frontal part of glabella most highly vaulted. Glabellar furrows quite indistinct. Occipital furrow almost straight, narrow, and relatively deep. Occipital ring bent gently backwards with a small median tubercle. Dorsal furrow shallow, slowly tapering forward about halfway, then distinctly contracted inward, subparallel and more discernible in frontal half. Anterior border wide; anterior border furrow broad and distinct. Cheeks slightly convex, gently sloping laterally. No eyes. Posterior border narrow, markedly narrower than anterior and lateral borders.

Pygidium semi-elliptic, without markedly delimited border, broader than long. Axis conical, highly vaulted, tapering backward, passing posteriorly into a conspicuous postaxial ridge, 7 to 8 axial rings separated by sharp transverse furrows. Pleural region wide and moderately convex, distinctly declining toward the outside. Both pleural and interpleural furrows markedly visible, with 7 ribs. Border undifferentiated, showing no border furrow. Test smooth.

Measurement of the holotype: Cephalon 4.7 mm, long; glabella 3.5 mm long, 2.5 mm wide at base; frontal part of glabella 1.8 mm wide; occipital ring 0.6 mm wide.

Variability: Some minor differences exist whereby broad specimens show their pygidial axes to be rapidly tapering (T1258) and relatively slender representatives show otherwise (T1259).

Remarks: Both cephalon and pygidium agree in main features with the genus *Typhloproetus*. The new species is closely related to *T. subcarintiacus* Richter (1913) from the upper Famennian, but differs in having an entirely smooth glabella, which extends more anteriorly to the border and becomes slightly expanded; here the dorsal furrow is more distinct. This new species may be compared with *T. sinensis* Chang (1955) from the Upper Devonian of Guangxi, but it differs in that it displays a slender bottle-shaped glabella without a preglabellar field, the presence of a pronounced postaxial ridge, 7 to 8 axial rings, and ribs in the pygidium. *T. sinensis* has a flat and narrow preglabellar field in the cephalon, a broad border, and a pygidium with fewer segments.

Horizon and locality: Daihua Formation, Upper Devonian, about 1.5 km east of Daihua, Changshun, Guizhou (Field No. 33).

Genus *Chaunoproetus* R. and E. Richter, 1919
***Chaunoproetus daihuanensis* sp. nov.**
Plate 1(5–11)

Material: 4 cranidia, 5 pygidia.
Holotype: Cranidium (T1263) figures in Plate 1(7).

Description: Cranidium subtrapezoidal with rounded anterior margin. Glabella convex, stout, widely tapering gradually forward and reaching almost the anterior border furrow; glabellar furrows absent. Occipital furrow narrow, sharply incised, slightly bent forward at the middle. Occipital ring broader in the median part with a minute tubercle. Dorsal furrows well defined, confining the glabella. Preglabellar field very narrow, almost absent. Anterior border slightly elevated, border furrow incised. Fixed cheeks narrow. Eyes lacking. Facial sutures subparallel with dorsal furrows. Posterior border furrow pronounced, widening sharply laterally. Posterior border relatively narrow.

Pygidium semi-elliptic or suboval in outline with pronounced relief. Axis short, strongly convex, tapering backward and quickly terminated. Axial transverse furrows appear as shallow ill-defined grooves. Postaxial region very wide. Pleural region smooth with segmentation only in frontal part. Lateral and posterior regions down-curved steeply to margin. Border furrow absent or very weak. Doublure broad.

Test smooth, showing only a shallow parallel line in anterior border of cranidium (T1261).

Holotype is a cranidium, with a length of 4.8 mm, width of 4.4 mm in the middle; a glabella with a length of 3.1 mm, width of 2.9 mm at base, and an occipital ring with a width of 0.9 mm

Variability: In general, the three cranidia look very similar among themselves, but the pygidia are somewhat different in outline; nevertheless, such differences are regarded as varitions within the same species. In one specimen (T1264) the pygidium is suboval, its anterior margin arched forward with strongly rounded anterolateral corners, which curve rapidly backward. In other pygidia the pleural region is comparatively wide with rounded corners, forming a semielliptic outline.

Remarks: This form is assigned to the genus *(Chaunoproetus* on account of its stout glabella, very narrow preglabellar field, absence of eyes, and particularly the short but highly vaulted axis, wide postaxial region, border coalescing with sloping pleural region, and broad doublure in the pygidium. The new species differs from *Chaunoproetus palensis* Richter in having a comparatively narrow and entirely smooth glabella, characteristic preglabellar field, and slightly elevated anterior border, in addition to the general outline and the more faintly segmented pygidium.

Horizon and locality: Same as the preceding.

Genus *Dianops* R. and E. Richter, 1923
***Dianops quizhouensis* sp. nov.**
Plate 1(12–15)

Material: 6 cephala.
Holotype: Cephalon (T1268) figures in Plate 1(12).

Description: Cephalon nearly semicircular in outline. Glabella strongly convex, elevated above cheeks, with widely rounded anterolateral corners. First pair of glabellar furrows (preoccipital furrow) distinct, uninterrupted throughout; second pair of glabellar furrows very faint, short and shallow, close to first glabellar pair, but separate from dorsal furrow; third pair of glabellar furrows divided into two sections—posterior and anterior rami, also slightly impressed; both second and third glabellar furrows almost indiscernible externally. Intercalating ring narrow and distinct, contracting medially, delimited anteriorly by the pronounced occipital furrow. Occipital furrow sharp, slightly bent forward; occipital ring relatively broad, widened medially, devoid of occipital tubercle or node. Dorsal furrows deep, distinctively divergent. Eyes absent. Cheeks convex, relatively narrow, declined laterally and limited by pronounced dorsal and border furrows. Border furrows run continuously along lateral and posterior margins of cheeks, forming a rounded genal angle. Lateral and posterior borders narrow and uniform.

Surface of cephalon sculptured with very fine and dense granulations.

In the holotype, the glabella including intercalating ring has a length of 6.6 mm width of glabella 3.5 mm at base, maximum width of glabella 7.0 mm, maximum width of cheek 3.5 mm.

Remarks: The new species is characterized by a broad glabella, continuous and complete intercalating ring as well as very fine granulations. It resembles *D. typhlops* Gürich from the Famennian of Europe, but is distinguished from the latter mainly by a continuous first glabellar furrow, which in *D. typhlops* is unconnected medially, so that the frontomedial lobe of the glabella coalesces with the central part of the intercalating ring. Our species is also closely allied to *D. vicarius* Chlupáč (1966) from the Famennian Krtiny limestone, Czechoslovakia, as the first glabellar furrow of both species is open throughout; however, in *D. vicarius* the cheek is wide, the posterolateral border furrow at the genal angle is faint, in the new species the cheek is comparatively narrow and the border furrow is open throughout and visible.

Horizon and locality: Same as the preceding.

Genus *Phacops* Emmrich, 1839
Subgenus *Phacops (Omegops)* Struve, 1976
Plate 2(1–10)

Material: 3 complete carapaces, 4 thoraces and attached pygidia, 4 cephala, 22 pygidia, 1 hypostome.

Holotype: Nearly complete carapace (T1271) figures in Plate 2(1).

Description: Carapace elliptical. Cephalon semicircular in outline without conspicuous genal spines. Glabella expanding forward, subtrapezoid, slightly curved in front. First pair of glabellar furrows deep laterally, broadening and shallowing toward the inside; anterior glabellar furrows weak or absent. Intercalating ring narrow. Occipital ring broader than intercalating ring. Dorsal furrows deep and prominent. Eyes large, composed of many lenses, near to glabella. Cheeks narrow, with bluntly rounded genal angle, where the width of border is at its maximum, narrowing toward front. Vincular furrow well developed.

Thorax composed of 11 segments. Axis slender, articulating half-ring fairly developed. Pleural region wide, inner half elevated, declining outwardly. Pleural furrows shallow. Articulating facets well developed.

Pygidium short, semi-elliptical. Axis conical, tapering quickly backward, to reach almost the margin, its 7 to 8 rings subdivided by conspicuous transverse furrows. Lateral region weakly vaulted, with 5 distinctly visible ribs marked out by broad pleural furrows. Border undifferentiated, without border furrow.

Sculpture present on glabella only, consists of pronounced and coarse tubercles.

Hypostome large, trigonal in outline with anterior wings extending transversely and their ends pointed; border furrow undifferentiated, posterior margin pointed with three denticles within a short distance.

Measurement of the holotype: length of cranidium 11 mm, length of thorax 19.5 mm, width of thorax 20 mm, length of pygidium 7.2 mm, frontal width of pygidium 15 mm, frontal width of axis 4 mm.

Remarks: Since the species *Calymene accipitrina* was erected by Phillips (1841), it was described and discussed in detail by Richter and Richter (1933). Chlupáč (1966) provided supplemental information about its hypostome in specimens from the Moravian Karst. Pillet and Lapparent (1969) examined trilobites from Afghanistan, among them many specimens of this species, including complete carapaces, so they obtained the first knowledge of thorax. They did not mention the number of thoracic segments in the text, but according to a figure (Pl. 39, fig. 9), the thorax is composed of 10 segments. Recently, Struve (1976) proposed the subgenus *Omegops* using *Calymene accipitrina* as its type species.

Among the numerous Upper Devonian trilobites collected from the West Junggar region, there are many dissociated cranida and pygidia, complete carapaces, and enrolled individuals. Being spheroidally enrolled in the typical phacopid fashion, the posterior margin of its pygidium fits into a vincular furrow on the cephalic doublure, which runs parallel and close to the cephalic margin. We base the new subspecies on these characteristics.

Phcacops (Omegops) accipitrinus mobilis subsp. nov. agrees with *P. (O.) accipitrinus (accipitrinus)* in main features, but the former bears three denticles on its trigonal hypostome, its thorax is composed of 11 segments instead of 10, the pygidial pleura are divided by 5 instead of 6 ribs. This characteristic of pygidium resembles that of Pillet and Lapparent's new, unnamed subspecies from the Famennian of Afghanistan. In the Afghanistan subspecies, however, the cephalon and hypostome are unkown, its thoracic pleura are narrower, and numerous prominent granulations cover both the thorax and pygidium.

Horizon and locality: Hongguleleng Formation, Upper Devonian, Hongguleleng (Field Nos. V-1 V-1–V-9, 303–6, 6G–1, 304–1) and Bulong (Field No. 306–3), Hoxtolgay in West Junggar, Xinjiang.

Phacopid, gen. et sp. indet.
Plate 2(11)

Description: This form is represented only by a nearly complete cephalon and a few fragments, but it bears distinct characteristics. Cephalon semicircular, moderately convex. Glabella large and vaulted, subhexagonal, reaching the anterior margin; maximum breadth of glabella almost equal to length of the same including occipital ring. First glabellar furrow marked and deep, opening throughout; other glabellar furrows absent. Intercalating ring prominent, depressed medially but convex laterally. Occipital furrow broad and deep on both sides, narrowing and shallowing abruptly into the center, and bending forward. Occipital ring very wide and strong. Dorsal furrows distinct, rapidly diverging forward for a distance of over two-thirds of their length, subparallel reaching the anterior border, then turning back to converge and connect together. Cheeks relatively narrow and gently convex. Eyes small, degenerated, composed of lenses, placed at the anterolateral parts of cephalon, and situated close to the anterolateral corners of glabella and border. Border narrow. Lateral border furrow very shallow and weakly impressed, but posterior border furrow strongly incised. Test smooth.

Remarks: These specimens are characterized by subhexagonal glabella, intercalating occipital rings, size and position of eyes, by which they can be distinguished from other representatives of the phacopids. Phylogenetically, our form is possible intermediate between *Cryphops* and *Trimerocephalus* along the phacopid evolutionary line. The above-mentioned features suggest a new genus or subgenus. However, the erection of a new taxon must wait until sufficient better-preserved material are available.

Horizon and locality: Shangdamingshan Formation, Upper Devonian, Damingshan, Baiku, Mianduhe of Hulun Buir Meng, Nei Mongol.

ACKNOWLEDGMENTS

I express my sincere thanks to Professor Yang Zun-yi for his critical reading of the manuscript.

The specimens described in this paper are in the Institute of Geology, Chinese Academy of Geological Sciences, Beijing.

SELECTED REFERENCES

Alberti, H., 1974, Neue Trilobiten (Chaunoproetiden, Mirabolen) aus dem Ober-Devon IV–VI (Nord-Afrika und Mittel-Europa), Beitrag 2: Neues Jahrbuch für Geologie und Paläontologie Abhandlungen, Band 146, Heft 2, P. 221–261, fig. 6–10.

Chang An Chi [Chang An-zhi], 1958, Stratigraphy, palaeontology, and palaeogeography of the ammonite fauna of the Clymeneenkalk from Great Khingan with special reference to the post-Devonian break (hiatus) of South China: Acta Palaeontologica Sinica, v. 6, no. 1, p. 71–89 (in Chinese and English).

——1960, New late Upper Devonian ammonite-faunas of the Great Khingan and its biological classification: Acta Palaeontologica, Sinica, v. 8, no. 2 p. 180–193, Pl. 1 (in Chinese and English).

Chang Wen Tang [Zhang Wen-tang], 1955. Note on a new proetid from Upper Devonian of Kwangsi: Acta Palaeontologica Sinica, v. 3, no. 3, p. 189–192, Pl. 1 (in Chinese and English).

Chlupáč, I., 1966, The Upper Devonian and Lower Carboniferous trilobites of the Moravian Karst: *Sbornik Geologický Ved, Palaeontologie,* 7, 143 p., 24 pls.

Durkoop, A., Mensink, H., and Plovdowski, G., 1967, Central and Western Afghanistan and Southern Iran, *in* Oswald, H. F., ed., Proceedings of the International Symposium on the Devonian System, Volume 1: Calgary, Alberta, p. 529–544.

Kobayashi, T., and Hamada, T., 1966, A Devonian proetid from Malaysia (Malaya): Japanese Journal of Geology and Geography, v. 37, nos. 2–4, p. 87–92, Pl. 2

——1973, Cyrtosymbolids (Trilobita) from the Langgon red beds in northwest Malaya, Malaysia: Geology and Palaeontology of Southeast Asia, v. 12, p. 1–28, Pls. 1–3.

——1977, Devonian trilobites of Japan: Palaeontological Society of Japan, Special Paper No. 20, 202 p., 13 pls.

Maximova, Z. A., 1955, Trilobity srednego i verkhnego Devona Urala i severnykh Mugodzhar: Trudy VSEGEI, v. 3, 263 p., 18 pls.

Phillips, J., 1841, Figures and descriptions of the Palaeozoic fossils of Cornwall, Devon, and West Somerset: 231 p., 60 pl.

Pillet, J., and Lapparent, A. F., 1969, Description des trilobites Ordoviciens, Siluriens et Dévoniens d'Afghanistan: Annales de la Société Géologique du Nord, v. 89, no. 4, p. 323–333, Pls. 34–39.

Richter, R., 1913, Beiträge zur Kenntnis devonischer Trilobiten. Oberdevonische Proetiden: Abhandlungen der Senckenbergischen Naturforschenden Gesellschaft, 31, no. 4, p. 341–424, Pls. 22–23.

Richter, R., and Richter, E., 1926, Die Trilobiten des Oberdevons. Beiträge zur Kenntnis devonischer Trilobiten, IV: Abhandlungen der Preussischen Geologischen Landesanstalt, N. F. 99, 313 p., 12 pls.

——1933, Die letzten Phacopidae: Bulletin du Muséee Royal d'Histoire Naturelle de Belgique, v. 9, no. 21, p. 1–19, Pls. 1–2.

Struve, W., 1976, *Phacops (Omegos)* n. sg. (Trilobita, Ober-Devon): Senckenbergiana Lethaea, v. 56, no. 6, p. 429–452, Pls. 1–2.

Sun Yun Chu [Sun Yun-zhu], 1937, Two new species of Devonian trilobites from Hunan: Bulletin of the Geological Society of China, v. 17, p. 349–354, Pl. 1.

Sun Yun Chu [Sun Yun-zhu] and Shen Yao Ting [Shen Yao-ting], 1965 [On the late Upper Devonian ammonite fauna of the *Wocklumeria* beds of South Kweichow and its stratigraphical significance]: Beijing, Geological Publishing House, Professional Papers of Academy of Geological Sciences, sec. B, no. 1, p. 33–109, Pls. 1–5 (in Chinese).

MANUSCRIPT ACCEPTED BY THE SOCIETY APRIL 22, 1981

PLATE 1

Figure

1–4.	***Typhloproetus laevis*** **sp. nov.**
1.	**Cephalon, ×6, holotype (T1257).**
2–4.	**Pygidia, ×8, ×6, ×6 (T1258, T1259, T1260).**
5–11.	***Chaunoproetus diahuaensis*** **sp. nov.**
5–6.	**Cranidia, ×7 (T1261, T1262).**
7.	**Cranidium, ×4.5, holotype (T1263).**
8–11.	**Pygidia, ×5.5, ×4.5, ×5, ×5.5 (T1264, T1265, T1266, T1267).**
12–15.	***Dianops guizhouensis*** **sp. nov.**
12.	**Cephalon, ×3.5, holotype (T1268).**
13–14.	**Cephala, ×5, ×5.5 (T1269, T1270).**
15.	**Cephalon, associated with the pygidium of *Typhloproetus laevis* (fig. 2), ×5, (T1258).**

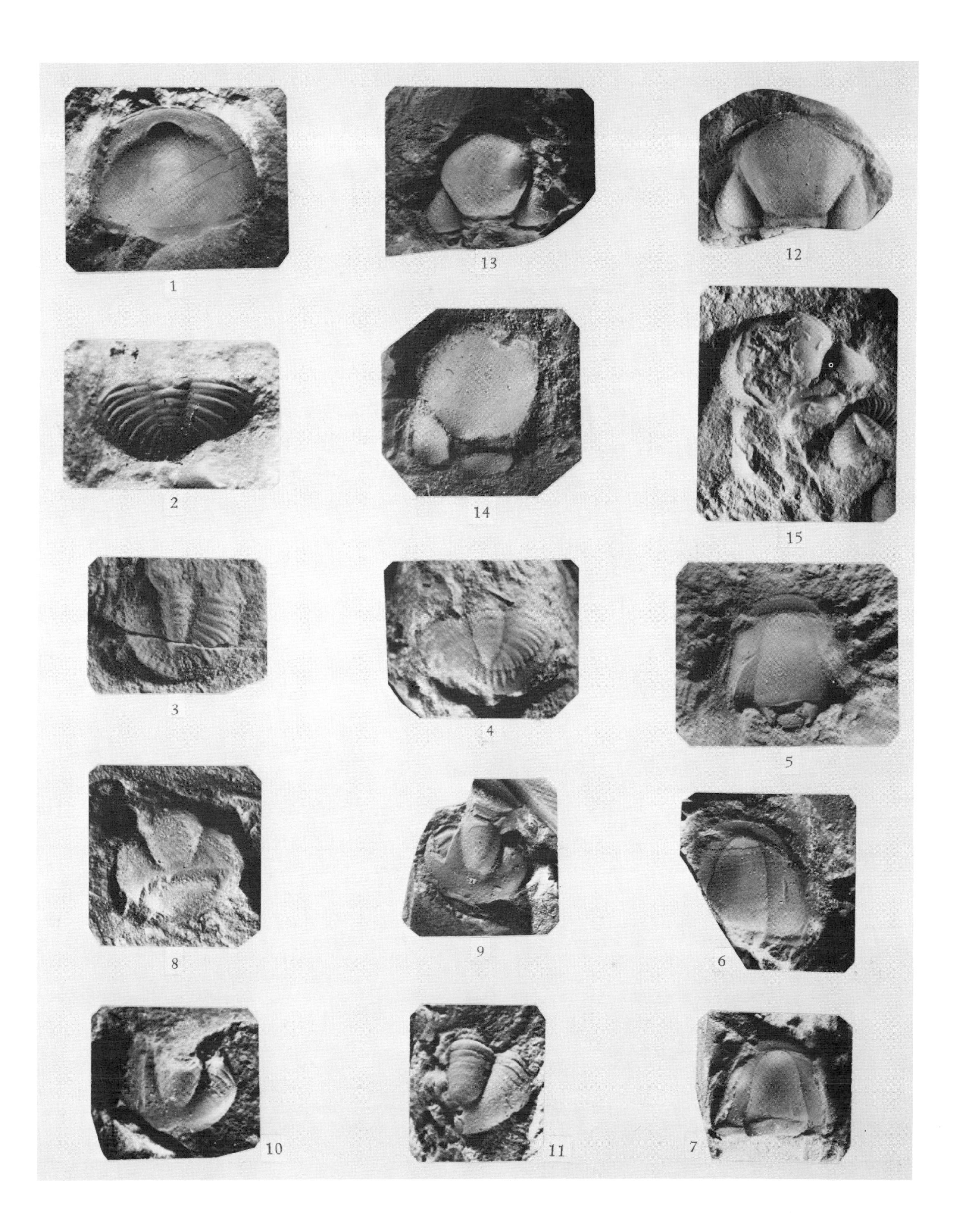
1
13
12
2
14
15
3
4
5
8
9
6
10
11
7

PLATE 2

Figure

1–10.	***Phacops (Omegops) accipitrinus mobilis*** **subsp. nov.**
1.	**Nearly complete carapace, ×1.5, holotype (Field No. V–7, Cat. No. T1271).**
2.	**Thorax, pygidium and hypostome, ×2 (Field No. V–7, Cat. No. T1272).**
3.	**An enrolled individual, ×2.5. 4a, cephalon; 4b, pygidium (Field No. V–3, Cat. No. T1273).**
4–6.	**Cephala, ×2.5, ×3.5, ×2.5 (Field Nos. 306–3, 303–6, 306–3, Cat. Nos. 1274, T1275, T1276).**
7–9.	**Pygidia and attached thoracic segments, ×3, ×2, ×2 (Field Nos. V–9, 304–1, 6G–1, Cat. Nos. T1277, T1278, T1279).**
10.	**Pygidium, ×2.5 (Field No. V–5, Cat. No. 1280).**
11.	**Phacopid, gen. et sp. indet. Nearly complete cephalon, ×4 (T1281).**

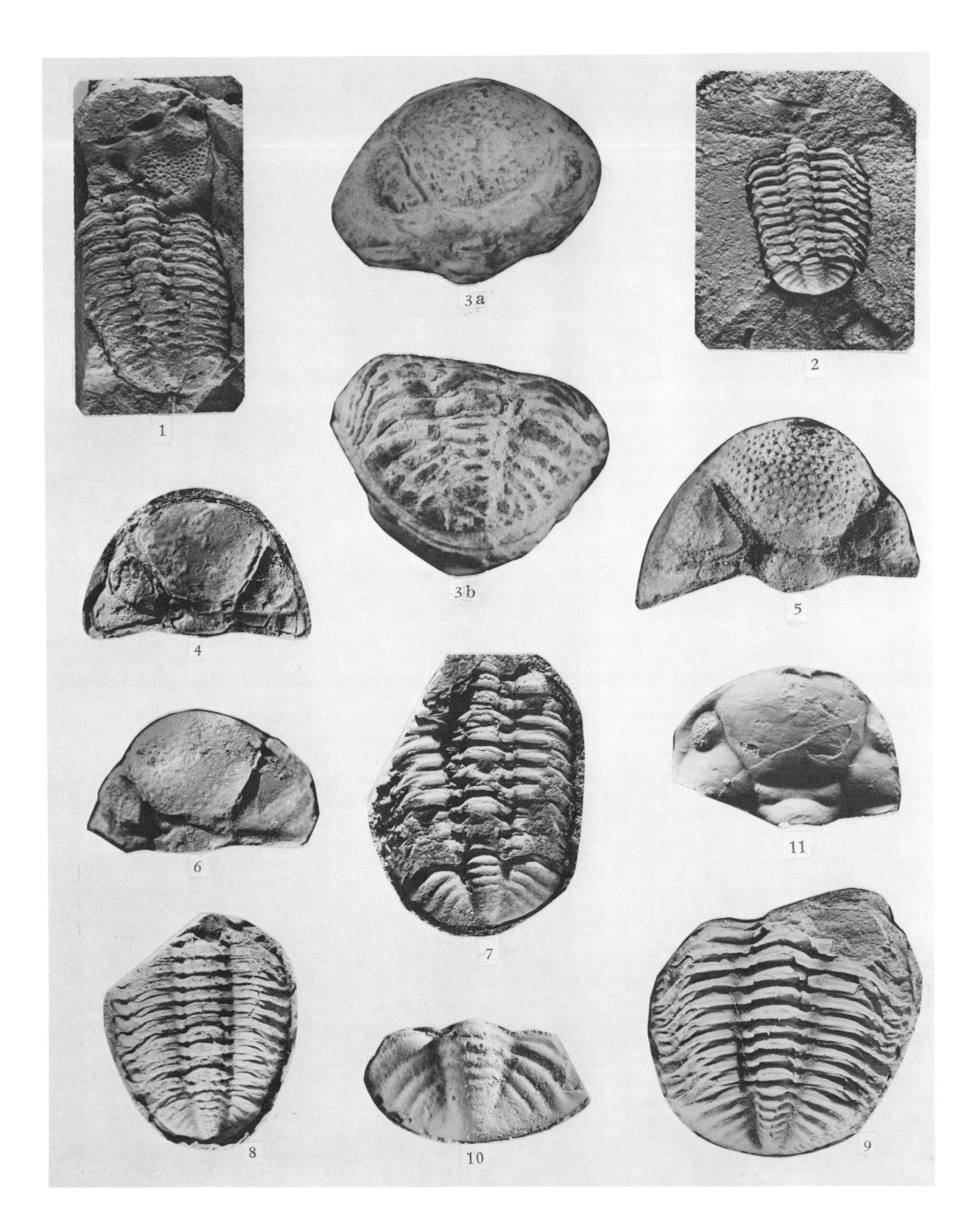
1
3a
2
3b
5
4
11
6
7
8
10
9

Printed in U.S.A.

Geological Society of America
Special Paper 187
1981

Paleogene conchostracan faunas of China

Chen Pei-ji
Shen Yan-bin
Nanjing Institute of Geology and Palaeontology, Academia Sinica, Nanjing 210008, People's Republic of China

ABSTRACT

Tertiary conchostracan fossils are rare in the world. In this paper the *Perilimnadia* fauna, *Fushunograpta changzhouensis* fauna, and *Paraleptestheria menglaensis* fauna from the Paleogene of China are discussed; 11 species (in 6 genera) are described systematically.

INTRODUCTION

It is generally known that the Conchostraca made their first appearance in the Devonian and reached the acme twice during the late Paleozoic and Mesozoic. They declined in the Cenozoic Era. Up to the present, more than 200 genera of fossil conchostracans have been reported, but only 14 living genera are known. Because no fossil conchostracans were known from the Tertiary for a long time, Kobayashi (1972) held that a discontinuity exists between fossil and living conchostracans.

Recently, valuable collections have been obtained from many localities in China. Based on the available material, the sequence of the Tertiary conchostracan faunas is tentatively established in descending order as follows:

Paleogene
Eocene
Paraleptestheria menglaensis fauna
Fushunograpta changzhouensis fauna
Paleocene
Perilimnadia fauna

The *Perilimnadia* fauna occurs in the second member of the Funing Formation of Jiangsu Province as well as in the Chuxian District of Anhui Province. *Perilimnadia,* with a large protoconch and a strong adductor muscle scar, includes *P. jiangsuensis* Chen, *P. gaoyouensis* Chen, *P. lingtangqiaoensis* Chen, and *P. taizhouensis* Chen. The first three species were obtained from a drill core in Gaoyou District of Jiangsu Province, the fourth species, together with a few Fushunograptidae (Xia Shu-fang and other, 1979), was collected from a drill core in Taizhou District (Chen Pei-ji, 1975) and from an outcrop in Donggou of the Liuhe District of Jiangsu Province.

Yunmenglimnadia is very similar to *Perilimnadia* in the protoconch, but its adductor muscle scar is surrounded by the characteristic shell gland. This genus was ascribed to the *Perilimnadia* fauna (Zhang Wen-tang and others, 1976; Chen Pei-ji and Shen Yan-bin, 1979). It was first obtained from the Jinshan and Yingchen Districts of Hubei Province and includes three species: *Y. hubeiensis* Chen, *Y. rhombica* Chen, and *Y. yingchengensis* Chen. According to the collector Sun Li-sen, the estheriid-bearing bed may belong to the fourth member of the late Late Cretaceous Gonganzhai Formation. Later, *Y. gansuensis* Chen and *Y. qinanensis* Chen were also found in the Qinan District of Gansu Province together with numerous Tertiary ostracodes, as has also been recorded in the above-mentioned localities (Chen Pei-ji, 1975). It is uncertain whether *Yunmenglimnadia* belongs to the *Perilimnadia* fauna, because the two genera have not yet been found together at the same locality.

The *Fushunograpta changzhouensis* fauna includes species of *Fushunograpta* and *Cixiella,* both of which are characterized by having slender and curvate radial striae on the growth bands. In this fauna, the index-species is *Fushunograpta changzhouensis* (Chen). *Fushunograpta ovata* Wang, *F. subcyclaria* Wang, and *F. brachysellipsa* Wang occur in green calcareous mudstone of the Guchengzi Formation in the Fushun coal field, Northeast China (Hong You-chong and others, 1974). *Fushunograpta changzhouensis* (Chen) was found in the fourth member of the Funing Formation or corresponding horizons in many localities of Jiangsu and Anhui Provinces (Chen Pei-ji, 1975). *Fushunograpta changheensis* Chen and Shen, *F. changzhouensis* (Chen), and *Cixiella serrula* Chen and Shen are known to occur from the middle part of the Changhe Formation in the Cixi District near the Gulf of Hangzhou (Shen Yan-bin and Chen Pei-ji, 1979). In addition, *Fushunograpta* is known from the Paleogene deposits in Gansu and Henan Provinces. On the whole, this fauna is widespread geographically, but confined to a certain horizon, thus being a useful indicator for stratigraphic division and correlation.

The Fushunograptidae, characterized by bearing various lirae on the growth bands of the carapace, first appeared in the Permian but gradually declined in the early and middle Mesozoic. They rose again in the Cretaceous, abounding both in number and variety and branching out into some new family stocks. Subsequently, it rapidly fell away in the Tertiary with only one genus, *Fushunograpta,* remaining. Finally, this family became extinct.

The forms of *Altanestheria* and some others from the Nemegetin basin of Mongolia (Trussova and Badamgarav, 1976) are thought to be *Fushunograpta* by the characteristics of their carapaces and sculptures.

The *Paraleptestheria menglaensis* fauna is very widely distributed in the Tertiary deposits from Guandong in South China to Heilongjiang in the north. Except for the index-species, the fauna includes *Paraleptestheria lanpingensis* Chen, *P. mohanensis* Chen, *P. triangularis* Chen, *P. yunlongensis* Chen, *P. jintanensis* (Chen), *P. aquilonaria* Zhang and Chen, *P. baoyuensis* Shen and X. Zhang and *Nanhaiestheria sanshuiensis* Shen and X. Zhang.

In the Loxomegaglyptidae, the Paleozoic and Mesozoic forms bear on the growth bands a large polygonal reticulate sculpture, which is composed of more regular meshes. The Tertiary *Paraleptestheria menglaensis* carries on the growth bands in the anterior and middle parts of the valve a large, transversely elongated, polygonal, reticulate sculpture, which is closely similar to that in living forms of Leptestheriidae. It seems most probable that some forms of this living family with such a sculpture on the valve have arisen from Loxmogaglyptidae.

The genus *Paraleptestheria* was found in the Guolang Formation of western and southern Yunnan, the Nongshan Formation of the Lofochai Group in the Nanxiong basin of the northern Guangdong, the Xiawanpu Formation in the Xiangxiang District of Hunan, the Dainan Formation of Jiangsu, the upper part of the Changhe Formation in the Cixi District of Zhejiang, the third member of the Dinyuan Formation in the Hefei basin of Anhui, the Minshui Formation of Heilongjiang, and the Paleogene beds in Dapinggou near Zhangyi in Ningxia.

In addition, *Paraleptestheria* is found in association with the genus *Nanhaiestheria* and a few forms of *Fushunograpta* in the Buxing Formation in the Sanshui basin of Guangdong. *Paraleptestheria* often occurs in red beds, in which gypsum or rock salt are common, but in Heilongjiang it was found in gray mudstone. This indicates that the animals could live in saline or brackish water. This fauna is considered to be of Eocene age, whereas the *Perilimnadia* fauna belongs to the Paleocene (Chen Pei-ji and Shen Yanbin, 1980).

In China and Mongolia, the *Fushunograpta changzhouensis* fauna is found together with fossil vertebrates, ostracodes, and charophytes, but the age of the fossil-bearing horizon is still in dispute. The writers are inclined to regard it as early Eocene (Zheng Jia-jian and Qiu Zhan-xian, 1979).

Neogene conchostracans have not yet been reported from any part of the world.

SYSTEMATIC DESCRIPTIONS

Order CONCHOSTRACA Sars, 1867
Suborder ESTHERITINA Kobayashi, 1972
Superfamily VERTEXOIDEA Kobayashi, 1954
Family PERILIMNADIIDAE Chang and Chen, 1975

Protoconch large, carrying an adductor muscle scar or shell gland; growth bands few in number, not recurved at terminal part nor protruded into serration on dorsal margin; late Permian–Paleogene.

Genus *Perilimnadia* Chen, 1975
Type species: *Perilimnadia jiangsuensis* Chen

Diagnosis: Carapace elliptical or subovate in outline; growth bands few; protoconch large with strong subcircular adductor muscle scar near the antero-dorsal angle.

Comparison: Many Paleozoic forms of Vertexiidae such as *Cornia, Pemphicyclus, Bipemphigus, Tripemphigus, Curvacornutus,* and *Gabonestheria* bear various processes, spines, or tubercles on the protoconch instead of adductor muscle scars. The genus *Perilimnadia* shows no shell gland on the large protoconch, but has an adductor muscle scar similar to that in modern *Limnadia.* Therefore, the former genus may be an extinct form, probably somehow related with the second one.

Distribution: Jiangsu, Anhui; Paleocene.

***Perilimnadia jiangsuensis* Chen**
Plate 1(1)

1975 *Perilimnadia jiangsuensis* Chen, *Scientia Sinica,* no. 6, p. 623, Pl. 1, figs. 3, 5b
1976 *Perilimnadia jiangsuensis* Chen, Fossil Conchostraca of China, p. 106–107, Pl. 9, figs. 5–7

Description: Carapace moderate in size; anterior margin somewhat protruded in upper part, becoming straight near ventral margin; ventral margin nearly rounded; posterior end of valve strongly contracted; dorsal margin long, slightly arched upward; protoconch large, about two-thirds to three-fourths the size of whole valve, with strong subcircular adductor muscle scar in antero-dorsal part; growth band few in number, only four distinct ones can be counted.

Discussion: This species is found along with *P. gaoyouensis* and *P. lingtangqiaoensis,* but the sculpture of the growth band could not be observed because of poor state of preservation.

Horizon and locality: Lingtangqiao, Gaoyou, Jiangsu; second member of Funing Formation; Paloecene.

Cat. No. 22912-3.

***Perilimnadia gaoyouensis* Chen**
Plate 1(2)

1975 *Perilimnadia gaoyouensis* Chen, *Scientia Sinica,* no. 6, p. 623, Pl. 1, figs. 2, 5c
1976 *Perilimnadia gaoyouensis* Chen, Fossil Conchostraca of China, p. 107, Pl. 10, fig. 9c

This species is found in association with *P. jiangsuensis,* and distinguished by the lanceolate-elliptical shape and the less contracted posterior end of valve.

Horizon and locality: Same as the preceding.

Cat. No. 22912-5.

***Perilimnadia lingtangqiaoensis* Chen**
Plate 1(3)

1975 *Perilimnadia lingtangqiaoensis* Chen, *Scientia Sinica,* no. 6, p. 623, Pl. 1, figs. 1, 5a
1976 *Perilimnadia lingtangqiaoensis* Chen, Fossil Conchostraca of China, p. 107, Pl. 10, fig. 8

This species is also found along with *P. jiangsuensis* and *P.*

gaoyouensis, but it differs from these two forms in the dorsal margin, which is strongly arched upward in the middle part.

Horizon and locality: Same as the preceding.

Cat. Nos. 22912-6, 22913.

Genus *Yunmenglimnadia* Chen, 1975
Type species: *Yunenglimnadia hubeiensis* Chen

Diagnosis: Carapace elliptical or ovate in outline; growth lines narrowly sulcate; growth bands few; protoconch large, with large and vertically elongate adductor muscle scar and shell glands; both growth bands and protoconch with shallow and crowded honeycomb sculpture.

Comparison: This genus is more closely allied to modern forms of Limnadiidae than *Perilimnadia,* because it not only bears an adductor muscle scar, but also carries loops of shell glands enclosing it.

Distribution: Hubei and Gansu; Paleogene?

***Yunmenglimnadia hubeiensis* Chen**
Plate 1(4–5)

1975 *Yunmenglimnadia hubeiensis* Chen, *Scientia Sinica,* no. 6, p. 624, Pl. 1, figs. 10–11; Pl. 2, fig. 10

1976 *Yunmenglimnadia hubeiensis* Chen, Fossil Conchostraca of China, p. 108, Pl. 9, fig. 8; Pl. figs. 2–5.

Description: Carapace moderate in size, subelliptical in outline, 6.6 to 7 mm in length, 5 mm in height; dorsal margin long and straight; anterior margin straight, both ventral margin and posterior margin nearly rounded; anterior height greater than posterior height; protoconch very large, about eight-tenths to nine-tenths of the entire valve, with large, feeble, vertically elongated adductor muscle scar and loops of fine shell gland in anterior part; shell glands elliptical in shape, more or less slightly contracted in middle, 2 growth bands, broader and flat, located near the ventral margin; both protoconch and growth bands with shallow and crowded honeycomb sculptue; crowded pustulate structures present on external mold of valve.

Discussion: In this species, the indistinct end sac and the faint and fine shell gland are probably due to poor state of preservation. This species is found to be infested by ostracodes (*Eucypris subtriangularis*) on the same specimen.

Horizon and locality: Yingcheng, Hubei; Gaoyan Formation of Yingcheng Group?, Paleogene?

Cat. Nos. 22918-22920.

***Yunmenglimnadia gansuensis* Chen**
Plate 1(6)

1975 *Yunmenglimnadia gansuensis* Chen, *Scientia Sinica,* no. 6, p. 628, Pl. 2, fig. 10

1976 *Yunmenglimnadia gansuensis* Chen, Fossil Conchostraca of China, p. 109, Pl. 12, figs. 1–2

Description: Carapace moderate in size, subelliptical in outline; 5 mm high; anterior margin nearly straight, both posterior and ventral margins rounded; anterior height greater than posterior height; protoconch large, elliptical in shape, about nine-tenths of the whole valve, carrying a thick and large adductor muscle scar with about 4 shell glands enclosing anterior part; shell gland vertically elongated and slightly contracted in middle part; 3 distinct growth bands visible near ventral margin; both protoconch and growth bands bear shallow and crowded honeycomb sculpture.

Comparison: This species closely resembles *Y. hubeiensis* in size, outline of the valve, and shell gland, but differs in the more distinct adductor muscle scar and the greater number of growth bands.

Horizon and locality: Dasi of Wangfu Commune, Qinan, Gansu; Paleogene.

Cat. No. 22924.

Superfamily LIOESTHERIOIDEA Raymond, 1946
Family LOXOMEGAGLYPTIDAE Novojilov, 1958
Genus *Paraleptestheria* Chen, 1976
Type species: *Paraleptestheria menglaensis* Chen

Diagnosis: Carapace moderate in size; elliptical, oval, or triangular in outline; growth lines stout, with median groove; interspace broad and flattened, ornamented with large horizontal polygons that gradually weaken toward posterior part of carapace.

Distribution: Yunnan, Guangdong, Zhejiang, Jiangsu, Anhui, Hunan, Ningxia, Henan, and Heilongjiang; early Eocene.

***Paraleptestheria menglaensis* Chen**
Plate 2(7, 8)

1976 *Paraleptestheria menglaensis* Chen, Fossil Conchostraca of China, p. 149, Pl. 36, figs. 6–9; Pl. 37, figs. 1–8; Pl. 39, fig. 11

1977 *Paraleptestheria menglaensis* Chen, Mesozoic Fossils from Yunnan, Fasc. 2, p. 348, Pl. 4, figs. 1–5.

1979 *Paraleptestheria menglaensis* Chen, Mesozoic and Cenozoic red beds of South China, p. 296–297, Pl. 1, figs. 1–6; p. 301, Pl. 1, figs. 1–11; Pl. 2, figs. 10–11

1980 *Paraleptestheria menglaensis* Chen, *Acta Palaeontologica Sinica, 19, no. 3, Pl. 1, figs. 1–8; Pl. 2, figs. 1–8.*

Description: Carapace moderate in size, oval in outline; dorsal margin long and straight; umbo small, anterior; anterior margin broadly rounded, posteror margin contracted; anterior height longer than posterior height; concentric ridges stout, with median groove; 10 to 16 broad and even growth bands, ornamented with large horizontal polygons that gradually weaken to posterior part of carapace.

Horizon and locality: Near Mengla, Yunnan; Paleogene; Guolang Formation. Hekou, Sanshui, Guangdong; Ecocene; Buxin Formation.

Cat. Nos. 26971–26979, 29870, 29872–29874, 44514–44515, 44527–44533, 44538, 49651–49661.

Family FUSHUNOGRAPTIDAE Wang, 1974

Carapace usually small, circular, elliptical, ovate, rhomboid, triangular, or subquadrate in outline; growth bands show simple or complicated striated sculpture; striae coarse or fine, straight or curvate, regularly or unevenly arranged, sometimes bifurcated upward and downward, or repeatedly bifurcated into a dendroitic sculpture; fine lines like cross bars occasionally present between striae; late Permian–Paleogene.

Genus *Fushunograpta* Wang, 1974

1975 *Curvestheria* Chen, *Scientia Sinica,* no. 6, p. 629

Type species: *Fushunograpta ovata* Wang

Diagnosis: Carapace, subcircular in outline, many growth bands with slender curvate, striated sculpture; striae closely spaced, occasionally bifurcated upward and downward; fine cross bars usually visible between the striae.

Comparison: This genus is very similar in sculpture and shape of valve to *Eremograpta,* recorded from the People's Republic of Mongolia, but the latter is distinguished by a striated sculpture which is composed of comparatively straight striae without fine cross bars in between as shown in the drawing attached in the publication.

The present genus also differs from *Nemestheria* in that the latter has unevenly arranged striae. However, if somewhat resembles Cretaceous *Rhombograpta* and *Deltostracus* in the shape of the valve and in the sculpture of the growth bands, but differs from *Rhombograpta* in the crowded and curvate striae and from *Deltostracus* in the subcircular shape of the valve.

Distribution: Liaoning, Jiangsu, Anhui, Zhejiang, Henan, and Guangdong; Paleogene.

Fushonograpta ovata Wang
Plate 2(1–2)

1974 *Fushunograpta ovata* Wang, *Acta Geologica Sinica,* no. 2, Pl. 8, figs. 1–5.

Description: Carapace flat, oval-elliptical in outline; dorsal margin straight; umbo small, situated between middle part and anterior end of dorsal margin; marked with 22 to 28 growth bands covered with fine radial ridges; ridges simple or somewhat branching; radial ridges from ventral to middle part of valve, long and robust, but from middle part upward relatively fine and short; near umbo growth bands not marked by radial ridges, but instead smooth or slightly punctate. Cross bars intercalated between ridges.

Horizon and locality: Fushun, Liaoning; Guchengzi Formation of Fushun Group; Paleogene.

Cat. Nos. 10003, 10004, 10007.

Fushunograpta changzhouensis (Chen)
Plate 2(3–6)

1975 *Curvestheria changzhouensis* Chen, *Scientia Sinica,* no. 6, p. 629, Pl. 2, figs. 3–7.

1976 *Curvestheria changzhouensis* Chen, Fossil Conchostraca of China, p. 193, Pl. 75, figs. 1–5

1979 *Fushunograpta changzhouensis* (Chen), Mesozoic and Cenozoic red beds of South China, p. 297–298, Pl. 1, fig. 13

Description: Carapace, subcircular in outline, 5.6 to 6.8 mm long, 4.2 to 5.2 mm high; dorsal margin straight; umbo small, located between interior end and middle part of dorsal margin; anterior, posterior, and ventral margins nearly rounded; anterior height greater than posterior height; growth lines stout, with concave sulcus on median axis called "median groove"; growth bands flat, about 25, or at most 30, in number, ornamented with slender and curvate striae, closely spaced, occasionally bifurcated upward and downward; fine cross bars usually intercalated between striae; impressed on external mold are many minutely closely spaced, discontinued, and curvate lines.

Comparison: This species differs from *F. ovata* in its subcircular carapace.

Horizon and locality: Changzhou, Jintan, and Xuyi of Jiangsu; Nanling, Wuhu, and Lai'an of Anhui; fourth member of Funing Formation; early Eocene.

Cat. Nos. 22928-22951.

Fushunograpta changheensis Chen and Shen
Plate 2(7–8)

1979 *Fushonograpta changheensis* Chen and Shen, Mesozoic and Cenozoic red beds of South China, p. 297, Pl. 1, figs. 7–12.

Description: Carapace of small to moderate size, oval in outline; holotype 5 mm long, 3.5 mm high; dorsal margin long, more or less straight, umbo small, subanterior; anterior margin oblique backwards, ventral margin arched downward, posterior margin rounded and slightly dilated; anterior height less than posterior height; more than 20 growth lines, stout and convex, with a median groove; growth bands broad and flattened, ornamented with slender and curvate radial striae, frequently bifurcated upward or downward, cross bars in between.

Comparison: This species resembles *F. ovata* in the outline of the carpace, but differs in the median groove of stout growth lines and in the striae on growth bands being more irregular. it differs from *F. changzhouensis* in the relatively longer dorsal margin and in the outline of the valve.

Horizon and locality: Dapaitang village, near Changhe town, Cixi, Zhejiang; middle part of Changhe Formation: Paleogene.

Cat. Nos. 4516–44519.

Superfamily AFROGRAPTIOIDEA Novozhilov, 1957
Family AFROGRAPTIDAE Novozhilov, 1957
Genus *Nanhaiestheria* Shen and X. Zhang, 1979
Type species: *Nanhaiestheria sanshuiensis* Shen and X. Zhang

Diagnosis: Carapace of moderate size, elliptical in outline; shallow polygonal reticulate sculpture seen on growth bands, mesh of reticulate sculpture moderate in size; growth lines stout, with a row of serrations in ventral margin. Among Afrograptidae, this is the only genus with polygonal reticulate sculpture on growth bands.

Distribution: Guangdong; Paleogene.

Nanhaiestheria sanshuiensis Shen and X. Zhang
Plate 1(11–12)

1979 *Nanhaiestheria sanshuiensis* Shen and X. Zhang, Mesozoic and Cenozoic red beds of South China, p. 302, Pl. 2, figs. 1–4

Description: Carapae moderate in size, elliptical in outline; dorsal margin long; umbo small, subcentral; anterior margin broadly rounded, ventral and posterior margins rounded; anterior height less than posterior height; growth bands broad and flattened, about 13 in number, ornamented with irregular reticulate mesh which is shallow and moderate in size; serration in the ventral margin of the growth lines gradually weakened to dorsal part of carapace valve.

Horizon and locality: Hekou, Sanshui, Guangdong; Buxin Formation; Eocene.

Cat. No. 44535.

Genus *Cixiella* Chen and Shen, 1979
Type species: *Cixiella serrula* Chen and Shen

Diagnosis: Carapace small or moderate in size, oval in outline; growth bands with crowded irregular striae; striae frequently bifurcated and connected by cross bars, rare on the umbonal region; external mold of the carapace showing irregular discontinuous linear, minute serration in the lower margin of growth lines on ventral region of carapace valve.

Discussion: This genus is similar to *Fushunograpta* in sculpture of growth bands, but differs chiefly in the serration on the ventral margin of the growth lines.

Distribution: Zhejiang, Paleogene.

Cixiella serrula Chen and Shen
Plate 2(9–12)

1979 *Cixiella serrula* Chen and Shen, Mesozoic and Cenozoic red beds of South China, p. 298, Pl. 2, figs. 1–8

Description: Carapace small in size, quadrate to circular in outline; holotype 6.5 mm long, 5 mm high; dorsal margin short and arched; umbo small, subcentral; anterior and posterior margins relatively straight, ventral margin arched downward; posterior height slightly greater than anterior height; growth bands flattened, about 20 in number, with slender and irregular curvate striated sculpture, striae closely spaced, frequently bifurcated and connected by cross bars, rare on the umbonal region; growth lines stout and convex, ornamented by serration in lower margin on ventral part of carapace.

Horizon and locality: Dapaitang village near Changhe town, Cixi, Zhejiang; middle part of the Chenghe Formation; Eocene.

Cat. Nos. 44521-44526.

ACKNOWLEDGMENTS

The writers wish to express their sincere thanks to Professor Zhang Wen-tang and Mr. Qi Bao-ji for their critical reading of the manuscript.

All specimens except Cat. Nos. L0003, L0004 are stored in the Nanjing Institute of Geology and Palaeontology, Academia Sinica.

SELECTED REFERENCES

Chen Pei-ji [Chen Pei-chi], 1975, Tertiary Conchostraca from China: Scientia Sinica, no. 6, p. 618–630, Pls. 1–2 (in Chinese).

Chen Pei-ji and Shen Yan-bin, 1979. [Mesozoic and Cenozoic conchostracan faunas of China, with notes on their distributions in the red beds of South China]: Beijing, Science Press, Mesozoic and Cenozoic red beds of South China, p. 79–97, text-figs. 1–3 (in Chinese).

—— 1980, On the *Paraleptestheria menglaensis* fauna, with reference to the age of the Lofochai group: Acta Palaeontologica Sinica, v. 19, no. 3, p. 182–189, Pls. 1–2 (in Chinese with English abstract).

Hong You-chong, Yang Zhu-chiang, Wang Shih-tao, Wang Sze-en, Li Yu-kuei, Sun Meng-rung, Sun Hsiang-chun, and Tu Nai-chiu, 1974, Stratigraphy and palaeontology of Fushun coal-field, Liaoning Province: Acta Geologica Sinica, no. 2, p. 113–149, Pls. 1–8 (in Chinese with English summary).

Kobayashi, T. 1972, On the two discontinuities in the history of the order Conchostraca; Proceedings of the Japan Academy, v. 48, no. 10, p. 725–729.

Qiu Zhan-xiang, Li Chuan-kui, Huang Xue-shi, Tang Ying-jun, Xu Qin-qi, Yan De-fa, and Zhang Hong, 1977, [Continental Paleocene stratigraphy of Qianshan and Xuancheng basins, Anhui]: Vertabrata PalAsiatica, v. 15, no. 2, p. 85–93 (in Chinese).

Shen Yan-bin and Chen Pei-ji [Chen Pei-chi], 1979, [A discovery of the Early Tertiary conchostracans near the Gulf of Hangzhou]: Beijing, Science Press, Mesozoic and Cenozoic red beds of South China, p. 295–299, Pls. 1–2, (in Chinses).

——Shen Yan-bin and Zhang Xian-qiu, 1979, [Early Tertiary conchostracans from Sanshui basin of Guangdong]: Beijing, Science Press, Mesozoic and Cenozoic red beds of South China, p. 300–304, Pls. 1–2 (in Chinese).

Trusova, E. K., and Badamgarav, D., 1976, O pervoy nakhodke kaynozoyskikh dvustvorchatykh listonogikh rakoobraznykh (Conchostraca): in Kramarenko, N. N. (ed.), Paleontologiya i biostratigrafiya Mongolii, Sovmestnaya Sovetsko-Mongolskaya Paleontologicheskaya Ekspeditsiya, Trudy, v. 3, p. 162–168, Pl. 1

Xia Shu-fang, Liu Guan-bang, Yin Pei, and Miao De-sui, 1979, [Early Tertiary fish from the western region of the North Jiangsu plain]: Beijing, Science Press, Mesozoic and Cenozoic red beds of South China, p. 321–329, Pls. 1–2 (in Chinese).

Zhang Wen-tang [Chang Wen Tang], Chen Pei-ji [Chen Pei-chi], and Shen Yan-bin, 1976, [Fossil Conchostraca of China]: Beijing, Science Press, 325 p., 138 pls. (in Chinese).

Zheng Jia-jian and Qiu Zhan-xiang, 1979, [Cretaceous and Tertiary continental strata in South China, with discussion on some problems concerned]: Beijing, Science Press, Mesozoic and Cenozoic red beds of South China, p. 1–57, text-figs. 1–18 (in Chinese).

MANUSCRIPT ACCEPTED BY THE SOCIETY APRIL 22, 1981

PLATE 1

Figure

1. *Perilimnadia jiangsuensis* Chen
   Internal mold of right valve; ×5; holotype; Cat. No. 22912-3. Lingtangqiao, Gaoyou, Jiangsu; second member of Funing Formation; Eocene.

2. *Perilimnadia lingtangqiaoensis* Chen
   External mold of left valve; ×5; holotype; Cat. No. 22913. Horizon and locality same as preceding.

3. *Perilimnadia gaoyouensis* Chen
   External mold of left valve, ×5; holotype; Cat. No. 22912-5. Horizon and locality same as preceding.

4–5. *Yunmenglimnadia hubeiensis* Chen
   4. Internal mold of open carapace; ×5; holotype; Cat. No. 22918.
   5. External mold of left valve; ×5; Cat. No. 22920. Yingcheng, Hubei; Gaoyan Formation of Yingcheng Group?, Paleogene?

6. *Yunmenglimnadia gansuensis* Chen
   Left valve; ×10; holotype; Cat. No. 22924. Dasi of Wangfu commune, Qinan, Gansu; Paleogene.

7–10. *Paraleptestheria menglaensis* Chen
   7. Left valve; ×10; holotype; Cat. No. 26971. Near Mengla, Yunnan; Paleogene Guolang Formation.
   8. Sculpture on growth bands of external mold of left valve; ×40; paratype; Cat. No. 26972. Near Shangyong, Mengla, Yunnan; Paleogene Guolang formation.
   9. External mold of right valve; ×17; Cat. No. 44527.
   10. Sculpture on growth bands of middle-ventral region of fig. 9; ×40. Hekou, Sanshui, Guangdong; Buxin Formation; Eocene.

11–12. *Nanhaiestheria sanshuiensis* Shen and X. Zhang
   11. External mold of left valve; ×10; holotype; Cat. No. 44534.
   12. Sculpture on growth bands of antero-ventral region of fig. 11. Hekou, Sanshui, Guangdong; Buxin Formation; Eocene.

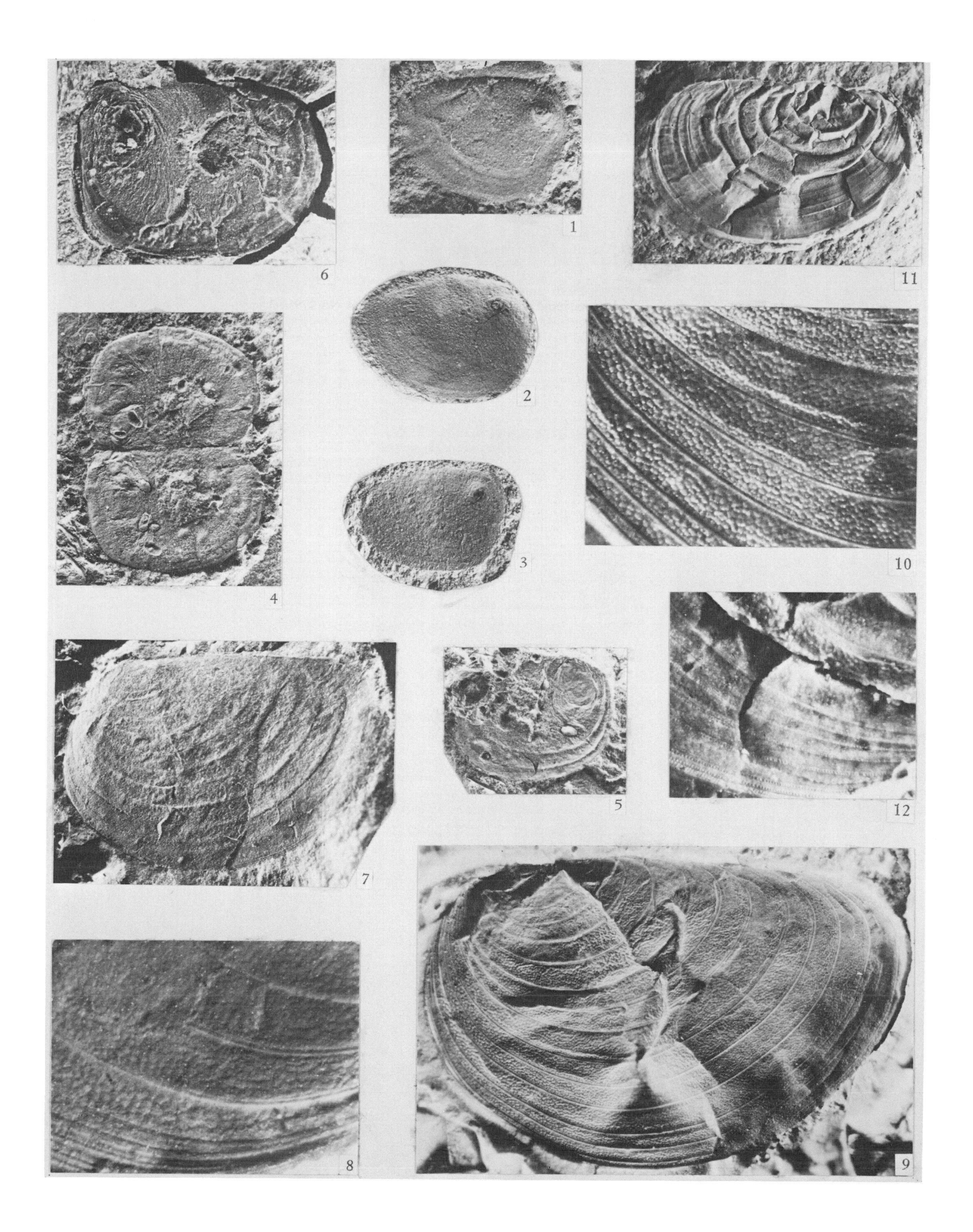

6
1
11
2
3
10
4
5
12
7
8
9

PLATE 2

Figure

1–2. *Fushunograpta ovata* Wang

1. Right valve; ×9; holotype; Cat. No. L0003.
2. Radial striated sculptue on growth bands of postero-ventral region of right valve; ×28; Cat. No. L0004. Fushun, Liaoning; Guchengzi Formation of Fushun Group.

3–6. *Fushunograpta changzhouensis* (Chen)

3. Internal views of left valve: ×5; holotype; Cat. No. 22928. Changzhou, Jiangsu; fourth member of Funing Formation; Eocene.
4. Internal views of left valve; ×5; Cat. No. 22933.
5. Sculpture on growth bands of fig. 4; ×40.
6. Sculpture on chitinous growth bands of another carapace; ×40; Cat. No. 22934. Xuyi, Jiangsu; fourth member of Funing Formation; Eocene.

7–8. *Fushunograpta changheensis* Chen and Shen

7. External mold of right valve; ×20; holotype; Cat. No. 44517.
8. Sculpture on growth bands of middle-ventral region of external mold of another carapace; ×40; Cat. No. 44518. Dapaitang village near Changhe town, Cixi, Zhejiang; middle part of Changhe Formation; Eocene.

9–12. *Cixiella serrula* Chen and Shen

9. Internal mold of right valve; ×4; holotype; Cat. No. 44521.
10. Sculpture on remaining chitinous growth bands in anterior region of fig. 9; ×40.
11. Sculpture on growth bands of postero-ventral region of external mold of holotype; ×40; Cat. No. 44522.
12. Sculpture on chitinous growth bands in postero-ventral region of another carapace; ×40; Cat. No. 44524. Horizon and locality same as the preceding.

11
3
12
9
10
1
4
5
6
2
8
7

Printed in U.S.A.

Geological Society of America
Special Paper 187
1981

Two new graptolite genera from the Ningkuo Formation (Lower Ordovician) of Wuning, North Jiangxi

Ni Yu-nan
Nanjing Institute of Geology and Palaeontology, Academia Sinica, Nanking 210008, People's Republic of China

ABSTRACT

The new generic name *Protabrograptus* is proposed for biramous abrograptids with sicula inclining and merging into the ventral filament of the second stipe. A new genus of three- or four-stiped kinnegraptids, *Wuninograptus,* is also described.

INTRODUCTION

The materials dealt with in this paper were collected by Miss Chang Yin-nan and me in the autumn of 1963 from the uppermost part of the Ningkuo Formation at Xinkaling, about 5 km northwest of Wuning in Jiangxi. Four species belonging to two new genera may be recognized as follows: *Protabrograptus sinicus* gen. et sp. nov.; *Wuninograptus quadribrachiatus* gen. et sp. nov.; *Wuninograptus erectus* gen. et sp. nov.; and *Wuninograptus tribrachiatus* gen. et sp. nov.

Associated with these new graptolites are *Reteograptus delicatus* Mu, *Reteograptus* spp., *Glossograptus* spp., and a number of pendent didymograpti. Together they constitute the *Didymograptus jiangxiensis* Zone, lying above the *Pterograptus elegans* Zone and being considered as the highest graptolite zone of the Lower Ordovician in Wuning and the adjacent districts; this may correspond to the *Didymograptus clavulus* Zone of Scania.

SYSTEMATIC DESCRIPTIONS

Family ABROGRAPTIDAE MU, 1958
Genus *Protabrograptus* gen. nov.
Type species: *Protabrograptus sinicus* gen. et sp. nov.

Diagnosis: Rhabdosome minute with a generally rounded base, consisting of two reclined stipes, which are composed of two longitudinal filaments (ventral and dorsal) and apertural rings or "cross-bars." Sicula periderm normal, inclining and merging into the ventral filament of second stipe originates from top of sicula; that of first stipe originates from apertural portion of sicula. Two dorsal filaments of stipes connected in the middle of the rhabdosome. Development probably of modified dichograptid type.

Remarks: The new genus *Protabrograptus* closely resembles *Abrograptus* Mu in the main characters, but differs from the latter in having a sicula inclining and merging into the ventral filament of the second stipe. Stratigraphically, it occurs at a lower horizon, and may be considered as the ancestor of *Abrograptus.* In the position of the sicula, *Protabrogaptus* is similar to *Janograptus* Tullberg, but the sicula of the latter merges into the dorsal edge of the second stipe.

Generally speaking, the reduction of thecal periderm is not unusual in the axonophorous graptoloids, several stages of which can be recognized from Early Ordovician to Late Silurian. Probably, *Reteograptus* is the most primitive representative with reduced thecal periderm. The occurrence of *Protabrograptus* reveals that the reduction of the periderm is usual in axonolipous graptoloids as well as in axonophorous graptoloids. In this connection, I am disposed to accept Mu's (1974) view that the family Abrograptidae Mu, 1958, and the family Teteograptidae Mu, 1974, represent two different lines of the reduction of the thecal periderm in Graptoloidea.

Protabrograptus sinicus gen. et sp. nov.
Plate 1(2–4,9); Figure 1

The rhabdosome is very small with a generally rounded base, consisting of two reclined stipes, about 6 mm in length. The sicula is slender, with normal periderm, measuring 0.5 to 0.6 mm in heighth and 0.1 mm in width, inclining and merging into the ventral filament of the second stipe. In the proximal portion of the first stipe there are three radiated short filaments. The thecal periderm is reduced. The ventral filament of the second stipe originates from the top of the sicula, and that of the first stipe originates from the apertural portion of the sicula. Two dorsal filaments of stipes are connected in the middle of the rhabdosome. The distance between the ventral filament and the dorsal one is variable, generally 0.3 to 0.4 mm. There are about 4 or 5 thecae in a length of 5 mm.

In the collection, there are a number of young individuals with

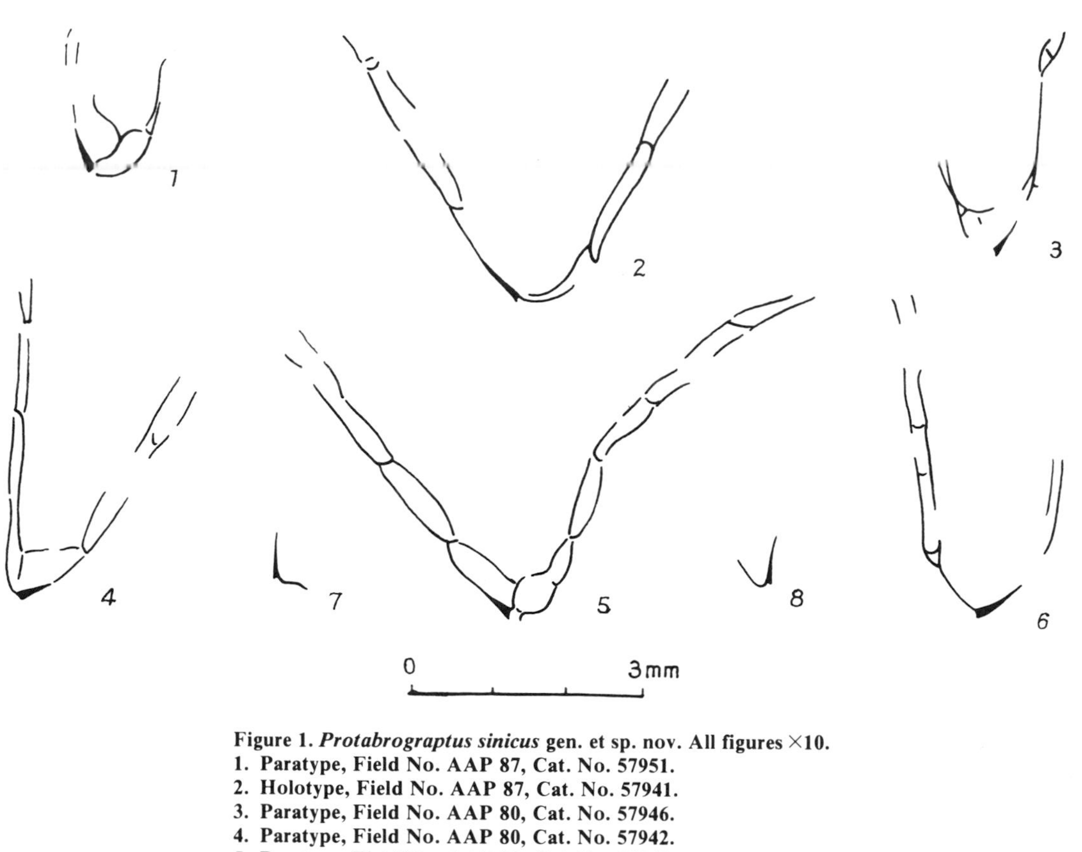

Figure 1. ***Protabrograptus sinicus*** **gen. et sp. nov. All figures ×10.**
1. Paratype, Field No. AAP 87, Cat. No. 57951.
2. Holotype, Field No. AAP 87, Cat. No. 57941.
3. Paratype, Field No. AAP 80, Cat. No. 57946.
4. Paratype, Field No. AAP 80, Cat. No. 57942.
5. Paratype, Field No. AAP 80, Cat. No. 57943.
6. Paratype, Field No. AAP 81, Cat. No. 57944.

at the most one pair of thecae. It is of interest to note that the thecae of the young rhabdosome are nearly equal in size to those in the correspondent part of the adult individuals, though their filaments are finer. The shape and size of thecae remain unchanged in the course of successive development, only their internal tissues being strengthened.

Comparison: This species is similar to *Abrograptus formosus* Mu in the main characters. It differs from the latter in having a sicula inclining and merging into the ventral filament of the second stipe.

Horizon and locality: Zone of *Didymograptus jiangxiensis* of the Ningkuo Formation at Xinkaling, Wuning.

Family KINNEGRAPTIDAE Mu, 1974

Diagnosis: Rhabdosome composed of two to four or more stipes, declined, horizontal or reclined. Sicula and thecae with a long tongue-shaped apertural process. Development of Isograptid type and Dichograptid type.

Distribution: Lower Ordovician; China, Sweden, Norway, Canada, Soviet Union, Australia.

Genus *Wuninograptus* gen. nov.
Type species: *Wuninograptus quadribrachiatus* gen. et sp. nov.

Diagnosis: Rhabdosome consisting of three to four reclined stipes. Sicula and thecae with a long, tongue-shaped apertural process and slight thecal overlapping. Development of probably dichograptid type.

Remarks: According to thecal specialization, the new genus *Wuninograptus* has been referred to the Kinnegraptidae, characterized by a peculiar tongue-shaped process on the ventral side of the apertural part of thecae. The apertural process supported soft tissue increasing the size of the apertural region for floating and catching food.

Generally, most of the axonolipous graptoloids in the Lower Ordovician have simple straight-walled thecae. It appears clear, therefore, that the families Kinnegraptidae and Sinograptidae represent two different evolutionary trends of the elaboration of the thecae in Axonolipa.

Wuninograptus may be distinguished from *Kinnegraptus* Skoglund by the number and the direction of growth of the stipes. It is difficult to compare *Wuninograptus* with *Prokinnegraptus* Mu exactly, because no complete rhabdosome of *Prokinnegraptus* has yet been found.

In the general aspect of the rhabdosome, this genus is similar to *Tetragraptus,* but differs strikingly in the character of the thecae. Based on the feature of the thecae, *Tetragraptus(?) insuetus* Keble and Benson and the four-stiped *"Kinnegraptus"* from Kirghizia (Skevington, 1974, p. 64) may belong to the present new genus.

Kinnegraptus(?) gracilis described by Chen from the *Glyptograptus austrodentatus* Zone of the Meitan Shale, Guizhou, is another special form of Kinnegraptidae with isolated apertures. *Kinnegraptus* sp. described by Archer and Fortey (1974) is similar to *Kinnegraptus(?) gracilis* Chen.

Wuninograptus is known to occur in Kalpin, Xinjiang, and Zotzeshan, Inner Mongolia. It appears that the Ordovician grap-

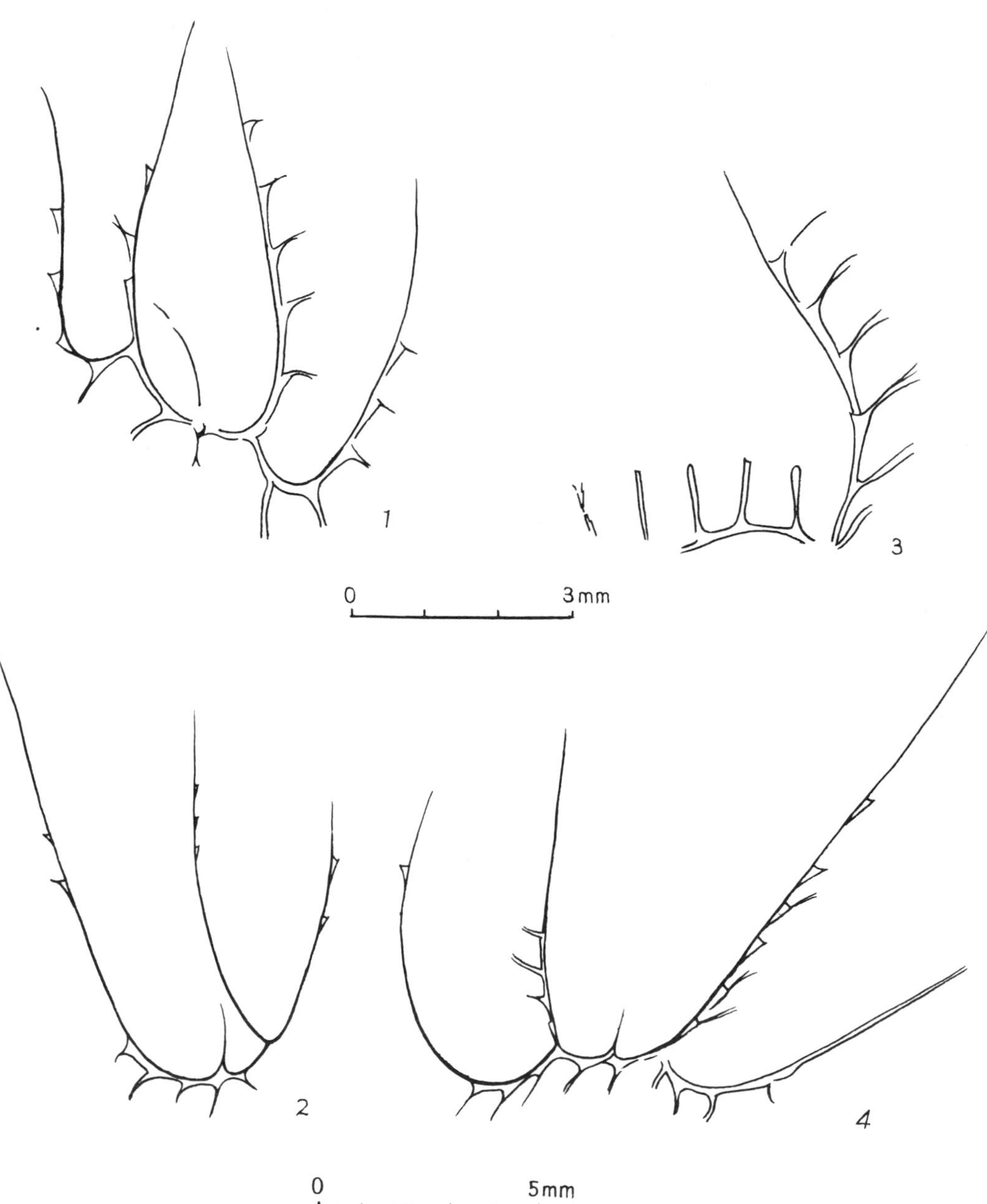

Figure 2. *Wuninograptus* gen. nov.
1. *Wuninograptus erectus* gen. et sp. nov., holotype, ×10, Field No. AAP 86, Cat. No. 54075.
2. *Wuninograptus tribrachiatus* gen. et sp. nov., holotype, ×6, Field N. AAP 86, Cat. No. 54076.
3. *Wuninograptus quadribrachiatus* gen. et sp. nov., paratype, ×10, Field No. AAP 88, Cat. No. 54077.
4. *Wuninograptus quadribrachiatus* gen. et sp. nov., holotype, ×6, Field No. AAP 88, Cat. No. 54074.

tolite faunas of the Northwest China Region belong to the South China type.

Wuninograptus quadribrachiatus gen. et sp. nov. Plate 1(1, 8); Figure 2(3, 4)

The rhabdosome comprises four reclined stipes of second order. The length of the stipes is about 14 mm, their width 0.2 to 0.3 mm. The funicle is short, consisting of two thecae. The sicula is 1 mm in height and 0.1 mm in width. The first theca (th_1^1) originates from the middle part of the sicula, having probably one crossing canal. The thecae are slender; the distance between two neighboring thecal apertures is 0.7 to 0.8 mm. The apertural process is about 0.4 to 0.8 mm in length. The torsion of the thecae seems to be pronounced. There are 5 to 6 thecae in a length of 5 mm.

Horizon and locality: Zone of *Didymograptus jiangxiensis* of the Ningkuo Formation at Xinkaling, Wuning.

Wuninograptus erectus gen. et sp. nov. Plate 1(7); figure 2(1)

The rhabdosome is small, its length being about 5.5 mm, nearly equal to its width. The four reclined stipes of second order are parallel to each other distally, 5 mm in length, 0.2 mm in width.

Th_1^3 and th_1^4 originate from th_1^1 and th_2^2 respectively. The sicula is 0.6 mm in height and 0.1 mm in width. The length of the nema is 1.4 mm. The process is 0.4 mm in length. The thecae are slender, only overlapping a little. The process is 0.3 to 0.5 mm in length. The torsion of the thecae is obvious. There are 5 thecae in a length of 5 mm.

Comparison: The new species closely resembles the type species of the genus in having four stipes, but has a small rhabdosome, nearly erect and asymmetrically branched stipes.

Horizon and locality: Same as the preceding form.

Wuninograptus tribrachiatus gen. et sp. nov.
Plate 1(5–6); Figure 2(2)

The rhabdosome consists of two reclined stipes of second order and one of first order. The stipe is 11 mm in length, 0.2 mm in width. The th_1^3 originates from the th_2^2. The sicula is minute, about 0.6 mm in height, 0.2 mm in width. The length of the nema is 1 mm. The process of the sicula is 0.4 mm in length. The th_1^1 originates from the middle part of the sicula, having a crossing canal. On account of the torsion of the thecae, the proximal thecae are more obvious than the distal one. The thecae are slender, overlapping very little. The distance between two neighboring thecal apertures is about 0.8 mm. The length of thecal process reaches 0.4 mm. There are about 4 thecae in a length of 3 mm.

Comparison: This form differs from *Wuninograptus quadribrachiatus* and *Wuninograptus erectus* in having two second order and one first order stipes.

Horizon and locality: Same as the preceding form.

ACKNOWLEDGMENTS

I thank Professor Mu En-zhi [A. T. Mu] for his valuable suggestions and careful reading of the manuscript.

REFERENCES CITED

Archer, J. B., and Fortey, R. A., 1974, Ordovician graptolites from the Valhallfonna Formation, Northern Spitsbergen: London, Special Papers in Palaeontology, no. 13, p. 87–97.

Bulman, O.M.B., and Cowie, C. M., 1962, On the occurrence of *Kinnegraptus* Skoglund in Norway: Norsk Geologisk Tidsskrift, Bd. 42, p. 253–260, Pl. 1.

——1970, Graptolithina (second edition), *in* Teichert, C., ed., Treatise on invertebrate paleontology, Part V: Boulder, Colorado, Geological Society of America (and University of Kansas Press), p. 1–163.

Jackson, D. E., 1966, On the occurrence of *Parabrograptus* and *Sinograptus* from the Middle Ordovician of western Canada: Geological Magazine, v. 103, p. 263–268.

Keble, R. A., and Benson, W. N., 1928, Ordovician graptolites of North-west Nelson: Transactions and Proceedings of New Zealand Institute, v. 59, p. 840–863, Pls. 104–107.

Kozlowski, R., 1951, Sur un remarquable graptolithe Ordovicien: Acta Geologica Polonica, v. II, p. 86–93.

Lu Yen-hao [Lu Yan-hao], Chu Chao-ling [Zhu Zhao-ling], Chein Yi-yuan [Qian Yi-yuan], Zhou Zhi-yi, Chen Jun-yuan, Liu Geng-wu, Yu Wen, Chen Xu, and Xu Han-kui, 1976, [Ordovician biostratigraphy and palaeozoogeography of China]: Memoirs of Nanjing Institute of Geology and Palaeontology, Academia Sinica, no. 7, 83 p., 14 pls. (in Chinese).

Mu An-tze [Mu En-zhi], 1958, *Abrograptus,* A new graptolite genus from the Hulo Shale (Middle Ordovician) of Kiangshan, Western Chekiang: Acta Palaeontologica Sinica, v. 6, no. 3, p. 259–265, Pl. 1 (in Chinese and English).

Mu An-tze [Mu En-zhi] and Qiao Xin-dong, 1962, New material of Abrograptidae: Acta Palaeontologica Sinica, v. 10, no. 1, p. 1–8, Pls. 1–2 (in Chinese and English).

Mu En-zhi [Mu An-tze], 1974, Evolution, classification and distribution of Graptoloidea and Graptodendroids: Scientica Sinca, v. 17, no. 2, p. 227–238.

Mu En-zhi [Mu An-tze], Ge Mei-yu [Geh Mei-yu], Chen Xu, Ni Yu-nan, and Lin Yao-kun, 1979, Lower Ordovician graptolites of Southwest China: Palaeontologica Sinica, whole no. 156, new ser. B, no. 13, 192 p., 48 pls. (in Chinese with English abstract).

Skevington, D., 1974, Controls influencing the composition and distribution of Ordovician graptolite faunal provinces: London, Special Papers in Palaeontology, no. 13, p. 59–73.

Skoglund, R., 1961, *Kinnegraptus,* a new graptolite genus from the Lower *Didymograptus* Shale of Västergötland, Central Sweden: Bulletin of the Geological Institutions of the University of Uppsala, v. 40, p. 389–400, Pl. 1.

Skwarko, S. K., 1962, Graptolites of Cobb River–Mount Arthur area, Northwest Nelson, New Zealand: Transactions of the Royal Society of New Zealand: Transactions of the Royal Society of New Zealand, Geology, v. 1, no. 15, p. 215–247.

MANUSCRIPT ACCEPTED BY THE SOCIETY APRIL 22, 1981

PLATE I

Photographs by Deng Dong-xing and Liang Xiao-yun. All figures unretouched. All type specimens are deposited in the Nanjing Institute of Geology and Palaeontology, Academia Sinica.

Figure

1, 8. ***Wuninograptus quadribrachiatus*** **gen. et sp. nov.**
- **1. Paratype, ×, Field No. AAP88, Cat No. 54077.**
- **8. Holotype, ×10, Field No. AAP88, Cat No. 54074.**

2–4, 9. ***Protabrograptus sinicus*** **gen. et sp. nov.**
- **2. Paratype, ×10, Field No. AAP80, Cat Nos. 57946, 57944.**
- **9. Holotype, ×10, Field No. AAP87, Cat No. 57941.**

5, 6. ***Wuninograptus tribrachiatus*** **gen. et sp. nov.**
- **Holotype, 5 ×10; 6, ×6, Field No. AAP86, Cat No. 54076.**

7. ***Wuninograptus erectus*** **gen. et sp. nov.**
- **Holotype, ×10, Field No. AAP86, Cat No. 54075.**

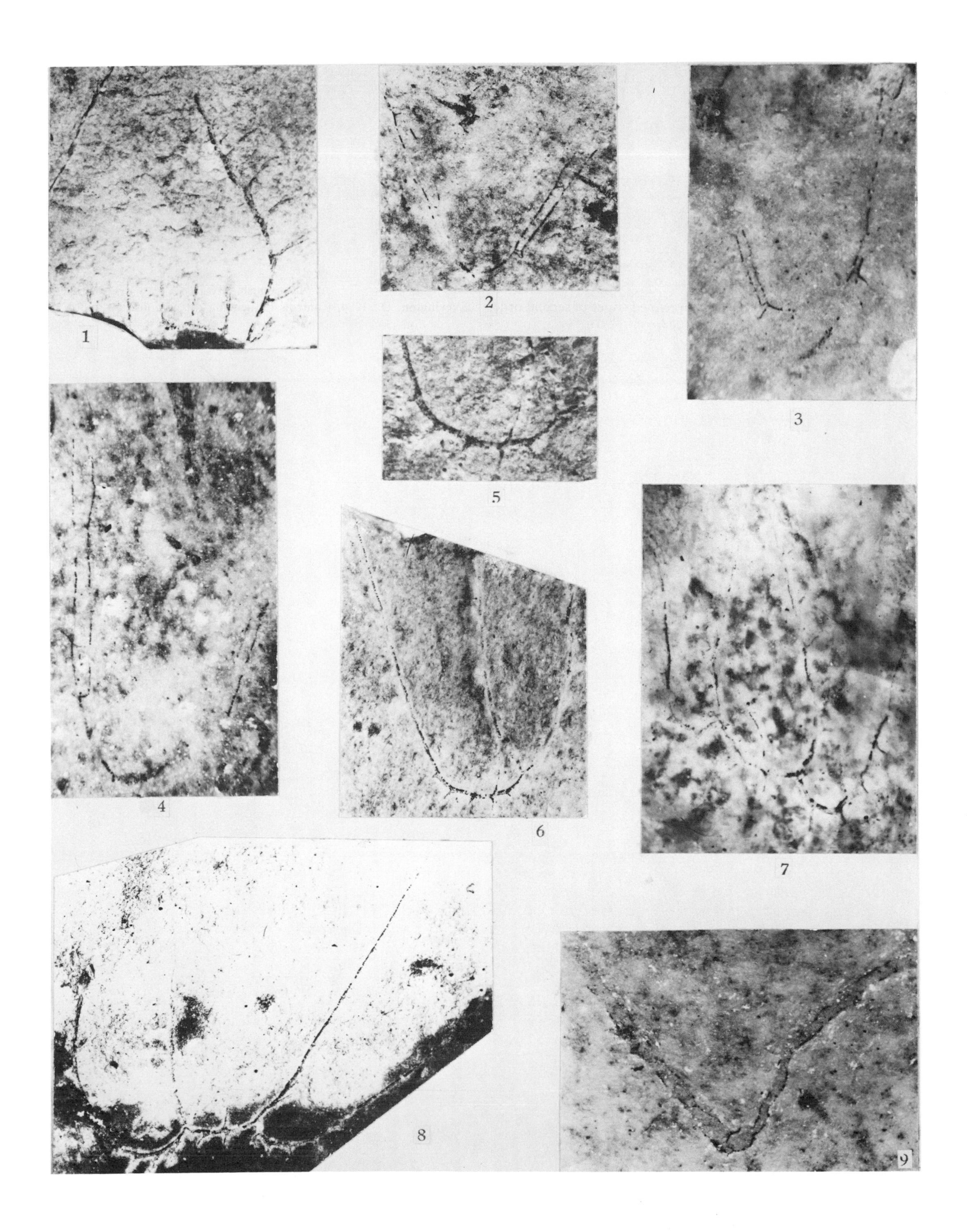
1
2
3
4
5
6
7
8
9

Printed in U.S.A.

Geological Society of America
Special Paper 187
1981

Recent progress in Cambrian and Ordovician conodont biostratigraphy of China

An Tai-xiang
Department of Geology, Beijing University, Peking (Beijing), People's Republic of China

ABSTRACT

Study of the Cambrian and Ordovician conodont biostratigraphy of China is mainly confined to the eastern parts of China, where the conodont faunas are very rich and varied. There are many stratigraphic sections with continuous conodont sequences that are ideal for biostratigraphic research.

Many Upper Cambrian conodonts show limited vertical ranges and are very widespread. Near the Cambrian and Orodovician boundary, 4 conodont zones are established and the boundary is designated at the bottom of the *Drepanodus simplex* Zone.

Ordovician conodonts are useful in biostratigraphy and show evident provincial differentiation. In the North China province, some 8 zones have been recognized, and in the South China province some 15 zones are defined. Many of these zones are also recognized in other parts of the world.

INTRODUCTION

The study of conodont biostratigraphy has developed rapidly since the 1960s, but reports in this field in China are extremely few. In 1966 and 1967, two articles were published by Nogami Yasuo, describing some conodonts from the Gushan and the Fengshan Formations at Dawekou in Shandong Province and from Changxing Island and Benxi in Liaoning Province, but without exact stratigraphical data.

The study of Cambrian and Ordovician conodont biostratigraphy in China developed in the 1970s, largely as a result of oil exploration and exploitation. To date, the general features of the conodont faunas have been studied in eastern China only, where two Ordovician biogeographic provinces have been recognized. The conodont sequences were preliminary established in both provinces.

BIOGEOGRAPHY OF CAMBRIAN AND ORDOVICIAN CONODONTS OF CHINA

Study of the Cambrian and Ordovician conodont biostratigraphy of South China is confined mainly to the middle and lower Yangtze region, where carbonate rocks are widespread. Fossil conodonts were found in Yunnan, Sichuan, Guizhou, Hubei, Anhui, Jiangsu, and Zhejiang Provinces. Among these the Yichang section of Hubei may be regarded as the standard sequence of Ordovician conodonts in southern China (Fig. 1).

Cambrian and Ordovician conodont faunas are found widely distributed also in northern China. Important localities are Western Hills of Beijing, Tangshan, Fengfeng, and Pingquan in Hebei Province, Yichang in Hubei Province, Xintai and Boshan in Shandong Province, and Benxi in Liaoning Province. Among these the Tangshan and Fengfeng sections have been systematically studied and may be regarded as the standard sequence in Northern China.

The Ordovician deposits developed along the southwest border of the Erdousi Basin are transitional between platform type in the east and geosynclinal type in the west. Study in conodont biostratigraphy of this region is of special significance, and investigation has been begun in Zhouzishan, Nei Monggol, in Huanxian and Longxian, Gansu Province, and in Yaoxian and Qishan, Shaanxi Province.

Up to 1979, little is known about Cambrian and Ordovician conodonts of western China, only a few having been found in Houcheng, Kalping, and Bachu of Xinjiang.

While biogeographic differentiation of conodonts in Cambrian time is not clear, two Ordovician conodont provinces can be recognized: a South China province and a North China province. The boundary of the two provinces runs roughly from Lianyungang through the Huaiyang and Wudang Mountains to Zhouzishan, mainly along the southern border of the Sino-Korean plat-

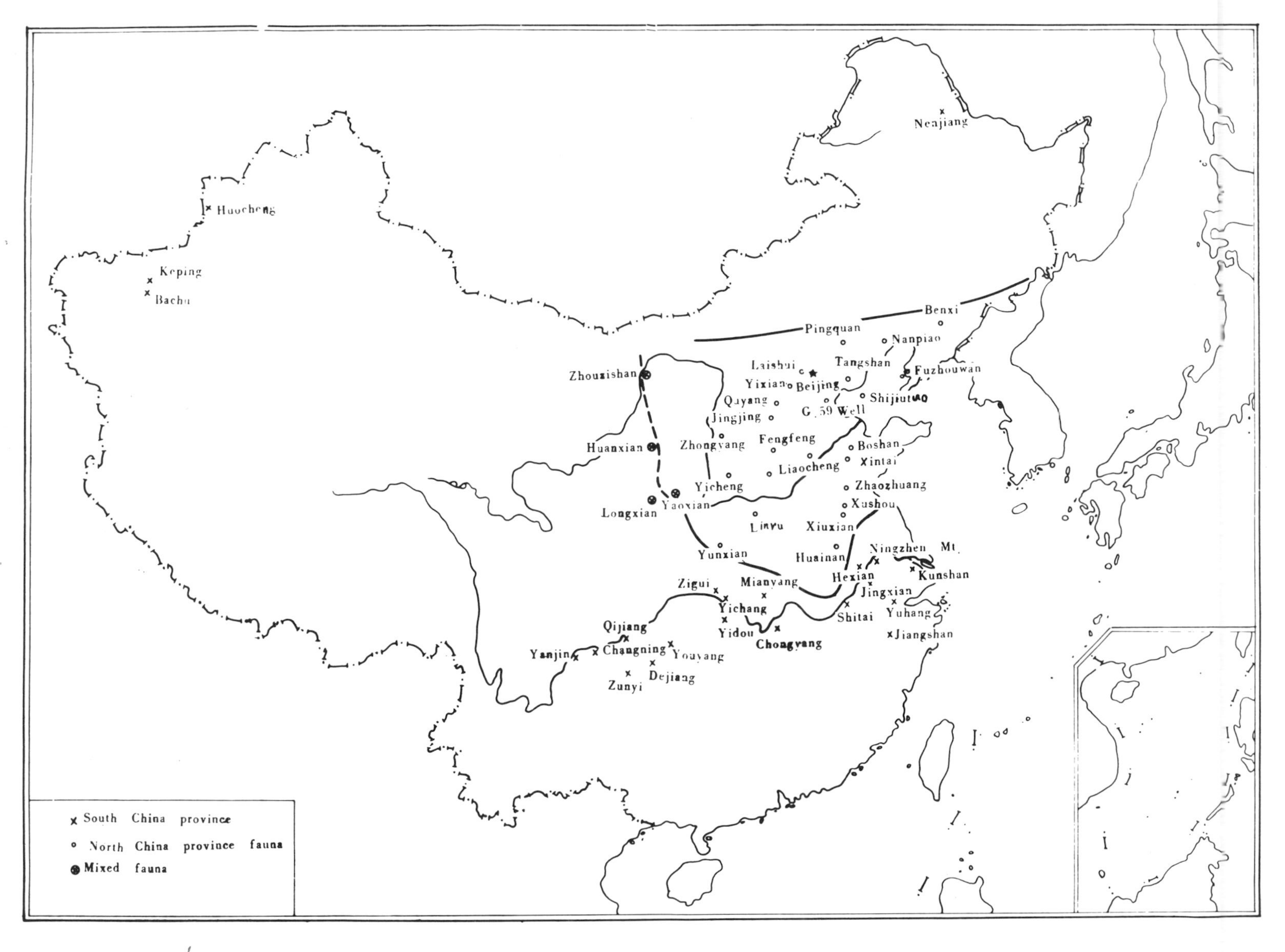

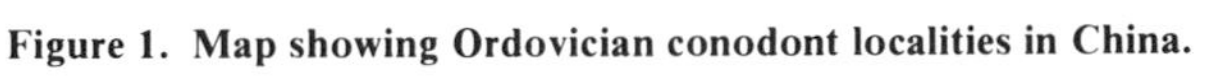

Figure 1. Map showing Ordovician conodont localities in China.

form, which became more clearly defined since the Dawan (Arenigian) epoch. The two provinces may be compared with the North American midcontinent conodont province and the North Atlantic conodont province, respectively.

It appears that biostratigraphic provinciality of the north and the south developed with time. So far as we know, there seems to be no bigeographic differentiation in Cambrian conodont faunas. This is due mainly to the primitive state of evolution of the conodont faunas. Tremadocian conodonts are less well studied in Europe, and their biogeographic differentiation is not clear. Judging from the worldwide distribution of *Cordylodus angulatus* Pander and *C. rotundatus* Pander, we think provinciality may not have developed, and there are many species common to the Tremadocian of North and South China. However, conditions changed after the Tremadocian, and the conodonts in the Liangjianshan Formation in the north and the Honghuayuan Formation in the south, both lower Arenigian, have only few forms in common, such as *Serratognathus* and *Bergstroemognathus,* and are difficult to correlate with each other. To sum up, the conodonts of the Tremadocian and lower Arenigian series in the north (Yeli Formation and Liangjianshan Formation) are related to the North America midcontinent faunas and those in the south (Nanjinguan, Fengxiang, and Honghuayuan Formations) to the North Atlantic faunas. The number of common forms in the northern and in the southern faunas decreases upwards, and since Dawan time few species remain common to both areas. But despite these differences, the evolutionary trend is from simple through compound to plate conodonts; the sequence of the occurrence and time of abundance in some genera or species are almost the same and might be used in chronostratigraphy. Before Arenigian, the southwestern margin of the Eerdosi Basin belonged to the North China province and since Llanvirnian, to the South China province. Little is known of conodonts in West China in Cambrian and Ordovician, though Llanvirnian and Llandeilian conodonts of Keping and Bachu obviously belong to the southern province.

STRATIGRAPHIC DISTRIBUTION

Cambrian

Some doubtful conodonts, *Protohertzina robusta* Qian and *P. anabarina* Missarzhevsky, were found in the Maidiping Formation at Emei in Sichuan Province, the Huangshandong Formation at Yichang in Hubei, and the Kuanchuapu Formation in Shaanxi. The formations are basal Cambrian and can be compared with the Tommotian Stage of Siberia.

Up to 1979, the following Middle Cambrian conodonts have been found: *Furnishina primitiva* (Müller) [Pl. 1(16)] from the Zhangxia Formation of Shandong, and *Westergaardodina* sp. and *Furnishina* sp. from the Zhangxia Formation of Tangshan.

Upper Cambrian conodonts are rather widespread, varied, well preserved, and mostly of worldwide occurrence. Notable localities are the Western Hills of Beijing, Xintai and Laiwu in Shandong, Tangshan, and Quyang in Hebei, and Benxi in Liaoning.

No important differences exist between the conodonts of the Gushan Formation and the Changshan Formation, the main forms being *Furnishina primitiva* Müller, *F. asymmetrica* Müller, *F. quadrata* Müller, *F. furnishi* Müller, *Hertzina americana* Müller, *H. tricarinata* Nogami, *Prosagittodontus dunderbergiae* (Müller), *P. eureka (Mü*ller), *Prooneotodus tenuis* (Müller), *P. gallatini* (Müller), *P. rotundatus* (Druce and Jones), *P. terashimai* (Nogami), *Westergaardodina mössebergensis* (Müller), [Pl. 1(9)], *W. muelleri* Nogami [Pl. 1(1)], *W. tricuspidata* Müller [Pl. 1(2)], *W. bicuspidata* Müller [Pl. 1(3)], *W. matsushitai* Nogam [Pl. 1(4)], *Proacodus sinensis* Nogami, *Prodistacodus palmeri* (Müller), and *Problematoconites perforatus* Müller[Pl. 1(12)]. They belong to the Paraconodont type.

Different stages of conodont evolution are present in the Fengshan Formation of North China. The conodonts at the base of the Fengshan Formation bear a close relation to those of the Changshan Formation. Higher up, however, they are replaced by dominant species such as *Proconodontus mülleri serratus* Miller [Pl. 1(11)], *P. notchpeakensis* Miller [Pl. 2(7)], and *P. compressus* An. By this time, the Euconodont type, for example *Oneotodus nakamurai* Nogami [Pl. 1(8, 20)], seems to make its first appearance and increases in the upper part of this zone, which is correlated with the *Ptychaspis-Tsinania* to *Quadraticephalus* Zone. The upper part of the Fengshan Formation is characterized by the primitive compound conodont *Cordylodus proavus* Müller [Pl. 2(18)], and *C. oklahomensis* Müller [Pl. 1(5, 6)] with deep basal cavities. There are also *Hirsutodontus bulbosus* (Miller) [Pl. 1(14)], *Hertzina* cf. *H. bisulcata* Müller [Pl. 1(17)], *Proconodontus carinatus* Miller [Pl. 2(10)], *Oistodus minutus* Miller, and *O. cabricus* Miller. This zone is approximately equivalent to the *Mictosaukia-Calvinella* Zone. Sequences similar to those mentioned above are found in Iran (Müller, 1973), Queensland, Australia (Druce and Jones, 1971), Utah, United States (Miller, 1969, 1978), and Newfoundland (Landing and others, 1978).

Few Cambrian conodonts have been reported from South China. At Yichang, Hubei, *Hirsutodontus primitivus* An and others, *Cordylodus proavus* Müller, *Oneotodus nakamurai* Nogami, and *Oneotodus* spp. were found in the upper 10 to 60 m of the Sanyoudong Dolomite. Obviously, this Upper Cambrian conodont assemblage may correspond to the *Cordylodus proavus* Zone or to the *Mictosaukia-Calvinella* Zone of North China.

The Cambrian-Ordovician Boundary

Sections across the Cambrian-Ordovician boundary containing conodonts are well developed in Tangshan, Benxi, Jiawang, the Western Hills of Beijing, and Yichang, where the conodont successions and assemblages are generally equivalent. The conodont assemblages may be grouped into 4 zones (in ascending order): (1) *Proconodontus* Zone, (2) *Cordylodus proavus* Zone, (3) *Drepanodus simplex* Zone, and (4) *"Acodus" oneotensis–Acanthodus costatus* Zone (Fig. 2). The two lower zones are of the Paraconodont type and the two upper zones of the Euconodont type. I consider the transition from the paraconodonts to the euconodonts as an important event which is critical in the designation of the Cambrian-Ordovician boundary. At Tangshan, Beijing, and Xuzhou in North China, characteristic latest Cambrian trilobites such as *Mictosaukia, Calvinella, Kingstonia, Plethopeltella, Andersonella,* and *Coreanocephalus* were found in the *Cordylodus proavus* Zone. Higher up in the sequence, typical Ordovician trilobites *Onychopyge, Leiostegium (Alloleiostegium) punctatum,*

China			Iran (Müller, 1973)		Queensland, Australia (Druce and Jones, 1971)			Utah, U.S.A. (Miller, 1976)		
Ordovician	Yehli Formation	"Acodus" oneotensis-Chosonodina herfurthi Zone	Ordovician	7	Ordovician	Warentian	Chosonodina herfurthi-Acodus Zone Cordylodus rotundatus-C. angulatus Zone Cordylodus prion-Scolopodus Zone	Ordovician	Cordylodus proavus Zone	Clavohamulus hintzei Subzone
				6		Datsonian	Cordylodus oklahomensis-C. lindstroemi Zone			
		Drepanodus simplex Zone		5			Drepanodus simplex-Oneotodus bicuspatus Zone Cordylodus proavus Zone			Hirsutodontus simplex Subzone Clavohamulus primitus Subzone Fryxellodontus inornatus Subzone
Cambrian	Fengshan Formation	Cordylodus proavus Zone	Cambrian	4	Cambrian	Payntonian		Cambrian		Hirsutodontus hirsutus Subzone
		Proconodontus Zone		3					Proconodontus Zone	Oistodus minutus Subzone Proconodontus notchpeakensis Subzone Proconodontus muelleri Subzone

Figure 2. Approximate correlation of Cambrian-Ordovician conodont zones.

Eotrinucleus solus (in Tangshan) and Ordovician graptolites *Dictyonema flabelliforme sociale* and *Callograptus* sp. (in Benxi) occur in the *Drepanodus simplex* Zone. Therefore, I consider that the boundary between Cambrian and Ordovician as determined by conodonts is coincident with that based on trilobites and graptolites.

Ordovician of South China

In South China, an Ordovician conodont sequence is well developed in the Huanghuachang section of Yichang. The Zigui section of Hubei Province and the Hexian section of Anhui Province may be regarded as supplementary.

Conodonts are generally of small size in the Nanjinguan Formation. An and others (in press) have divided the conodont assemblages into 3 zones (in ascending order): (1) *Drepanodus simplex* Zone, (2) *"Acodus" oneotensis-Acanthodus costatus* Zone, and (3) *Scolopodus quadraplicatus* Zone (Fig. 3). The *Drepanodus simplex* Zone is taken unanimously as the first Ordovician conodont zone in South as well as North China. The Cambrian-Ordovician strata from northeast Yunnan to the lower Yangtze region are usually composed of dolomites. Study of the conodonts shows that the top part of the dolomite at different localities ranges in age from basal Ordovician to Honghuayuan, getting younger westward. The *"Acodus" oneotensis-Acanthodus costatus* Zone is widely distributed and has been found in the Ningzhen Mountains of Jiangsu Province and in western Hubei and Guizhou Provinces. The *Scolopodus quadraplicatus* Zone includes most of the Nanjiguan Formation, which has a high percentage of dolomite and bears a rather rare and uniform conodont fauna.

Triangulodus proteus [Pl. 2(19, 20)] is peculiar to the Fenxiang Formation, which also yields *Drepanodus deltifer, Scolopodus paucicostatus [Pl. 3(3)], and S. barbatus.* The former is a drepanodiform element of *Paroistodus deltifer* (Lindström) of late Tremadocian age in the North Atlantic conodont province, and may be compared with the *Paroistodus deltifer* Zone of western Europe. The appearance of *Paroistodus proteus, Serratognathus diversus* [Pl. 2 (23, 27, 30)], and *Scolopodus asperas* marks the top of the Fengxiang Formation, which may belong to the basal part of Arenigian.

The conodonts of the Honghuayuan Formation are very widely distributed in South China. Simple conodonts are still dominant, among which *Serratognathus* is the most characteristic, found only in East Asia. *Baltoniodus communis* (Ethington and Clark) [Pl. 4(20–29)], *Juanognathus variabilis, Drepanodus perlongus, Reutterodus depressus,* and *Protopanderodus gradatus* are characteristic elements in the upper Honghuayuan Formation. Thus, the Hunghuayuan Formation in Yichang can be divided into 2 subzones. An and others have compared the upper subzone with the

Stage	Great Britain graptolite zones	North Atlantic conodont zones and subzones (Bergström and Sweet, 1971)		Forma- tion	South China conodont zones (After An and others, 1979)	Forma- tion		North China conodont zones (After An, 1976)	Midcontinent faunas (From Harris and others, 1979) Series	Fauna
Ashgillian	Dicellograptus anceps	Amorphognathus ordovicicus		Lin-hsiang Wu-feng				Hiatus	Cincinnatian	12
	Dicellograptus complanatus									
	Pleurograptus linealis	Amorphognathus superbus		Pagoda	Protopanderodus insculputus	Fengfeng				11
Carndocian	Dicranograptus clingani				Hamarodus europaeus				Champlainian	10
	Diplograptus multidens	Amorphognathus tvaerensis	Prioniodus alobatus,,					Belodina-Badoudus-Microcoelodus Assemblage		9 8 7 6
			P. gerdae							
			P. variabilis							
Llandeilian	Nemagraptus gracilis	Pygodus anserinus		Miaopo	Polyplacognathus P. friendsvillensis	Machia Kou	Upper	Scandodus sp.		5
		Pygodus serrus	Eoplacognathus lindstroemi,							
	Glyptograptus teretiusculus		E. robustus	Kuniu-tang	Eoplacognathus reclinatus,			Eoplacognathus reclinatus		
	Didymograptus murchisoni		E. reclinatus, E. foliaceus,		E. foliaceus,			Eoplacognathus foliaceus		
			E. suecicus		E. pseudoplanus					
					Amorphognathus antivariabilis					4
Llanvirnian	Didymograptus "bifidus"	Amorphognathus variabilis		Dawan	Baltoniodus aff. B. navis					
Arenigian	Didymograptus hirundo	Microzarkodina parva			Paroistodus originalis		Lower	Tangshanodus tangshanensis		3
		Paroistodus originalis			Oistodus multicorrugatus-Periodon flabellum					2
	Didymograptus extensus	Baltoniodus navis								1
		B. triangularis								
		Oepikodus evae			Oepikodus evae			Scolopodus flexilis-S. filiformis		E
		Prioniodus elegans		Hung-hua-yuan	Baltoniodus communis	Liang-chiashan		Serratognathus bilobatus	Canadian	
	Tetragraptus approximatus	Paroistodus proteus			Serratognathus diversus					D
Tremadocian	Anisograptidae	Paltodus deltifer		Feng-siang	Paltodus deltifer-Triangulodus proteus	Yehli		"Acodus" oneotensis-Chosonodina herfurthi		C
					Scolopodus paucicostatus-S. barbatus					B
	Clonograptus tenellus	Cordylodus angulatus		Nantsin-kuan	Scolopodus quadraplicatus Acanthodus costatus-"Acodus" oneotensis					
	Dictyonema flabelliforme				Drepanodus simplex			Drepanodus simplex		

Figure 3. Correlation of Ordovician conodont zones in China, the North Atlantic, and the midcontinent province, United States.

Paroistodus elegans Zone of the North Atlantic and the lower subzone with the *Paroistodus proteus* Zone. The upper subzone of the Honghuayuan Formation is variable in thickness, thinning generally from east to west. From Qijiang in Sichuan Province westward, the upper subzone passes into shale of the lower member of the Meitan Formation. In Yichang, the boundary between the Honghuayuan Formation and the Dawan Formation is marked by 1 m of limestone rich in glauconite. Both above and below this limestone, *Oepikodus evae* (Lindström) is found. Thus, this glauconite limestone may be correlated with the *O. evae* Zone of the North Atlantic province.

The conodont fauna of the Dawan Formation is very widespread and very rich. The most important conodonts in this formation are *Drepanoistodus forceps* (Lindström) [Pl. 3(7, 12, 13)], *Paroistodus originalis* (Sergeeva) [Pl. 3(8, 14)], *P. parallelus* (Pander) [Pl. 3(20)], *Scolopodus rex* Lindström [Pl. 3(10)], *Drepanodus arcuatus* Pander [Pl. 3(22)], *Baltoniodus triangularis* (Lindström) [Pl. 4(12, 13, 19)], *B. navis* (Lindström) [Pl. 4(15–18, 24)], *Oepikodus evae* (Lindström), *Bergstroemognathus extensus* (Graves and Ellison), and *Oistodus multicorrugatus* (Harris). In Yichang, conodonts of the Dawan Formation may be divided into (in ascending order): (1) *Oepikodus evae-Bergstroemognatus extensus* Zone, (2) *Oistodus multicorrugatus-Periodon flabellum* Zone, (3) *Paroistodus orginalis* Zone, and (4) unnamed zone. The Dawan Formation is generally divided into three members: the lower and the upper shale members and the "middle limestone" member, in which *Protocycloceras deprati* was found. Zones 1 and 2 were found, respectively, in the lower and middle parts of the lower shale member; zone 3 was found in the upper part of the lower shale member and in the "middle limestone" member; and zone 4 occurred in the upper shale member. In the middle limestone and the top of the lower shale, *Paroistodus originalis* was found, which can be compared with *P. originalis* Zone of the North Atlantic province, and equals the lower part of the *Didymograptus hirundo* Zone. As there are abundant clastic rocks in the upper part of the Dawan Formation in Yichang, conodonts are rather rare. Conodonts of the upper shale member of the Dawan Formation were also found in the lower member of Xiaotan Formation in Anhui Province, which is referred to the *Baltoniodus* aff. *B. navis* Zone. *Amorphognathus antivariabilis* An was found in the upper part of the Dawan Formation at Yichang and Zigui in Hubei Province.

Conodonts are very abundant in the Guniutan Formation, where they are characterized by platform type forms. In the Huanghuachang area they have been divided into (in ascending order): (1) *Amorphognathus antivariabilis* [Pl. 4(10, 11)] Zone, (2) *Eoplacognathus pseudoplanus* [Pl. 4(14)] Zone, (3) *E. foliaceus* [Pl. 4(5, 6)] Zone, and (4) *E. reclinatus* Zone. The genera and species mentioned above are all elements of Llanvirnian conodont zones or subzones in Western Europe. *Eoplacognathus reclinatus* also occurs at the bottom of the Miaopo Shale. In Western Europe, this zone is generally considered to be at the top of the *Didymograptus murchisoni* Zone, so it is possible that the basal part of the Miaopo Formation contains the upper part of the *Didymograptus murchisoni* Zone. Conodont species known from the Guniutan Formation were also found in the Kelimuli Formation of Nei Mongol. *Eoplacognathus foliaceous* and *Pygodus serrus* were obtained from the Keping and Bachu regions of Xinjiang.

In South China, the Miaopo Formation comprises three facies: shelly facies, mixed facies, and graptolitic facies. Conodont faunas in the shelly and mixed facies are approximately the same. *Polyplacognathus* aff. *P. friendsvillensis, Pygodus serrus* [Pl. 4(1–3)], *P. anserinus, Complexodus pugionifer [Pl. 4(4)], Eoplacognathus miaopoensis,* and *Baltoniodus variabilis* are found in the lower part of the Miaopo Formation. These elements are found in the *Glyptograptus teretiusculus* Zone in other parts of the world. As platform conodonts are rare in the *Nemagraptus gracilis* Zone in the upper part of the Miaopo Formation, it is difficult there to establish conodont zones.

In the Pagoda Formation, conodonts are abundant but poor in species and similar at different localities. The main elements are *Amorphognathus superbus* (Rhodes) [Pl. 4(7)], *Hamarodus europaeus* [Pl. 3(32, 33)], *Protopanderodus insculptus* [Pl. 3(28)], *Panderodus gracilis, Dapsilodus mutatus* [Pl. 3(18)], *D. similaris* (Rhodes) [Pl. 3(4, 5)], *Belodella fenxiangensis, Icriodella baotanensis,* and *Protopanderodus liripipus* [Pl. 3(29)]. Two zones have been established, an upper *P. insculputus* Zone and a lower *H. europaeus* Zone, the former corresponding to the Meijiang Member (upper part of the Pagoda Formation) and easily recognized at Jiancaogou in Guizhou Province, and at Zigui and Yichang in Hubei Province.

The conodont fauna of the Linxiang Formation is close to that of the Pagoda Formation, especially the simple conodonts. Up to 1979, no platform conodont types have been found and no fossil zones established in the Linxiang Formation.

In South China, the Wufeng Formation consists mainly of dark shale in which conodonts are rare. Only *Drepanodus* sp. and *Acontiodus* sp. were found at Huanghuachang in Yichang.

Ordovician of North China

Ordovician conodont faunas have been found at many localities in North China, and systematic studies have been made of collections from Tangshan, Fengfeng, the Western Hills of Beijing, Laiwu, and Xintai.

Yeli Formation. The conodonts in this formation were studied in detail. The main elements are *Oneotodus gracilis* (Furnish) [Pl. 2(3)], *O. erectus* Druce and Jones [Pl. 1(10), Pl. 2(6)], *O. datsonensis* Druce and Jones) [Pl. 1(13)], *Acontiodus iowensis* Furnish, *A. transitans* (Druce and Jones) [Pl. 2(11)], *A. staufferi* Furnish, A. *(Semiacontiodus) bicostatus* Miller [Pl. 2(9)], *A. (S.) nogamii* Druce and Jones [Pl. 2(13, 14)], *Scolopodus primitive* An, *Chosonodina fisheri* Druce and Jones [Pl. 2(21)], *C. herfurthi* Mü [Pl. 2(22)], *Paltodus parvus* An, *P. variabilis* Furnish [Pl. 2(29)], *Drepanodus simplex* Branson and Mehl, [Pl. 1(15)], *Acanthodus costatus* Druce and Jones, *Scolopodus quadraplicatus* (Branson and Mehl) [Pl. 2(1, 8)], *S. bassleri* (Furnish) [Pl. 2(2)], *S. restrictus* An, *S. quinquecostatus* Müller, *Drepanodus subarcuatus* (Branson and Mehl), and *"Acodus" oneotensis* Furnish [Pl. 1(21, 22)]. The fauna can be easily correlated with that of the Nanjinguan Formation in South China and bears a strong resemblance to the conodont fauna of the Warentian in Queensland, Australia. The *Drepanodus simplex* Zone occurs near the bottom and the *"Acodus"* *oneotensis—Chosonodina herfurthi* Zone in the middle and upper part of the Yeli Formation.

Liangjiashan Formation. Conodonts of this formation are *Serratognathus bilobatus* Lee [Pl. 2(26)], *S. extensus* Yang, *Bergstroemognathus hubeiensis* An and others, *Paraserratognathus paltodiformis* An, *Triangulodus proteus* An and others, *Scolopodus huolianzhaiensis* An and Xu, *Scolopodus asperus* An and others, *Cornuodus longibasis* [Pl. 2(24, 25)], *Paltodus variabilis* Furnish, *P. deltifer* (Lindström) [Pl. 3(15, 19)], and *Drepanodus* spp. Of these *Serratognathus, Bergstroemognathus hubeiensis, Scolopodus asperus, Cornuodus longibasis,* and *Paltodus variabilis* are also common in the Honghuayuan Formation in South China. Without doubt, the Liangjiashan Formation and Honghuayuan Formation were deposited roughly contemporaneously, but the designation of their upper and lower limits needs further study. Because graptolites are rare in the Liangjiashan Formation, the appearance of *Archaeoscyphia, Coreanoceras, Scolopodus houlianzhaiensis,* and *Serratognathus* is commonly considered to indicate the lower boundary of the Liangjiashan Formation. In the Houlianxhai area of Benxi, Liaoning, the above-mentioned fossils occur with *Callograptus? taitzihoensis.* The thickness of the Liangjiashan Formation in Pingquan is 250 m, but is not more than 150 m in Thangshan, decreasing gradually toward the south, finally tapering off in this direction. The Yeli and Liangjiashan Formations in the southern part of North China consist mainly of dolomite, in which conodonts are very rare.

Lower Majiagou Formation. The formation contains abundant conodonts and is therefore easily correlated throughout the whole area. The most important conodonts in this formation are *Tangshanodus tangshanensis* An, *Scolopodus nogamii* Lee [Pl. 3(21)], *S. flexilis* An [Pl. 3(1, 2)], *S. filiformis* An [Pl. 3(11)], *Histiodella infrequens* An, *Loxodus disectus* An,*Rhipitognathus laiwuensis* Zhang, and *Paracordylodus orientalis* An, which on the whole indicate the stage of compound conodonts.

As is the case with cephalopods, the conodont fauna shows many differences from those of other parts of the world. At present, *Polydesmia* and *Ordosoceras,* two important cephalopods in the Lower Majiagou Formation, have not been found outside Asia. From that time onward, North and South China obviously began to form separate conodont biogeographic provinces. The conodont fauna of the Lower Majiagou Formation resembles that found in North Korea (Lee, 1975).

Upper Majiagou Formation. The major conodonts occurring in this formation are *Belodella rigida* An, *Plectodina onychodonta* An and Xu, *Tricladiodus aulilobus* Lee [Pl. 3(30,31)], *Panderodus exilis* An, *Eoplacognathus foliaceus* (Fåhraeus), *E. reclinatus* (Fåhraeus), *Scolopodus nogamii* Lee, and *Scandodus* spp. The appearance of platform conodonts is important. *Eoplacognathus foliaceus* and *E. reclinatus* are index fossils of the upper part of the Llanvirian series of the North Atlantic province and of the upper part of the Guniutan Formation in South China. They are roughly equivalent to the middle and lower parts of the *Didymograptus murchisoni* Zone. In addition, *Panderodus* and *Belodella* began to appear in the Guniutan Formation. Therefore, the beds of the upper Majiagou Formation containing these platform conodonts can be safely correlated with the upper Guniutan Formation. Above the *E. recinatus* Zone is a thick sequence of upper Majiagou Formation which may be higher than the Guniutan Formation. Obviously, the conodonts described by Lee (1975) include forms of the upper Majiagou Formation.

Fengfeng Formation. This formation is only distributed in the southern part of North China and was studied at Boshan in Shangdong and at Fengfeng, Hebei. The most important conodonts of this formation are *Panderodus gracilis* (Branson and Mehl) [Pl. 3(23, 27)], *Microcoelodus asymmetricus* Branson and Mehl [Pl. 3(24, 25)], *M. symmetricus* (Branson and Mehl) [Pl. 3(26)], *Belodina compressa* (Branson and Mehl) [Pl. 3(17)], *Badoudus badouensis* Zhang, *Plectodina praevara* An, *Leptochirognathus irregularis* An, *Dapsilodus mutatus* (Branson and Mehl), and *Scandodus* sp. The appearance of abundant *Belodina* and *Microcoelodus* is characteristic, which makes the fauna comparable to that of the Joachim and Plattin Formations in North America, belonging to the Black River Stage. The same conodonts are also found at Yaoxian in Shaanxi and at Yunxian in Hubei Province.

The correlation of the above-mentioned fossil zones is summarized in Figure 3.

DESCRIPTION OF NEW CONODONTS

Amorphognathus Branson and Mehl, 1933

Type species: _Amorphognathus ordovicicus_ Branson and Mehl, 1933

Amorphognathus antivariabilis An sp. nov.

Plate 4(10, 11)

The apparatus formed by ambalodiform and polyplacognathiform elements, with a large cusp and deep basal cavity; all processes pointed distally.

The ambalodiform element has fairly large posterior process and shorter anterior and anterior-lateral process; posterior process about twice as long as anterior process; anterior and lateral processes equal in length; angle between posterior and anterior processes generally 180°; angle between anterior process and anterior-lateral process about 30°. Aborally, unit is deeply excavated.

The polyplacognathiform element has four processes; posterior, posterior-lateral, anterior, and anterior-lateral; all processes confluent; angle between posterior-lateral and posterior processes vary between about 100° and 110°. Anterior and anterior-lateral processes shorter than posterior and posterior-lateral processes. In cross section, outline of base nearly rectangular.

Comparisons: This species resembles *Amorphognathus variabilis* Sergeeva in general aspects, but differs from the latter in the confluence of all processes, and in having a large cusp and a deep basal cavity. Apart from the distinct morphological differences between *Amorphognathus variabilis* and *A. antivariabilis,* their ranges are also different: *A. variabilis* is younger and occurs mainly in the Gunituan Formation, whereas *A. antivariabilis* occurs mainly in the uppermost Dawan Formation.

Type locality: Shizipu in Zunyi County, Guizhou, China

Type stratum: *Glyptograptus sinodentatus* var. *minor* Zone of Upper Dawan Formation (late Arenigian).

Holotype: BUC-8091 [Pl. 4(10)], a polyplacognathiform element.

Regional occurrence: The species has been found in the basal Guniutan Formation and Upper Dawan Formation of South

China, for example, at Nanjing, Hexian, Yichang, Youyang, Zunyi, and Changning.

Material: Ambalodiform elements 7, polyplacognathiform elements, 9 specimens.

Scolopodus Pander, 1856
Type species: _Scolopodus sublaevis_ Pander, 1856
Scolopodus filiformis An sp. nov.
Plate 3(11)

Description: Unit consists of a clear base and a cusp; base shallow with subcircular cross section; basal cavity conical with its apex developed near the anterior face at point of flexure. Cusp massive, reclined, and generally circular in cross section; surface ornamented by numerous threadlike costae that extend along the entire length of cusp.

Comparisons: In its general morphological features, this species somewhat resembles *Scolopodus filiosus* Ethington and Clark, but differs in the expanded base and the thicker and stronger unit.

Type locality: Zhaogezhuang, Tangshan City, Hebei, China.

Type stratum: *Wutinoceras* Zone, basal lower Majiagou Formation (late Arenigian).

Holotype occurrence: The geographical distribution of *Scolopodus filiformis* is not yet clear, but it seems limited to the basal Lower Majiagou Formation in Tangshan.

Material: 8 specimens.

Scolopodus flexilis An sp. nov.
Plate 3(1, 2)

Description: Recurved, roughly biconvex cusp with a few lateral costae that appear to run over the length of the cusp. One costa is more prominent and extends along entire length of cusp; shallow grooves parallel to costae. Transverse section normally narrowly elliptic; anterior and posterior keels straight. Outline of aboral surface of cusp elliptic with anterior part narrower; basal cavity low-conical with the apex situated anteriorly.

Type locality: Zhaogezhuang, Tangshan City, Hebei, China.

Type stratum: *Wutinoceras* Zone of basal lower Majiagou Formation (late Arengian).

Holotype: BUC-8052 [Pl. 3(2)].

Regional occurrence: This species is widely distributed in North China, for example, Hebei, Shandong, Shanxi, and Liaoning.

Material: About 100 specimens.

Scolopodus paucicostatus An sp. nov.
Plate 3(3)

Description: Recurved scolopodontids which are regularly curved above the base and distally straight; cross section of lower half of cone nearly rhombic; anterior and posterior keels sharp. Lateral costae start just above the aboral margin and extend along entire length of cusp, base rather small and slightly expanded posteriorly. Aboral surface elliptic with anterior and posterior parts narrower; basal cavity triangular and its apex pointing along axis of cone.

Comparison: The present species is somewhat similar to *Scolopodus flexilis,* but is distinguished from the latter by its smaller base and fewer costae.

Type locality: Huanghuachang, Yichang District, Hubei, China.

Type stratum: Uppermost Nanjinguan Formation (Tremadocian).

Holotype: BJ-9-12 [Pl. 3(3)].

Regional occurrence: Upper Nanjinguan Formation and lower Fenxiang Formation of Yichang, Zigui, and Nanjing.

Material: 30 specimens.

Serratognathus Lee, 1970
Type species: _Serratognathus bilobatus_ Lee, 1970
Serratognathus diversus An sp. nov.
Plate 2(23, 27, 30)

This species is comparable to *Serratognathus bilobatus* Lee (1970), but is distinguished from the latter by having a cusp that may be large or small, by the divergence of both lobes.

Type locality: Huanghuachang, Yichang District, Hubei, China.

Type stratum: *Manchuroceras* Zone, upper Honghuayuan Formation (early Arenigian).

Holotype: BJ-79-10 [Pl. 2(30)].

Regional occurrence: The species is a most important element in the Honghuayuan Formation, but it also occurred in the uppermost Fenxiang Formation.

Material: About 40 specimens.

Triangulodus Van Wasmel, 1974
Type species: _Triangulatus brevibasis_ Sergeeva, 1965
Triangulodus proteus An and others, sp. nov.
Plate 2(19, 20)

A species of *Triangulodus* with asymmetrical cones bearing 2 to 5 costae formed by a proclined cusp and long base; basal cavity moderately deep. Elements with two costae have sharp anterior and posterior edges and acodiform cross sections; elements with three costae have sharp posterior edges and both anterior-lateral costae; four-costae elements have two anterior-lateral and two posterior-lateral costae; five-costae elements have sharp posterior and anterior edges, other costae in lateral faces.

I have not found the oistodiform element of *Triangulodus* that normally occurs together with oistodiform elements.

This species is unlike any previously described *Triangulodus.*

Type locality: Huanghuachang, Yichang, Hubei, China.

Type stratum: *Acanthograptus sinensis* Zone, Fenxiang Formation (late Tremadocian).

Holotype: BUC-8043 [Pl. 2(20)].

Regional occurrence: Yichang, Zunyi, Nanjing.

Material: Two-costae element 5, three-costae element 1, four-costae element 3, five-costae element 4.

ACKNOWLEDGMENTS

I am greatly indebted to Professor Yue Sen-xun and Professor Wang Hong-zhen for reading the manuscript and for much helpful advice. I extend my heartfelt thanks to the Ministry of Petroleum Industry in defraying the expenses for the present research. Finally, I am much indebted to Du Gou-qing and Zhang Fang for their valuable assistance. All figured specimens are deposited in the Paleontology Collection of the Department of Geology, Beijing University.

SELECTED REFERENCES

An Tai-xiang, Du Gup-qing, Gao Qin-qin, Chen Qin-bao, and Li Wei-tong, [Conodont biostratigraphy of the Ordovician System of Yichang, Hubei, China]: (in Chinese) (in press).

Barnes, C. R., and Poplowski, M.L.S., 1973, Lower and Middle Ordovician conodonts from the Mystic Formation, Quebec, Canada: Journal of Paleontology, v. 47, p. 760–790, Pls. 1–5.

Bengtson, S., 1977, The structure of some Middle Cambrian conodonts, and the early evolution of conodont structure and function: Lethaia, v. 9, p. 188–206.

Bergström, S. M., 1977, Early Paleozoic biostratigraphy in the Atlantic borderlands, *in* Swain, F. M., ed., Stratigraphic micropaleontology of Atlantic basin and borderlands: Amsterdam, Elsevier Scientific Publishing Company, p. 85–110.

Druce, E. C., and Jones, P. J., 1971, Cambro-Ordovician conodonts from the Burke River structure belt, Queensland: Bureau of Mineral Resources Geology and Geophysics Bulletin 110, 159 p., 20 pls.

Dzik, J., 1978, Conodont biostratigraphy and paleogeographical relation of the Ordovician Mojcza Limestone: Acta Paleontologica Polonica, v. 23, no. 1, p. 12–14.

Ethington, R. L., and Clark, D. L., 1964, Conodonts from the El Paso Formation of Texas and Arizona: Journal of Paleontology, v. 38, p. 685–704, Pls. 113–115.

Fähraeus, L. E., 1966, Lower Viruan conodonts from the Gullhögen Quarry, south-central Sweden: Stockholm, Sveriges Geologiska Undersökning C, 610, p. 1–40.

Harris, A. G., Bergström, S. M., Ethington, R. L., and Ross, R. J., Jr., 1979, Aspects of Middle and Upper Ordovician conodont biostratigraphy of carbonate facies in Nevada and Southeast California and comparison with some Appalachian successions: Brigham Young University Geological Studies, v. 26, pt. 3, p. 7–44, Pls. 1–5.

Landing, E. D., Taylor, M. G., and Erdtmann, B. D., 1978, Correlation of the Cambrian-Ordovician boundary between the Acado-Baltic and North American faunal provinces: Geology, v. 6, no. 2, p. 75–78.

Lee Ha Yong, 1970, Conodonten aus der Chosen-Gruppe von Korea: Neues Jahrbuch für Geologie and Paläontologie, Abh., v. 136, p. 303–344, 2 pls.

——1975, Conodonten aus dem unteren and mittleren Ordovicium von Nordkorea: Palaeontographica, Abt. A, v. 150, Leif. 4–6, pl 161–186, 2 pls.

Lindström, M., 1954, Conodonts from the lowermost Ordovician strata of south-central Sweden: Geologiska Föreningens i Stockholm Förhandlingar 76, p. 517–603.

Löfgren, A., 1978, Arenigian and Llanvirnian conodonts from Jämtland, northern Sweden: Fossils and Strata, no. 13, 129 p., 16 pls.

Miller, J. F., 1969, Conodont fauna of the Notch Peak Limestone House Range, Utah: Journal of Paleontology, v. 43, p. 413–439, Pls. 63–66.

——1978, Upper Cambrian and lowest Ordovician conodont faunas of the House Range, Utah, *in* Miller, J. F., ed., Upper Cambrian to Middle Ordovician conodont faunas of western Utah: Southwestern Missouri State University Geoscience Series, no. 5, p. 1–33, 4 figs., 2 tables.

Müller, K. J., 1959, Kambrische Conodonten: Zeitschrift Deutsche Geologische Gesellschaft, 111 (2), p. 434–485, 5 pls.

——1964, Conodonten aus dem unteren Ordovicium von Südkorea: Neues Jahrbuch für Geologie and Paläaontologie, Abh., v. 119, p. 93–102, Pls. 12–13.

——1973, Late Cambrian and Early Ordovician conodonts from Northern Iran: Geological Survey of Iran Report, v. 30, p. 1–53, 11 pls.

Nogami, Y., 1966, Kambrische Conodonten von China, Teil I: Memoirs of the College of Science, University of Kyoto, ser. B., v. 32, no. 4, p. 351–366, Pls. 9–10.

——1967, Kambrische Conodonten von China, Teil 2: Memoirs of the College of Science, University of Kyoto, ser. B, v. 33, p. 211–216, Pl. 1.

Sweet, W. C., and Bergström, S. M., (eds.), 1971, Symposium on conodont biostratigraphy: Geological Society of America Memoir 127, 499 p.

Viyra, V., 1974, Konodonty Ordovika Pribaltiki: "Valgus," Tallin, 142 p. 13 pls. (in Russian).

MANUSCRIPT ACCEPTED BY THE SOCIETY APRIL 22, 1981

PLATE 1

All illustrations are scanning electron photomicrographs except as otherwise indicated.

Figure

1. *Westergaardodina muelleri* Nogami, Gushan Formation, Upper Cambrian, Quyang, western Hebei Province; ×160; repository BUC-8001.
2. *W. tricuspidata* Müller, Gushan Formation, Upper Cambrian, Quyang, western Hebei Province; ×160; repository BUC-8002.
3. *W. bicuspidata* Müller, Changshan Formation, Upper Cambrian, western Hills of Beijing; collector Liu Zhan; ×65; repository BUC-8003.
4. *W. matsushitai* Nogami, Gushan Formation, Upper Cambrian, Quyang, western Hebei Province, ×65; repository BUC-8004.

5, 6. *Cordylodus oklahomensis* Müller, Fengshan Formation, uppermost Cambrian, Zhaogezhuang village, Tangshan area, northeastern Hebei Province, 5. ×240; repository BUC-8005. 6. ×170; repository BUC-8006.

7. *Muellerina oelandica* (Müller), Fengshan Formation, uppermost Cambrian, Jingjing, western Hebei Province; ×250; repository BUC-8007.

8, 20. *Oneotodus nakamurai* Nogami, Fengshan Formation, uppermost Cambrian. 8. Xuzhou, northern Jiangshui Province; collector Yang Chang-sheng; ×200; repository BUC-8008. 20. Zhaogezhuang village, Tangshan area, northeastern Hebei Province; × 200; repository BUC-8009.

9. *Westergaardodina moessebergensis* Müller, Gushan Formation, Upper Cambrian, Quyang, western Hebei Province; ×24; stereophotograph; repository BUC-8010.

10. *Oneotodus erectus* Druce and Jones, Yehli Formation, Lower Ordovician, Zhaogezhuang village, Tangshan area, northeastern Hebei Province; ×160; repository BUC-8001.

11. Cf. *Proconodontus muelleri serrus* Miller, Fengshan Formation, uppermost Cambrian, Laishui, western Hebei Province; ×144; repository BUC-8013.

12. *Problematoconites perforatus* Müller, Fengshan Formation, uppermost Cambrian, Laishui, western Hebei Province; ×144; repository BUC-8013.

13. *Oneotodus datsonensis* Druce and Jones, Fengshan Formation, uppermost Cambrian, Tangshan area, northeastern Hebei Province; ×170; repository BUC-8014.

14. *Hirsutodontus* aff. *H. bulbosus* (Miller), Fengshan Formation, uppermost Cambrian, Xuzhou, northern Jiangshui Province; collector Yang Chang-sheng; ×240; repository BUC-8015.

15. *Drepanodus simplex* Branson and Mehl, Yehli Formation, Lower Ordovician, Tangshan area, northeastern Hebei Province; ×130; repository BUC-8016.

16. *Prooneotodus tenuis* (Müller), Fengshan Formation, uppermost Cambrian, Tangshan area, northeastern Hebei Province; ×240; repository BUC-8017.

17. *Hertzina* cf. *H. bisulcata* Müller, Fengshan Formation, uppermost Cambrian, Tangshan area, northeastern Hebei Province; ×65; repository BUC-8018.

18, 19. *Cordylodus rotundatus* Pander. 18. Yehli Formation, Lower Ordovician, Xingtang, Hebei Province; ×96; repository BUC-8019. 19. Tongze Formation (Tremadocian), Zunyi, Guizhou Province; ×40; stereophotograph; repository BUC-8020.

21, 22. *Acodus oneotensis* Furnish. 21. Yehli Formation, Lower Ordovician, Tangshan area, northeastern Hebei Province; ×220; repository BUC-8021. 22. Sanshanzi Formation, Lower Ordovician, Xuzhou, northeastern Jiangshui Province; collector Yang Chang-sheng; ×180; repository BUC-8022.

23. *Cordylodus* aff. *C. prion* Lindström, Fengshan Formation, uppermost Tangshan area, northeastern Hebei Province; ×220; repository BUC-8023.

1
2
3
4
5
6
7
8
9
10
11
12
13
14
15
16
17
18
19
20
21
22
23

PLATE 2

Figure

1, 8. *Scolopodus quadraplicatus* (Branson and Mehl), Yehli Formation, Lower Ordovician, Quyang, western Hebei Province. 1. ×75; repository BUC-8024. 8. ×65; repository BUC-8025.

2. *S. bassleri* (Furnish), Yehli Formation, Lower Ordovician, Tangshan area, northeastern Hebei Province; ×120; repository BUC-8026.

3. *Oneotodus gracilis* (Furnish), Yehli Formation, Lower Ordovician, Tangshan area, northeastern Hebei Province; ×170; repository BUC-8027.

4. *Acanthodus costatus* Druce and Jones, Lunshan Formation, Lower Ordovician, Ningzhen Mountains, Nanjing area; collector Ding Lian-sheng, ×60; repository BUC-8026.

5. Cf. *Proconodontus muelleri serrus* Miller, Sanshanze Formation, Lower Ordovician, Xuzhou, northern Jiangshui Province; collector Yang Chang-sheng; ×220; repository BUC-8029.

6. *Oneotodus erectus* Druce and Jones, Yehli Formation, Lower Ordovician, Tangshan area, northeastern Hebei Province; ×75; repository BUC-8030.

7. *Proconodontus notchpeakensis* Miller, Fengshan Formation, uppermost Cambrian, Tangshan area, northeastern Hebei Province; ×240; repository BUC-8031.

9. *Acontiodus (Semiacontiodus) bicostatus* Miller, Yehli Formation, Lower Ordovician, Tangshan area, northeastern Hebei Province; ×420; repository BUC-8032.

10. *Proconodontus carinatus* Miller, Fengshan Formation, uppermost Cambrian, Tangshan area, northeastern Hebei Province; ×324; repository BUC-8033.

11. Cf. *Acontiodus transitans* (Druce and Jones), Yehli Formation, Lower Ordovician, Quyang, western Hebei Province; ×165; repository BUC-8034.

12. A. *(Semiacontiodus) nogamii* Miller, Yehli Formation, Lower Ordovician, Tangshan area, northeastern Hebi Province; ×165; repository BUC-8035.

13, 14. *Scolopodus warendensis* Druce and Jones, Yehli Formation, Lower Ordovician. 13. Quyang, western Hebei Province; ×165; BUC-8036. 14. Thangshan area, northeastern Hebei Province; ×73; repository BUC-8037.

15. *Cordylodus* aff. *C. prion* Lindström, Fengshan Formation, uppermost Cambrian, Zhaozhuang, southern Shandong Province; ×150; repository BUC-8038.

16. *C. intermedius* Furnish, Fengshan Formation, uppermost Cambrian, Tangshan area, northeastern Hebei Province; ×144; repository BUC-8039.

17. *C. angulatus* Pander, Yehli Formation, Lower Ordovician, Tangshan area, northeastern Hebei Province; ×250; repository BUC-8040.

18. *C. proavus* Miller, Fengshan Formation, uppermost Cambrian, Tangshan area, northeastern Hebei Province; ×240; repository BUC-8041.

19, 20. *Triangulodus proteus* An and others, sp. nov. Fenxiang Formation, Lower Ordovician, Yichang, Hubei Province; ×220. 19. Four-costate element, innerview; repository BUC-8042. 20. Five-costate element; and holotype, outer view; repository BUC-8043.

21. *Chosonodina fisheri* Druce and Jones, Yehli Formation, Lower Ordovician, Tangshan area, northeastern Hebei Province; ×70; steropotograph; repository BUC-8044.

22. *Ch. herfurthi* Müller, Wanjuanshu Formation, early Tremadocian, Changning, Sichuan Province; ×40; stereophotograph; repository BUC-8045.

23, 27, 30. *Serratognathus diversus* An, sp. nov., Honghuayuan Formation, Lower Ordovician, 23. Youyang, Sichuan Province; ×30; repository BUC-8046. 27. Nanjing; collector Ding Liansheng; ×65; stereophotograph; holotype, BJ79-10. 30. Yichang, Hubei Province; ×220; BJ79-10.

24, 25. *Cornuodus longibasis* (Lindström). 24. Honghuayuan Formation, Lower Ordovician, Tongzi, northern Guizhou Province; × 40; stereophotograph; repository BUC-8047. 25. Guniutan Formation, Hexian, Anhui Province, Lower Ordovician; ×42; repository BUC-8048.

26. *Serratognathus bilobatus* Lee, Liangjiashan Formation, Lower Ordovician, Tangshan area, northeastern Hebei Province; ×30; stereophotograph; repository BUC-80108.

28. *Oistodus* aff. *O. lanceolatus* Pander, Honghuayuan Formation, Lower Ordovician, Tongzi, northern Guizhou Province; ×40; stereophotograph; repository BUC-8049.

29. *Paltodus variabilis* Furnish, Yehli Formation, Lower Ordovician, Tangshan area, northeastern Hebei Province; ×65; repository BUC-8050.

1
2
3
4
5
6
7
8
9
10
11
12
13
15
16
17
18
19
20
21
22
23
24
25
26
27
28
29
30

PLATE 3

All illustrations are scanning electron photomicrographs except as otherwise indicated.

Figure

1, 2. ***Scolopodus flexilis*** **An, sp. nov., Lower Majiagou Formation, Lower Ordovician, Tangshan area, northeastern Hebei Province. 1. ×45; holotype; repository BUC-8051. 2. ×56; repository BUC 8052.**

3. ***S. paucicostatus*** **An, sp. nov., Fenxiang Formation, Lower Ordovician, Yichang, western Hubei Province; ×50; holotype.**

4, 5. ***Dapsilodus similaris*** **Rhodes, Pagoda Formation, Middle Ordovician, Qijiang, southern Sichuan Province; ×30; stereophotograph; repository BUC-8053, BUC-8054.**

6. ***Scolopodus gracilis*** **Ethington and Clark, Xiaotan Formation, late Arenigian-Llanvirnian, Hexian, Anhui Province; ×96; collector Ding Liang-sheng; repository BUC-8055.**

7, 12, 13. ***Drepanoistodus forceps*** **(Lindström), Xiaotan Formation, late Arenigian-Llanvirnian, Hexian, Anhui Province; collector Ding Lian-sheng; ×96. 7. Homocurvatiform element; repository BUC-8056. 12. Suberectiform element; repository BUC-8057. 13. Oistodontiform element; repository BUC-8058.**

8, 14. ***Paroistodus originalis*** **(Sergeeva), Xiaotan Formation, late Arenigian-Llanvirnian, Hexian, Anhui Province; collector Ding Lian-sheng; ×98. 8. Drepanodontiform element; repository BUC-8059. 14. Oistodontiform element; repository BUC-8060.**

9. ***Protopanderodus gracilis*** **Serpagli, Shizipu Formation, Llanvirnian-Llandelian, Zunyi, Guizhou Province, stereophotograph; ×30; repository BUC-8061.**

10. ***Scolopodus rex*** **Lindström, Dawan Formation, Lower Ordovician, Nanjing; repository BUC-8062.**

11. ***Scolopodus filiformis*** **An, sp. nov., Lower Majiagou Formation, Lower Ordovician, Tangshan area, northeastern Hubei; ×70; stereophotograph; holotype; repository BUC-8063.**

15, 19. ***Paltodus deltifer*** **(Lindström), Liangjiashan Formation, Lower Ordovician, Pingquan, Hebei Province; stereophotograph; ×50. 15. Oistondontiform element; repository BUC-8064. 19. Drepanodontiform element; repository BUC-8065.**

16. ***Walliserodus ethingtoni*** **Fähraeus, Xiaotan Formation, late Arenigian-Llanvirnian, Hexian, Anhui Province; collector Ding Lian-sheng; ×44; oblique-posterior view; repository BUC-8066.**

17. ***Belodina compressus*** **Branson and Mehl, Fengfeng Formation, Middle Ordovocian, Xiyang, Shanxi Province; ×145; repository BUC-8067.**

18. ***Dapsilodus mutatus*** **Branson and Mehl, Pagoda Formation, Middle Ordovician, Changning, Sichuan Province; stereophotograph; ×30; repository BUC-8068.**

20. ***Paroistodus parallelus*** **(Pander), Xiaotan Formation, late Arenigian-Llanvirnian, Hexian, Anhui Province; collector Ding Lian-sheng; ×96; drepanodontiform element; repository BUC-8069.**

21. ***Scolopodus nogamii*** **Lee, Upper Majiagou Formation, Middle Ordovician, Tangshan area, northeastern Hebei Province; stereophotograph; ×50; repository BUC-8070.**

22. ***Drepanodus arcuatus*** **Pander, Dawan Formation, late Arenigian, Hexian, Anhui Province; collector Ding Lian-sheng; ×86; oistodontiform element; repository BUC-8071.**

23, 27. ***Panderodus gracilis*** **Branson and Mehl, Fengfeng Formation, Middle Ordovician, Fengfeng, western Hubei Province; ×70; compressiform element. 23. Repository BUC-8072. 27. Repository BUC-8073.**

24, 25. ***Micorcoelodus symmetricus*** **Branson and Mehl, Fengfeng Formation, Middle Ordovician, Fengfeng, western Hebei Province; ×22; posterior view; repository BUC-8074. 25. Anterior view; × 40; repository BUC-8075.**

26. ***M. asymmetricus*** **Branson and Mehl, Fengfeng Formation, Middle Ordovician, G59 Well, Hebei Province; ×100; posterior view; repository BUC-8076.**

28. ***Protopanderodus insculptus*** **Branson and Mehl, Pagoda Formation, Middle Ordovician, Yanjin, Yunnan Province; stereophotograph; ×40; repository BUC-8077.**

29. ***P. Liripipus*** **Kennedy and others, Pagoda Formation, Middle Ordovician, Hexian, Anhui Province; collector Ding Lian-sheng; ×67; repository BUC-8078.**

30, 31. ***Tricladiodus? auliobus*** **Lee, Upper Majiagou Formation, Middle Ordovician, Tangshan area, northeastern Hebei Province; ×40; stereophotograph. 30. Cordylodiform element; repository BUC-8079. 31. Roundiform element; repository BUC-8080.**

32, 33. ***Hamarodus europaeus*** **Serpagli, Pagoda Formation, Middle Ordovician. 32. Qijiang, Sichuan Province; ×65; repository BUC-8081. 33. Yichang, western Hubei Province; ×70; repository BUC-8082.**

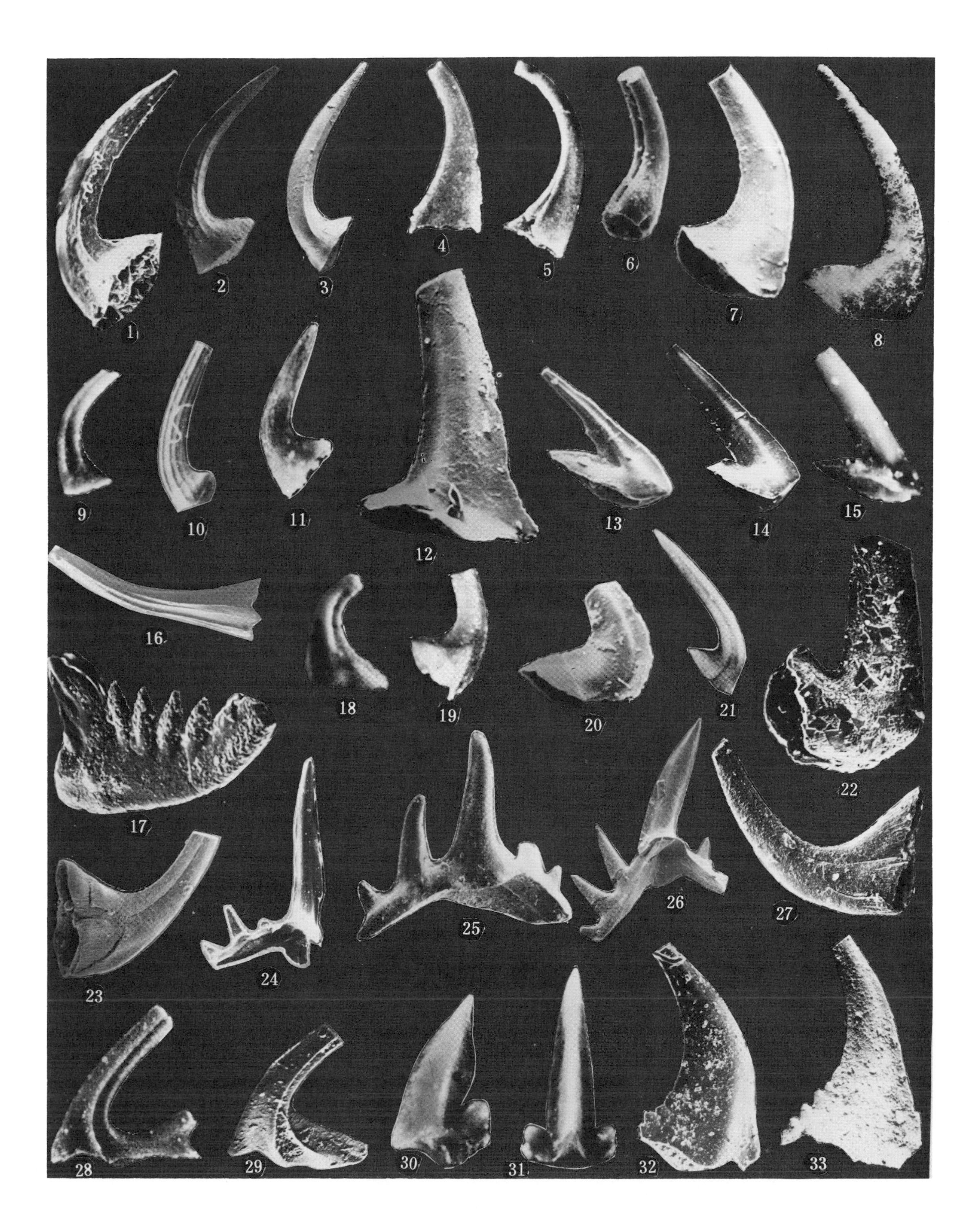
1
2
3
4
5
6
7
8
9
10
11
12
13
14
15
16
17
18
19
20
21
22
23
24
25
26
27
28
29
30
31
32
33

PLATE 4

All illustrations are scanning electron photomicrographs except as otherwise indicated.

Figure

1, 2, 3. *Pygodus serrus* Hedding. 1. Datianba Formation, Llandelian, Nanjing; collector Ding Lian-sheng; ×220; repository BUC-8083. 2, 3. Qiulitake Formation, Arenigian, Kalpin, Xinjiang Province; collector Zhao Zhi-xiz; ×90. 2. Haddingodiform element; repository BUC-8084. 3. Pygodiform element, upper view; repository BUC-8085.

4. *Complexodus pugionifer* Drygent, Miaopo Formation, Middle Ordovician, Zigui, western Hubei Province; ×60; repository BUC-8086.

5, 6. *Eoplacognathus foliaceus* Fahraeus, Guniutan Formation, Llanvirnian, Yichang, Yunnan Province; collector Du and others. 5. Polyplacognathiform element; repository BUC-8087. 6. Ambalodiform element; repository BUC-8088.

7. *Amorphognathus superbus* Rhodes, Pagoda Formation, Middle Ordovician, Yanjin, western Hubei Province; stereophotograph; ×30; dextral amorphognathiform element; repository BUC-8089; BUC-8090.

8. *Amorphognathus kieleensis* Dzik, Guniutan Formation, Llanvirnian, Yichang, western Hubei Province; ×75.

9a, 9b. *Prioniodus* cf. *P. alobatus* Bergström, Datianba Formation, Llandelian, Nanjing; collector Ding Lian-sheng; ×60; lateral and upper views; repository BUC-8091.

10, 11. *Amorphognatus antivariabilis* An, sp. nov., Meitan Formation, late Arenigian. 11. Zunyi, northern Guizhou Province; ×30; stereophotograph; aboral-lateral view; holotype; repository BUC-8092. 10. Xiaotan Formation, Lower Ordovician, Hexian, Anhui Province; ×110; oblique-upper view, repository BUC-8011.

12, 13, 19. *Baltoniodus triangularis* (Lindström), Xiaotan Formation, Arenigian, Hexian, Anhui Province; collector Ding Lian-sheng; ×96. 12. Gothodontiform element; repository BUC-8093. 13. Paracordylodiform element; repository BUC-8094. 19. Prioniodiform element; repository BUC-8095.

14. *Eoplacognathus pseudoplanus* Viira, Guniutan Formation, Llanvirnian, Nanjing; collector Ding Lian-sheng; ×60; sinistral amorphognathiform element; repository BUC-8093.

15–18, 24. *Baltoniodus navis* (Lindström), Xiaotan Formation, Arenigian, Hexian, Anhui Province; collector Ding Lian-sheng; ×89. 15. Paracordylodiform element; repository BUC-8096. 16. Prioniodiform element; repository BUC-8097. 17. Tetraprioniodontiform element; repository BUC-8098. 18. Trichonodelliform element; repository BUC-8099. 24. Oistodiform element; repository BUC-8100.

20–23, 25–29. *Baltoniodus communis* Ethington and Clark, Xiaotan Formation, Arenigian, Hexian, Anhui Province. 20, 25. Oepikodiform elements; ×89; repository BUC-80101, BUC-80102. 23. Paracordylodiform element; ×89; repository BUC-80105. 21, 22. Trichonodelliform elements; ×89; Repository BUC-80103, BUC-80104. 26. Oistodiform element; ×89; repository BUC-80106. 27, 28, 29. Prioniodiform elements; ×40; ×89; ×89; repository BUC-80107, BUC-80108, BUC-80109.

1
2
3
4
5
6
7
8
9a
9b
10
11
12
13
14
15
16
17
18
19
20
21
22
23
24
25
26
27
28
29

Printed in U.S.A.

Geological Society of America
Special Paper 187
1981

Permian conodont biostratigraphy of China

Wang Cheng-yuan
Wang Zhi-hao
Nanjing Institute of Geology and Palaeontology, Academia Sinica, Nanjing 210008, People's Republic of China

ABSTRACT

In this article, seven conodont assemblage zones are discussed. They represent all four stages of the Permian—Chihsian, Maokouan, Wuchiapingian, and Changhsingian—and lend support to the view that all the Permian stages are rather complete in China.

On the basis of conodont evolution, we suggest that the lower boundary of the Permian in China is drawn at the base of the *Neogondolella bisselli–Sweetognathus whitei* Assemblage Zone and approximately corresponds to the base of the Baomoshan Formation in South China and the base of the Shensi Formation in North China.

INTRODUCTION

In his article entitled "Conodonts from the Kufeng Formation of Lungtan, Nanking," Jin Yu-gan [Ching Yü Kan] (1960) reported for the first time on Permian conodonts in China. But, as has been pointed out by some authors (Bender and Stoppel, 1965; Szaniawski and Malkowski, 1979), some species that he determined were inadequately described. The species *Neogondolella nankingensis* may be valid, although some problems concerning it were still left to be solved. After publication of this article, studies on conodonts were at a standstill for more than 10 years. During this time, no one paid any attention to conodonts in China except Wang Cheng-yuan, who began his fundamental work on conodonts of different ages. In 1974, he described and figured some Permian conodonts from the Chihsia and Maokou Formations of Sichuan, such as *Neogondolella serrata* and *N. idahoensis*. In 1978, Wang Zhi-hao published an article on Permian conodonts from the Maokou and Wuchiaping Formations of Shanxi, making a first report on *Sweetognathus hanzhongensis (Gnathodus hanzhongensis), Neogondolella liangshanensis, N. aserrata (N. idahoensis),* and some other important Permian conodont species in China. Recently, we collected systematically many samples containing many conodont specimens from the Chihsia, Maokou, Lungtan, and Changhsing Formations in Zhejiang and Guizhou. Additional information on Permian conodonts was obtained from our colleagues working in Xinjiang and Jiangxi. As a result, a preliminary Permian conodont succession or zonation has been established, though still rather incomplete. There are many gaps in our knowledge with reference to the Permian conodont sequences as well as their regional differences.

So far as we are aware, the first report on Permian conodonts was given by C. C. Branson (1932), and the first conodont zonation of the Permian system was proposed by Clark and Behnken (1971). Later on, somewhat more detailed Permian conodont zonations were published by Behnken (1975), Kozur (1975), and Clark and others (1979). Compared with the studies in the United States and Germany, our work is too incomplete to determine the range of some species. There are guide fossils in some horizons and localities. The conodont faunas in Permian rocks of China have proved to be extremely rich and they should provide an excellent basis for the establishment of a complete Permian conodont zonation. We believe that with our joint efforts we will make more contributions to the study of Permian conodonts in the days ahead.

Based on the newest data concerning the Permian conodont faunas in China, a brief summary of the Permian conodont biostratigraphy follows.

LOWER PERMIAN

Chihsia Stage

The names of Permian stages used in China are the Chihsia, Maokou, Wuchiaping, and Changhsing, in ascending order. According to traditional correlation, the lower boundary of the Chihsia Stage of the Permian corresponds to the upper boundary of the Sakmarian Stage, but this conclusion was revised recently by some authors who placed the lower boundary of the Chihsia Stage within the Sakmarian. Many conodont samples have been collected from the type section of the Chihsia Formation of Nanjing, but no conodont guide fossils were recognized except *Anchignathodus minutus*.

Neogondolella bisselli–Sweetognathus whitei Assemblage Zone. This assemblage characterizes an interval of the upper Sterlitamakian in the Soviet Union and the Sakmotozawan in Japan. The name-giving species are known in the lower part of the Chihsia Formation in Ziyun, Guizhou, in association with *Neostreptognathodus pequopensis* and *Anchignathodus minutus. Sweetognathus whitei* has also been found in the Chihsia Formation of Pingxiang, Jiangxi, and in the Kecüligiman Formation of Xinjiang. There is not enough material, however, to determine the range of the name-giving species in China. The horizon above this assemblage has not been recognized. Clark and Behnken (1971) established in ascending order two conodont assemblage zones in Nevada, namely, *Idiognathodus ellisoni* and *Neogondolella bisselli–Sweetognathus whitei.* The former represents the lower Wolfcampian and is identical with Pennsylvanian species. According to Clark and Behnken (1971), *Neogondolella bisselli* differs from the typical Pennsylvanian *Neogondolella* and is rather similar to Permian and later neogondolellids. Therefore, they expressed their view that the faunal changes from the Pennsylvanian to the Permian type conodonts started in later Wolfcampian.

Streptognathodus elongatus, S. wabaunsensis, and other Upper Carboniferous species are known to occur in the upper part of the Taiyuan Formation of Shanxi and the Tahagi Formation of Xinjiang, associated with the fusulinids *Occidentoschwagerina texa* and *Schwagerina richthofeni.* From the point of view of conodont evolution, the lower boundary of Permian is best placed at the base of the *Neogondolella bisselli–Sweetognathus whitei* Assemblage Zone. This boundary probably coincides approximately with the lower boundary of the Baomoshan Formation in South China and of the Shanxi Formation in North China.

Neogondolella idahoensis Assemblage Zone. The name-giving species is characteristic of an assemblage in the upper Chihsia Formation in Ziyun, Guizhou, where it is associated with *Merrillina oertlii* and *Anchignathodus minutus.* These are guide fossils of the Leonardian in North America and the Sakamotozawan in Japan and are known also from the Chihsia Formation of Sichuan.

Maokou Stage

The Maokou Stage is assigned to the upper Lower Permian. Although it is known to occur widely in South China, our knowledge of its conodonts is rather inadequate as compared with that in other Permian stages. So far as we know, only one conodont assemblage can be recognized.

Neogondolella serrata–N. aserrata Assemblage Zone. Besides the name-giving species, two more species, *Sweetognathus hanzhongensis* (Wang) and *Anchignathodus minutus,* have been found in the middle and upper Maokou Formation of Liangshan, Shaanxi, and of Pingxiang, Jiangxi. Ching described a new species, *Neogondolella nankingensis,* from the Kufeng Formation that is equivalent to the lower Maokou Formation in stratigraphic position. Unfortunately, the holotype of *N. nankingensis* was lost by Ching Yü-kan. Having reexamined a specimen of the species (Ching Yü-kan, 1960, Pl. II, fig. 5, Cat. No. 10488) housed in our Institute, we found there was no serration or nodulation at the posterior platform as sketched by Ching Yü-kan. In fact, the surface of the posterior platform was smooth. It is, therefore, possible that *Neogondolella serrata* is a junior synonym of *N. nankingensis,* as suggested by Kozur.

One specimen of *Neostreptognathodus prayi* is known to occur in the middle Maokou Formation. According to Clark and others (1979), this species occurs mainly in the Leonardian, while *Neogondolella serrata* occurs in the Wordian. Kozur considered that the occurrence of *N. serrata* and *Sweetognathus hanzhongensis* in the middle and upper Maokou was in good agreement with a Wordian to Capitanian age.

UPPER PERMIAN

Wuchiaping Stage

Neogondolella liangshanensis–N. bitteri Assemblage Zone. The name-giving species characterize an assemblage of the Wuchiaping Formation in Liangshan of Shaanxi, associated with *Anchignathodus minutus, Enantiognathus ziegleri,* and *Xaniognathus elongatus. Neogondolella liangshanensis* in the Wuchiaping Formation is transitional between *N. bitteri* Kozur and *N. leveni* Kozur, Mostler, and Pjatakova. The age of this species is therefore uppermost Abadehian.

Neogondolella orientalis Assemblage Zone. The uppermost part of the Lungtan Formation at the type locality of the Changhsing Formation is well defined by the prolific occurrence of *Neogondolella orientalis.* This species is also known in the Dzhulfian Stage of Transcaucasia, the Soviet Union, and in the Ali Bashi Formation of northwestern Iran.

Changhsing Stage.

Neogondolella subcarinata subcarinata–N. s. elongata Assemblage Zone. At the stratotype of the Changhsing Stage, *Neogondolella subcarinata subcarinata* and *N. s. elongata* are abundant in the lower part of the Changhsing Limestone, along with *Anchignathodus minutus, Enantiognathus ziegleri, Prioniodella decrescens,* and *Xaniognathus elongatus. Neogondolella subcarinata changxingensis* is also known in the upper part of this assemblage. *N. subcarinata subcarinata* has been found in the Dorashamian Formation of Transcaucasia of the Soviet Union and in the Ali Bashi Formation of Iran.

Neogondolella deflecta–N. subcarinata changxingensis Assemblage Zone. The upper part of the Changhsing Formation is characterized by *Neogondolella deflecta* and *N. subcarinata changxingensis* known in Changxing, Zhejiang. This assemblage also contains *Neogondolella carinata, Anchignathodus minutus, Ellisonia teicherti, Enantiognathus ziegleri, Prioniodella decrescens,* and *Xaniognathus elongatus.* A small number of *N. subcarinata subcarinata* can be found in this assemblage. This assemblage is transitional with the *N. s. subcarinata–N. s. elongata* assemblage zone. It is a rather characteristic fauna. The upper part of the Changhsing Formation in which the assemblage occurs may be higher than the Ali Bashi Formation or the Dorasham Formation, and may occupy the uppermost position in the Permian stratigraphic sequence.

SYSTEMATIC DESCRIPTIONS

The specimens described here are preserved in the Nanjing Institute of Geology and Palaeontology, Academia Sinica. The generic name *Neogondolella* Bender and Stoppel (1965) is regarded by Kozur (1975) as a junior synonym of *Gondolella* Stauffer and Plummer (1932), but this problem is not within the scope of this paper.

Anchignathodus Sweet, 1970
Type species: _Anchignathodus typicalis_ Sweet, 1970
Anchignathodus minutus (Ellison), 1941
Plate 2(6, 7)

Gnathodus sichuanensis Wang, 1974, p. 315, Pl. 166, figs. 18–22
Spathodus minutus Ellison, 1941, p. 120, Pl. 20, figs. 50–52
Anchignathodus typicalis Sweet, 1970, p. 222, 223, Pl. 1, figs. 13, 20; Wang and Wang, 1976, p. 398, 399, Pl. 5, fig. 13; Wang Zhi-hao, 1978, p. 215, Pl. 1, figs. 26–28
Anchignathodus minutus (Ellison) Kozur, 1977, p. 1118–1120, Pl. 1, figs. 1–16.

Description: Blade short, straight, broadly arched with aboral edge in anterior half nearly straight. Anteriormost cusp large and erect or slightly inclined posteriorly, from which about 13 denticles tend to diminish regularly toward the posterior tip of unit. Outline of basal cavity lachrymiform, broadly rounded anteriorly with posterior point reaching near posterior end of blade; deepest point near anterior and slightly back of superior cusp; cavity transversed by a longitudinal groove, extending anteriorly and posteriorly.

Occurrence: Chihsia Formation of Ziyun, Guizhou.

Merrillina Kozur, 1975
Type species: _Spathognathodus divergens_ Bender and Stoppel, 1965
Merrillina oertlii Kozur, 1976

Merrilina oertlii Kozur, 1976, p. 12, Pl. 3, fig. 10
Sweetocristatus oertlii (Kozur), Szaniawski and Malkowski, 1979, p. 254

Description: Only one broken specimen. Bladelike element arched. Anterior blade with 6 fused long denticles, posterior blade with 4 fused posteriorly inclined short denticles; main cusp large, inclined posteriorly; basal cavity large, expanding laterally. Basal groove narrow under anterior blade, expanded under posterior blade.

Occurrence: Chihsia Formation of Ziyun, Guizhou.

Neogondolella Bender and Stoppel, 1965
Type species: _Gondolella mombergensis_ Tatge, 1956
Neogondolella aserrata Clark and Behnken, 1979
Plate 1(9, 10); Plate 2(12, 14)

Neogondolella idahoensis (Youngquist, Hawley, and Miller), Z. H. Wang, 1978, Pl. 2, figs. 23–26
Neogondolella aserrata Clark and Behnken, 1979, pl 271, 272, Pl. 1, figs. 1–11.

Description: Platform long and narrow as in *N. serrata,* tapering from approximate midpoint of platform to anterior tip. Anterior serration on platform margin indistinct or absent; platform asymmetrical at posterior end. Carina composed of nodose denticles, with a distinct cusp in the posteriormost part; keel of lower surface narrow.

Remarks: This species is characterized by tapering from approximate midpoint of platform to anterior tip and by general absence of anterior serrations on the platform margins.

Occurrence: Maokou Formation of Liangshan area, Southern Shaanxi.

Neogondolella bisselli (Clark and Behnken, 1971)
Plate 2(16, 17)

Gondolella bisselli Clark and Behnken, 1971, p. 429, Pl. 1, figs. 12–14
Neogondolella bisselli (Clark and Behnken) Behnken, 1975, p. 306, Pl. 1, figs. 27, 31

Description: Long, tear-drop shaped; posterior end round; margins tapering gradually to pointed anterior. Surface of platform smooth or pitted; whole unit slightly arched. Carina with 14 discrete, small, round, and pointed denticles; lower surface smooth; basal opening of basal cavity type; occupying half of total lower surface, with shallow V-shaped cross section; a large loop on posterior end.

Remarks: This species is distinguished from other species by the gracefully arched, narrow, tear-drop-shaped platform.

Occurrence: Lower Chihsia Formation of Ziyun, Guizhou.

Neogondolella bitteri (Kozur, 1975)
Plate 2(9, 13)

Gondolella bitteri Kozur, 1975, p. 19, 20; Kozur, 1978, Pl. 5, figs. 30, 31; Pl. 6, figs. 17, 19, 20

Description: Platform moderately broad. Greatest width at about middle part, abruptly tapering anteriorly about anterior third of unit. Outline of posterior end subcicular. Length of free blade about a third of unit; carina high at anterior and low at middle and posterior. Oral surface smooth; keel moderately broad.

Remarks: This species is closely similar to *Neogondolella liangshanensis,* but has parallel margins at the mid-posterior end of platform and tapers abruptly anteriorly at about anterior third of unit, giving a long free blade.

Occurrence: Wuchiaping Formation of Liangshan area, southern Shaanxi.

Neogondolella deflecta sp. nov.
Plate 1(22, 23)

Diagnosis: A species of *Neogondolella* characterized by a wide platform with a truncated posterior end and a gradually tapering anterior end and by a sharp inward deflection of the carina at the posterior end of the platform.

Description: Most of specimens here referred to *Neogondolella deflecta* sp. nov. are faintly bent inward and slightly arched, but a few are straight. Inconspicuous cusp in median position or situated slightly to inner side of the midline in position. In most specimens, cusp fused with 2 or 3 adjacent low nodes to form a low short carina that extends to inner-posterior angle of platform, giving an impression that carina is sharply deflected inwardly at posterior end of unit. Nodes of carina low and fused, increasing in size and height anteriorly, middle nodes commonly showing most fusion in some specimens, and forming a smooth carina; anterior nodes or denticles high and separated. Adcarina grooves well developed. Platform wide, smooth, and having a truncated posterior end, gradually narrowing and tapering anteriorly for whole length of unit, free blade absent. Lower attachment surface wide, having a truncated posterior margin, and a pit surrounded by a slightly elevated loop posteriorly; pit extending anteriorly as a narrow median groove bordered by a slightly elevated keel.

Remarks: This species is similar to *Neogondolella rosenkrantzi* from which it may have evolved, but differs from the latter by having a generally wide platform and sharp carina deflection at its posterior end and a low fused middle carina. It is also very similar to *N. aserrata,* but the new species has a straight truncated posterior margin and a stronger deflection at posterior carina. In addition, their stratigraphic horizons are also different. *Neogondolella rosenkrantzi* has been suggested as the ancestor of *N. deflecta.*

Occurrence: Upper Changhsing Formation of Changxing, Zhejiang.

***Neogondolella liangshanensis* Z. H. Wang, 1978**
Plate 1(1–3, 11, 12, 14, 15)

Neogondolella liangshanensis Z. H. Wang, 1978, p. 221, Pl. 2, figs. 9–11, 19, 27–33.

Description: Platform wide, widest at mid-posteror part; gradually tapering anteriorly. Posterior platform wide and round. Anterior free blade absent or very short. Upper surface of platform smooth, with indistinct nodes at anterior platform. Mid-posterior carina very low, extending to posterior end of platform without cusp. Adcarina grooves distinct; keel wide; posterior edge of keel round or square.

Remarks: *Neogondolella liangshanensis* is closely similar to *N. bitteri,* but it can be distinguished by its platform shape, absent or short free anterior blade, indistinct nodes at anterior platform, and very low mid-posterior carina.

Occurrence: Wuchiaping Formation of Liangshan area, Southern Shaanxi.

***Neogondolella idahoensis* (Youngquist and others, 1951)**
Plate 2(15, 23)

Gondolella idahoensis Youngquist and others, 1951, p. 361, Pl. 54, figs. 1–3, 14, 15; Clark and Behnken, 1971, p. 431, Pl. 1, fig. 9

Gondolella phosphoriensis Youngquist and others, 1951, p. 362, Pl. 54, figs. 10–12, 27, 28

Description: Platform long and narrow, tapering only in anteriormost portion; upper surface of platform smooth. Denticles of carina discrete and low; suberect posterior cusp distinct. Keel on lower surface narrow, bearing a V-shaped basal groove.

Remarks: We agree with Behnken's (1975) discussion of this species. According to Behnken, it differs from other Permian species of *Neogondolella* in (1) the narrow, long platform which tapers only in anteriormost portion; (2) the narrow keel of the lower surface which bears a V-shaped basal groove; (3) the surface of the platform which bears no other ornamentation; and (4) the suberect posterior cusp.

Occurrence: Upper Chihsia Formation of Ziyun, Guizhou.

***Neogondolella orientalis* (Barskov and Koroleva), 1970**
Plate 1(16, 17)

Gondolella orientalis Barskov and Koroleva, 1970, p. 933, 934, fig. 1a–1c.

Neogondolella orientalis (Barskov and Koroleva) Sweet, 1973, p. 143, Pl. Neogondo-Plate 1, fig. 7

Description: Platform wide and oval in outline, abruptly tapering toward anterior at about anterior third of unit; surface of platform smooth. Anterior carina high, decreasing posteriorly, without cusp; end of posterior carina surrounded by wide platform brim. Keel on lower surface wide.

Remarks: This species differs from *N. subcarinata* in absence of cusp and wide platform brim at posterior end of carina.

Occurrence: Uppermost Lungtan Formation of Changxiang, Zhejiang.

***Neogondolella serrata* (Clark and Ethington, 1962)**
Plate 2(1–4)

Gondolella serrata Clark and Ethington, 1962, p. 108, 109, Pl. 1, figs. 10, 11, 15, 19; Pl. 2, figs. 1, 6, 8, 9, 11–14; Clark and Mosher, 1966, p. 389, Pl. 47, figs. 13–15; Clark and Behnken, 1971, p. 431, Pl. 1, fig. 10

Neogondolella serrata serrata (Clark and Ethington), Behnken, 1975, p. 308, Pl. 2, figs. 21–24, 37

Neogondolella serrata (Clark and Ethington), Z. H. Wang, 1978, p. 222, Pl. 2, figs. 6–8, 14, 15, 20–22

Description: Platform narrow, gradually tapering anteriorly, widest at middle-posterior part; whole unit slightly arched. Upper surface of platform smooth at posterior and serrated or ridged at anterior platform margins. Carina with more than 9 denticles, anteriormost 4-5 being the highest; denticles at mid-posterior. Keel of lower surface high and narrow; groove V-shaped. Pit distinct.

Remarks: This species is characterized by a narrow platform, prominent serrate ridges at anterior platform, and a narrow elevated keel.

Occurrence: Middle-upper Maokou Formation of Liangshan area, Southern Shaanxi.

***Neogondolella subcarinata* (Sweet, 1973)**
***Neogondolella subcarinate changxingensis* subsp. nov.**
Plate 1(6, 7)

Neogondolella carinata subcarinata Sweet *in* Teichert and others, 1973, P. 13, figs. 14, 15

Diagnosis: A subspecies of *Neogondolella subcarinata* characterized by a narrow and elongate platform and by the lack of a distinct notch and buttress at posterior platform.

Description: Platform slightly arched, straight, narrow, lanceolate, symmetrical or subsymmetrical, abruptly tapering posteriorly and gradually narrowing anteriorly. Greatest width at about middle or near mid-posterior of unit; its posterior end characterized by indistinct posterior platform buttress and lacking distinct notches. Carina composed of many denticles, forming a low ridge, but becoming larger anteriorly; posterior cusp obvious. Juvenile keel narrow and high, gerontic ones wider; a single median groove extending from anterior and ending in a pit beneath cusp.

Remarks: The elements of *Neogondolella subcarinata changxingensis* are distinct from those of the *N. subcarinata subcarinata* established by Sweet (1973) which included two morphotypes, one with wide platform, the other with narrow, elongate platform. According to Sweet's description, the characteristics of *N. s. subcarinata* is the wide platform, the ratio of length to width varying from 2.2 to 3. We assign the group with wide platform to *N. subcarinata subcarinata,* and establish a new subspecies *N. subcarinata changxingensis* for forms with a narrow platform.

Occurrence: Upper Changhsing Formation of Changxing, Zhejiang.

Neogondolella subcarinata elongata subsp. nov.
Plate 1(21, 25)

Diagnosis: A subspeices of *Neogondolella subcarinata* characterized by an elongated and posteriorly inclined cusp and by a wide platform that abruptly narrows anteriorly.

Description: Platform wide, arched, and slightly curved laterally, its greatest width near middle or mid-posterior. Cusp commonly projects beyond posterior platform, forming a posterior separate wedge, commonly forming part of margin of platform. Denticles on carina partly fused, increasing in size and spacing anteriorly. Platform abruptly tapering at posterior end and at anterior one-third to one-fourth of unit. Free blade not well developed. Lower attachment surface wide and having a pit, surrounded by elevated loop posteriorly; keel gradually elevated from pit.

Remarks: This new subspecies can be distinguished from *Neogondolella subcarinata subcarinata* by a large inclined cusp and a posterior platform elongation. It is also closely similar to *N. elongata,* but differs from the latter in having a very narrow platform at both sides of anterior blade-carina, and having lower and closer denticles on the anterior blade.

Occurrence: Lower Changhsing Formation of Changxing, Zhejiang.

Neogondolella subcarinata subcarinata (Sweet, 1973)
Plate 1(4, 5, 8)

Neogondolella carinata subcarinata Sweet *in* Teichert and others, 1973, p. 436, Pl. 13, figs. 12, 13, 16, 17; Textfigs. 16E–16H.
Gondolella carinata subcarinata, Kozur, 1976, p. 19, Pl. 2, figs. 9, 10
Gondolella subcarinata (Sweet), Kozur, 1978, Pl. 8, figs. 16–18

Description: Unit symmetrical or subsymmetrical; platform wide and short, the ratio of length to width varying from 2.2 to 3. Posterior platform buttress indistinct and lacking a distinct notch at posterior end of platform. Carina composed of 7 to 18 fused denticles, forming a low ridge with a cusp. Keel of lower surface narrow and elevated in juvenile specimens, but low and wide in gerontic specimens.

Occurrence: Lower Changhsing Formation of Changxing, Zhejiang.

Neogondolella sp.
Plate 1(18, 24)

Description: Platform wide, greatest width at about middle, tapering anteriorly at anterior one-third to one-fourth of unit, without free blade. Carina composed of many fused denticles, ridged at mid-posterior platform and bladelike at anterior portion. Posterior cusp distinct. Adcarina grooves well developed. Lower attachment surface wide and a pit surrounded by an elevated loop posteriorly.

Occurrence: Upper Chihsia Formation of Ziyun, Guizhou.

Neostreptognathudus Clark, 1972
Type species: _Streptognathudus sulcoplicatus_ Youngquist and others, 1951
Neostreptognathodus pequopensis Behnken, 1975
Plate 1(13, 19)

Neostreptognathodus pequopensis, Behnken, 1975, p. 310, Pl. 1, figs. 19–22, 25

Description: Only one gerontic specimen. Two rows of equidimensional nodes on platform, separated by sharply defined medial groove, and each composed of 7 nodose denticles; free blade composed of 4 denticles, about a third of unit. Basal cavity centrally located under platform and extending anteriorly as a narrow groove on lower surface of free blade.

Remarks: This species is very similar to *N. sulcoplicatus.* Behnken (1975) considered that later growth stages of this species show a single row of denticles divided into two rows of equidimensional nodes by a sharply defined medial groove, but mature forms of *N. sulcoplicatus* possess transverse to oblique ridges on at least the posterior portion of the two carinas.

Occurrence: Lower Chihsia Formation of the Ziyun, Guizhou.

Neostreptognathodus prayi Behnken, 1975
Plate 2(18, 19)

Neostreptognathodus prayi Behnken, 1975, p. 310, 311, Pl. 2, figs. 14, 17–20

Description: Only one specimen, closely similar to gerontic specimens described by Behnken (1975). Large, bowed asymmetrical unit, with 9 transverse low ridges on mid-posterior platform, extending across a barely discernible medial groove; an elevated ridgelike structure distinct at anterior-lateral margins of platform. Free blade long, about a third of unit, its denticles broken. About 5 anteriormost ridges on platform indistinct by diminution, so that they become smooth and structureless. Basal cavity large, extending throughout entire length.

Remarks: This is a gerontic specimen. It is characterized by diminution of 5 anteriormost ridges on the platform.

Occurrence: Middle Maokou Formation of Sichuan.

Sweetognathus Clark, 1972
Type species: _Spathognathodus whitei_ Rhodes, 1963
Sweetognathus whitei (Rhodes, 1963)
Plate 1(20); Plate 2(8, 10, 11, 22)

Spathognathodus whitei Rhodes, 1963, p. 404, Pl. 47, figs. 4, 9, 10, 25, 26; Clark and Behnken, 1971, Pl. 1, figs. 2–6; Merrill, 1973, Pl. 3, figs. 1–9
Sweetognathus whitei (Rhodes), Clark, 1972, p. 155; Behnken, 1975, p. 312, fig. 26

Description: Unit straight or gently curved, oval in outline, blade-carina bearing 6 to 11 denticles. In free blade, these laterally compressed denticles may be as many as 7 in number. Denticles of carina on the cup greatly expanded laterally, forming 6 to 11 transverse ridges, each one composed of 2 to 5 small nodes. Transverse ridges discrete, being separated from each other by an interval of the same width as ridges. In these intervening depressions they are joined only by a low median bladelike ridge, whose

PLATE 1

Figure

1–3, 11, 12, 14, 15. *Neogondolella liangshanensis* **Z. H. Wang**

1–3 Lateral, aboral and oral views of paratype, ×60, Wuchiaping Formation of Liangshan area, southern Shaanxi. Cat. No. 45442.

11, 12. Oral and aboral views, ×60, Wuchiaping Formation of Liangshan area, southern Shaanxi. Cat. No. 45446.

14, 15. Aboral and oral views of holotype, ×60, Wuchiaping Formation of Liangshan area, southern Shaanxi, Cat. No. 45448.

4, 5, 8 *Neogondolella subcarinata subcarinata* **(Sweet)**

4, 5. Aboral and oral views, ×60, lower Changhsing Formation of Zhejiang. Cat. No. 53213.

8. Oral view, ×60, lower Changhsing Formation of Zhejiang. Cat. No. 53216.

6, 7. *Neogondolella subcarinata changxingensis* **subsp. nov. Aboral and oral views of holotype, ×45, upper Changhsing Formation of Zhejiang. Cat. No. 53217.**

9, 10. *Neogondolella aserrata* **Clark and Behnken. Aboral and lateral views, ×60, Maokou Formation of Liangshan area, southern Shaanxi. Cat. No. 45453.**

13, 19. *Neostreptognathodus pequopensis* **Behnken. Aboral and oral views, ×80, lower Chihsia Formation of Guizhou. Cat. No. 63323.**

16, 17. *Neogondolella orientalis* **(Barskov and Koroleva). Oral and aboral views, ×45, upper Lungtan Formation of Zhejiang. Cat. No. 53229.**

18, 24. *Neogondolella* **sp. Oral and aboral views, ×60, Chihsia Formation of Guizhou. Cat. No. 63324.**

20. *Sweetognathus whitei* **(Rhodes). Oral view, ×60, lower Chihsia Formation of Guizhou. Cat. No. 63325.**

21, 25. *Neogondolella subcarinata elongata* **subsp. nov. Oral and aboral views of holotype, ×60, lower Changhsing Formation of Zhejiang. Cat. No. 53226.**

22, 23. *Neogondolella deflecta* **sp. nov. Oral and aboral views of holotype, ×45, upper Changhsing Formation of Zhejiang. Cat. No. 53228.**

1
2
3
4
5
6
8
10
11
12
13
14
15
16
17
18
19
20
21
22
23
24
25

PLATE 2

Figure

1–4. *Neogondolella serrata* (Clark and Ethington).
- 1, 2. Aboral and oral views, ×60, Maokou Formation of Liangshan area, southern Shaanxi. Cat. No. 45450.
- 3, 4. Aboral and oral views, ×60, Maokou Formation of Liangshan area, southern Shaanxi. Cat. No. 45452.

5. *Merrillina oertlii* Kozur. Lateral view, ×80, Chihsia Formation of Guizhou. Cat. No. 63326.

6, 7. *Anchignathodus minutus* (Ellison)
- 6. Lateral view, ×80, Chihsia Formation of Guizhou. Cat. No. 63327.
- 7. Lateral view, ×80, Chihsia Formation of Guizhou, Cat. No. 63328.

8, 10, 11, 22. *Sweetognathus whitei* (Rhodes)
- 8. Oral view, ×80, lower Chihsia Formation of Guizhou. Cat. No. 63329.
- 10, 11. Oral and aboral views, ×60, lower Chihsia Formation of Guizhou. Cat. No. 63330.
- 22. Oral view, ×80, lower Chihsia Formation of Guizhou. Cat. No. 63331.

9, 13. *Neogondolella bitteri* (Kozur). Oral and aboral views, ×60, Wuchiaping Formation of Liangshan area, southern Shaanxi. Cat. No. 45444.

12, 14. *Neogondolella aserrata* Clark & Behnken. Aboral and oral views, ×60, Maokou Formation of Liangshan area, southern Shaanxi. Cat. No. 45454.

15, 23. *Neogondolella idahoensis* (Youngquist and others). Oral and aboral views, ×60. Upper Chihsia Formation of Guizhou. Cat. No. 63332.

16, 17. *Neogondolella bisselli* (Clark and Behnken). Aboral and lateral-oral views, ×60, lower Chihsia Formation of Guizhou. Cat. No. 63334.

18,19. *Neostreptognathodus prayi* Behnken. Oral and aboral views, ×60, Maokou Formation of Sichuan. Cat. No. 63335.

20, 21. Form transitional between *Neogondolella bisselli* and *Neogondolella idahoensis.* Aboral and oral views, ×80, upper Chihsia Formation of Guizhou. Cat. No. 63333.

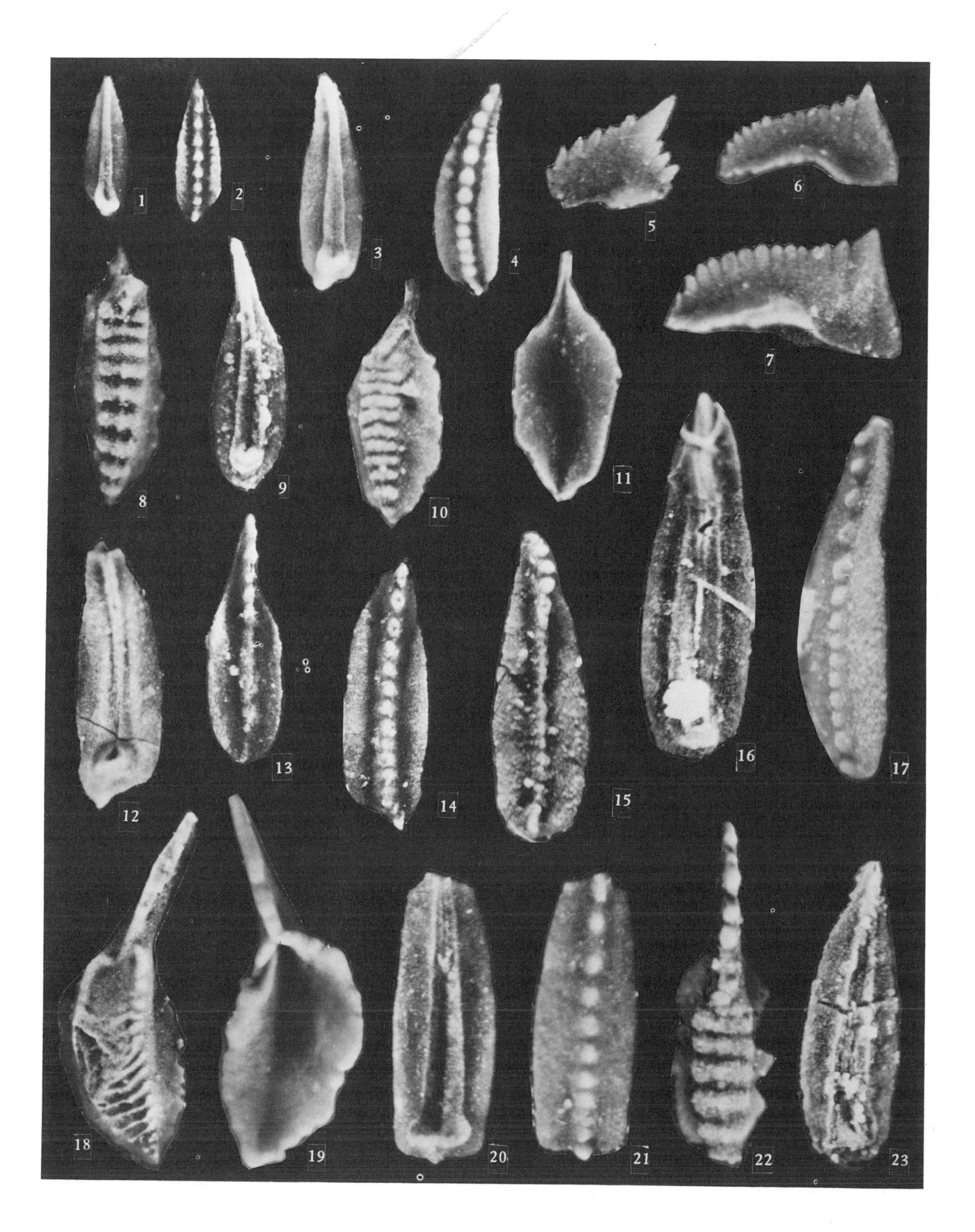

1
2
3
4
5
6
7
8
9
10
11
12
13
14
15
16
17
18
19
20
21
22
23

oral surface lies well below that of denticles, so that it is conspicuous in oral but not in lateral view. Basal cavity large, occupying entire aboral surface of cup, and extending anteriorly as a narrow groove on lower surface of free blade.

Remarks: This species is characterized by laterally expanded, transverse ridges on oral surface connected by a low median bladelike ridge.

Occurrence: Lower Chihsia Formation, Guizhou, Jiangxi, and Xinjiang.

ACKNOWLEDGMENTS

All specimens are in the Nanjing Institute of Geology and Palaeontology, Academia Sinica.

REFERENCES CITED

Barskov, I. S., and Koroleva, N. V., 1970, Pervaya nackodka verck-nepermskick konodontov na territorii SSSR: Moskva, Doklady Akademii Nauk SSSR, v. 194, no. 4, p. 933–934, Pl. 1.

Behnken, H. F., 1975, Permian conodont biostratigraphy: Journal of Paleontology, v. 49, no. 2, p. 284–315, Pls. 1–2.

Bender, H., and Stoppel, D., 1965, Perm-Conodonten: Geologisches Jahrbuch, v. 82, p. 331–364, Pl. 14–16.

Branson, C. C., 1932, Origins of phosphate in the Phosphoria Formation (abs.): Geological Society of America Bulletin, v. 43, p. 284.

Ching Yü-kan [Jin Yu-gan], 1960, Conodonts from the Kufeng Suite (formation) of Lungtan, Nanking: Acta Palaeontologica Sinica, v. 8, no. 3, p. 242–248, Pls. 1–2 (in Chinese and English).

Clark, D. L., and Behnken, F. H., 1971, Conodonts and biostratigraphy of the Permian: Geological Society of America Memoir 127, p. 415–439, Pls. 1–2.

——1979, Evolution and taxonomy of the North America Upper Permian *Neogondolella serrata* complex: Journal of Paleontology, v. 53, no. 2, p. 263–275, Pls. 1–2.

Clark, D. L., and Ethington, R. L., 1962, Survey of Permian conodonts in western North America: Brigham Young University Geology Studies, v. 9, pt. 2, p. 102–114, Pl. 2.

Clark, D. L., and Mosher, L. C., 1966, Stratigraphic, geographic, and evolutionary development of the conodont genus Gondolella: Journal of Paleontology, v. 40, p. 376–394, Pls. 45–47.

Clark, D. L., Carr, T. R., Behnken, F. H., Wardlaw, B. R., and Collinson, J. W., 1979, Permian conodont biostratigraphy in the Great Basin: Brigham Young University Geology Studies, v. 26, pt. 3, p. 143–150, Pl. 1.

Ellison, S. P., 1941, Revision of the Pennsylvanian conodonts: Journal of Paleontology, v. 15, p. 107–143, Pls. 20–23.

Kozur, H., 1975, Beiträge zur Conodontenfauna des Perm: Geologisch-Paläontologische Mitteilungen Innsbruk, v. 5/4, p. 1–44, Pl. 1–4.

——1976, Neue Conodonten aus dem Jungpaläozoikum und der Trias: Geologisch-Paläontologische Mitteilungen Innsbruck, v. 6/3, p. 1–33, Pl. 1–4.

——1978, Beiträge zur Stratigraphie des Perms Teil II: Die Conodontenchronologie des Perms: Freiberger Forschungsheft C 334, p. 85–161, Pl. 1–8.

Merrill, G. K., 1973, Pennsylvanian nonplatform conodont genera, I: *Spathognathodus:* Journal of Paleontology, v. 47, no. 2, p. 289–314.

Rhodes, F.H.T., 1963, Conodonts from the topmost Tensleep Sandstone of the eastern Bighorn Mountains: Journal of Paleontology, v. 37, p. 401–408, Pl. 47.

Stauffer, C. R., and Plummer, H. J., 1932, Texas Pennsylvanian conodonts and their stratigraphic relations: Texas University Bulletin 3201, p. 13–50, 4 pls.

Sweet, W. C., 1970, Uppermost Permian and Lower Triassic conodonts of the Salt Range and Trans-Indus ranges, West Pakistan, *in* Kummel, B., and Teichert, C., eds., Stratigraphic boundary problems: Permian and Triassic of West Pakistan: Department of Geology, University of Kansas Special Publication 4, p. 207–275, Pls. 1–5.

Sweet, W. C., 1973, *Neogondolella orientalis* (Barskov and Koroleva), *in* Ziegler, W., ed., Catalogue of conodonts: Schweizerbart, Stuttgart, p. 143, Neogondo-Pl. 1, fig. 7.

——1973, Late Permian and Early Triassic conodont fauna, *in* Logan, A., and Hills, L. V., eds., The Permian and Triassic Systems and their mutual boundary: Calgary, Alberta, Canadian Society of Petroleum Geologists Memoir 2, p. 630–646.

Szaniawski, H., and Malkowski, K., 1979, Conodonts from the Kappstarostin Formation (Permian) of Spitsbergen: Acta Palaeontologica Polonica, v. 24, no. 2, p. 231–264, Pls. 4–10.

Teichert, Curt, Kummel, Berhard, and Sweet, Walter, 1973, Permian-Triassic strata, Kuh-e-Ali Bashi, Northwestern Iran: Museum of Comparative Zoology Bulletin, v. 145, no. 8, p. 359–472, 14 pls.

Wang Cheng-yuan, 1974, [Early Permian conodonts from Sichuan, *in* A handbook of the stratigraphy and palaeontology of SW China]: Beijing, Science Press, p. 314–315, pl. 166 (in Chinese).

Wang Cheng-yuan and Wang Zhi-hao, 1976, [Triassic conodonts from the Mount Jolmo Lungma region, *in* A report of scientific expedition in the Mount Jolmo Lungma Region (1966–1968)]: Beijing, p. 387–416, 5 pls. (in Chinese).

Wang Zhi-hao, 1978, Permian-Lower Triassic conodonts of the Liangshan area, Southern Shaanxi: Acta Palaeontologica Sinica, v. 17, no. 2, p. 213–232, Pls. 1–2 (in Chinese with English abstract).

Wardlaw, B. R., and Collonson, J. W., 1979, Youngest Permian conodont faunas from the Great Basin and Rocky Mountain Regions: Brigham Young University Geology Studies, v. 26, pt. 3, p. 151–164, Pl. 1–2.

MANUSCRIPT ACCEPTED BY THE SOCITY APRIL 22, 1981

Geological Society of America
Special Paper 187
1981

On the occurrence of cycadophytes with slender growth habit in the Permian of China

Zhang Shan-zhen
Mo Zhuang-guan
Nanjing Institute of Geology and Palaeontology, Academic Sinica, Nanjing 210008, People's Republic of China

ABSTRACT

Fronds of cycadophytes in organic connection with slender axes from the Lower Permian of North China are described and assigned to new genera. *Procycas* is characterized by *Nilssonia*-like leaves and a slender smooth axis. *Procycas densinervioides* gen. et sp. nov. is derived from the upper part of Shansi Formation of Henan. The species described as *Dioonites densinervis* by Halle in the Lower Shihhotse Formation of central Shansi, transferred to the genus *Nilssonia* and recorded by Stockmans and Mathieu in the upper part of the Chao Ko Chuang Series of Kaiping Basin, is thought to be closely allied to *Procycas densinervioides* and is here transferred to the genus *Procycas*. The new genus *Cladotaenipteris* is monotypic, typified by *C. shaanxiensis* gen. et sp. nov. from the lower Shihhotse Formation of Shaanxi. It is characterized by *Taeniopteris*-like leaves that are loosely disposed on a slender axis which is lacking persistent leaf bases and ornamented with round leaf scars, with clustering of leaves at the tip.

INTRODUCTION

The cycadophytes form by far the most conspicuous section of the extant gymnosperms. There are only nine genera scattered in the tropics and subtropics, four in the Western and five in the Eastern Hemisphere. Extinct members of the cycadophytes, however, played a very important role in the Mesozoic flora. Distributed worldwide, they were highly diversified at that time, and many of their organs have been described in different taxa. The foliages or pinnate fronds of the Mesozoic cycadophytes are very characteristic and agree fairly well with those of the extant ones in general appearance, but show considerable variations in the microscopic structures and fructifications. They are rarely preserved with the fructifications in organic connection with the stems. Because of the apparent similarities of the foliages, authors have assumed that they had a growth habit similar to that of the extant cycads. In the older literature, the Mesozoic cycadophytes were generally reconstructed as having stout, unbranched, erect stems with a cluster of spirally disposed, closely spaced foliages or pinnate fronds at the tip.

Early in 1961, Harris expressed his disagreement with these imaginary stout stems and restored a *Nilssonia* with an interesting, quite different growth habit. It differed considerably from the extant cycads in having repeatedly branched stems with leaves borne on the apical parts of the young stems which were not very thick (Harris, 1961, Text-fig. 2). Harris was also of the opinion that the stem was bare and not covered with the bases of dead leaves like most of the cycads, because the leaf bases were cut off cleanly like the leaves of ordinary dicotyledon trees, and he considered that the modern cycad was unlikely to offer a good model (Harris, 1961, p. 320). For want of an actual stem, the reconstruction was more or less theoretical.

The suggestion of the prevalence of a slender growth habit in Mesozoic cycadophytes is supported by the recent discovery of leaf-bearing stems in *Leptocycas gracilis* Delevoryas and Hope (1971), *Nilssoniocladus nipponense* (Yokoyama) Kimura and Sekiko (1975), and *Ischnophyton iconicum* Delevoryas and Hope (1976). *Leptocycas gracilis* and *Ischnophyton iconicum* were discovered in the Late Triassic Pekin Formation of central North Carolina, U.S.A. The former is a cycadalean with the leaves of *Pseudoctenis;* the latter is a cycadeoidalean with leaves of *Otozamites* to *Zamites* type. The two contemporary forms are distantly related and represent a remarkably similar growth habit in that their leaves are loosely spaced on slender stems and that the stems lack persistent leaf bases. According to Delevoryas and Hope (1976, p. 97), *Ischnophyton iconicum,* if found without attached leaves, would be virtually indistinguishable from stems of *Leptocycas gracilis. Nilssoniocladus nipponense* is a cycadalean too, discovered in the early Lower Cretaceous Itoshiro Group of Ishakawa Prefecture, Japan, characterized by its slender long shoot with spirally arranged dwarf shoots. The internodes of the long shoots are long and smooth. Dwarf shoots are covered with spirally placed rhomboidal leaf scars and with terminal clusters of

Nilssonia leaves. Leafy long shoots and brachyblasts, with clusters of leaves of Nilssoniales, have also been reported by Krassilov (1972) in the Mesozoic of the Soviet Union.

The morphology of the Mesozoic cycadophytes is now well known, as are the diversity and evolutionary advancement of them. The known Jurassic genera, as pointed out by Harris (1961, p. 322), are no more primitive than the living ones. It seems more than likely that the cycadophytes existed as a distinct group, or cluster of groups, by Late Paleozoic times, but precise details are obscure below the Triassic, as held by Andrews (1961, p. 312).

From Paleozoic strata, petrified wood with cycadlike anatomical features has been described as *Cycadoxylon* by Andrews (1940). *Taeniopteris* and other foliages with more or less cycadophyte-like external morphological appearance have been repeatedly recorded from the Permian and Carboniferous of the world. For want of further information, these pieces of evidence were not always regarded as indisputable.

The recent discovery of Permian fructifications by Mamay (1969, 1973, 1976) furnished decisive evidence for the existence of Paleozoic cycadophytes. Most of them are cycadalean megasporophylls derived from the Leonardian of Kansas and Texas, U.S.A. Six species have been recognized, attributed to the genera *Phasmatocycas* Mamay and *Archaeocycas* Mamay. They are characterized by stout axes with two lateral rows of naked ovules. In the former genus, the ovules are attached to the axis. The latter genus is characterized by the attachment of the ovules to the basal lamina, instead of the axis proper, and by the fact that the lamina is inrolled and at least partly encloses the ovules. In the distal part of *Archaecocyas,* there is a sterile lamina, and the distal part of *Phasmatocycas* is broken off and thought to have been provided with a sterile lamina too. The well-preserved double cuticles of *Phasmatocycas kansana* resemble the Jurassic cycad *Beania,* and the small resinoid spherules alternating with the ovules on the fertile axis are identical with the interneural sperules of the associated taeniopterid leaves. *Phasmatocycas* has been reconstructed as a cycadlike megasporophyll with an elaminate, fertile base and a taeniopterid distal lamina. Associated with them are *Cycadospadix yochelsoni* Mamay with cycadalean or bennettitalean affinities and specimens of probably cycadean male cones. The discovery of fructifications, the most indisputable evidence of a Paleozoic origin of the cycadophytes, favors strongly the cycadean affinities of the Paleozoic cycadophyte-like foliage.

In the more or less contemporary deposits of China, the pinnate frond fragments of *Pterophyllum* spp., *Dioonites densinervis, Nilssonia* spp., and a number of *Taeniopteris* species bearing strong resemblance to the fronds of certain known cycadophytes have generally been accepted as possibly of cycadean affinities. However, further information was obscured by fragmentary preservation.

It is interesting to note that the recent study of collections of fossil plants from the Permian and Carboniferous of China by the authors has revealed several well-preserved specimens with slender axes to which a number of fronds are still attached. Two species are recognized. One of them bears fronds of *Nilssonia*-type, and the other is provided with *Taeniopteris*-type fronds. Presenting some insight as to the habit of the whole plants, they are regarded as new genera rather than belonging to any existing form genus, as advocated by Delevoryas and Hope (1976, p. 95).

MATERIAL AND METHOD

The plant fossils studied in this paper are impressions or coalified impressions. *Procycas densinervioides* derived from the upper part of the Shansi Formation of Dajiancun, Yuxian District, Henan Province, is preserved in a buff-colored siltstone. *Cladotaeniopteris shaanxiensis* from the lower Shihhotse Formation of Sangsjuping, Hancheng District, Shaanxi Province, is preserved in a grayish-green siltstone. The main technique employed in this study is the careful uncovering of the specimens by removal of the matrix with a fine needle and hammer. Transfers and maceration were attempted, but no information was gained in this way, because the carbon fragmented during the preparation.

SYSTEMATIC DESCRIPTIONS

Division CYCADOPHYTA
Order UNCERTAIN (CYCADIALES or BENNETTITALES)
Genus *Procycas* gen. nov.
Type-species: *Procycas densinervioides* (gen. et sp. nov.)

Stem slender in ultimate part which are the only ones known. Frond spirally disposed, distant, of *Nilssonia* type. Cuticular structure unknown.

Procycas densinervioides (*gen. et sp. nov.*)
Figure 1; Plate 1(1–6); Plate 2(1–4)

Branch slender, rigid, with a breadth of 7 mm and a length of more than 16 cm, tapering gradually to the apex; surface smooth, lacking persistent leaf bases. Frond spirally disposed, simple compound of unknown length, attaining a breadth of 3 to 4 cm; rachis rounded, 1 to 2 mm broad on the impression. Pinnae inserted on the upper surface of the rachis, nearly perpendicular to it, usually distant, linear, with slightly widened base and blunt or rounded apex, of unequal length and breadth, with short pinnae proximally, larger distally, about 1.5 cm in length, 2 mm in breadth in the middle of the frond, about 4 to 5 mm in length, 0.6 mm in breadth in the basal part of the frond, and confluent to a continuous lamina in the distal part of the frond. Veins simple or bifurcate entering the pinna in a number of more than two, according to the breadth of the pinna, with minute projecting black dots, probably representing glands, between the veins. Axis of juvenile frond bent, with the pinnae straight. Cuticular structure unkown.

Of this species, several leaf-bearing shoots are known. The fragments shown on Plate 2(1,2) are parts of a fractured shoot. Its counterpart being more complete, with its posterior end broken, is shown on Plate 1(1). It measures about 16 cm in length without being complete, about 7 mm in breadth, tapering gradually upward, with spirally disposed leaves. The rachis of the frond is fairly thin. Those in the posterior end are slightly thicker, measuring scarcely 2 mm in breadth. The fronds are unfortunately broken off distally. Their length is unknown. The left one in the middle of the shoot shown on Plate 1(1) is the longest ever preserved, measuring about 8.5 cm in length. The surface of the rachis is finely striated. To the side of the rachis are distantly attached small narrow pinnae, though these are not very clearly shown in the figure. Of special interest is a juvenile frond preserved at the tip of the shoot as shown on Plate 2(2), enlarged 5 times in Plate 2(3). It is rather small, only a few millimetres in

Figure 1. Reconstruction of *Procycas* showing the leaves spirally disposed on the slender branches.

length. It arises from the tip of the branch, runs forward for 4 mm, then bends downward and recurves to form an open ring of 2 mm diameter. The pinnae of the juvenile frond are straight.

Another shoot shown on Plate 1(2, 3) is also a distal part. The fronds appear to be more densely disposed as compared with the above-mentioned specimen. The pinnae are well preserved as shown in the enlarged figure, linear in shape, measuring about 4 to 5 mm in length and 0.6 mm in breadth, with two parallel veins. Between the veins are minute black dots.

The frond shown on Plate 1(4, 5), though fragmentarily preserved, is no less important. It bears a thin rachis, with linear, larger pinnae, about 1.5 cm in length and 2 mm in breadth. Eight parallel veins are minute black dots. In the same locality, same horizon, or even in the same slab are fronds with broader pinnae or pinnae with contiguous laminae, assuming a more or less taeniopteroid aspect, but with the same venation and minute dots. Furthermore, in no case are these contiguous ones preserved for more than 5 cm in length. They are all fragmentarily preserved with very thin rachis. One of them in shown on Plate 1(6).

The specimens described above are seemingly varied, but the venation and the presence of black dots between the veins would indicate that the specimens may be closely related, or belong to one and the same species. Similar variations may also be met with in the pinnae of *Dioonites densinervis* Halle (1927, p. 196, Pl. 56, figs. 1–5). *D. densinervis* is an interesting species created by Halle, from the lower Shihhotse Formation of Early Permian age in Shanxi [Shansi], China, which was transferred to the genus *Nilssonia* and recorded by Stockmans and Mathieu (1939, p. 91, Pl. 21, fig. 1) in the upper part of the Chao Ko Chuang Series of the Kaiping Basin, which is correlated with the lower Shihhotse Formation. According to Halle, this species is characterized by the varying breadth of the pinnae. A frond figured by Halle (1927, p. 196, Pl. 56, fig. 1927, Pl. 56, fig. 5).

More or less comparable undivided distal segment with irregularly laciniated pinnae can also be met with in the Jurassic cycadean *Anomozamites nilssoni* (Phillips) Seward (1917, p. 557, Fig. 615; Harris, 1969, p. 83, Fig. 38A). Another Jurassic cycadean of Yorkshire has been figured by Harris (1969, p. 70, Fig. 32A) as *Nilssoniopteris vittata* (Brongniart) Nathorst, although it has as a rule an entire taeniopterid lamina and an unusual leaf with two small basal segments. it is obvious that in fossil cycadophyte frond with deeply laciniated pinnae a distal entire segment is not a very unusual feature.

In this respect, the writers are of the opinion that it would not be too rash to assign the present specimens to one species. A schematic restoration can thus be proposed. Although there are good examples for the fronds disposed on slender axes, it is very difficult to determine on what base the axes arose. However, judging from the very slender axes, together with their long and rigid, large fronds, it seems to the writers that the frond-bearing axes may have branched from a tree rather than arisen directly from a root. Though further information concerning the stem is as yet wanting, the writers would accept an imaginary stem or branch system as proposed by Harris for his restoration of a plant composed of *Bucklandia pustulosa* Harris and *Ptilophyllum pecten* (Phillips) Halle (Harris 1969, p. 138, Fig. 59C) for the present restoration (Fig. 1).

As stated above, the present species bears a considerable resemblance to *D. densinervis* in the general appearance of the long and delicate fronds and dissection of pinnae, but the pinnae of the latter are strongly widened at the base and united with the next pinnae as a rule, instead of being rather distant and only faintly expanded at the base as in the present species. Furthermore, the present species may also be distinguished from *D. densinervis* by the presence of minute projecting black dots between the veins. The writers are thus of the opinion that there seems to be no question about their having distinct specific identities, though the two species are very closely allied. The specific name *P. densinervioides* is thus proposed in allusion to its resemblance to *D. densinervis*.

In this connection, it is interesting to note that several incomplete leaves of the latter species arising from a small piece of branch have also been figured by Halle (1927, Pl. 56, fig. 5) in Shanxi [Shansi]. The specimen appears to represent the tip of a branch, and may be correlated with the distal end of a branch in the present species. It seems to the present writers that *D. densinervis* had a growth habit similar to that of the present species. It would be convenient to assume that the branch pattern of *D.*

densinervis may also be similar to that of the present species, and that the species may be transferred to the new genus *Procycas*.

Procycas densinervioides is of considerable morphological interest because it reminds us of the important discovery of a slender growth habit in the Upper Triassic *Leptocycas gracilis* (Delevoryas and Hope, 1971) and *Ischnophyton iconicum* (Delevoryas and Hope, 1976) which were also characterized by loosely spaced leaves and fairly smooth stems. But their fronds, especially the pinnae, are too different from those of the present form to warrant any generic identification.

The presence of well-preserved juvenile fronds in the present species seems to deserve special notice. It is known that the venation of cycadophytes exhibits less uniformity than that of ferns. It differs considerably among different extant genera, of which *Zamia* and *Stangeria* are characterized by their straight pinnae on a bent rachis. In extinct genera, a bent rachis of juvenile fronds with straight pinnae has been figured by Nathorst (1909, Pl. 1, figs. 2–20) in *Nilssonia brevis* Brongniart.

Genus *Cladotaeniopteris* (gen. nov.)

Type species: *Cladotaeniopteris shaanxiensis* (gen. et sp. nov.)

Diagnosis: Same as for type species.

***Cladotaeniopteris shaanxiensis* (gen. et sp. nov.)**

Figure 2; Plate 3(1–4)

Stem or branch known in part, about 1 cm wide, covered with spirally arranged leaves or rounded leaf scars, lacking persistent leaf bases. Phyllotaxy lax, with the exception of clustering leaves at the tip. Leaves taeniopterid, oblanceolate, entire, more than 12 cm in length, about 3 cm in breadth, tapering gradually to the apex and base, abscissed at the base. Petiolate, petiole short, 5 mm in length, 3 mm in breadth, expanded to 4 to 5 mm at the base. Rachis about 2.5 mm broad, tapering gradually to the apex, secondary veins rather indistinct, probably arising at an open angle and pursuing a straight course to the margin. Cuticular structure unkown.

A slab of grayish-green siltstone about 21 cm long and 13.2 cm wide with two frond-bearing shoots together with its counterpart is derived from the lower Shinhotse Formation of Hancheng, Shaanxi, and is shown on Plate 3(1, 2). The shoots are incomplete, with their posterior ends broken off, and are marked A and B in the figure. The preserved part of axis A is about 1 cm in breadth and 4.5 cm in length, and is covered irregularly with faint, round, small scars, which seem to be leaf scars. In the middle of each scar is a round dot representing the vascular bundle scar. In addition to the scar, there are also small wrinkles about 1 mm in length. At the summit of the axis is a cluster of fronds, four of which are exposed on the rock surface as shown in Plate 3(1). Three of the fronds overlap, preserved approximately in their original position. About 1.2 and 2.4 cm below the cluster, two fronds arise from the right side of the axis. In the lower half of the figure, another shoot marked B is situated below, measuring about 9 mm in breadth and 1.2 cm in length. There are at least five leaves or fronds in the tip of the shoot. The posterior part of the frond, preserved at the left of the shoot, slightly above the level of the shoots, also seems to arise from the tip of shoot B. According to the orientation of the shoots as well as the fronds, shoot B appears to have spread from A, but the specimen is fractured at the place where the shoots are attached.

The present form is characterized by *Taeniopteris*-type leaves on a slender axis which is covered with leaf scars. The leaves are clustering at the tip and are widely spaced below it. The mode of leaf attachment is thus of considerable interest, inasmuch as it bears a certain resemblance to that of *Nilssoniocladus nipponensis* (Yokoyama) Kimura (1975) of the Lower Cretaceous of Japan on the one hand and *Ischnophyton iconicum* Delevoryas and Hope (1976) on the other. The tip of the branch of the present species closely resembles that of the former species in the spiral disposition of leaf scars and terminal cluster of leaves. But the pinnae of the Japanese Mesozoic fossil are triangular in shape and the leaf-bearing shoots are dwarf shoots which are spirally placed on smooth, internoded, long shoots. The present form recalls *Ischnophyton iconicum* in that the leaves are widely spaced on the branch. But the leaves of the American Triassic fossil are of the *Otozamites* or *Zamites* type, and the branch is found to be smooth. It seems to the present writers that the present form resembles more closely the former in general habit, as shown in the restoration Pl. 3(2).

DISCUSSION

The materials here described as *Procycas* and *Cladotaeniopteris* from the Lower Permian of China are interesting in their geological distribution and growth pattern. Formerly, fossil cycadophytes were considered to be characterized by stout stems with closely spaced persistent leaf bases similar to those of the extant *Cycas*. Through the recent works of Harris (1969), Delevo-

Figure 2. Reconstruction of *Cladotaeniopteris* showing the apical leaf clusters and the loosely disposed leaves below them on the slender shoots.

ryas and Hope (1971, 1976), and Kimura (1975), the pattern of growth of fossil cycadophytes of Mesozoic age is known to be usually of a slender growth habit. The old conceptions and many reconstructions based on them are no longer regarded as reliable. The slender growth of the Paleozoic plants shows almost no essential differences from that of the Mesozoic cycadophytes. It is true that the cuticular structure of the present material is as yet unkown. The striking resemblance of the fronds, however, to certain known cycadophytes renders it permissible to suggest a cycadean affinity, though the precise position of the genera within the class has not yet been decided. Without the aid of further information, it is often impossible to determine whether a cycadean foliage belongs to a cycad or bennettitite, or whether it borders on both. The same is also true of branches. According to Delevoryas and Hope (1976, p. 99), the growth habit of the Late Triassic cycadeoidalean *Ischnophyton iconicum* is remarkably similar to that of its contemporary cycadalean *Leptocycas gracilis.* Until knowledge of the cuticular structure or other information is available, the writers prefer to leave open the question as to the ordinal classification of the new genera. At any rate, the discovery of such different kinds of foliage-bearing shoots in the Lower Permian deserves more than passing notice.

The information supplied by the diversification of shoots is consistent with that derived from the fructifications. According to Mamay (1976), six species of cycadean fructifications in probably four genera have been described from the upper Lower Permian Leonardian Series of southwestern United States. On the basis of a series of critical studies, Mamay has demonstrated a hypothetical evolutionary development of a primitive *Cycas*-like *Phasmatocycas* megasporophyll from a *Spermopteris*-like pteridosperm (Mamay, 1976, Fig. 6) and a hypothetical evolution of a *Cycas*-like megasporophyll from *Archaeocycas* (Mamay, 1969, Fig. 1; 1976, Fig. 7). The writers are in complete accord with these conclusions.

Concerning the foliages, Mamay is of the opinion that the pinnate habit in cycadalean foliages has evolved from the gradual incision or lacination of an originally *Taeniopteris*-like lamina. He suggested, for instance, that

> *Taeniopteris serrulata,* from the lower Shihhotse Series (Lower Permian) of Shansi, China (Halle, 1927, p. 160–161, Pl. 42, figs. 13–18; Pl. 64, fig. 13), is a species that well might represent an early evolutionary stage in such a series. *T. serrulata* has a simple leaf and is quite ordinary in most morphological aspects; however, it has a finely serrate margin. The marginal teeth are very small, but each one is occupied by a single vein ending. [Mamay, 1976, p. 40].

The present writers are in complete accord with Mamay as to the supposition about the gradual incision of the *Taeniopteris*-like lamina leading to the pinnate frond in cycadophytes. From the presence of *Procycas* in the Lower Permian of China they are inclined to conclude that the dissected frond of *Procycas* already represents an advanced evolutionary stage, for the incision is so deep as to enable the pinnae to be separate and free down to the rachis. The incision might have taken place long before the appearance of *Procycas* or even in the Pennsylvanian. Furthermore, since both *Procycas densinervis* and *P. densinervioides* have no serrate margins in their pinnae, they seem to be rather different from and only distantly related to *Taeniopteris serrulata.* The suggestion that *T. serrulata* represents an early evolutionary stage seems to have to be reconsidered. Judging from the evolutionary advancement of the foliages and the broad variety of fructifications in the Lower Permian, the present writers think that probably a considerable number of by no means primitive cycadophytes existed in the Early Permian. If one is to find progressive evolution, it must have taken place before the Permian.

The writers are also inclined to think that, in view of the resemblance in the general appearance of the fronds as well as juvenile stages of Paleozoic *Procycas* and Mesozoic *Nilssonia,* it is not improbable that there would be certain similarities in their fructifications, and the Paleozoic primitive cycadean megasporophylls or *Spermopteris*-like pteridosperms might have shown a trend toward evolving a *Beania*-like Mesozoic cone.

Regarding the evolutionary trends of the cycadophytes, it should be noted that the incision theory advocated by Mamay seems to be convincing in the interpretation of cycads with parallel or radiated veins. In extant cycads, a group distinguished by its fernlike habit and venation that should be mentioned is *Stangeria.* Being so nearly like a common tropical fern, *Lomaria,* Stangeria was not even described as a separate genus (Chamberlain, 1935, p. 75). The large fronds of *S. paradoxa* bear broadly linear acuminate pinnae with entire, unevenly lobed, serrate, or pinnatifid margins. Some leaflets are so deeply dissected as almost to justify the appellation pinnate (Seward, 1917, p. 19, 20). Each pinna has

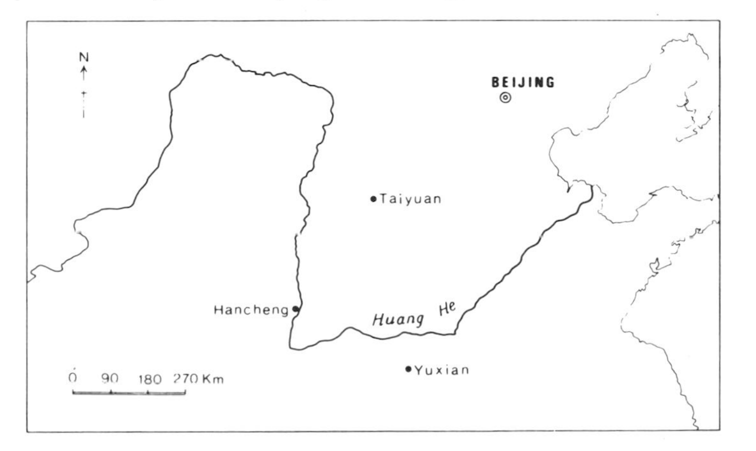

Figure 3. Map of northern part of China showing the fossil localities.

a pronounced midvein with a dense system of transverse, parallel, dichotomously divided secondary veins (Mamay, 1976, p. 49). Mamay is of the opinion that the pinnately compound taeniopterid leaf could evolve with only relatively minor modifications into a *Stangeria* leaf (Mamay, 1976, p. 43). It seems to the present writers that the fernlike habit and venation of *Stangeria* are peculiar enough to invite speculation as to whether alternative interpretations may be drawn from these morphological features. In this connection it is interesting to recall some seed-bearing fronds of pteridosperms described by the writers from the Lower Permian of Henan, China (Zhang and Mo, 1979). One of the species, *Sphenopecopteris beaniata* Zhang and Mo, with a slender flexuous midrib and more or less freely forked lateral veins, bears ovoid seeds (ovules) on the ultimate pinnae rachis. It seems to the present witers that it would be possible too that a *Stangeria* leaf could evolve from a *Sphenopecopteris* leaf by fusion, inasmuch as a coherent leaf of the latter can match a *Stangeria* leaf fairly well.

ACKNOWLEDGMENTS

The writers are grateful to Professors Li Xing-xue and Liu Bin for having read the manuscript. The reconstructions were drawn by Zhang Wu-cong, staff artist of the Institute. The photographs are by Deng Dong-xing. The writers were assisted in the field by Ding Hui, Meng Feng-yuan, and Feng Bao-dong of the Ministry of Coal Industry of China. To them the writers are indebted.

REFERENCES CITED

Andrews, H. N., 1940, On the stelar anatomy of the pteridosperms, with particular reference to the secondary wood: Missouri Botanical Garden Annual Report, v. 27, no. 1, p. 51–118.

——1961, Studies in paleobotany: New York, John Wiley & Sons, Inc., 487p.

Chamberlain, C. J., 1935, Gymnosperms: Structure and evolution: University of Chicago Press, 484p.

Delevoryas, T., 1968, Some aspects of cycadeoid evolution: Journal of the Linnean Society of London (Botany), v. 61, no. 384, p. 137–146, 4 figs.

Delevoryas, T., and Hope, R. C., 1971, A new Triassic cycad and its phyletic implications: Postilla, no. 150, p. 1–31 (not seen by the authors).

——1976, More evidence for a slender growth habit in Mesozoic cycadophytes: Review of Palaeobotany and Palynology, v. 21, no. 1, p. 93–100, 2 pls.

Halle, T. G., 1927, Palaeozoic plants from Central Shansi: Palaeontologia Sinica, Ser. A, no. 2, fasc. 1, p. 1–316, 64 pls.

Harris, T. M., 1961, The fossil cycads: Palaeontology, v. 4, pt. 3, p. 313–323, 2 figs.

——1969, The Yorkshire Jurassic flora. III. Bennettitales: London, British Museum (Natural History), Publication no. 675. p. 1–186, 69 figs., 7 pls.

Kimura, T., and Sekido, S., 1974, *Nilssoniocladus* n. gen. (Nilssoniaceae N. Fam.), newly found from the early Lower Cretaceous of Japan: Palaeontographica, Abteilung B, Bd. 153, Lfg. 1–3, p. 111–118, 2 pls.

Krassilov, V. A., 1975, Paleoecology of terrestrial plants, basic principles and techniques, translated from Russian by Hilary Hardin: New York, John Wiley & Sons, p. 1–29.

Mamay, S. H., 1969, Cycads: Fossil evidence of Late Paleozoic origin: Science, v. 164, no. 3877, p. 295–296, 1 fig.

——1973, *Archaeocycas* and *Phasmatocycas*—new genera of Permian cycads: Journal of Research of the U.S. Geological Survey, v. 1, no. 6, p. 687–689, Fig. 1a–g.

——1976, Paleozoic origin of the cycads: U.S. Geological Survey Professional Paper 934, p. 1–48, 5 pls.

Nathorst, A. G., 1909, Ueber die Gattung *Nilssonia* Brongn. mit besonderer Berücksichtigung schwedischer Arten: Kungliga Svenska Vetenskapsakademiens Handlingar, Bd. 43, no. 12, p. 1–40, 8 pls.

Seward, A. C., 1917, Fossil plants, volume 3: London, Cambridge University Press, 656 p.

Stockmans, F., and Mathieu, F. F., 1939, La flore paléozoique du Bassin houiller de Kaiping (Chine): Bruxelles, Patrimoine du Musée royal d'Histoire naturelle de Belgique, p. 49–165, Pls. 1–34.

Zhang, Shan-zhen [Chang, Shanchen J.], and Mo Zhuang-guan, 1979, New forms of seed-bearing fronds from the Cathaysia flora in Henan, China: Paper for the 9th International Congress of Carboniferous Stratigraphy and Geology, Urbana: Nanjing Institute of Geology and Palaeontology, Academia Sinica, Nanjing, China, 4 p., 3 pls.

MANUSCRIPT ACCEPTED BY THE SOCIETY APRIL 22, 1981

PLATE 1

Photographs are unretouched and unless otherwise stated are in natural size.

Figure

1–6. ***Procycas densinervioides*** **(gen. et sp. nov.).**

1. **Specimen showing a slender axis with spirally disposed fronds of which only posterior parts are preserved. Syntype, PB 8766.**

2. **Tip of a shoot showing small linear pinnae in the basal part of the fronds. Syntype, PB 8767.**

3. **Part of PB 8767, enlarged, showing parallel veins of the linear pinnae and minute black dots between the veins, ×5.**

4. **Frond fragment showing the linear pinnae. Syntype, PB 8768.**

5. **PB 8768, enlarged, showing the linear pinnae and pinnae with broad, contiguous laminae, ×3.**

6. **Distal part of a frond showing contiguous lamina. Syntype, PB 8769.**

6
2
4
3
5
1

PLATE 2

Photographs are unretouched and unless otherwise stated are in natural size.

Figure

1–4. ***Procycas densinervioides*** **(gen. et sp. nov.).**

1,2. **Fractured parts of a shoot, counterpart shown in Plate 1(1). Syntype, PB 8770, PB 8771.**

3. **Tip of PB 8770, enlarged, showing juvenile frond, ×5.**

4. **Frond fragment showing fronds with small linear pinnae. Syntype, PB 8772.**

5. ***Procycas densinervis*** **(Halle) (comb. nov.) (reproduced from Halle, 1927, Pl. 56, figs. 1–5).**
Note that broad expanse of continuous laminae in the distal part of a frond shown in figure 5. The figure is upside down as figured by the original author.

9. **Several incomplete fronds on the tip of a branch.**

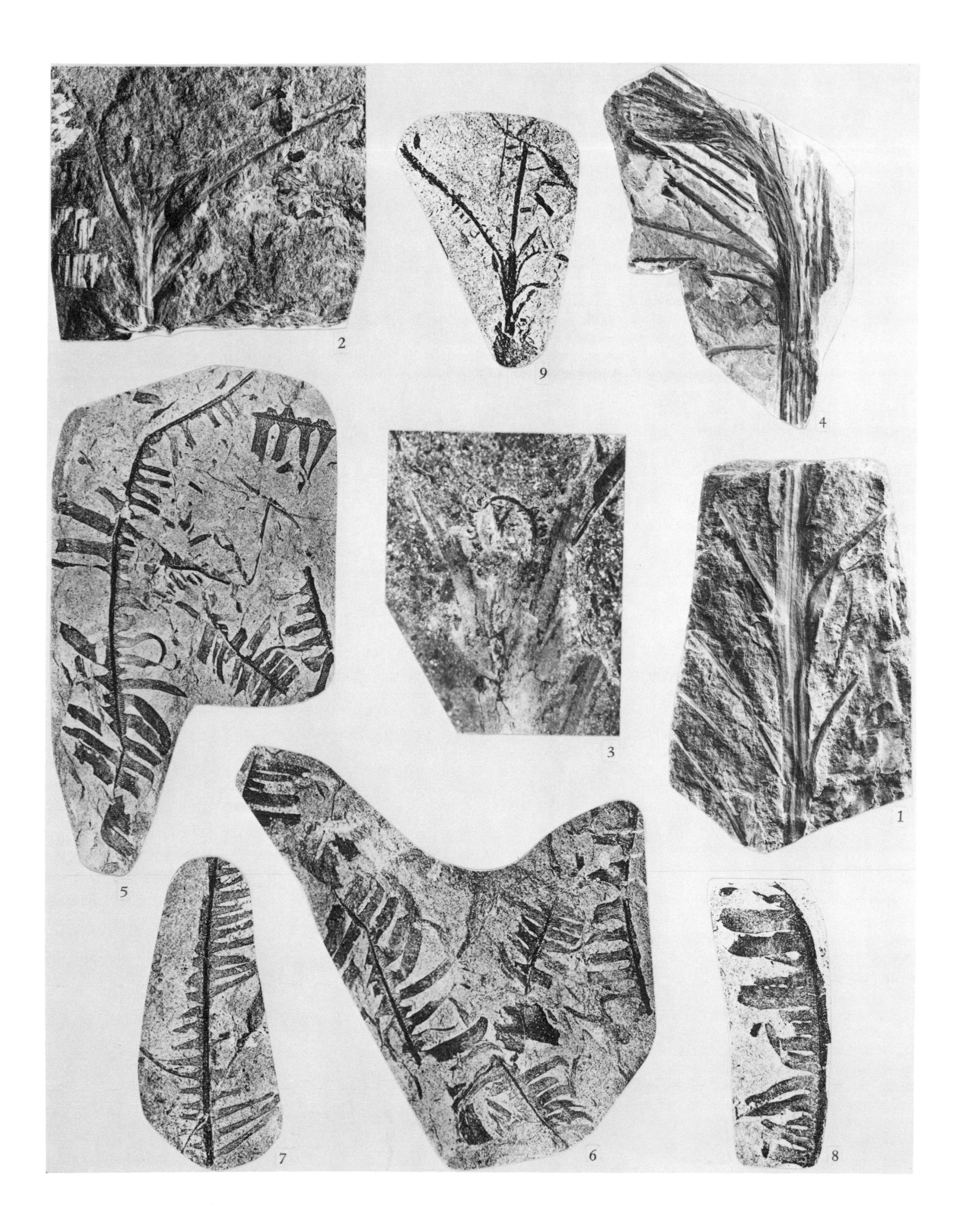

2
9
4
5
3
1
7
6
8

PLATE 3

Photographs are unretouched and unless otherwise stated are in natural size.

Figure

1–4. ***Cladotaeniopteris shaanxiensis*** **(gen. et sp. nov.).**

1. **Specimen showing two clusters of fronds on axes. At the summit of axis A, four fronds (arrows) of the cluster are exposed. Below the cluster, on the right side of axis A, two fronds (arrows) are disposed loosely. Holotype, PB 8773.**

2. **Part of specimen shown in figure 1, enlarged, showing leaf scars on axis B, ×2.**

3. **Counterpart of the specimen shown in figure 1, slightly reduced, ×0.73. Below the scale is a fragment of** ***Gigantonoclea*** **sp. S51–39.**

4. **Detached frond. Paratype, S51–19.**

A
B
1
2
3
4

Printed in U.S.A.

Geological Society of America
Special Paper 187
1981

Miocene floristic regions of China

Song Zhi-chen
Li Hao-min
Zheng Ya-hui
Liu Geng-wu
Nanjing Institute of Geology and Palaeontology, Academia Sinica, Nanjing 210008, People's Republic of China

ABSTRACT

In the light of the palynological and paleobotanical information available, three Miocene floristic regions may be recognized in China, namely, (1) the Qinghai-Xizang *Quercus-Betula*-Thicket Floristic Region, (2) the Inland Forest–Meadow and Steppe Floristic Region, and (3) the Eastern Monsoon Broad-Leaved Floristic Region. Each may be further subdivided into two or three provinces. Their floral characteristics and regional boundaries are discussed in some detail.

INTRODUCTION

The study of floristic regions is an important branch of botany. It is important as a guide in interpreting and remolding nature and, what is more, in exploiting and utilizing plant resources. The purpose of the paleofloristic studies is to look into paleofloristic successions and paleoclimatic changes as well as to make correct correlations among different floristic regions. For a long time, paleobotanists have been paying attention to this subject with great interest. During the past two decades, the growing accumulation of information of both plant megafossils and microfossils has made possible the systematic study of the post-Cretaceous floras. Earlier, more attention had been paid to studies of Late Cretaceous to Paleogene floras. Based on botanical and palynological evidence, a number of works on this topic have been published since 1960 by A. N. Kryshtofovich, B. A. Vokhramcev, D. I. Axelrod, R. W. Chaney, and E. D. Zaklinskaya. However, so far as the late Tertiary floras are concerned, only a small number of papers have appeared, which is in part due to the higher differentiation and diversity of these floras. For the Miocene, we have only the distribution map of early and late Miocene floras of the Soviet Union, drawn on the basis of palynological evidence by Pokrovskaya (1956).

Not a single article, dealing specially with this subject, has so far been written by Chinese authors, although characteristics of Tertiary floras of China have been mentioned both in the *Cenozoic Plants from China* (in Chinese) published in 1978 and the "Paleogeography" (manuscript) of *China's Natural Geography* compiled by Peking Normal University. In this paper, an attempt has been made to present a preliminary idea on the division of Miocene floristic regions of China, in the light of both palynological and paleobotanical evidence available.

In general, the formation and development of floristic regions are controlled by the physical environment, especially by the climate, as well as the historical background of floral successions (*Cenozoic Plants from China,* 1978; Chien Sung-shu and others, 1956).

In China, especially in its western part, recognition of the influence of the Himalayan orogenesis is of great importance. By the latest Paleogene, as a result of the uplift caused by this orogenesis, the eastern Tethys disappeared and gave way to a combined continent of Europe-Asia-India, forming in China the Qinghai-Xizang Plateau. The natural environment (including the climate) and the development of the floras in the late Tertiary were seriously affected by these tremendous changes, making them remarkably different from conditions in the early Tertiary.

According to the differences of composition of the Miocene floras and the environments concerned, three floristic regions might be recognized in China (Fig. 1). Although the data available are rather scattered and incomplete, and some relevant ages have not been exactly dated, it is on the whole feasible to make a division of Miocene floristic regions in China.

QINGHAI-XIZANG *QUERCUS-BETULA*-THICKET FLORISTIC REGION

This region is situated in the Hengduan Mountains and the Qinghai-Xizang Plateau to the south of the Kunlun Mountains. Probably because of the complexity of the topography, a primary vertical differentiation of vegetation had begun here in the Miocene.

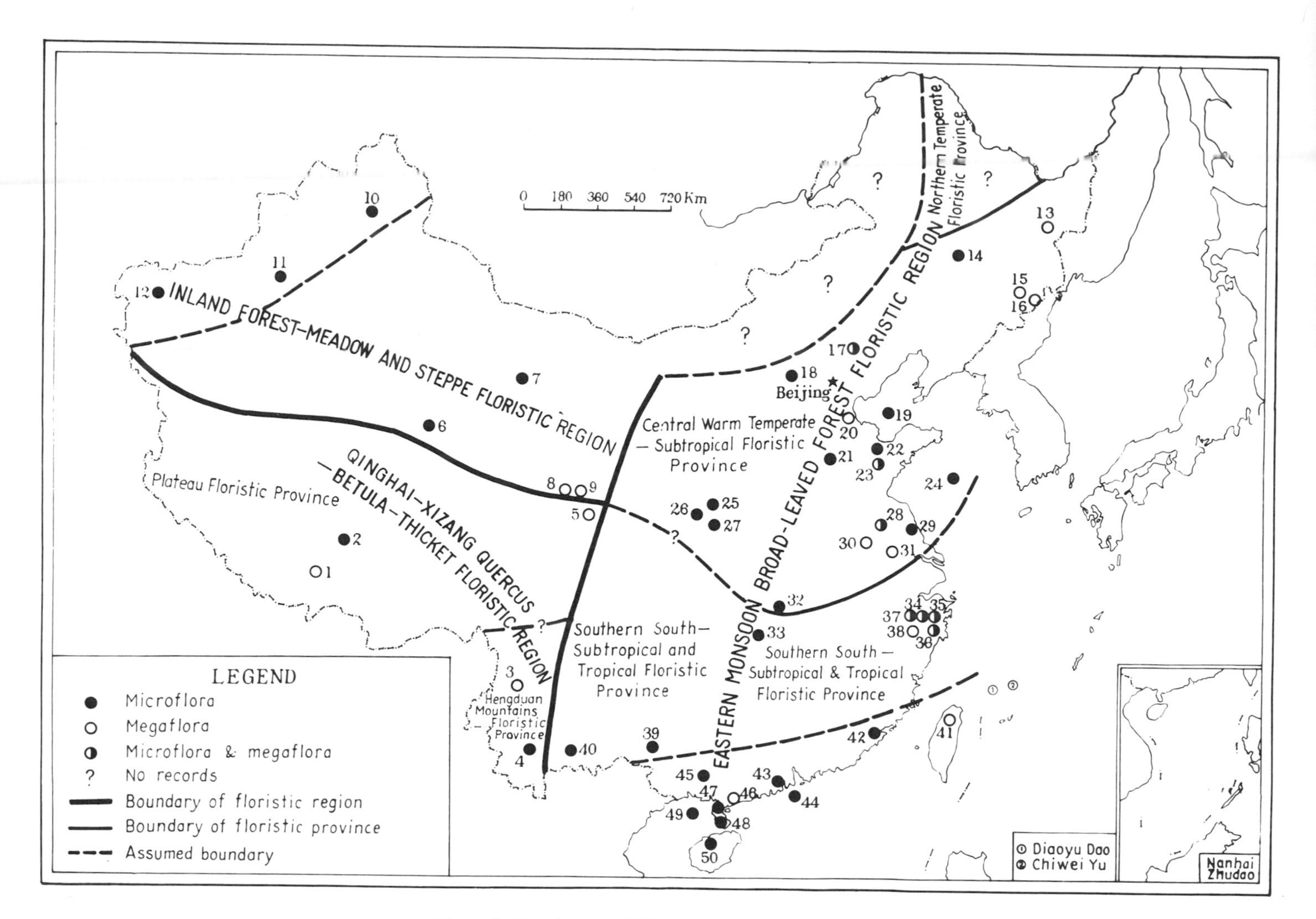

Figure 1. Sketch map of Miocene floristic regions of China.

1. Namling
2. Bagon
3. Jianchuan [Chienchuan]
4. Jinggu
5. Zoige
6. Qaidam Pendi [Tsaidam Basin]
7. Jiuquan [Chiuchuan]
8. Zeko [Zekog]
9. Caka [Chaka]
10. Junggar Pendi [Dzungar Basin]
11. Kuqa [Kuche]
12. Kashi [Kaxgar]
13. Huanan
14. Fuyu
15. Dunhua [Tunhua]
16. Helong [Holung]
17. Weichang
18. Tianzhen [Tienchen]
19. Bohai [Pohai Sea]
20. Huanghua
21. Guantao [Kuantao]
22. Dongying [Tungyin]
23. Shanwang
24. Huanghai [Yellow Sea]
25. Lintong [Lintung]
26. Huxian [Huhsien County]
27. Lantian [Lantien]
28. Sihong [Szuhung]
29. Yancheng [Yencheng]
30. Mingguang
31. Puzhen
32. Jianghan Pingyuan [Chianghan Plain]
33. Changde [Changteh]
34. Shengxian [Shenghsien County]
35. Xinchang
36. Ninghai
37. Xianju [Hsienchu]
38. Linhai
39. Bose [Pose]
40. Kaiyuan
41. Taibei [Taipei]
42. Zhangpu [Changpu]
43. Xinhui [Hsinhui]
44. Zhujiangkou [Pearl River Mouth]
45. Nanning
46. Maoming
47. Zhanjiang [Chanchiang]
48. Xuwen [Hsuwen]
49. Weizhoudao [Weichou Island]
50. Changpo

The most important characteristic of the vegetation of this region is that after the dominance of Betulaceae in the early-middle Miocene came such representatives as *Quercus, Rhododendron,* and *Picea* in the late Miocene. Owing to the preponderance of broad-leaved species of forest with a number of shrubby plants, we name this region the Qinghai-Xizang *Quercus-Betula*-Thicket Floristic Region. If the differences of paleogeography and paleoclimate as well as of composition of vegetation between its western and eastern parts are considered, this region may be further subdivided into two provinces, namely, the Plateau Floristic Province in the west and the Hengduan Mountain (or western Yunnan) Floristic Province in the east.

On the Qinghai-Xizang Plateau to the south of the Kunlun Mountains and to the west of the Hengduan Mountains lies the Plateau Floristic Province. Palynoflorals of the Wulong Formation in Namling County and of the lower member of the Lunpola Group (formerly the Tingching Formation) in Bagon County may be considered as representative. This assemblage in this area is characterized by an abundance of grains of *Quercus, Pinus,* and monolete spores of Polypodiaceae, with a certain number of the herbaceous grains. Toward the north of this province, *Salix* grains become common, while Polypodiaceae spores tend to decline and herbaceous grains are more abundant.

Of the three kinds of palynomorphs in the Wulong Formation, angiospermous pollen occupies the dominant position, constituting generally about 50% or more of the total number of the assemblage. Gymnospermous pollen forms about 10% to 30% of the total content, and the pteridophyte percentage is approximately equal to that of the gymnosperm. In the upper part of this formation, however, the percentage of angiospermous pollen drops down to 30%, and the gymnospermous rises to 50% of the total number of palynomorphs.

In the spectrum of the angiospermous pollen, the notably highest percentage goes to genus *Quercodites,* including the following species: *Quercoidites henrici, Q. microhenrici, Q. minutus, Q. densus, and Q. asper,* which all together account for 20% to 30% of the total number of palynomorphs. The Betulaceae pollen consists of such genera as *Betulaceoipollenites, Ostryoipollenites,* and *Momipites,* representing 2% to 10% of the entire assemblage. In the upper part of this formation, the percentage of pollen of the herbaceous families, such as Ranunculaceae and Polygonaceae, increases and pollen of Liliaceae and Gramineae also appear.

The Wulong Formation contains such gymnospermous pollen as *Podocarpidites, Cedripites, Taxodiaceaepollenites,* and *Keteleeria* grains, besides the pine grains which constitute 3% to 18% of the total. At the top of this formation, *Piceapollenites* and *Abiespollenites* have been found for the first time. Among the pteridophyte spores, the dominant genera are *Polypodiisporites, Polypodiaceaesporites;* the others are *Echinatisporis, Deltoidospora,* and *Polypodiceoisporites.*

In the pollen and spore assemblage of the lower member of the Lunpola Group in Bagon County, according to the datum reported by Wang and others (1975), angiospermous grains dominate over gymnospermous grains. The former may be identified as *Quercus* (the dominant genus), *Salix, Juglans, Alnus,* and *Corylus,* with a considerable number of herbaceous grains of Chenipodiaceae, Compositae, Leguminosae, Cruciferae, Liliaceae, and others. The gymnospermous pollen is composed mainly of *Pinus, Picea,* and *Abies.* Pteridophytes, such as the spores of Polypodiaceae, *Pteris* and *Adiantum,* are very scarce.

The plant megafossils of the Plateau Floristic Province may be represented by that of Namling (Li Hao-min and Guo Shuangxing, 1976). In the lower part of the Wulong Formation of this area, Betulaceae, in addition to the subordinate families Salicaceae, Ulmaceae, and others, play the leading role. The important species of these families are *Betula parautilis, Carpinus grandis, C. wulongensis, Populus latior, Salix* sp., *Ulmus hedini, Ribes* sp., and *Crataegus* sp., whereas those of the upper part of this formation are mainly of alpine species of *Quercus* and the various species of *Rhododendron,* such as *Quercus wulongensis, Q. prespathulata, Q. namlingensis, Rhododendron sanzugawaense,* and *R. namlingense.* Other species in this bed are *Thermopsis prebarbata, Phragmites* sp., and *Cyperacites* sp. The microleaved fossil *Thermopsis prebarbata* Li and Guo, which appeared at the upper part of the Wulong Formation, was also discovered in the Zoige area of northwestern Sichuan.

The megafossil and microfossil records mentioned above indicate that a rather warm and humid climate was prevalent in the middle Miocene and tended to become cooler and less moist in the late Miocene, which was a favorable time for herbaceous cover.

The Hengduan Mountains Floristic Province is located in the western part of Yunnan. As the valleys extended in a south-north direction with broader cols opening southward during the Miocene, the trade winds of the Indian Ocean made the climate there temperate and humid. The florules reported from many sites of this province may be represented by that of the Shuanghe Formation (Coal Bed) in Jianchuan County, western Yunnan. The florule of this formation, as shown in *Cenozoic Plants from China* (1978), includes *Picea* sp., *Cupressus* sp., *Dryophyllum yunnanense, D.* cf. *puryearense, Quercus scottii, Q. spathulata, Sassafras paratsuma, Phoebe megaphylla, Cinnamomum* sp., *Zelkova ungeri, Pistacia miochinensis, Rhus trifolia,* and *Palirus* sp. (fruit).

This list provides convincing evidence that during Miocene time the tropical-subtropical evergreen woody plants were still growing in the warm and moist valleys of the Hengduan Mountains, while pines and other alpine plants were living at the mountain top.

INLAND FOREST-MEADOW AND STEPPE FLORISTIC REGION

To the west of the Helan Mountains and to the north of the Kunlun Mountains is the vast Inland Floristic Region. The moisture-bearing winds of the Indian Ocean are diverted by the increasing altitude of the Qinghai-Xizang Plateau resulting from the movement of the Himalayas; the monsoon of the Donghai (the East China Sea) could hardly travel for such a long distance. Thus, a rather arid climate prevails in this region.

Palynological taxa found in the Hongxiakou Formation and the Yiushashan Formation in the Qaidam Pendi (the Qaidam Basin) (Hsü Jen and others, 1958) and the Baiyanghe Formation in the Jiuquan area (Sung Tze-chen, 1958) are characterized by abundant herbaceous and shrubby pollen (*Artemisia,* Compositae, Chenopodiaceae, *Nitraria,* and others) and a small number of woody angiospermous pollen (Betulaceae, Ulmaceae, Juglandaceae, Fagaceae, and other), sometimes with more pollen of Pinaceae.

The palynomorph assemblages of the Hongxiakou Formation and the Yiushashan Formation are characterized by abundant Pinaceae pollen and a great increase of the herbs and shrubs (*Nitraria, Chenopodium, Artemisia, Ephedra,* and others). The palynoflora of these formations shows an affinity mainly to the following genera or families: conifers: *Pinus, Tsuga, Cedrus,* and *Picea;* woody angiosperms: Betulaceae, *Quercus, Castanea,* and *Fagus;* and shrubs and herbs: Compositae, Liliaceae, *Chenopodium,* Gramineae, Leguminosae, *Potamogeton, Nitraria,* and others.

Such an aspect of the assemblage suggest a landscape of probably broadly open steppe, locally forest-meadows, with some pine forests growing on the mountains.

In the easternmost part of Qinghai Province, the absence of the south-north oriented folded ranges gave access to the East China monsoon which influenced slightly the flora of this district. Therefore, the woody angiospermous plants were well developed and even formed some massive woodlands under suitable environment. In the Miocene deposits of Caka, Zeko, for example, the megaflorule contains such species as *Taxus qinghaiense, Salix* sp., *Ranunculus* sp., *Cercus miochinensis, Podogonium oehningense, Acer pseudocarpinifolium, A. subginnala, Typha latissima,* and *Phragmites oehningensis.*

In the Junggar Pendi (Dzungar Basin), western Xinjian, the Miocene microflora may be typified by the assemblages of the upper Green Rock Formation and the Brown Rock Formation of the Horgus section and Manas section (Sung Tze-chen and others, 1964a, 1964b). The sporo-pollen flora of the upper Green Rock Formation is composed as follows:

The woody angiospermous pollen belongs mainly to Betulaceae, *Alnus,* Ulmaceae, Juglandaceae, and Tiliaceae, with a few Anacardiaceae, Aquifoliaceae, Myrtaceae, Ericaceae, Elaegnaceae, and others. The diagnostic herbaceous pollen are those of Chenopodiaceae that constitutes 20% to 25% of the total number of the palynomorphs, others are related to Compositae, Liliaceae (all together below 2%), *Sparganium, Potamogeton,* and others. The gymnospermous pollen is as important as the herbaceous pollen. Pollen of *Ephedra* (in places 15%) and Pinaceae and other elements (20%) are commonly present, but pollen of Taxodiaceae and Podocarpaceae are scarcely present. Spores of Polypodiaceae and fungi are also found.

The palynological assemblage of the Brown Rock Formation is composed of various angiospermous pollen referable to *Betula, Alnus, Corylus,* Ulmaceae, Juglandaceae, *Tilia,* Fagaceae and Chenopodiaceae (43% to 52%), Compositae (30%), *Sparganium, Potomogeton,* and a few subtropical representatives, besides plenty of gymnospermous pollen belonging to *Pinus, Picea* (the dominant genera) and *Cedrus,* Podocarpaceae, and *Ephedra;* spores of clubmoss affinity appear in this assemblage only very rarely.

The general features of the palynological complexes of both the Upper Green Rock Formation and the Brown Rock Formation are that plenty of herbaceous pollen related to *Chenopodium,* Compositae, and aquatic *Potomageton* exist together with a moderate number of hardwoods classified as Betulaceae, Ulmaceae, Juglandaceae, Fagaceae, Liliaceae, and others, with a few subtropical elements of Aquifoliaceae, Myrtaceae, and others. Rather similar assemblages have been discovered in the lower Miocene series of Kazakhstan, Soviet Union (Pokrovskaya, 1956) where lacustrine Miocene deposits are widespread. It follows, therefore, that in the Junggar Basin, Xinjiang, rocks of the same age were probably deposited in a closely comparable environment. Recently, it was reported that a number of woody angiospermous pollen of this period had been found in borehole samples in southern Xinjiang, findings that compared well with data from the Junggar Basin.

Because the floral content in western Xinjiang is evidently different from that of the rest of this region, it may be regarded as a subregion, named Western Xinjiang Floristic Province. Comparison of microfloras of the Western Xinjiang Province and the inland province shows that they are alike in the abundance of the herbaceous pollen, unlike in the quantity of the woody angiospermous pollen. In the Western Xinjiang Province, the presence of more woody angiospermous elements, especially some subtropical elements, may be the result of favorable paleoclimatic influences from the Tethys. By the latest Miocene and later, as this influence gradually abated and even disappeared altogether, the woody angiospermous plants began to fade out accordingly, and subtropical members became nearly extinct. The appearance of the flora in this province was henceforth also entirely similar to that of the inland province.

EASTERN MONSOON BROAD-LEAVED FOREST FLORISTIC REGION

This floristic region occupies almost the whole eastern part of China, including all the islands in the Donghai (East China Sea) and the Nanhai (South China Sea). It is composed of plains and hilly countries stretching to the eastern Hengduan Mountains in the south. Owing to the powerful influence of the Pacific trade winds, a climate with considerable warmth and high humidity had prevailed in general, which caused extensive distribution of broad-leaved forest (deciduous or evergreen) or of mixed forest consisting mainly of broad-leaved angiospermous types.

In consideration of the differences in floristic composition in the latitude belts from north to south, this floristic region may be subdivided into three floristic provinces.

Northern Temperate Floristic Province

This floristic province lies to the north of the Songhua River of Helongjiang Province. On the basis of presently available data, this floristic province cannot yet be assigned with certainty to the Miocene age. From palynological information from Primorye Krai (Maritime Territory of the Soviet Union) and the area to the north of the Helong River (Amur River) (Pokrovskaya, 1956), it may be concluded that there might have existed a broad-leaved (deciduous oaks), coniferous (mainly pines), mixed forest consisting mainly of angiosperms, still intermingling with a few subtropical elements of such genera as *Carya, Liquidambar,* and *Ilex.* Climatically, this might have been comparable to the conditions prevailing in the area of northern China today.

Central Warm-Temperate-Subtropical Floristic Province

This floristic province occupies a large tract from the middle and lower Yangtze Valley and from the valleys of the Jing River and the Wei River eastward to the littoral areas of the Bohai and Huanghai [Yellow Sea]. The eastern part of Ningxia and Neimongol [Inner Mongolia] to the east of the Helan Mountain is also included because of occurrence of beds equivalent to the Hannuoba Formation, which is intercalated with basalt at Tianzhen County, Shanxi Province.

Some highly fossiliferous localities were discovered in this province. The characteristics of both megaflora and microflora of these localities are that temperate angiosperms, such as Betulaceae, Juglandaceae, Ulmaceae, and deciduous oaks, existed together with subtropical members, such as chestnut and evergreen oaks (Fagaceae), *Carya, Liquidambar, Rhus,* Oleaceae, Meliaceae, Sterculiaceae, and Lauraceae, along with *Certopteris, Salvinia,* and *Trapa.* All of these tend to prove that the vegetation was a deciduous broad-leaved forest intermixed with a number of evergreen members, and some aquatic plants in pools.

Judging from the described taxa, we think the climate in this province was similar to that of today's north-middle subtropical type prevailing in the area between the lower and middle Yangtze Valley and the Nanling Mountains. The well-known Shanwang flora may be regarded as a typical representative of this province, which seems to reflect, in our opinion, a climate of the southern northern-subtropical type rather than the warm-temperate or north-subtropical type inferred by Hu and Chaney (1940). Chaney was inclined to believe that the aspect of this flora might be comparable with that of the middle Yangtze Valley of today.

The Shanwang flora is the richest in number of genera and species in this province. As shown in *Cenozoic Plants from China* (1978), the Shangwang flora consists of 88 genera and 127 species. Following are some additional important fosiliferous sites containing leaf remains: Mengjiagang in Huanan County, Heilongjiang Province; Sanhe in Yanbian Autonomous District, and Qiuligou in Dunhua County, Jilin Province; Zhangzhuang in Sihong County, and Puzhen in Nanjing, Jiangsu Province; Mingguang in Anhui Province. The following genera and species are most frequently represented in these floras: *Betula, Carpinus, Corylus, Quercus sinomiocenica, Q. miocrispula, Q. protodentata, Magnolia, Liriodendron, Ailanthus, Podogonium oehningense, Glyptostrobus europaeus, Magnolia, Liriodendron, Ailanthus, Podogonium oehningense, Glyptostrobus europaeus, Metasequoia distichia, Pseudolarix,* and *Pinus.* These floras have some members in common with the Shanwang flora, despite the differences in their complexity and in the size of their collections. They should be considered to be roughly contemporaneous, that is, middle Miocene or slightly later. One of them, the Sanhe florula of the Yanbian Autonomous District, Jilin Province, has rather more Taxodiaceae and other older forms, for example, *Dryophyllum dewalquei* and *Engelhardtia korianica* Oishi. Its geological age should thus be slightly older than that of the Shanwang flora, although still in the Miocene.

On the basis of differences in their composition, the palynological assemblages of this province may be divided into three subassemblages: (1) The early Miocene subassemblage is well represented by the floras of the Guantao Formation, Shandong Province (Sung Tze-Chen and others, 1974a, 1964b), and of the Hannuoba Formation, Shanxi Province (Wang Hsian-tzeng, 1978). It is characterized by the frequent occurrence of Pinaceae and mild, broad-leaved tree grains and the occasional occurrence of herbaceous grains, sometimes with a number of *Trapa* grains and spores of *Ceratopteris* and *Salvinia.* The palynoflora of the Guanta Formation of Shandong, the Hannuoba Formation of Tianzhen County, Shanxi, and the Fengshan Formation of Sihong County, Jiangsu, may represent this stage. In the microflora of the Guantao Formation, for example, spores of *Certopteris* (dominant element) and Polypodiaceae and *Salvinia* are most numerous. Gymnospermous pollen which has an affinity to *Picea* (dominant member), *Pinus, Tsuga,* and a few Taxodiaceae, *Podocarpus, Eohedra,* and others are subordinate; angiospermous pollen is not as rich as the previous forms, consisting mainly of grains of *Carya, Ulmus,* as well as grains of *Juglans, Quercus, Betula, Alnus, Corylus,* and a few *Fagus, Rhus, Liquidambar, Magnolia,* and others. The recognizable herbaceous grains belong to Convolvulus, Polygonum, Gramineae, Compositae, *Potamogeton,* and others.

(2) The middle Miocene subassemblage is represented by the Shanwang palynoflora. It is marked by very abundant thermic tree grains, with rare but definite spores of *Salvinia,* whereas Pinaceae pollen is more poorly represented. (Sung Tze-chen and others, 1964a, 1964b).

(3) The late Miocene subassemblage is palynologically similar to the second, except for the higher content of Pinaceae and herbaceous grains. The microflora of the lignite beds of Liushanxiang, Linju County, Shandong Province, may serve to illustrate the features of the late Miocene assemblage (Sung Tze-chen and others, 1964a, 1964b).

During the Miocene, inasmuch as the east monsoon had weakened in the area to the north of the Yinshan Mountains and the Daxinganling Mountains, there might possibly have been a forest-meadow or steppe developing, but more reliable information is needed to corroborate this suggestion.

Southern South-Subtropical and Tropical Floristic Province

This province includes the vast area south of the middle and lower Yangtze Valley extending westward to the eastern part of Yunnan, eastward to the Donghai (East China Sea) and to the Nanhai (South China Sea).

In the northern part of this floristic province, the fossil remains of both leaf impressions and palynomorphs were obtained mainly from the Xiananshan Formation of the Shengxian Group, Zhejiang Province. In comparison with its northern neighbor, two distinctive features stand out: (1) the impoverishment or absence of such temperate forms as Betulaceae, poor in species as well as in specimens; and (2) an obvious rise of evergreen species, consisting of forms related to Hamanelidaceae, Buxaceae, Aquifoliaceae, Rutaceae, and evergreen types of Fagaceae which are the dominant members of the flora. Notable is the discovery of subtropical Magnoliaceae, Lauraceae, and Annonaceae whose living representatives are nowadays growing in the low latitudes of the subtropical-tropical zone. Palynologically, this province is noted for the rich presence of forms assigned to *Quercus* (the evergreen), Carya, Ulmaceae, and *Trapa.* Other common elements are *Castanea, Liquidambar,* Leguminosae, and *Certopteris.* Betulaceae pollen is rare. Many Pinaceae pollen have been discovered, possibly coming from the contemporaneous adjacent alpine coniferous forest. With the help of the floristic data mentioned above, the Xiananshan flora may be easily distinguished from the Shanwang flora by its comparatively fewer temperate forms and more abundant subtropical evergreen forms. This flora probably suggests climate conditions of a southern-subtropical type.

In the south of this province, that is, in Eastern Yunnan Province, Guangdong Province, and the Guangxi Autonomous Region, more records of megafossils and microfossils are known. The Xiaolongtan flora *(Cenozoic Plants from China,* 1978) of the Kaiyuan District, Yunnan, and the plant remains of the Yongning Group of the Nanning Basin, for example, are characterized by the abundant species of lauraceae and certain additional tropical forms.

Many samples for maceration were collected and examined from many localities, such as outcrop samples from the Yongning Group of the Nanning Basin, the Xiayang Formation of Weizhou Island and the Leizhou Penninsula, the Changpo Formation of Hainan Island, and borehole samples from Xinhui County, the estuary of the Zhujiang (Pearl River). Of the palynoflora, the dominant components are tricolpate and tricolporate grains related to such families as Leguminosae, Flacourtiaceae, Theceae, Hamamelidaceae, and the genus *Trochodendron.* Polypodiaceae monolete spores are also richly represented. Pinaceae pollen generally is rare, and the *Cacrydium* grains are rich only in some localities. Here it is worth emphasizing that the interesting *Gothanipollis,* which may be referable to the Loranthaceae, and the grains of *Sonneratia* are always found in bore cores on Weizhou Island, Hainan Island, and in the estuary of the Zhujiang River. These plants exist in the tropical zone today. So far as we know, there is every reason to believe that the area to the south of Nanning-Guangzhou (Canton) had a tropical climate during Miocene time.

Some fossiliferous beds of the lower member of the Fotan Group have also been discovered in southern Fujian Province. Despite the smaller number of specimens, it is obvious that each florule is composed predominantly of species of Annonaceae, with some rarer temperate forms. If this florule is of Miocene age, southern Fujian should also fall into the tropical zone. As for Taiwan Provinces, the flora of the Shihti coal field, about 15 km southeast from Taipei, is of middle Miocene age. In this flora, Lauraceae are very plentiful; the evergreen trees such as *Castanopsis* and *Pasania* of Fagaceae are abundant, but the typically temperate members of this family are absent; in the meantime *Coniogramme, Bambusa,* and *Phyllostachys,* restricted today to low latitudes, also have many representatives, all of which point to the subtropical, or even tropical, character of the Shihti flora. Taiwan Province thus may also be classified as a tropical region.

SUMMARY

During the Miocene period, three main floristic regions came into existence in China: (1) the Qinghai-Xizang *Quercus-Betula*-Thicket Floristic Region, (2) the Inland Forest-Meadow and Steppe Floristic Region, and (3) the Eastern Monsoon Broad-Leaved Floristic Region. Each may be subdivided into two or three floristic provinces.

REFERENCES CITED

[Cenozoic Plants from China]: Compiled by the Beijing Institute of Botany and the Nanjing Institute of Geology and Palaeontology, Academia Sinica, Science Press, 232 p., 149 pls., Beijing, 1978 (in Chinese).

Chien Sung-shu, Wu Cheng-yih, and Chen Chang-tu, 1956, The vegetation types of China: Acta Geographica Sinica, v. 22, no. 1, p. 37–92 (in Chinese with English Summary).

Hsü Jen [Xu Ren], Sung Tze-chen [Song Zhi-chen], and Chou Ho-i [Zhou He-yi], 1958, Sporo-pollen assemblages from the Tertiary deposits of the Tsaidam Basin and their geological significance: Acta Palaeontologica Sinica, v. 6, no. 4, p. 429–440, 6 pls. (in Chinese with English summary).

Hu Hsen-hsu and Chaney, R. W., 1940, A Miocene flora from Shantung Province, China: Palaeontologia Sinica, new ser. A, no. 1, 147 p., 57 pls.

Li Hao-min and Guo Shuang-xing, 1976, The Miocene flora from Namling of Xizang: Acta Palaeontologica Sinica, v. 15, no. 1, p. 7–20, 3 pls. (in Chinese with English abstract).

Pokrovskaya, I. M., ed., 1956, Atlas Miotsenovykh Sporovo-Pyltsevykh Kompleksov Razhichnykh Rayonov SSSR: 461 p., 32 pls., Materialy VSEGEI, novaya seriya, vyp. 13, Paleontologiya i Stratigrafiya, Gosudarstvennye Nauchno-Tekhnicheskoe Izdatelstvo Literatury po Geologii i Okhrane nedr, Moskva.

Sung Tze-chen [Song Zhi-chen], 1958, Tertiary spore and pollen complexes from the red beds of Chiuchuan, Kansu, and their geological and botanical significance: Acta Palaeontologica Sinica, v. 6, no. 2, p. 159–166, 7 pls. (in Chinese with English summary).

——1959, Miocene sporo-pollen complex of Shanwang, Shantung: Acta Palaeontologica Sinica, v. 7, no. 2, p. 99–110, 3 pls. (in Chinese with English abstract).

Sung Tze-chen [Song Zhi-chen], Tsao Liu [Cao Liu], and Li Man-ying, 1964a, Tertiary sporo-pollen complexes of Shantung: Memoirs of the Nanjing Institute of Geology and Palaeontology, Academia Sinica, no. 3, p. 179–291, 28 pls. (in Chinese with English abstract).

Sung Tze-chen [Song Zhi-chen], Zhang Lu-jin [Chang Lu-chin], Liu Jin-ling, Huang Feng-bao, Yuan Feng-xiang, Zheng Ya-hui, Ouyang Shu, Zhang Chun-bin, and Li Wen-ben, 1964b, [Sporo-Pollen Analysis]: Beijing, Science Press, 250 p., 36 pls. (in Chinese).

Wang Hsian-Tzeng [Wang Xian-zeng], 1978, On the discovery of late Tertiary sporo-pollen assemblage in the brown-coal beds of Tianchen Region and its significance: Acta Scientiarum Naturalium Universitatis Pekinensis, 1978, no. 4, p. 89–105, 3 pls. (in Chinese with English abstract).

Wang Kai-fa, Yang Jiao-wen, Li Zhe, and Li Zeng-rui, 1975, On the Tertiary sporo-pollen assemblages from Lunpola Basin of Xizang, China, and their paleogeographical significance: Scientia Geologica Sinica, v. 4, no. 4, p. 366–374, 3 pls. (in Chinese with English abstract).

MANUSCRIPT ACCEPTED BY THE SOCIETY APRIL 22, 1981

Printed in U.S.A.

Geological Society of America
Special Paper 187
1981

Chinese personal and geographic names in Wade-Giles and Pinyin romanization

Curt Teichert
Liu Lu*
Department of Geological Sciences, University of Rochester, Rochester, New York 14627

Although it is not possible to render Chinese words in any of the European alphabets, especially the Roman alphabet, in a way to ensure their correct pronunciation, systems of transcribing Chinese sounds as combinations of the Roman alphabet have been proposed; this procedure is called "romanization." About 1860, an English diplomat, Sir Thomas Wade, proposed a system of romanization which was later modified by Herbert Giles of Cambridge University and became known as the Wade-Giles system which was used almost exclusively to romanize Chinese words in English-language literature for more than 100 years.

On November 1, 1957, the State Council of the People's Republic of China adopted a considerably modified system which became known as Pinyin (literally "transcription") and was made official on January 1, 1979. It is now used in all official Chinese communications and publications and is finding increasing application in England and North America. Pinyin is used by most authors of this volume.

The following list of names has been prepared to assist readers in the identification of names of Chinese authors and places used in this volume and to be found in the earlier literature (books, periodicals, maps). For authors' names as well as geographic names, two lists are given in each category—Wade-Giles to Pinyin and Pinyin to Wade-Giles. In addition, a list of conventional, non-Chinese names has been added, and their equivalents in Pinyin are given. In order to find the Wade-Giles equivalents of such names, one of the appropriate Pinyin to Wade-Giles lists should be consulted.

In using these lists the following additional points should be noted:

1. Names mentioned in this volume in Pinyin for which there are no Wade-Giles equivalents have not been listed.
2. Names whose spellings are identical in both systems have not been listed.
3. Chinese personal names consist of either two or three words, the first of which is the surname or family name. This is followed, without intervening comma, by the given name which may consist of one or two words which may or may not be hyphenated. According to Wade-Giles, if the given name consisted of two words, these were both written with initial capital letters and were not hyphenated; for example, Yin Tsan Hsun. In Pinyin, the given name may be either written as two hypenated words or united into one word; for example, Yin Zan-xun or Yin Zanxun.
4. During the 1920s and through the 1940s, Chinese authors, when publishing in English either in China or abroad, often adopted the western style of placing the given name before the family name, at the same time abbreviating it by printing initials only; for example, T. H. Yin. Such names are listed by placing the initials after the surname, separated from it by a comma; for example, Yin, T. H.

In alphabetizing the lists, we have been guided by the *Gazetteer of the People's Republic of China,* published by the Defense Mapping Agency, Washington, D.C. (1979).

*Present address: Amoco Orient Petroleum Company, P.O. Box 4381, Houston, Texas 77210.

AUTHORS' NAMES FROM WADE-GILES TO PINYIN

Wade-Giles	Pinyin
Chang An Chi	Chang An-zhi
Chang, H. C.	Zhang Xi-zhi
Chang Lu Chin	Zhang Lu-jin
Chang Shan Jean	Zhang Shan-zhen
Chang Wen Tang	Zhang Wen-tang
Chang Wen You	Zhang Wen-you
Chao King Koo	Zhao Jin-ke
Chao, Ya Tseng (Y. T. Chao)	Zhao Ya-zeng
Chen Pei Chi	Chen Pei-ji
Chen, S.	Chen Xu
Chen Te Chiung	Chen De-quiong
Chi, Y. S.	Ji Rong-sen
Chien Yi Yuan	Qian Yi-yuan
Ching Yü-kan	Jing Yu-gan
Chou Ho I	Zhou He-yi
Chow, M. C.	Zhou Ming-zhen
Chow Tse Yen	Zhou Zhi-yan
Chu Chao Ling	Zhu Zhao-ling
Chu Hao Jan	Zhu Hao-ran
Geh Mei Yu	Ge Mei-yu
Hsiang Lee Wen	Xiang Li-wen
Hsü Jen	Xu Ren
Hsü, Singwu C.	Xu Jie
Hsü, T. Y.	Xu De-you
Huang, T. K.	Huang Ji-qing
Lee, Hsing Hsüeh	Li Xing-xue
Lee, Jonquei S. (J. S. Lee)	Li Si-guang
Lee Pei Chuan	Li Pei-juan
Liu Kui Zhih	Liu Gui-zhi
Lu Yen Hao	Lu Yan-hao
Ma, T. Y.	Ma Ting-ying
Mu An Tze	Mu En zhi
Pan Chuang Hsiang	Pan Zhong-xiang
Ping, C.	Bing Zhi
Qi Dun-luan	Qi Dun-lun
Shen Kuang-lung	Shen Guang-long
Sheng Jing Chang	Shen Jing-zang
Sheng Si Fu	Shen Xin-fu
Sin Yu Sheng	Xing Yu-shen
Sun Kou	Shen Gua
Sun, Y. C.	sun Yun-zhu
Sung Tze Chen	Song Zhi-chen
Sze Hsin Chien	Si Xing-jian
Tien, C. C.	Tian Qi-jun
Tsai Chung Yang	Cai Chong-yang
Tsao Cheng Yao	Cao Zheng-you
Tsao Liu	Cao Liu
Tsao Rui Chi	Cao Rui-ji
Tsou Si-ping	Zou Xi-ping
Wang Hong Chen	Wang Hong-zhen
Wang Hsian Tseng	Wang Xian-zeng
Wu Chun-ching	Wu Shun-qing
Wu Wang Shih	Wu Wang-shi
Xu Guei Yong	Xu Gui-rong
Yang Chi'i	Yang Qi
Yang King Chih	Yang Jing-zhi
Yang Tsun Yi	Yang Zun-yi
Yeh Mei Na	Ye Mei-na
Yin Tsan Hsun (T. H. Yin)	Yin Zan-xun
Yoh, S. S.	Yue Sen-xun
Young Chong Ching	Yang Zhong-jian
Yü, C. C.	Yu Jian-zhang
Yü Wen	Yu Wen

AUTHORS' NAMES FROM PINYIN TO WADE-GILES

Pinyin	Wade-Giles
Bing Zhi	Ping, C.
Cai Chong-yang	Tsai Chung Yang
Cao Liu	Tsao Liu
Cao Rui-ji	Tsao Rui Chi
Cao Zheng-you	Tsao Cheng Yao
Chang An-zhi	Chang An Chi
Chen De-quiong	Chen Te Chiung
Chen Pei-ji	Chen Pei Chi
Chen Xu	Chen, S.
Ge Mei-yu	Geh Mei Yu
Huang Ji-qing	Huang, T. K.
Ji Rong-sen	Chi, Y. S.
Jing Yu-gan	Ching Yü-kan
Li Pei-juan	Lee Pei Chuan
Li Si-guang	Lee, Jonquei S. (J. S. Lee)
Li Xing-xue	Lee, H. H.
Liu Gui-zhi	Liu Kui Zhih
Lu Yan-hao	Lu Yen Hao
Ma Ting-ying	Ma, T. Y.
Mu En-zhi	Mu An Tze
Pan Zhong-xiang	Pan Chuang Hsiang
Qian Yi-yuan	Chien Yi Yuan
Qi Dun-lun	Qi Dun-luan
Shen Gua	Sun Kou
Shen Guang-long	Shen Kuang-lung
Shen Xin-fu	Sheng Si Fu
Sheng Jing-zang	Sheng Jing Chang
Si Xing-jian	Sze Hsin Chien
Song Zhi-chen	Sung Tze Chen
Sun Yun-zhu	Sun, Y. C.
Tian Qi-jun	Tien, C. C.
Wang Hong-zhen	Wang Hong Chen
Wang Xian-zeng	Wang Hsian Tseng
Wu Shun-qing	Wu Chun Ching
Wu Wang-shi	Wu Wang Shih
Xiang Li-wen	Hsiang Lee Wen
Xing Yu-shen	Sin Yu Sheng
Xu De-you	Hsü, T. Y.
Xu Gui-rong	Xu Guei Yong
Xu Jie	Hsü, Singwu C.
Xu Ren	Hsü Jen
Yang Jing-zhi	Yang King Chih
Yang Qi	Yang Chi'i
Yang Zhong-jian	Young Chong Ching
Yang Zun-yi	Yang Tsun Yi
Ye Mei-na	Yeh Mei Na
Yin Zan-xun	Yin Tsan Hsun (T. H. Yin)
Yu Jian-zhang	Yü, C. C.
Yu Wen	Yü Wen
Yue Sen-xun	Yoh, S. S.
Zhang Lu-jin	Chang Lu Chin
Zhang Shan-zhen	Chang Shan Jean
Zhang Wen-tang	Chang Wen Tang
Zhang Wen-you	Chang Wen You
Zhang Xi-zhi	Chang, H. C.
Zhao Jin-ke	Chao King Koo
Zhao Ya-zeng	Chao Ya Tseng (Y. T. Chao)
Zhou He-yi	Chou Ho I
Zhou Ming-zhen	Chow, M. C.
Zhou Zhi-yan	Chow Tse Yen
Zhu Hao-ran	Chu Hao Jan
Zhu Zhao-ling	Chu Chao Ling
Zou Xi-ping	Tsou Si-ping

GEOGRAPHIC NAMES WADE-GILES TO PINYIN

Wade-Giles	Pinyin
Ai-lao Shan	Ailaoshan
Anhwei	Anhui
An-yüan	Anyuan
Ch'ai-ta'mu P'en-ti	Qaidam Pendi (Chaidamu Pendi)
Ch'ang-chou	Changzhou
Chang-ho	Changhe
Ch'ang-hsing	Changxing
Chang-i	Zhanyi
Ch'ang-ning	Changning
Ch'ang-shan	Changshan
Ch'ang-tu Ti-chü	Qamdo Diqu
Chang-yeh	Zhangye
Chao-chuang	Zhaozhuang
Ch'ao Hsien	Chaoxian
Chao-ko-chuang	Zhaogezhuang
Chekiang	Zhejiang
Chengchow	Zhengzhou
Cheng-chiang	Chengjiang
Ch'eng-k'ou	Chengkou
Chengtu	Chengdu
Chen-p'ing	Zhengping
Chen-siung	Zhenxiong
Chia-ling Chiang	Jialing jiang
Chiang-ning	Jiangning
Chiang-p'u	Jiangpu
Chiang-shan	Jiangshan
Chiang-ta	Jiangda
Chiang-yu	Jiangyou
Chian-shan	Qianshan
Chiao-ho	Jiaohe
Ch'i Chiang	Qijiang
Ch'ien-wei	Qianwei
Chih-chin	Zhijin
Chih-wei (Islet)	Chiwei
Ch'i-lien Shan	Qilianshan
Ch'in-an	Qinan
Ching-ching	Jingjing
Ch'ing-feng	Qingfeng
Ching Hsien	Jingxian
Ch'ing Ling	Qining
Ching-shan	Jingshan
Ch'ing-shui Ho	Qingshui He
Ching-si	Jingxi
Ch'ing-tao (Tsing-tao)	Qingdao
Ching-yüan	Jingyuan
Chin-hua	Jinhua
Chin-sha	Jinsha
Chin-t'an	Jintan
Chin-yang	Jingyang
Chiu-ch'üan	Jiuquan
Cho-tzu Shan	Zhuozishan
Ch'uan-chiao	Quanjiao
Chu Chiang	Zhujiang
Ch'ü-ching	Qujing
Ch'u Hsien	Chuxian
Chu-ko-erh Pen-ti	Junggar Pendi
Chu-mu-long-ma Feng	Qomolongma Feng
Ch'ung-ch'ing	Chongqing
Ch'ung-yang	Chongyang
Chung-yang	Zhongyang
Ch'ü-yang	Quyang
Dabashan	Ta-pa Shan
Dali	Ta-li
Da-tian-ba	Datianba
Er Hai	Erhai
Erh-lien-hao-t'e	Erenhot
Erh-yüan	Eryuan
Feng-feng	Fengfeng
Fen-hsiang	Fenxiang
Fu-chou-wan	Fuzhouwan
Fukien	Fujian
Fu-ning	Funing
Fu-shun	Fushun
Hai-p'o-wan	Haibowan
Han-cheng	Hancheng
Hangchow	Hangzhou
Heilungkiang	Heilongjiang
Heng-nan	Hengnan
Heng-tuang Shan	Hengduan Shan
Ho-chin	Hejin
Ho-fei	Hefei
Ho-hsiang	Hexiang
Ho Hsien	Hexian
Ho-lan Shan	Helan Shan
Honan	Henan
Hopei	Hebei
Ho-shih-t'o-lo-kai	Hoxtolgay
Hsiang-chou	Xiangzhou
Hsiang-hsiang	Xiangxiang
Hsiao-ksing-an Ling	Xiao Hinggan Ling
Hsiao-shui	Xiaoshui
Hsien (County)	Xian
Hsi-hsia-pang-ma Feng	Xixiabangma Feng
Hsi-kuang Shan	Sikuangshan
Hsing-an	Hinggan
Hsing-t'ang	Xingtang
Hsiu Hsien	Xiuxian
Hsin-t'ai	Xintai
Hsin-yü	Xinyu
Hsi-shui	Xishui
Hsi-yang	Xiyang
Hsuan-cheng	Xuangcheng
Hsü-chou	Xuzhou
Hsü-i	Xuyi
Hsun-tien	Xundien
Hsü-shan	Xushan

GEOGRAPHIC NAMES WADE-GILES TO PINYIN

Wade-Giles	Pinyin
Huai-nan	Huainan
Huang Ho	Huang He
Huang-shan-tung	Huangshandong
Huan-hsien	Huanxian
Hua-ning	Huaning
Hu-chou	Huzhou
Hui-li	Huili
Hui-tung	Huidong
Hui-tze	Huize
Hu-lun-pei-erh Meng	Hulum Buir Meng
Huo-ch'eng	Houcheng
Hupeh (Hupei)	Hubei
Hupei (Hupeh)	Hubei
I-chang	Yizhang
I-ch'ang	Yichang
I-cheng	Yicheng
I Hsien	Yixian
I-tou	Yidou
Jen-huai	Renhuai
Kai-li	Kaili
K'ai-p'ing	Kaiping
Kansu	Gansu
Kao-yen	Gaoyan
Kao-yu	Gaoyou
K'e-p'ing	Keping
Kiangsi	Jiangxi
Kiangsu	Jiangsu
Kirin	Jilin
Kuan-chung	Guanzhung
Kuang-teh	Guangde
Kuang-yüan	Guangyuan
Kuanti	Guandi
K'u-lu-k'o-ta-ka	Kuruktag
Ku--niu-tang	Guniutan
K'un-lun Shan	Kunlunshan
Kun-ming	Kunming
K'un-shan	Kunshan
Kuo-long	Guolong
Ku-shan	Gushan
Kwangchow (Canton)	Guangzhou
Kwangsi Chuang Autonomous Region	Guangxi Zhuangzu Zizhiqu
Kwangtung	Guangdong
Kweichow	Guizhou
Kweilin	Guilin
Kweiyang	Guiyang
Lai-an	Lai'an
Lai-shui	Laishui
Lia-yang	Laiyang
Lanchow	Lanzhou
Lang-shan	Langshan
Lang-ya-Shan	Langyashan
Lan-t'ien	Lantian
Lei-po	Leibo
Le-ping (Loping)	Leping
Liang Shan	Liangshan
Lian-shan	Liangshan
Liao-ch'eng	Liaocheng
Liaoning	Liaoning
Li Chiang	Lijiang
Lien Hsien	Lianxian
Lien-yuan	Lianyuan
Lin-ch'ü	Linqu
Ling-ch'uan	Lingchuan
Lin-ju	Linru
Ling-ling	Lingling
Liu-yang	Liuyang
Lo-shan	Leshan
Lu-ch'uan	Luquan
Lu-ho	Luhe
Lung Hsien	Longxian
Lung-t'an	Longtan
Lun-shan	Lunshan
Lu-shi	Luxi
Ma-chia-kou	Majiagou
Meng-la	Mengla
Meng-tzu	Mengzi
Men-t'ou-kou	Mentougou
Miao-po	Miaopo
Mien Hsien	Mianxian
Mien-tu-ho	Mianduhe
Mien-yang	Mianyang
Ming-shui	Mingshui
Mu-fu Shan	Mufushan
Nan-cheng	Nanzheng
Nan-chiang	Nanjiang
Nan-hsiung	Nanxiong
Nanking	Nanjing
Nan-ling	Nanling
Nan-p'iao	Nanpiao
Nan-tan	Nandan
Nen-chiang	Nenjiang
Ning-chen Shan (Nanking Hills)	Ningzhen Shan
Ning-chiang	Ningqiang
Ningsia Hui Autonomous Region	Ningxia Huizu Zizhiqu
Nin-yüan	Ninyuan
Niu-hsin Shan	Niuxinshan
Nong Shan	Nongshan
O-me	Emei
O-pien	Ebian
Pa-ch'u	Bachu (Maralwexi)
Pai-k'u	Baiku
Pai-se	Bosa
Pan-tang	Bantang
Pao-ch'ing	Baoqing
Pei-p'iao	Beipiao
Pei San	Beishan
Peking	Beijing
Pen-hsi	Benxi
Pi-chieh	Bijie

GEOGRAPHIC NAMES WADE-GILES TO PINYIN

Wade-Giles	Pinyin
Ping-chuan	Pingquan
P'ing-hsiang	Pingxiang
P'ing-ting Shan	Pingdingshan
Po-shan	Boshan
Pu-erh-han-pu-ta Shan	Burhan Buda Shan
Pu-lung	Bulong
San-shui	Sanshui
Shansi	Shanxi
Shantung	Shandong
Shan-wang	Shanwang
Shensi	Shaanxi
Shen-yang	Shenyang
Shiao-shan	Xiaoshan
Shih-chiu-t'o	Shijiutuo
Shih-t'ai	Shitai
Shih-tzu-pu	Shizipu
Shou-ch'ang	Shouchang
Sian	Xi'an
Sinkiang Uighur Autonomous Region	Xinjiang Uygur Zizhiqu
Soochow	Suzhou
Ssu-nan	Sinan
Suchou	Suzhou
Szechuan	Sichuan
Ta-hsing-an Ling (Great Khingan Range)	Da Hinggan Ling
T'ai-chou	Taizhou
Tai-hua	Daihua
Tai-nan	Dainan
Taipei	Taibei
Tai-yüan	Taiyuan
Ta-li-mu P'en-ti	Tarim Pendi
Ta-ming Shan	Damingshan
Tang-ch'üan	Tanquan
Tang-shan	Tangshan
T'an-ku-la Shan	Tanggula Shan
Talien (Dairen)	Dalien
Ta-pa Shan	Daba Shan
Ta-ping-kou	Dapinggou
Ta-ssu	Dasi
Ta-t'ung	Datung
Ta-wan	Dawan
Ta-yü	Dayu
Te-chiang	Dejiang
Tiao-yu Island	Diaoyu Da
T'ien Shan	Tianshan
Tientsin	Tianjin
T'ing-t'sun	Dingcun
Ting-yuan	Dingyuan
Tsinan	Jinan
Tsinghai	Qinghai
Tsun-i	Zunyi
T'u-lu-p'an	Turpan
Tung-chuan	Dongchuan
Tung-kou	Donggou
Tung-tzu	Tongzi
Tze-yüan	Ziyun
Tzu-hsi	Cixi
Tzu-kuei	Zigui
Wang-fu	Wangfu
Wa-se	Wase
Wu-ch'ia	Wuqia
Wu-han	Wuhan
Wu-hsing	Wuxing
Wu-hu	Wuhu
Wu-lu-mu-ch'i	Ürümqi
Wu-ning	Wuning
Wuhsi	Wuxi
Yao Hsien	Yaoxian
Ya-tung	Yadong
Yeh-li	Yeli
Yenan	Yan'an
Yen-chin	Yanjin
Yin-ch'eng	Yincheng
Yin-chiang	Yinjiang
Yüan-mou	Yuanmou
Yü-hang	Yuhang
Yü-hsien	Yuxian
Yung-ning	Yongning
Yung-shan	Yongshan
Yün Hsien	Yunxian
Yu-yang	Youyang

GEOGRAPHIC NAMES PINYIN TO WADE-GILES

Pinyin	Wade-Giles
Ailoshan	Ai-lao Shan
Anhui	Anhwei
Anyuan	An-ÿan
Bachu (Maralxwexi)	Pa-ch'u
Bagon	Pa-kung
Baiku	Pai-k'u
Bantang	Pan-tang
Baode	Pao-te
Baoqing	Pao-ch'ing
Beijing	Peking
Beipiao	Pei-p'iao
Beishan	Pei Shan
Benxi	Pen-hsi
Bijie	Pi-Chieh
Bose	Pai-se
Boshan	Po-shan
Bulong	Pu-lung
Burhan Buda Shan	Pu-erh-han-pu-ta Shan
Chaidamu Pendi	Ch'ai-ta'mu P'en-ti
Changhe	Chang-ho
Changning	Ch'ang-ning
Changshan	Ch'ang-shan
Changxing	Ch'ang-hsing
Changzhou	Ch'ang-chou
Chaoxian	Ch'ao Hsien
Chengdu	Chengtu
Chengjiang	Cheng-chiang
Chengkou	Ch'eng-k'ou
Chiwei	Chi-hwei
Chongqing	Ch'ung-ch'ing
Chongyang	Ch'ung-yang
Chuxian	Ch'u Hsien
Cixi	Tzu-hsi
Dabashan	Ta-pa Shan
Da Hinggan Ling	Ta-hsing-an Ling (Great Khingan Range)
Daihua	Tai-hua
Dainan	Tai-nan
Dali	Ta-li
Dalian	Talien
Damingshan	Ta-ming Shan
Dapinggou	Ta-ping-kou
Dasi	Ta-ssu
Datianba	Da-tian-ba
Datong	Ta-t'ung
Dawan	Ta-wan
Dayu	Ta-yü
Dejiang	Te-chiang
Diaoyu Dao	Tiao-yu Island
Dingcun	Ting-ts'un
Dingyuan	Ting-yuan
Dongchuan	Tung-chuan
Donggou	Tung-kou
Ebian	O-pien
Emei	O-mei
Erenhot	Erh-lien-hao-t'e
Erhai	Er Hai
Eryuan	Erh-yüan
Fengfen	Feng-feng
Fengshan	Feng-shan
Fenxiang	Fen-hsiang
Fujian	Fukien
Funing	Fu-ning
Fushun	Fu-shun
Fushouwan	Fu-chou-wan
Gaoyan	Kao-yen
Gaoyou	Kao-yu
Gansu	Kansu
Guandi	Kuanti
Guangde	Kuang-teh
Guangdong	Kwangtung
Guangxi Zhuangzu Zizhiqu	Kwangsi Chuang Autonomous Region
Guangyuan	Kuang-yüan
Guangzhou	Kwangchow (Canton)
Guangzhung	Kuan-chung
Guilin	Kweilin
Guiyang	Kweiyang
Guizhou	Kweichow
Guniutan	Ku-niu-tang
Guolang	Kuo-lang
Gushan	Ku-shan
Haibowan	Hai-p'o-wan
Hancheng	Han-cheng
Hangzhou	Hangchow
Hebei	Hopei
Hefei	Ho-fei
Heilongjiang	Heilungkiang
Hejin	Ho-chin
Helan Shan	Ho-lan Shan
Henan	Honan
Hengduan Shan	Heng-tuang Shan
Hengnan	Heng-nan
Hexian	Ho Hsien
Hexiang	Ho-hsiang
Hinggan	Hsing-an (Khingan)
Houcheng	Hou-ch'eng
Hoxtolgay	Ho-shin-t'o-lo-kai
Huainan	Huai-nan
Huang He	Huang Ho
Huangshandong	Huang-shan-tung
Huaning	Hua-ning
Huanxian	Huan-hsien
Hubei	Hupei (Hupeh)
Huidong	Hui-tung
Huili	Hui-li
Huize	Hui-tse
Hulun Buir Meng	Hu-lun-pie-erh Meng
Huzhou	Hu-chou
Jialingjiang	Chia-ling Chiang

GEOGRAPHIC NAMES PINYIN TO WADE-GILES

Pinyin	Wade-Giles
Jiangda	Chiang-ta
Jiangpu	Chiang-p'u
Jiangning	Chiang-ning
Jiangshan	Chiang-shan
Jiangsu	Kiangsu
Jiangxi	Kiangsi
Jiangyou	Chiang-yu
Jiaohe	Chiao-ho
Jilin	Kirin
Jinan	Tsinan
Jingjing	Ching-ching
Jingshan	Ching-shan
Jingxi	Ching-si
Jingxian	Ching Hsien
Jingyuan	Ching-yüan
Jinhua	Chin-hua
Jinsha	Chin-sha
Jinquan	Chiu-ch'üan
Jintan	Chin-t'an
Jinyang	Chin-yang
Junggar Pendi	Chu-ko-er Pen-ti (Dzungar Basin)
Kaili	Kai-li
Kaiping	K'ai-p'ing
Kaplin	K'o-p'ing
Kunlunshan	K'un-lun Shan
Kunming	Kun-ming
Kunshan	K'un-shan
Kuruktag	K'u-lu-k'o-ta-ko
Laishui	Lai-shui
Laiyang	Lai-yan
Langyashan	Lang-ya Shan
Lanshan	Lang-shan
Lantien	Lan-t'ien
Lanzhou	Lanchow
Leibo	Lei-po
Leping	Le-ping
Leshan	Lo-shan
Liangshan	Lian-shan
Lianxian	Lien Hsien
Lianyuan	Lien-yuan
Liaocheng	Liao-ch'eng
Liaoning	Liaoning
Lijiang	Li Chiang
Lingchuan	Ling-ch'uan
Lingling	Ling-ling
Linqu	Lin-ch'ü
Linru	Lin-ju
Liuyang	Liu-yang
Longtai	Lung-t'an
Longxian	Lung Hsien
Luhe	Lu-ho
Lunshan	Lun-shan
Luquan	Lu-ch'uan
Luxi	Lu-shi
Majiagou	Ma-chia-kou
Mengla	Meng-la
Mengzi	Meng-tzu
Mentougou	Men-t'ou-kou
Mianduhe	Mien-tu-ho
Mianxian	Mien Hsien
Mianyang	Mien-yang
Miaopo	Miao-po
Mingshui	Ming-shui
Mufushan	Mu-fu Shan
Nandan	Nan-tan
Nanjiang	Nan-chiang
Nanjing	Nanking
Nanling	Nan-ling
Nanpiao	Nan-p'iao
Nanxiong	Nan-hsiung
Nanzheng	Nan-cheng
Nenjiang	Nen-chiang
Ningqiang	Ning-chiang
Ningxia Huizu Zizhiqu	Ningsia Hui Autonomous Region
Ningzhen	Ning-chen Shan (Nanking Hills)
Ninyuan	Nin-yüan
Niuxinshan	Niu-hsin Shan
Nongshan	Nong Shan
Pingdingshan	P'ing-ting Shan
Pingquan	Ping-chuan
Pingxiang	P'ing-hsiang
Qaidam Pendi	Ch'ai-ta'mu P'en-ti
Qamdo Diqu	Ch'ang-tu Ti-chü
Qianshan	Chian-shan
Qianwei	Ch'ien-wei
Qijiang	Ch'i Chiang
Qilianshan	Ch'i-lien Shan
Qinan	Ch'in-an
Qingdao	Ch'ing-tao (Tsing-tao)
Qingfeng	Ch'ing-feng
Qinghai	Ch'ing-hai (Tsinghai)
Qingshui He	Ch'ing-shui Ho
Qinling	Ch'in Ling
Qomolongma Feng	Chu-mu-long-ma Feng (Mount Jolmolongma; Mount Everest)
Quanjiao	Ch'uan-chiao
Qujing	Ch'ü-ching (Kütsing)
Quyang	Ch'ü-yang
Renhuai	Jen-huai
Sanjiang	San-chiang
Sanshui	San-shui
Shaanxi	Shensi
Shandong	Shantung
Shanwang	Shan-wang
Shanxi	Shansi
Shenyang	Shen-yang

GEOGRAPHIC NAMES PINYIN TO WADE-GILES

Pinyin	Wade-Giles
Shijiutuo	Shih-chin-t'o
Shitai	Shih-t'ai
Shizipu	Shih-tzu-pu
Shouchang	Shou-ch'ang
Sichuan	Szechuan
Sikuangshan	Hsi-kuang Shan
Sinan	Ssu-nan
Suzhou	Suchou, Soochow
Taibei	Taipei
Taiyuan	Tay-yüan
Taizhou	T'ai-chou
Tanggula Shan	T'an-ku-la Shan
Tangquan	Tang-ch'üan
Tangshan	Tang-shan
Tarim Pendi	Ta-li-mu P'en-ti
Tianjin	Tientsin
Tianshan	T'ien Shan
Tongzi	Tung-tzu
Turpan	T'u-lu-p'an
Ürümqi	Wu-lu-mu-chi
Wangfu	Wang-fu
Wase	Wa-se
Wuhan	Wu-han
Wuhu	Wu-hu
Wuning	Wu-ning
Wuqia	Wu-ch'ia
Wuxi	Wuhsi
Wuxing	Wu-hsing
Xian	Hsien (County)
Xi'an	Sian
Xiangxiang	Hsiang-hsiang
Xiangzhou	Hsiang-chou
Xiao Hinggan Ling	Hsiao-hsing-an Ling
Xiaoshan	Shiao-shan
Xiaoshui	Hsiao-shui
Xingtang	Hsing-t'ang
Xinjiang Uygur Zizhiqu	Sinkiang Uighur Autonomous Region
Xintai	Hsin-t'ai
Xinyu	Hsin-yü
Xishui	Hsi-shui
Xiuxian	Hsin Hsien
Xixiabangma Feng	Hsi-hsia-pang-ma Feng
Xiyang	Hsi-yang
Xizang Zizhiqu	Tibet Autonomous Region
Xuangcheng	Hsuan-cheng
Xundian	Hsun-tien
Xushan	Hsü-shan
Xuyi	Hsü-i
Xuzhou	Hsü-chou
Yadong	Ya-tung
Yan'an	Yenan
Yangzi Jiang	Yang-tze Chiang (Yangtze River)
Yanjin	Yen-chin
Yaoxian	Yao Hsien
Yeli	Yeh-li
Yichang	I-ch'ang
Yidou	I'tou
Yincheng	Yin-ch'eng
Yinjiang	Yin-chiang
Yixian	I Hsien
Yizhang	I-chang
Yongning	Yung-ning
Yongshan	Yung-shan
Youyang	Yu-yang
Yuanmou	Yuan-mou
Yuhang	Yu-hang
Yunxian	Yun Hsien
Yuxian	Yu-hsien
Zhangye	Chang-yeh
Zhanyi	Chang-i
Zhaogezhuang	Chao-ko-chuang
Zhaozhuang	Chao-chuang
Zhejiang	Chekiang
Zhengping	Chen-p'ing
Zhengzhou	Chengchow
Zhenxiong	Chen-siung
Zhijin	Chih-chin
Zhongyang	Chung-yang
Zhuozishan	Cho-tzu Shan
Zigui	Tzu-kuei
Ziyun	Tze-yun
Zhujiang	Chu Chiang (Pearl River)
Zunyi	Tsun-i

CONVENTIONAL GEOGRAPHIC NAMES TO PINYIN

Conventional	Pinyin
Amur River	Heilong Jiang
Canton	Guangzhou
Chamodo Prefecture	Qamdo Diqu
Chihli Gulf	Bohai Wan
Dairen	Dalian
Dzungar Basin	Junggar Pendi
East China Sea	Dong Hai
Everest, Mount (Mount Jolmolongma; Chu-mu-long-ma Feng)	Qomolongma Feng
Great Khingan Range	Da Hinggan Ling
Inner Mongolia Autonomous Region	Nei Mongol Zizhiqu
Jolmolongma, Mount (Mount Everest)	Qomolongma Feng
Khingan	Hinggan
Kütsing	Qujing
Lesser Khingan Range (Hsiao-hsing-an Ling)	Xiao Hinggan Ling
Mount Everest	Qomolongma Feng
Nanking Hills	Ning Shan
Pearl River (Chu Chiang)	Zhujiang
South China Sea	Nan Hai
South China Sea Islands	Nanhai Zhudao
Tibet Autonomous Region	Xizang Ziziqu
Tsaidam Basin	Qiadam Pendi (Chaidamu Pendi)
Tsinghai	Qinghai
Tsing-tao	Qindao
Yangtze River	Chang Jiang
Yellow River	Huang He
Yellow Sea	Huang Hai

MANUSCRIPT ACCEPTED BY THE SOCIETY APRIL 22, 1981

Typeset by WESType Publishing Services, Inc., Boulder, Colorado
Printed in U.S.A. by Malloy Lithographing, Inc.,
Ann Arbor, Michigan